Current Progress in Retrovirology

Current Progress in Retrovirology

Edited by Adeline Foley

AMERICAN
MEDICAL PUBLISHERS
www.americanmedicalpublishers.com

American Medical Publishers,
41 Flatbush Avenue,
1st Floor, New York,
NY 11217, USA

Visit us on the World Wide Web at:
www.americanmedicalpublishers.com

ISBN: 978-1-63927-261-7

Cataloging-in-Publication Data

Current progress in retrovirology / edited by Adeline Foley.
p. cm.
Includes bibliographical references and index.
ISBN 978-1-63927-261-7
1. Retroviruses. 2. Medical microbiology. 3. Microbiology. I. Foley, Adeline.
QR414.5 .C87 2022
616.019 4--dc23

Table of Contents

Preface

It is often said that books are a boon to mankind. They document every progress and pass on the knowledge from one generation to the other. They play a crucial role in our lives. Thus I was both excited and nervous while editing this book. I was pleased by the thought of being able to make a mark but I was also nervous to do it right because the future of students depends upon it. Hence, I took a few months to research further into the discipline, revise my knowledge and also explore some more aspects. Post this process, I begun with the editing of this book.

Retrovirology is the study of retroviruses. A retrovirus is a kind of RNA virus that inserts a copy of its genome into the DNA of a host cell, thus changing the genome of that cell. These viruses are either single-stranded RNA or double-stranded DNA viruses. The main viron components found in retroviruses are envelope, RNA and protein. Both animals and humans are affected by retroviruses. It causes tumor growth and certain cancers that are linked with slow infections of animals such as equine infectious anemia. Retroviruses such as HTLV-1 cause cancer called adult T-cell leukemia in human beings. Other retroviruses such as HIV cause AIDS in human beings. The treatment of diseases by retrovirus can be done with the help of antiretroviral drugs and medications. This book unravels the recent studies in the field of retrovirology. It traces the progress of this field and highlights some of its key concepts and applications. It is a vital tool for all researching or studying retrovirology, as it gives incredible insights into emerging trends and concepts.

I thank my publisher with all my heart for considering me worthy of this unparalleled opportunity and for showing unwavering faith in my skills. I would also like to thank the editorial team who worked closely with me at every step and contributed immensely towards the successful completion of this book. Last but not the least, I wish to thank my friends and colleagues for their support.

Editor

1

Potent and reversible lentiviral vector restriction in murine induced pluripotent stem cells

Franziska K. Geis[1,2], Melanie Galla[1,2], Dirk Hoffmann[1,2], Johannes Kuehle[1,2], Daniela Zychlinski[1,2], Tobias Maetzig[1,2], Juliane W. Schott[1,2], Adrian Schwarzer[1,2], Christine Goffinet[3], Stephen P. Goff[4,5,6] and Axel Schambach[1,2,7*]

Abstract

Background: Retroviral vectors are derived from wild-type retroviruses, can be used to study retrovirus-host interactions and are effective tools in gene and cell therapy. However, numerous cell types are resistant or less permissive to retrovirus infection due to the presence of active defense mechanisms, or the absence of important cellular host co-factors. In contrast to multipotent stem cells, pluripotent stem cells (PSC) have potential to differentiate into all three germ layers. Much remains to be elucidated in the field of anti-viral immunity in stem cells, especially in PSC.

Results: In this study, we report that transduction with HIV-1-based, lentiviral vectors (LV) is impaired in murine PSC. Analyses of early retroviral events in induced pluripotent stem cells (iPSC) revealed that the restriction is independent of envelope choice and does not affect reverse transcription, but perturbs nuclear entry and proviral integration. Proteasomal inhibition by MG132 could not circumvent the restriction. However, prevention of cyclophilin A (CypA) binding to the HIV-1 capsid via use of either a CypA inhibitor (cyclosporine A) or CypA-independent capsid mutants improved transduction. In addition, application of higher vector doses also increased transduction. Our data revealed a CypA mediated restriction in iPSC, which was acquired during reprogramming, associated with pluripotency and relieved upon subsequent differentiation.

Conclusions: We showed that murine PSC and iPSC are less susceptible to LV. The block observed in iPSC was CypA-dependent and resulted in reduced nuclear entry of viral DNA and proviral integration. Our study helps to improve transduction of murine pluripotent cells with HIV-1-based vectors and contributes to our understanding of retrovirus-host interactions in PSC.

Keywords: HIV-1-based vectors, Murine pluripotent stem cells, Anti-viral restriction, Cyclophilin A

Background

In addition to their function as gene transfer vehicles, retroviral vectors can be utilized as tools to investigate specific vector-host interactions. Retroviral vectors are derived from their respective wild-type (wt) viruses by removal of nucleotide sequences in the RNA genomes, which are not essential for transgene delivery. Retroviral vectors are packaged into viral particles composed of structural and enzymatic proteins (Gag-Pol) and follow the natural course of the early retroviral life cycle. This includes cell entry, followed by a poorly-characterized uncoating process of the viral capsid, reverse transcription (RT) of the viral RNA genome into double-stranded DNA, nuclear entry of the viral DNA and stable, proviral integration into the host genome [1].

In this study, we investigated vector-host interactions in pluripotent stem cells (PSC). These cells naturally exist in early embryonic development and give rise to all three germ layers. PSC represent the equivalent to the inner cell mass of the blastocyst and possess major differentiation

*Correspondence: schambach.axel@mh-hannover.de
[1] Institute of Experimental Hematology, Hannover Medical School, Carl-Neuberg-Str. 1, Hannover, Germany
Full list of author information is available at the end of the article

potential. Among PSC, induced pluripotent stem cells (iPSC) represent a prominent target cell type and cell source as they are easily generated by reprogramming of differentiated cells, a process mediated by ectopic expression of specific transcription factors [2]. After successful reprogramming, iPSC obtain full pluripotency capacity and can differentiate into cells of all three germ layers in vitro as well as in vivo [3]. Therefore, iPSC display a suitable alternative to embryonic stem cells (ESC) for the study of cellular differentiation and disease modeling. Additionally, iPSC are promising resources for cell and gene therapy applications, e.g. innovative cell transplants mediated by retroviral gene transfer.

Retroviral vector application and their gene transfer efficiency are closely dependent on virus-host cell interactions that either support or inhibit efficient infection. Retroviral replication is supported by host co-factors, which are crucial for successful infection. A prominent retroviral host co-factor is cyclophilin A (CypA), which directly binds to an exposed loop of the human immunodeficiency virus type 1 (HIV-1) capsid protein [4, 5]. CypA was described to support early events of HIV-1 infection, such as uncoating and RT initiation [6, 7]. In addition, it was shown that the interaction of CypA with other host co-factors determined the route of nuclear entry and led to alterations in integration site targeting [8]. Furthermore, capsid-associated CypA can shield HIV-1 against cellular recognition and prevent a type 1 interferon response [9]. In contrast, CypA-capsid binding can also inhibit HIV-1 infectivity in a species-specific manner by increasing the sensitivity to certain restriction factors [9–11]. Restriction factors are part of the cell autonomous immunity, are expressed by host cells and inhibit viral replication by inducing viral degradation, transcriptional silencing of integrated proviruses or initiation of anti-viral interferon responses. In this regard, previously identified restriction factors in PSC include the Trim28/ZFP809 complex [12, 13] and YY1 [14]. Both factors target murine leukemia virus (MLV) in murine ESC. However, the role of cell autonomous immunity in PSC is only incompletely understood and so far not described for HIV-1.

Here, we describe an HIV-1-based vector restriction found in iPSC as well as in ESC. Focusing on iPSC, we identified a CypA-dependent restriction mechanism that impaired nuclear entry and reduced proviral integration. The restriction could be decreased by application of high vector doses, by prevention of CypA binding to the HIV-1 capsid or by iPSC differentiation.

Results

Murine iPSC exhibit a potent block to HIV-1-based vectors

In this study we analyzed the susceptibility of murine iPSC for vectors derived from HIV-1 (LV; lentiviral vectors) or from Moloney MLV (GV; gammaretroviral vectors). Both retroviral vectors were designed with self-inactivating (SIN) architecture and expressed the enhanced green fluorescence protein (EGFP) reporter gene driven by a short version of the human elongation factor 1 alpha gene (EFS) promoter (Fig. 1a). If not indicated otherwise, we generated vector particles pseudotyped with the vesicular stomatitis virus glycoprotein (VSVg) envelope containing the vector RNA and the retroviral wt replication enzymes and structural proteins. Prior to iPSC transduction, vector supernatants were always titrated on permissive HT1080eCat cells in order to utilize comparable vector doses for following experiments. To compare retroviral transgene transfer into iPSC, LV and GV were applied to iPSC at defined multiplicities of infection (MOI). All iPSC used in this study were generated by ectopic expression of the octamer-binding transcription factor 4 (Oct4), sex determining region Y-box 2 gene (Sox2), Kruppel-like factor 4 (Klf4) and c-Myc in primary fibroblasts (Fig. 1b). A previously characterized and described iPSC clone (clone #2EX) was used throughout this study [15]. In this clone, the reprogramming cassette was excised by Flp recombinase to prevent interference with the integrated lentiviral reprogramming vector during PCR-based analyses. Transduced cells were analyzed for EGFP and anti-stage specific antigen marker 1 (SSEA1) (pluripotency marker in murine cells) co-expression by flow cytometry. Major differences in the susceptibility of murine iPSC for the tested retroviral vectors were observed. Notably, both GV and LV exhibited low transduction rates in iPSC using MOI 10 and 100 (Fig. 1c, left graph). However, LV transduction efficiency remained significantly reduced compared to GV at both MOI. Analyses of the mean vector copy number per cell (detecting EGFP) revealed significantly less integrated vector copies in LV transduced iPSC indicating a pre-integration block (Fig. 1c, right graph). To exclude iPSC clone-specific abnormalities as the basis for the poor permissiveness to LV, we compared clone #2EX (excised reprogramming cassette) to a variety of iPSC clones generated by different reprogramming approaches and derived from various mouse strains. Additionally, we incorporated clones with non-excised reprogramming cassettes, including the parental clone #2 for direct comparison to the #2EX excised version. Remarkably, all tested iPSC clones demonstrated significantly less susceptibility to LV transduction, suggesting that LV restriction is independent from the genetic background and the presence of the reprogramming cassette (Fig. 1d). To analyze whether naturally occurring PSC are susceptible to LV, we tested murine ESC. As similarly observed in iPSC, we measured decreased transduction rates for LV, in contrast to GV, caused by reduced integration events (Fig. 1e).

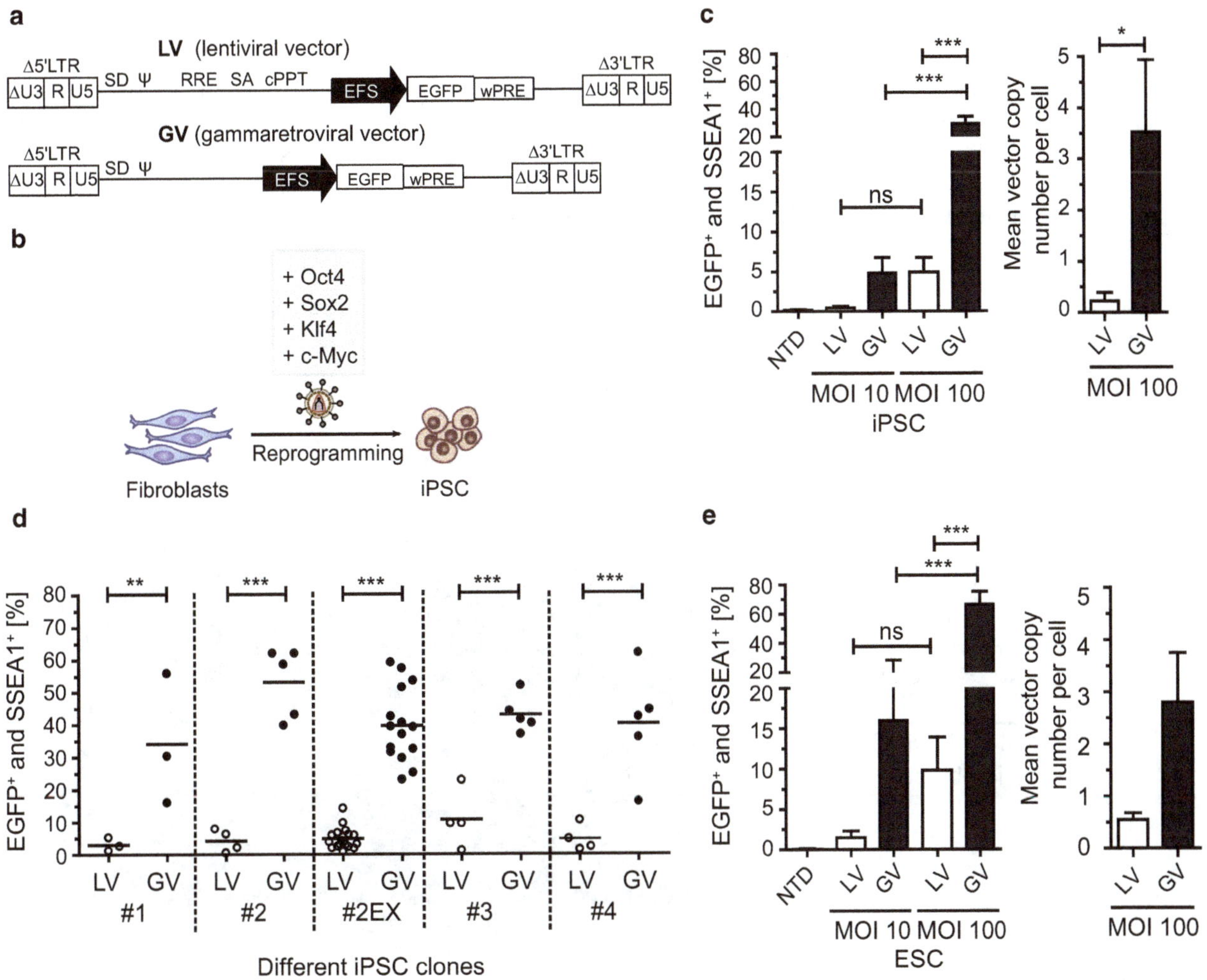

Fig. 1 Restriction of HIV-1-based vectors in murine iPSC. **a** Schematic overview of integrated LV and GV used in this study. Vectors contain two LTR with deleted U3 regions (ΔU3, self-inactivating design), flanking an EGFP expression cassette driven by an EFS enhancer/promoter. *SD* splice donor, *Ψ* retroviral packaging signal, *RRE* rev responsive element, *SA* splice acceptor, *cPPT* central polypurine tract, *wPRE* woodchuck hepatitis virus post-transcriptional regulatory element. **b** Scheme of reprogramming murine fibroblasts into iPSC by retroviral expression of Oct4, Sox2, Klf4 and c-Myc transcription factors. **c** iPSC transduced with three independently produced viral supernatants of LV and GV at an MOI of 10 and 100 ($n = 3$). Flow cytometry data of collected EGFP and SSEA1 (pluripotency marker) double positive cells are shown. *NTD* non-transduced control. One-way ANOVA with Tukey-Kramer post hoc test was used for statistical analyses. ns (not significant) $p = 0.258$; *** $p \leq 0.001$ (*left panel*). LV ($n = 6$) or GV ($n = 4$) applied to iPSC at an MOI of 100. Mean vector copy numbers per cell were determined 6–8 days after transduction with SYBR Green-based quantitative real-time PCR, based on EGFP copies, and normalized to endogenous PTBP2 DNA copies. The unpaired *t* test with Welch's correction was used for statistical analysis. * $p = 0.017$ (*right panel*). **d** LV and GV were applied to different iPSC clones at an MOI of 100. The percentage of EGFP and SSEA1 double positive cells was determined by flow cytometry. One-way ANOVA with Tukey-Kramer post hoc test was used for statistical analyses. #1 ** $p = 0.004$; #2, #2EX, #3 and #4 *** $p \leq 0.001$. **e** LV and GV from three independently produced supernatants were applied to ESC at an MOI of 10 and 100. Analyses were performed 6–8 days after transduction. The percentage of EGFP and SSEA1 double positive cells was measured by flow cytometry. One-way ANOVA with Tukey-Kramer post hoc test was used for statistical analyses. ns $p = 0.660$; *** $p \leq 0.001$ (*left panel*). Mean vector copy numbers per cell were determined with SYBR Green-based quantitative real-time PCR, based on EGFP copies, and normalized to endogenous PTBP2 DNA copies and a plasmid standard (*right panel*)

LV restriction is observed exclusively in the pluripotent state of a cell

Next, we asked, if inhibition of LV gene transfer appears only in iPSC or can also be observed in the parental cells used for reprogramming or in iPSC differentiated progeny. In contrast to iPSC, the parental fibroblasts used for reprogramming, did not show a LV restriction (Fig. 2a). To further test, whether LV restriction is directly correlated with the pluripotent state of the cells, we investigated if the differentiated progeny of iPSC regained permissiveness to LV. We differentiated iPSC by withdrawal of the leukemia inhibitory factor (LIF) and cultured the cells in the absence of supporting feeder cells (Fig. 2b, upper panel). Differentiated cultures were transduced with LV or GV encoding DsRedExpress (LV.DsRed and GV.DsRed). Transduced cells were stained with a green fluorescent SSEA1 antibody and live cell imaging was accomplished by fluorescence microscopy. Almost exclusively SSEA1 negative cells were transduced with LV, whereas SSEA1 and DsRedExpress double positive cells were observed more frequently after transduction with GV (Fig. 2b, lower panel). To better quantify transduction rates, we analyzed differentiated (SSEA1−) or remaining pluripotent (SSEA1+) cells by flow cytometry. In line with the fluorescence microscopy approach, only a small amount of SSEA1+ cells were susceptible to

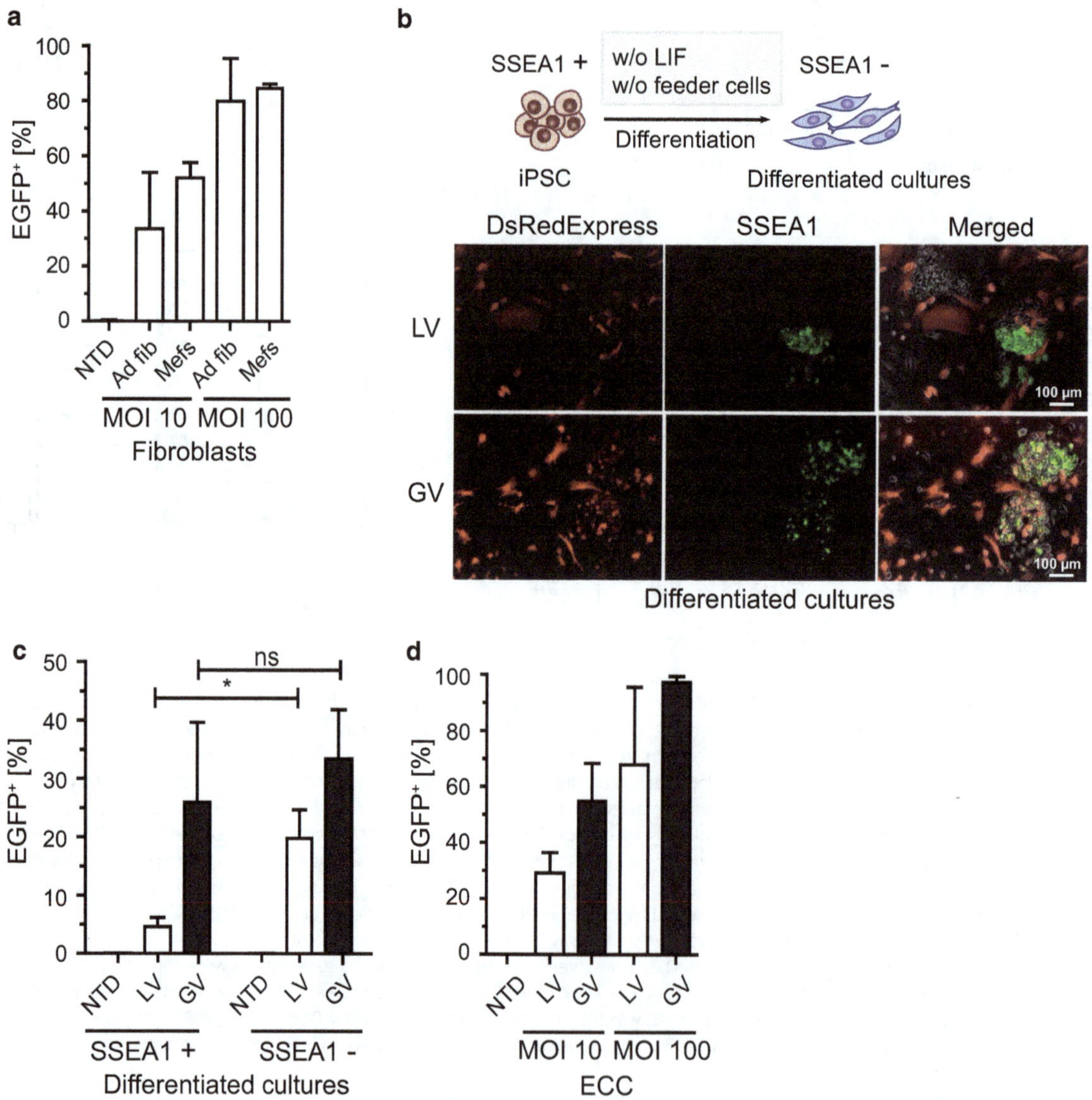

Fig. 2 Restriction is observed at the pluripotent cell state. **a** LV were applied to C57BL/6 p14f/f adult fibroblasts (Ad fib) (n = 8) or C3H Mefs (n = 3) at an MOI of 10 and 100 (independently produced viral supernatants). The percentage of EGFP positive cells was determined by flow cytometry. *NTD* non-transduced control. **b** Scheme of iPSC differentiation by withdrawal of LIF and feeder cell co-cultivation (*upper panel*). Representative fluorescence microscopy pictures of transduced differentiated cultures with LV.DsRed or GV.DsRed at an MOI of 100, SSEA1 live staining in *green* and an overlay of all channels are depicted (*lower panel*). **c** Differentiated cultures were transduced with LV (n = 5) and GV (n = 4) encoding EGFP at an MOI of 100 (independently produced viral supernatants). Pluripotent (SSEA1+) and differentiated (SSEA1−) cells were analyzed by flow cytometry. One-way ANOVA with Tukey-Kramer post hoc test was used for statistical analyses. * p = 0.049; ns p = 0.706. **d** ECC transduced with LV (n = 3) or GV (n = 2) at an MOI of 10 and 100 (independently produced viral supernatants). The percentage of EGFP positive cells was analyzed by flow cytometry

LV in contrast to differentiated SSEA1− cells (4.6 ± 1.6% vs. 19.8 ± 4.9%), while permissiveness of SSEA1+ and SSEA1− cells for GV particles was comparable (26 ± 13.7% vs. 33.5 ± 8.4%) (Fig. 2c). The impact of cell pluripotency on the LV block was further tested using embryonic carcinoma cells (ECC) as an additional pluripotent-like cell model [16]. Interestingly, ECC showed no LV restriction as observed in iPSC or ESC (Fig. 2d).

LV transduction rate in murine iPSC is enhanced by high vector doses and CSA treatment

It was previously shown that permissiveness of specific cell types to HIV-1 can be enhanced by altering the viral entry route [17, 18]. Therefore, we performed a side-by-side comparison of envelope pseudotypes used for LV particle generation. We tested VSVg-pseudotyped LV (LV-VSVg) and included particles pseudotyped with glycoproteins derived from amphotropic (LV-Ampho) or ecotropic MLV (LV-Eco). While LV-VSVg enters cells via clathrin-mediated endocytosis [19], LV-Ampho employs cell membrane fusion and/or macropinocytosis [20, 21], and LV-Eco enters cells by calveolae-dependent endocytosis and/or direct cell membrane fusion [22–24]. Although LV-VSVg, LV-Ampho and LV-Eco particles obtained comparable titers on permissive HT1080eCat cells (Fig. 3a, left graph), all vectors showed very low transduction rates of iPSC (less than 1%) at an MOI of 10 (Fig. 3a, right graph). Previous work demonstrated the possibility to apply high virus doses to overcome a restriction factor-mediated block [25–28]. Thus, we transduced murine iPSC with LV at increasing MOI. Transduction rates were enhanced from 0.84 ± 0.45% (MOI 10) to 25.83 ± 2.57% (MOI 2000) (Fig. 3b). Despite a 30-fold increase with the highest vector dose, the majority of iPSC remained untransduced, suggesting relatively inefficient abrogation of LV restriction. Next, we performed pre-transduction (Pre-Td) experiments followed by a second transduction to investigate a potential increase in transducability in a second round of transduction. This method is frequently used to provide evidence for a saturable restriction factor, when the factor can be abrogated at a certain MOI of the restricted incoming viral particles [28, 29]. Pre-Td was conducted with DsRedExpress encoding LV.DsRed and GV.DsRed at an MOI of 1000. GV.DsRed Pre-Td served as a control to exclude common cellular defense mechanisms after transduction with high MOI. Following published protocols, a second transduction was then performed with EGFP-encoding LV, at an MOI of 100, 6 h after Pre-Td [25, 28]. Although having conducted successful Pre-Td (Fig. 3c, right graph), Pre-Td resulted in no beneficial effect on LV transduction rates (Fig. 3c, left graph). Next, we tested the small molecules MG132 and cyclosporine A (CSA), which were previously described to enhance lentiviral transduction and counteract lentiviral restriction [11, 30–33]. Dose escalation experiments (using 0.05–0.4 μM) with the proteasome inhibitor MG132 failed to increase LV transduction rates in iPSC (Fig. 3d). Of note, at doses ≥ 0.2 μM, we observed a cytotoxic effect of the drug, resulting in decreased cell numbers as well as in increased AnnexinV and PI positive cells (see Additional file 1). To investigate if abrogation of the CypA-capsid interaction by CSA treatment could circumvent the restriction, as this has been described in other cell types in previous studies [34–37], we transduced iPSC in the presence or absence of different CSA doses. Strikingly, CSA treatment during transduction significantly enhanced LV transduction rates in a dose-dependent manner (Fig. 3e, left graph). In contrast, CSA had no statistically significant effect on GV transduction rates (Fig. 3e, right graph). Of note, while no apoptosis or cell death was observed for any tested CSA concentration, decreased cell numbers were observed with 20 μM CSA (see Additional file 2). Since doses ≤ 10 μM did not affect cell numbers, a concentration of 10 μM CSA was chosen for all further experiments.

CypA-capsid interaction mediates LV restriction in iPSC

To characterize the dependency of CSA on the time of its addition, we conducted a kinetic study and treated iPSC with CSA at different time points prior to and during transduction. LV transduction without CSA treatment contained equal amounts of DMSO as a solvent control. Interestingly, the strongest CSA effect on LV transduction rate enhancement in iPSC was achieved when adding CSA before (−0.5 h) or at the time of transduction (0 h) (Fig. 4a). Nevertheless, CSA addition still had a significant effect on iPSC susceptibility to LV even when applied 12 h after retroviral vector application. To test whether CypA-capsid interaction mediates the LV restriction or if CSA acts through different mechanisms, we prevented CypA binding to the HIV-1 capsid by using specific CypA-independent HIV-1 capsid mutants. We compared HIV-1 wt capsid and the mutant capsid variants G89V, P90A, H2.1 and A88T for their ability to transduce iPSC. The G89V, P90A and A88T capsid mutants harbor a point mutation causing a single amino-acid change in the CypA binding loop of the HIV-1 capsid, preventing CypA binding to the capsid [36, 38]. Of note, the A88T mutant was shown to prevent binding of CypA and also the myxovirus resistance 2 (Mx2) protein, which was identified as an HIV-1 restriction factor [34–36]. The H2.1 mutant contains a substitution of the equivalent CypA binding loop sequences of HIV-1 to HIV-2, which does not bind CypA [39]. Remarkably, all CypA-independent capsid mutants exhibited clearly improved LV transduction rates (Fig. 4b). In addition, we

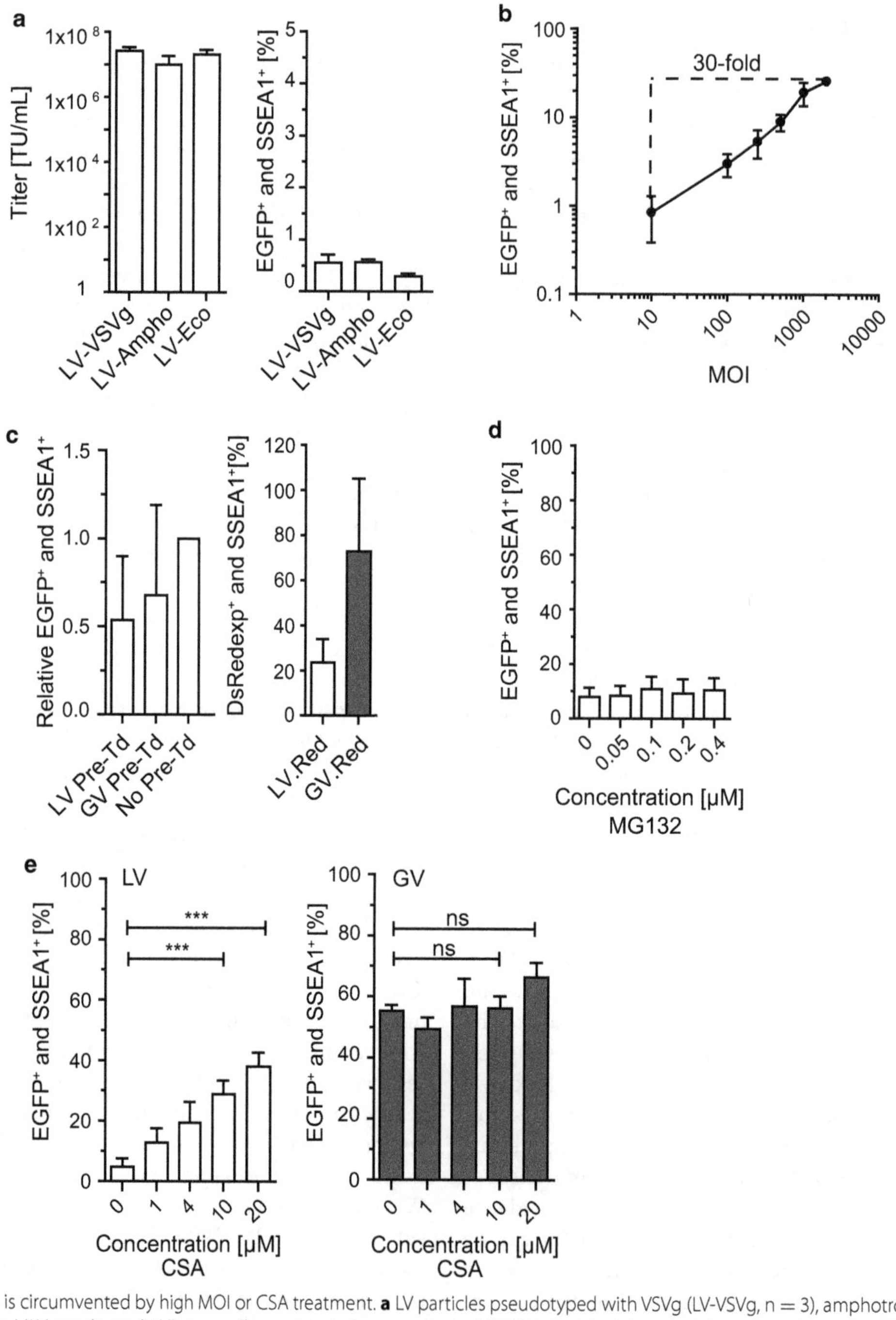

Fig. 3 Restriction is circumvented by high MOI or CSA treatment. **a** LV particles pseudotyped with VSVg (LV-VSVg, n = 3), amphotropic (LV-Ampho, n = 3) or ecotropic MLV envelopes (LV-Eco, n = 2) were titrated on permissive HT1080eCat fibroblasts and titers were determined by flow cytometry as transducing units per mL (TU/mL from independently produced viral supernatants) (*left panel*). The different pseudotyped LV particles were applied to iPSC at an MOI of 10 and the percentage of EGFP and SSEA1 double positive cells was assessed by flow cytometry (*right panel*). **b** iPSC were transduced at indicated MOI. EGFP and SSEA1 double positive cells from independent replicates were collected by flow cytometry. MOI 10, n = 5; MOI 100, n = 4; MOI 250, n = 5; MOI 500, n = 5; MOI 1000, n = 5; MOI 2000, n = 4. **c** Pre-transduction (Pre-Td) experiments in iPSC were performed by Pre-Td with LV.Red or GV.Red at an MOI of 1000 (n = 3; independently produced viral supernatants) followed by a second transduction with LV at an MOI of 100, 6 h after Pre-Td. Data of the second transduction are shown relative to no Pre-Td (*left panel*). Transduction of LV.Red and GV.Red are shown as the Pre-Td control (*right panel*). **d** iPSC transduced with LV at an MOI of 100 in the presence of MG132 at indicated concentrations. Data from 3 independently produced viral supernatants were determined by flow cytometry. **e** LV (*left graph*) or GV (*right graph*) were applied to iPSC at an MOI of 100 in the presence of CSA at the indicated concentrations. Data from 3 to 4 independently produced viral supernatants were determined by flow cytometry. One-way ANOVA with Dunnett post hoc test was used for statistical analyses. LV 0 versus 10, *** $p \leq 0.001$; LV 0 versus 20, ***$p \leq 0.001$; GV 0 versus 10, ns $p = 0.999$; GV 0 versus 20, ns $p = 0.112$

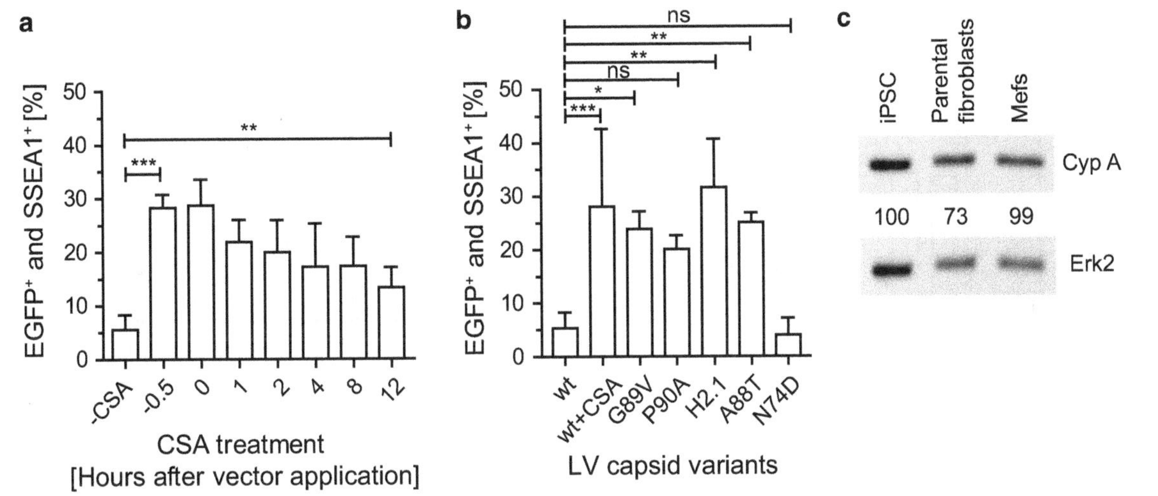

Fig. 4 CypA-capsid interaction mediates LV restriction independent from CypA expression. **a** iPSC were transduced with LV at an MOI of 100 in the presence of 10 µM CSA. CSA was added at indicated time points prior to or during transduction. The percentage of EGFP and SSEA1 double positive cells from 3 independently produced viral supernatants were assessed by flow cytometry. Repeated measures one-way ANOVA with Dunnett post hoc test was used for statistical analyses. *** $p \leq 0.001$; ** $p = 0.003$. **b** Transduction of iPSC with LV, harboring wt or mutated HIV-1 capsids. The percentages of EGFP and SSEA1 double positive cells were quantified by flow cytometry (wt $n = 9$, wt + CSA $n = 9$; G89V $n = 3$; P90A $n = 3$; H2.1 $n = 3$; A88T $n = 5$; independently produced viral supernatants). One-way ANOVA with Dunnett post hoc test was used for statistical analyses. wt versus wt + CSA, *** $p \leq 0.001$; wt versus G89V, * $p = 0.014$; wt versus P90A, ns $p = 0.066$; wt versus H2.1, *** $p \leq 0.001$; wt versus A88T, ** $p = 0.002$; wt versus N74D, ns $p = 0.9996$. **c** Western blot analysis showing cellular CypA protein levels in iPSC, C57BL/6 p14f/f adult fibroblasts and CF1-Mefs. The percentage of the CypA/Erk2 ratio is depicted for each cell type relative to iPSC

tested the capsid mutant N74D. N74D harbors a point mutation, which prevents binding of cleavage and polyadenylation specific factor 6 (CPSF6) to the capsid, binds CypA to a lower extent than wt capsids and favors a different nuclear entry route compared to the wt capsid [8, 40, 41]. It was previously reported that the early steps of infection by the N74D capsid mutant virus were inhibited prior to RT in macrophages [42]. Importantly, we did not observe this effect in iPSC, since late RT products using N74D mutant and wt capsid were comparable (see Additional file 3). Interestingly, the N74D capsid mutant did not improve LV transduction (Fig. 4b). To investigate whether the differences in LV transduction efficiencies observed between iPSC and differentiated cell types were due to different intrinsic CypA protein levels, Western blot analyses were used to compare CypA levels in iPSC with their parental fibroblasts and other primary fibroblasts. iPSC displayed comparable CypA levels to the permissive Mefs, indicating that LV restriction was not due to altered CypA protein expression (Fig. 4c).

Low efficiency of LV transduction in iPSC is associated with reduced levels of 2-LTR circles and proviral DNA integration

We observed strongly reduced vector copies in LV transduced iPSC (Fig. 1c). To characterize the observed LV restriction in greater detail and to identify the step of the early retroviral life cycle affected in iPSC, we analyzed the generation of viral intermediates between cellular entry and integration. iPSC were transduced with LV at an MOI of 10 and emerging RT products were analyzed with TaqMan-based quantitative real-time PCR at different time points. CSA treated cells served as a control for effective LV transduction. RT product levels were normalized to endogenous polypyrimidine tract binding protein 2 (PTBP2) DNA copies. Heat-inactivated supernatants (plasmid ctrl) served as a control for plasmid contamination in the viral vector preparations and DMSO as the solvent control for CSA treated cells. No differences were detected in early RT (Fig. 5a) or late RT product levels (Fig. 5b) when comparing DMSO to CSA treated iPSC. To analyze whether the reverse transcribed viral DNA is imported into the nucleus, we investigated the levels of 2-LTR circles, which are considered as a marker for nuclear entry and are formed in the nucleus by the host cell repair machinery [43–45]. The levels of 2-LTR circles were analyzed by TaqMan-based quantitative real-time PCR (normalized to PTBP2). Primers and probes were designed to detect the LTR-LTR junction region to avoid detection of LV autointegration [45]. In DMSO treated samples, 2-LTR circles were significantly reduced at 6 and 12 h after vector application when compared to CSA-treated cells (Fig. 5c). Next, we included the integrase inhibitor Raltegravir to impede

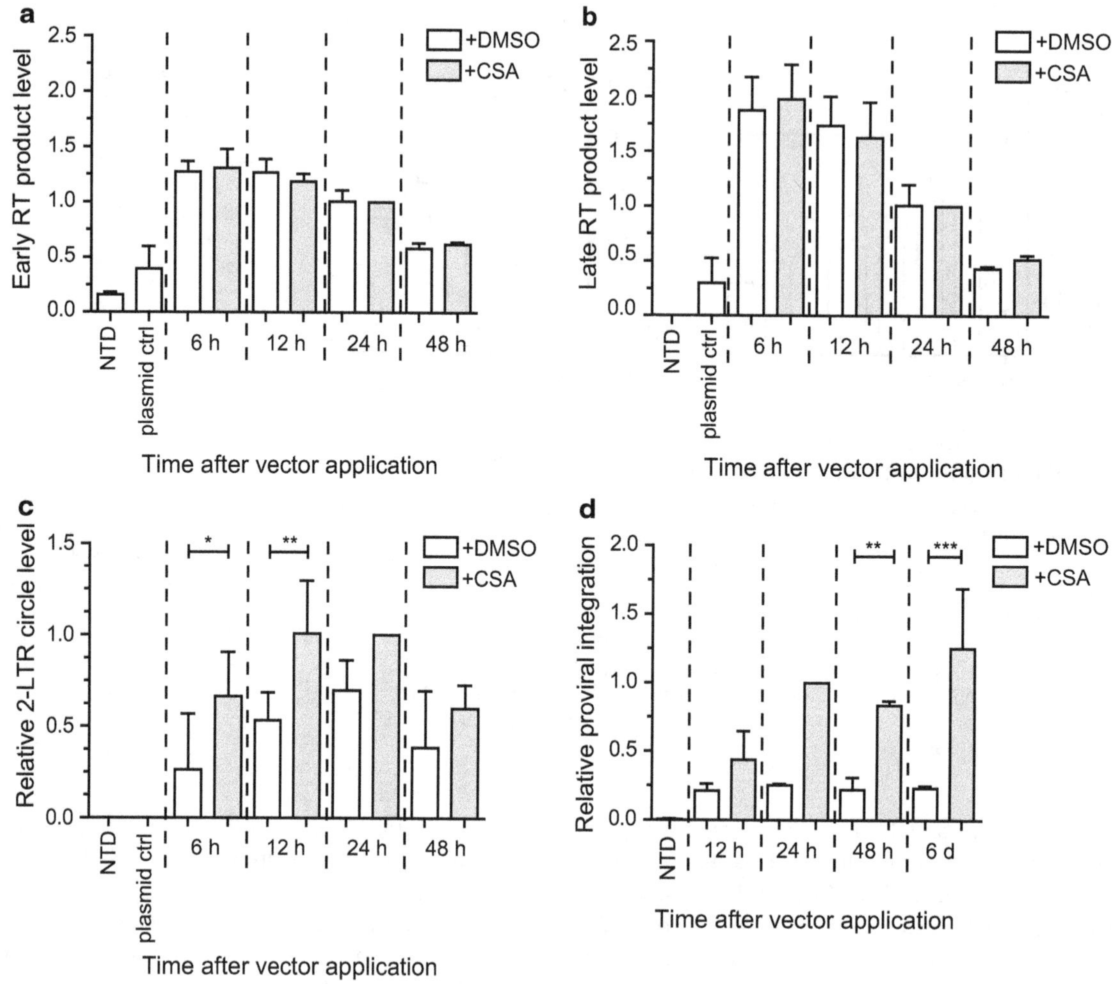

Fig. 5 Nuclear entry and integration of LV is reduced in iPSC. iPSC were transduced at an MOI of 10 (with independently produced viral supernatants as indicated) in the presence of 10 μM CSA or an equal volume of DMSO as solvent control. Cells were washed 6 h after vector application and harvested at indicated time points. Data are shown relative to the 24 h time point of LV transduced cells treated with CSA. Samples were analyzed for early RT (n = 3) (**a**), late RT products (n = 3) (**b**), 2-LTR circles (n = 6; * p = 0.030; ** p = 0.005) (**c**) and proviral vector copies (n = 3; ** p = 0.004; *** p ≤ 0.001) (**d**). RT products and 2-LTR circles were evaluated with TaqMan-based quantitative real-time PCR with the $2^{-\Delta\Delta Ct}$ method, normalized to endogenous PTBP2 copies. Proviral vector copies were determined by B1-LTR PCR and obtained values were corrected for plasmid contamination. Repeated measures one-way ANOVA with Tukey-Kramer post hoc test was used for statistical analyses. *NTD* non-transduced control; *plasmid ctrl* plasmid contamination control

integration and to determine the amount of 2-LTR circles in the presence or absence of CSA. The inhibition of integration by Raltegravir led to significantly higher levels of 2-LTR circles in CSA treated cells, but only a minor increase of 2-LTR circles in DMSO treated cells (see Additional file 4A). Interestingly, treatment of transduced iPSC with Raltegravir and CSA compared to Raltegravir alone led to significantly elevated 2-LTR circle levels. These findings further strengthen the hypothesis that nuclear entry of LV was impaired in iPSC. Strikingly, CSA treatment also significantly increased integrated vector copy numbers (see Additional file 4B). Transduction of Mefs treated under identical conditions served as permissive controls and showed no major effect of CSA on transduction rates in these cells (see Additional files 4A and B). To determine the time point of integration, we analyzed samples from Fig. 5a–c with the B1-LTR PCR, which is the mouse equivalent to the human Alu-LTR PCR. Briefly, proviral integrants were detected with primers detecting the LTR and the B1 repetitive elements in the genome (first stage of PCR), followed by a nested LTR amplification step (second stage PCR). Proviral integration started at 12 h and reached a plateau 24 h after vector application, with no major differences thereafter (as measured at 48 h and a late 6 day time point) (Fig. 5d).

Discussion

In this study, we transduced iPSC in order to investigate vector-host interactions in the early retroviral life cycle in pluripotent cell types. Our results revealed that murine iPSC exhibited a potent restriction against HIV-1-based vectors, which was observed after reprogramming fibroblasts into the pluripotent state. Dissecting retroviral intermediates to localize LV restriction revealed functional RT (including early and late RT products), impaired nuclear entry (as measured by 2-LTR circle formation) and reduced proviral integration (as determined by mean vector copy number per cell and proviral integrates). Transduction rates were improved by the use of high vector doses, prevention of CypA binding to HIV-1 capsid (by CSA or using CypA-independent capsid mutants) or differentiation into iPSC progeny (Fig. 6).

The fact that efficiently reverse transcribed proviral DNA of LV failed to integrate into the chromatin of iPSC (Figs. 1c, 5d) suggests inefficient cytoplasmic trafficking, nuclear entry and/or integration. Our 2-LTR circle analyses at different time points after transduction (6, 12, 24, 48 h) revealed significantly reduced levels in untreated compared to CSA-treated cells (Fig. 5c). In addition, cells treated with the CSA/Raltegravir combination exhibited a pronounced increase of 2-LTR circles when compared to cells treated with Raltegravir application alone (see Additional file 4A). These findings clearly support that the mechanism responsible for the LV restriction in iPSC involved inefficient nuclear entry and that CSA relieved this hindrance. This could be explained by unproductive nuclear entry due to misdirected cytoplasmic trafficking to the nucleus, trapping of the PIC or impaired translocation into the nucleus. Alternatively, it is possible that, in addition to the nuclear entry block, there is an intranuclear block (e.g. caused by altered nuclear trafficking or perturbed integration), which is supported by different ratios of 2-LTR circles and proviral integration levels (see Additional file 4A and Fig. 5d). However, we hypothesize that the main block affects the nuclear entry step.

We identified the CypA-capsid interaction as a key event that mediated the restriction. Prevention of CypA-capsid interaction by CSA or specific CypA-independent capsid mutants led to a significant increase in transduction (Figs. 3e, 4b). Our data provide evidence that this interaction is most crucial relatively early after cell entry, but we also observed supportive effects of CSA at the latest time point measured (12 h) (Fig. 4a). Interestingly, CypA expression levels were comparable in restricted iPSC and permissive cells, thus there was no correlation between CypA levels and the block (Fig. 4c). Furthermore, we were unable to saturate the block by pretransduction with high doses (MOI 1000) of LV in our abrogation experiments (Fig. 3c). Potential explanations for the lack of an abrogation phenotype include: (I) even higher concentrations are needed to saturate the potential factor; (II) the block is not saturable; (III) using our doses, other essential cellular pathways and receptors are blocked and are not recycled in a timely manner in the 6 h time window before the second transduction.

The restriction was relieved by use of CypA-independent capsid mutants (P90A, G89V, H2.1, A88T) (Fig. 4b). In contrast, the N74D capsid mutant, which binds CypA to a lower extent than the wt capsid, revealed no beneficial effect on transduction [42]. This suggests that the amount of capsid-bound CypA is important for the restriction. Furthermore, it was previously reported that

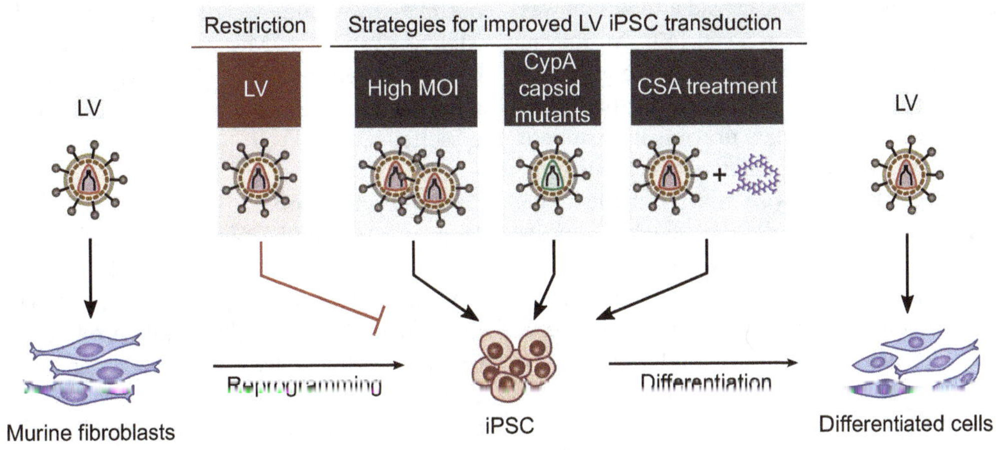

Fig. 6 Schematic overview of LV restriction and strategies to improve murine iPSC transduction. Primary fibroblasts of murine origin were efficiently transduced with LV. After reprogramming, iPSC exhibited a potent LV restriction phenotype, which was decreased by high vector amounts, HIV-1 CypA-independent capsid mutants, treatment with CSA during transduction or iPSC differentiation

the capsid mutants P90A and G89V utilize different nuclear entry routes compared to wt capsid [8]. We can therefore envision that, in addition to avoiding binding of CypA molecules to the capsid, also the different nuclear entry routes could play a role in relieving the restrictive phenotype of LV in iPSC.

In addition to a potential restriction factor, the block may also be explained by the lack of a necessary host co-factor. Thus, we performed a detailed microarray analysis of different iPSC clones and their parental fibroblasts and compared expression of previously reported HIV-1 host co-factors for nuclear entry and integration as well as of a set of murine nucleoporins (see Additional file 5). When compared to adult fibroblasts, iPSC exhibited similar or even higher expression levels of examined nucleoporins and important host co-factors for nuclear entry and integration. This suggests that differences in expression levels of the analyzed HIV-1 host co-factors and nucleoporins do not contribute to the underlying mechanism of the observed block.

As previously described, somatic and pluripotent stem cells evolved anti-viral defense mechanisms [12, 14, 31, 46–48]. The novel LV block we identified in iPSC is acquired during the process of reprogramming and relieved upon differentiation (Fig. 2a–c). In our experiments ESC and iPSC exhibited similarly low LV transduction rates and integration events compared to GV (Fig. 1c, e). In contrast, this phenomenon is less pronounced in ECC (Fig. 2d) [16]. This might be explained by the fact that ECC are supposed to represent a more differentiated status compared to PSC, including iPSC and ESC. Thus, we hypothesize that the LV restriction in iPSC may be associated with the pluripotent state.

Although, murine cells are not the natural host for HIV-1, lentiviruses were isolated from several species other than primates and the CypA-capsid interaction was also discovered to be highly conserved [49]. Moreover, the presence of endogenous lentiviruses in the genomes of a variety of different species demonstrates that lentiviruses are more widespread than previously assumed. Therefore, an active lentiviral defense mechanism in murine PSC is imaginable. Likewise, much remains to be discovered regarding anti-viral defense mechanisms in stem cells, especially in PSC. This knowledge can help to develop more efficacious anti-viral treatments in order to modulate cell autonomous immunity, and also increase gene transfer efficiencies for studies in basic biology and therapeutic approaches. Our findings document the characteristics of the restriction against HIV-1 found in PSC and contribute to our understanding of HIV-1-host interactions.

Conclusions

We identified CypA as a key player in LV restriction in iPSC and discovered impaired nuclear entry and proviral integration as underlying mechanisms for low transduction rates. To the best of our knowledge an HIV-1 block in PSC has not been previously described. Furthermore, the restriction was not observed in parental fibroblasts or differentiated iPSC progeny, which suggests that LV restriction is acquired through reprogramming into the pluripotent state. While our investigation demonstrated a novel HIV-1 restriction in murine PSC, it remained unclear whether the restriction is caused by a restriction factor or the absence of an essential host co-factor. While CypA was previously described to either support or inhibit wt HIV-1 infection in a cell type-dependent manner, the CypA-capsid interaction, we identified to be responsible for the restriction in iPSC, differs from previously described studies and clearly contributes to our understanding of the anti-viral activity in PSC.

Methods

Retroviral vector plasmids

Lentiviral vector plasmids pRRL.PPT.EFS.EGFP.pre [50–52] and pRRL.PPT.SFFV.DsRedExpress.pre [53] were referred to as LV and LV.DsRed, respectively. Gammaretroviral vectors pSERS11.EFS.EGFP.pre [52] and pSF91.DsRedExpress [53] were termed GV and GV.DsRed. Except for the LTR-driven GV.DsRed, all vectors had SIN designs with deletions of the viral U3 regions (ΔU3), instead containing the internal EFS promoter (short version/250 bp fragment of promoter/enhancer sequences from the human elongation factor 1 alpha gene) or, for LV.DsRed, the SFFV promoter (promoter/enhancer sequences from spleen focus forming virus). For reprogramming, the lentiviral "4-in-1" construct pRRL.PPT.SF.hOKSMco.id.Tom.pre.F+F3, co-expressing codon-optimized versions of the transcription factors Oct4, Klf4, Sox2 and wt c-Myc, was used [15, 54]. The construct contained heterospecific FRT sites (F and F3) to allow Flp recombinase-mediated excision of the reprogramming cassette [15, 55].

Cells and cultivation

Human 293T (embryonic kidney cell line) [56], human HT1080eCat (fibroblast cell line HT1080 [57] expressing the murine ecotropic receptor mCAT-1 [53]) and the murine ECC line F9 (kindly provided by S. P. Goff, Columbia University Medical Center, New York, NY, USA) were cultured in Dulbecco's modified Eagle's medium (DMEM) (Biochrom GmbH, Berlin, Germany) supplemented with 10% heat-inactivated fetal bovine serum (FBS) (GE Healthcare Europe GmbH, Freiburg,

Germany), 100 U/mL penicillin, 100 µg/mL streptomycin and 1 mM sodium pyruvate (all from PAN-Biotech, Aidenbach, Germany). CF-1 Mefs (MTI Global Stem, Gaithersburg, MD, USA) and C3H Mefs (kindly provided by T. Cantz, Hannover Medical School, Hannover, Germany) as well as freshly isolated murine adult fibroblasts (C57BL/6 p14f/f) were grown in low glucose DMEM (PAN-Biotech) supplemented with 15% heat-inactivated FBS, 100 U/mL penicillin, 100 µg/mL streptomycin, 2 mM L-glutamine, 1% MEM non-essential amino acids solution (Gibco Life Technologies GmbH, Darmstadt, Germany) and 100 µM beta-mercaptoethanol (Sigma-Aldrich, Munich, Germany) on gelatin pre-coated (0.1% gelatin in phosphate buffered saline (PBS)) wells or flasks. PSC were co-cultured with irradiated (30 Gy) C3H Mefs in ESC medium composed of Knockout DMEM (Gibco Life Technologies) with 15% ESC-tested and heat-inactivated FBS (GE Healthcare Europe GmbH), 2 mM L-glutamine, 100 U/mL penicillin, 100 µg/mL streptomycin, 1% MEM non-essential amino acids, 100 µM beta-mercaptoethanol and 10^3 U/mL LIF (kindly provided by the Department of Technical Chemistry, Leibniz University Hannover, Hannover, Germany). The following murine PSC were used in this study: iPSC clone #1 (C57BL/6 mouse strain), clones #2, #2EX (excised reprogramming cassette) and #3 (C57BL/6 p14f/f, wt version without conditional knockout of the p14 gene, reprogrammed from adult fibroblasts, mice kindly provided by D. Kotlarz and C. Klein, Ludwig Maximilian University, Munich, Germany), clone #4 (reprogrammed from C3H Mefs) and ESC (C57BL/6 mouse strain, kindly provided by I. Prinz, Hannover Medical School).

Retroviral particle production

One day before transfection, 5×10^6 293T cells were seeded per 10-cm dish. Transfection was performed based on the calcium phosphate precipitation method assisted by 25 µM chloroquine (Sigma-Aldrich). For LV or LV.DsRed, packaging cells were co-transfected with 5 µg LV plasmid, 5 µg pRSV-Rev (kindly provided by T. Hope, Northwestern University, Chicago, IL, USA), 1.5 µg pMD.G (VSVg) [58], 12 µg wt pcDNA3.gp.4xCTE (HIV-1 Gag-Pol) [59] or CypA-independent capsid mutant Gag-Pol P90A, G89V, A88T or H2.1 or the N74D mutant [36, 38, 39, 41]. Mutations were introduced by overlap extension PCR and confirmed by sequencing. For production of LV-Ampho(tropic) and LV-Eco(tropic) particles, 2 µg ecotropic [60] or 2 µg amphotropic [61] MLV envelope encoding plasmids were used. To package VSVg pseudotyped GV or GV.DsRed particles, 5 µg GV plasmid, 7 µg pcDNA3.MLV.GP (MLV Gag-Pol) [59] and 1.5 µg pMD.G were co-transfected. Supernatants were harvested 36 and 48 h after transfection, filtered through 0.22 µm pore-size-filters and concentrated (100×) by ultracentrifugation (2 h, 82,740×*g*, 4 °C) (SW32Ti rotor; Beckman Coulter GmbH, Krefeld, Germany). Viral pellets were resuspended in ESC medium and stored in aliquots at −80 °C until further usage.

Retroviral particle titration and transduction

One day before titration of vector supernatants, 7×10^4 HT1080eCat cells were seeded per well of a 12-well plate. The next day, the culture medium was removed and serial dilutions of vector supernatants were added to the cells. Viral transduction of cells was performed in the presence of 4 µg/mL protamine sulfate and spin-inoculation (1 h, 863×*g*, 37 °C). Three days after transduction, the percentage of fluorescent cells was analyzed by flow cytometry. Vector particle titers were calculated only including samples with < 30% transduced cells to avoid multiple proviral integration events per cell and false titer estimation. Mefs (3×10^4 cells per well of a 12-well plate) or ECC (5×10^4 cells per well of a 12-well plate) were seeded the day before transduction. Specific amounts of vector particles per cell, referred to as MOI, were applied and cells were treated as described above for titration. PSC were transduced in single cell suspension and separated with 0.5% Trypsin-EDTA in PBS (Gibco Life Technologies GmbH) on the day of transduction. Afterwards, PSC were centrifuged (138×*g*, 5 min) and depleted from feeder cells. Depletion was performed by incubating the cells on a 15-cm plate (40 min, 37 °C, 5% CO_2) to separate the faster attaching feeder from PSC. After applying vector supernatants at specific MOI to 3×10^4 cells (12-well plate) and adding 4 µg/mL protamine sulfate, a spin-inoculation step (1 h, 863×*g*, 37 °C) and an incubation step (1 h, 37 °C, 5% CO_2) were performed, before PSC were added to feeder cells. The next day, PSC had attached to the feeder cell layer, and the virus-containing medium was replaced by fresh ESC medium. For small molecule supplemented transductions, cells were incubated 30 min prior to and during transduction with indicated concentrations of CSA (Sigma - Aldrich), MG132 (Calbiochem/Merck Millipore, Darmstadt, Germany), Nevirapine (Sigma-Aldrich) or Raltegravir (Santa Cruz Biotechnology, Dallas, TX, USA). Depending on the cell type and experiment, cells were processed by flow cytometry 2 days (differentiated cultures), 3 days (HT1080eCat titration, C3H Mefs), 5–8 days (adult fibroblasts, ESC, ECC, iPSC) or at indicated time points after transduction.

Reprogramming murine primary fibroblasts to iPSC

Murine fibroblasts from the indicated mouse strains were reprogrammed and characterized as previously described by our group [15, 54].

Differentiation of iPSC

To induce differentiation, iPSC were depleted from feeder cells, seeded in gelatin pre-coated (0.1% gelatin in PBS) 12-well plates (5×10^4 per well) and cultured in ESC medium without LIF and feeder cells. Medium was exchanged daily and cells were split in the range of 1:5–1:20 on days 3–4 and 5–6. Prior to transduction of differentiated cultures, cells were stained for SSEA1 expression and analyzed by flow cytometry to investigate their stem cell status. Differentiated cultures were transduced with retroviral particles (MOI 100) encoding EGFP or DsRedExpress 6–9 days after starting differentiation. Cells transduced with DsRedExpress encoding vectors were additionally stained with Stain Alive SSEA1 Antibody DyLight 488 according to the manufacturer's protocol (Stemgent, Cambridge, MA, USA) and analyzed by fluorescence microscopy 2 days after transduction. Pictures were acquired with Axio Observer Z1 (Carl Zeiss AG, Jena, Germany) using Zeiss filter sets 43 (DsRedExpress) and 38 (DyLight 488) and AxioVision 4.8 software. Cells transduced with EGFP encoding vectors were analyzed by flow cytometry 2 days after transduction.

Flow cytometry

Cells were harvested and subsequently analyzed by flow cytometry (FACS Calibur, Becton-Dickinson, Heidelberg, Germany) using FlowJo software (Tree Star Inc, Ashland, OR, USA). In addition, PSC or differentiated cultures were stained with an Alexa 647 labeled SSEA1 antibody (1.5 ng per sample, 30 min, 4 °C) (eBioscience, San Diego, CA, USA). Samples were pre-gated for viable cells. Data from total viable cell populations with at least 50% SSEA1 positive cells were included in the analyses. The range of SSEA1 positive cells constituted mostly 70–100% throughout the study. The 70–100% range derives from differences between independent transduction experiments, but there were equivalent amounts of SSEA1 positive cells within each experiment. Cells, double positive for SSEA1 and EGFP, are shown in the graphs.

SYBR Green-based quantitative real-time PCR for detection of mean vector copy number per cell

iPSC were transduced with LV and GV and DNA was isolated 6–8 days after transduction. Vector copies per cell shown in Fig. 1c, e and Additional file 4B were determined by SYBR Green-based quantitative real-time PCR based on EGFP copies (Applied Biosystems, Darmstadt, Germany) using the QuantiTect SYBR-Green PCR Kit (Qiagen, Hilden, Germany) normalized to endogenous PTBP2 (EGFP for: 5′CTATATCATGGCCGACAAGCAGA3′, rev: 5′GGACTGGGTGCTCAGGTAGTGG3′; PTBP2 for: 5′GTCTCCATTCCCTATGTTCATGC3′, rev: 5′GTTCC CGCAGAATGGTGAGGTG3′). Master Mix preparation and PCR were performed as instructed by the manufacturer. Quantification of the mean vector copy number per cell were determined based on the comparative $2^{-\Delta\Delta Ct}$ method and based on a reference plasmid containing EGFP and PTBP2 sequences [62–64].

Analysis of RT products, 2-LTR circles and proviral integration by TaqMan-based quantitative real-time PCR

Vector supernatants were applied (MOI 10) to iPSC and CF-1 Mefs and remaining vector particles were removed by washing the cells twice with PBS 6 h after transduction. Cells were harvested at indicated time points. To test for plasmid contamination of vector supernatants, equivalent volumes of vector supernatants were heat-inactivated at 65 °C for 1 h and included in the analyses (plasmid ctrl). DNA isolation was performed with the QIAamp DNA Blood Mini Kit (Qiagen) according to the manufacturer's instructions. For TaqMan-based quantitative real-time PCR (Applied Biosystems), RT products were analyzed for early RT products (strong-stop DNA, for: 5′GCCTCAATAAAGCTTGCCTTGA3′, rev: 5′TGACTAAAAGGGTCTGAGGGATCT3′, probe: 5′AGAGTCACACAACAGACGGGCACACACTA3′), late RT products (U5/downstream PBS, for: 5′TA GTGTGTGCCCGTCTGTTG3′, rev: 5′GAGTCCTG CGTCGAGAGAG3′, probe: 5′TCCCTCAGACCCT TTTAGTCA3′) and 2-LTR circles (junction, for: 5′TAGTGTGTGCCCGTCTGTTG3′, rev: 5′CAGAGA GACCCAGTACAAGC3′, probe: 5′CTCTAGCAGTAAC TGGAAGGGCT3′). PTBP2 served as a housekeeping control (for: 5′TCTCCATTCCCTATGTTCATGC3′, rev: 5′GTTCCCGCAGAATGGTGAGGTG3′, probe: 5′ATGTTCCTCGGACCAACTTG3′). RT products were relatively quantified using the comparative $2^{-\Delta\Delta Ct}$ method [62, 63]. PCR was performed at 50 °C for 2 min and at 95 °C for 20 s followed by 60 cycles of 5 s at 95 °C, 20 s at 56 °C and 20 s at 65 °C.

B1-LTR PCR for analysis of proviral integration

Briefly, a first stage PCR was run with a LTR forward primer, which was designed to harbor a lambda-phage heel sequence at the 5′ end and LTR sequences at the 3′ end. In addition, two outward facing primers binding the highly redundant consensus sequence within the mouse B1 repetitive element were used in the first stage PCR. This was followed by a second stage TaqMan-based quantitative real-time PCR, which amplified the LTR using a lambda-specific forward and an LTR-specific reverse primer, scored using an LTR-specific probe. Primer sequences and protocol were used from Tervo et al. [65]. For data analyses, genomic DNA from a SC-1 standard cell line, harboring three HIV-1-based vector integrations [66], was diluted over a range of concentrations covering three logs. Slope and y-intercept of the standard curve were used to determine proviral copies. PTBP2 served as a housekeeping control. To exclude plasmid contamination of vector supernatants, 5 μM Nevirapine was used as an RT inhibitor, for every time point with equivalent volumes of vector supernatants, and signals were subtracted from the total signal. For each sample, PCR controls, omitting the B1 forward and reverse primers or the Lambda-LTR forward primer, were performed in parallel during the first PCR stage. The approach omitting the Lambda-LTR forward primer did not reveal any signal. The signal achieved by omission of B1 forward and reverse primers was subtracted from the total signal for each sample.

Western blot analysis

Cells were harvested, washed with PBS and lysed by using radioimmunoprecipitation assay buffer supplemented with proteinase inhibitors (Complete Mini, Roche, Mannheim, Germany). Samples (15 μg) were separated by sodium dodecyl sulfate-polyacrylamide gel electrophoresis (12.5%) and blotted onto nitrocellulose membranes (GE Healthcare Europe GmbH). Antibody probing was conducted with rabbit monoclonal anti-CypA (Cell Signaling Technology, Danvers, MA, USA) and rabbit polyclonal Erk2 (Santa Cruz Biotechnology) according to the manufacturer's instructions. Goat anti-rabbit IgG conjugated with horseradish peroxidase was used as a secondary antibody. Quantitative detection was carried out using the Fusion Fx system (Peqlab GmbH/VWR Life Science Competence Center, Erlangen, Germany).

Pre-transduction/abrogation experiments

iPSC were pre-transduced with LV.DsRed or GV.DsRed at an MOI of 1000. After spin-inoculation cells were incubated at 37 °C and 5% CO_2. After 6 h incubation time, iPSC were washed twice with PBS and a second transduction with LV was performed. LV encoding EGFP at an MOI of 100 were used for the second transduction. Flow cytometry analyses were conducted 4 days after transduction experiments.

Statistical analysis

Data were expressed as mean ± standard deviation. Where appropriate, we used one-way ANOVA with Tukey-Kramer post hoc test to adjust for multiplicity effects. When comparing all groups to one control group, we utilized one-way ANOVA with Dunnett post hoc test. For time course analyses, we utilized repeated measures one-way ANOVA with appropriate post hoc test. The unpaired t test was performed for comparison of two groups. In case of significantly different variances between the groups, the unpaired t test with Welch's correction was applied. p values of ≤ 0.05 were considered significant (*), $\leq$ 0.01 very significant (**), $\leq$ 0.001 extremely significant (***), and ns was considered not significant. Supplementary material and methods are described in Additional file 6.

Additional files

Additional file 1. MG132 exhibits cytotoxicity at doses ≥ 0.2 μM in iPSC. iPSC were treated with MG132 at indicated concentrations (n = 3). **(A)** Cell counts are shown after 12 hours. **(B)** Cells were stained for AnnexinV and PI after 12 hours and percentages of fluorescence positive cells are shown. Camptothecin served as a positive control for cytotoxicity.

Additional file 2. CSA exhibits no changes in cell growth or apoptosis at a concentration of 10 μM in iPSC. iPSC were treated with CSA at indicated concentrations and washed after 12 hours (n = 3). **(A)** Cell counts are shown after 48 hours. **(B)** Cells were stained for AnnexinV and PI after 48 hours and percentages of fluorescence positive cells are shown. Camptothecin served as a positive control for cytotoxicity.

Additional file 3. iPSC transduced with wt or N74D capsid mutants exhibit comparable late RT levels. iPSC were transduced with LV N74D capsid mutant and wt at an MOI of 100. Late RT products were analyzed with TaqMan-based quantitative real-time PCR with $2^{-\Delta\Delta Ct}$ method 24 hours after transduction. Data are shown from 3 independent retroviral supernatants (n = 3) and as a ratio of late RT product level and plasmid contamination control, for which Nevirapine (Nev) was used, relative to endogenous PTBP2 level. The unpaired t-test was used for statistical analysis. ns p = 0.669.

Additional file 4. LV nuclear entry is impaired in iPSC. LV were applied to iPSC and CF-1 Mefs at an MOI of 100 in the presence of 10 μM CSA and/or 50 μM Raltegravir or an equal volume of DMSO as solvent control. Data are shown from three independent retroviral supernatants (n = 3). **(A)** Relative 2-LTR circle levels were determined 48 hours after transduction and analyzed with TaqMan-based quantitative real-time PCR with the $2^{-\Delta\Delta Ct}$ method, and normalized to endogenous PTBP2 copies. Data are shown relative to Mefs treated with DMSO. One-way ANOVA with Tukey-Kramer post-hoc test was used for statistical analyses. ns p = 0.8338; ** p = 0.0013; *** p ≤ 0.001. **(B)** Relative vector copies were determined 21 days after transduction and analyzed with TaqMan-based quantitative real-time PCR with the $2^{-\Delta\Delta Ct}$ method, and normalized to endogenous PTBP2 copies. Data are shown relative to Mefs treated with DMSO. One-way ANOVA with Tukey-Kramer post-hoc test was used for statistical analysis. *** p ≤ 0.001.

Additional file 5. Microarray analysis comparison of iPSC and fibroblasts reveals similar or even higher expression of a set of HIV-1 host co-factors and nucleoporins. Heat map is shown for 2 independent preparations of primary adult fibroblasts (Ad fib I + II), which served as parental fibroblasts for reprogramming, and different murine iPSC clones (#3, #2, #2EX). **(A)** Log_2-intensity values for important HIV-1 host co-factors for nuclear entry and integration. **(B)** Log_2-intensity values for a set of murine nucleoporins.

Additional file 6. Supplementary material and methods.

Author details
[1] Institute of Experimental Hematology, Hannover Medical School, Carl-Neuberg-Str. 1, Hannover, Germany. [2] REBIRTH Cluster of Excellence, Hannover Medical School, Hannover, Germany. [3] Institute of Experimental Virology, TWINCORE, Centre for Experimental and Clinical Infections Research, Hannover, Germany. [4] Department of Biochemistry and Molecular Biophysics, Columbia University Medical Center, New York, NY, USA. [5] Department of Microbiology and Immunology, Columbia University Medical Center, New York, NY, USA. [6] Howard Hughes Medical Institute, Columbia University Medical Center, New York, NY, USA. [7] Division of Hematology/Oncology, Children's Hospital Boston, Harvard Medical School, Boston, MA, USA.

Authors' contributions

FG, MG, DH, SPG, A. Schambach designed the experiments and analyzed the data. FG performed the experiments Figs. 1, 2, 3, 4 and 5 and Additional files 1, 2, 3 and 4. TM cloned and validated the H2.1 capsid mutant. JWS illustrated Fig. 6 and helped to perform additional experiments for paper revision. JK generated and characterized iPSC clones #1, #2, #2EX and #3. DZ initially observed the LV block in iPSC. A. Schwarzer analyzed microarray data, performed bioinformatics and prepared Additional file 5. CG provided B1-LTR PCR protocol and contributed her expertise analyzing the data. FG, MG, DH, TM, JWS, SPG, and A. Schambach wrote the manuscript. All authors read and approved the final manuscript.

Acknowledgements

We thank Michael Morgan, Hildegard Büning and Verena Labenski for helpful discussion and assistance with the manuscript. We furthermore thank Thomas Neumann and Girmay Asgedom for technical assistance, Michael Rothe for consultation on quantitative PCR analysis and Steven Talbot for support and approval of statistical analyses. We are grateful to Eva Warlich for providing the C3H iPSC clone #4, to Daniel Griffin (Columbia Medical Center, New York, NY, USA) for the early RT TaqMan primer/probe, to Daniel Kotlarz and Christoph Klein (both Ludwig Maximilian University, Munich, Germany) for providing the C57BL/6 p14f/f mice as well as to Hanna-Mari Baldauf (Ludwig Maximilian University) for providing reference cell lines for the B1-LTR PCR. We would like to acknowledge the assistance of the Cell Sorting Core Facility of the Hannover Medical School supported in part by Braukmann-Wittenberg-Herz-Stiftung and Deutsche Forschungsgemeinschaft.

Competing interests

The authors declare that they have no competing interests.

Funding

This work was supported by grants from the Deutsche Forschungsgemeinschaft (SFB738, projects C4 and C9), Cluster of Excellence REBIRTH (EXC 62/1 SPP1230), the Bundesministerium für Bildung und Forschung (BMBF, IFB-Tx, PidNet), the DAAD (Modern Applications in Biotechnology, Short-Term USA Fellowship 2014), the European Union (FP7 project PERSIST) and NIH NCI (R01 CA30488). SPG is an investigator of the Howard Hughes Medical Institute. The funders had no role in study design, data collection and interpretation, or the decision to submit the work for publication.

References

1. Suzuki Y, Craigie R. The road to chromatin—nuclear entry of retroviruses. Nat Rev Microbiol. 2007;5(3):187–96.
2. Takahashi K, Yamanaka S. Induction of pluripotent stem cells from mouse embryonic and adult fibroblast cultures by defined factors. Cell. 2006;126(4):663–76.
3. Robinton DA, Daley G. The promise of induced pluripotent stem cells in research and therapy. Nature. 2012;481(7381):295–305.
4. Luban J, Bossolt KL, Franke EK, Kalpana GV, Goff SP. Human immunodeficiency virus type 1 Gag protein binds to cyclophilins A and B. Cell. 1993;73(6):1067–78.
5. Gamble TR, Vajdos FF, Yoo S, Worthylake DK, Houseweart M, Sundquist WI, et al. Crystal structure of human cyclophilin A bound to the amino-terminal domain of HIV-1 capsid. Cell. 1996;87(7):1285–94.
6. Braaten D, Franke EK, Luban J. Cyclophilin A is required for an early step in the life cycle of human immunodeficiency virus type 1 before the initiation of reverse transcription. J Virol. 1996;70(6):3551–60.
7. Shah VB, Shi J, Hout DR, Oztop I, Krishnan L, Ahn J, et al. The host proteins transportin SR2/TNPO3 and cyclophilin A exert opposing effects on HIV-1 uncoating. J Virol. 2013;87:422–32.
8. Schaller T, Ocwieja KE, Rasaiyaah J, Price AJ, Brady TL, Roth SL, et al. HIV-1 capsid-cyclophilin interactions determine nuclear import pathway, integration targeting and replication efficiency. PLoS Pathog. 2011;7(12):e1002439.
9. Rasaiyaah J, Tan CP, Fletcher AJ, Price AJ, Blondeau C, Hilditch L, et al. HIV-1 evades innate immune recognition through specific cofactor recruitment. Nature. 2013;503(7476):402–5.
10. Towers GJ, Hatziioannou T, Cowan S, Goff SP, Luban J, Bieniasz PD. Cyclophilin A modulates the sensitivity of HIV-1 to host restriction factors. Nat Med. 2003;9(9):1138–43.
11. Luban J, Cyclophilin A. TRIM5, and resistance to HIV-1 infection. J Virol. 2007;81:1054–61.
12. Wolf D, Goff SP. TRIM28 mediates primer binding site-targeted silencing of murine leukemia virus in embryonic cells. Cell. 2007;131(1):46–57.
13. Wolf D, Goff SP. Embryonic stem cells use ZFP809 to silence retroviral DNAs. Nature. 2009;458(7242):1201–4.
14. Schlesinger S, Lee AH, Wang GZ, Green L, Goff SP. Proviral silencing in embryonic cells is regulated by Yin Yang 1. Cell Rep. 2013;4(1):50–8.
15. Kuehle J, Turan S, Cantz T, Hoffmann D, Suerth JD, Maetzig T, et al. Modified lentiviral LTRs allow Flp recombinase-mediated cassette exchange and in vivo tracing of "factor-free" induced pluripotent stem cells. Mol Ther. 2014;22(5):919–28.
16. Alonso A, Breuer B, Steuer B, Fischer J. The F9-EC cell line as a model for the analysis of differentiation. Int J Dev Biol. 1991;35(4):389–97.
17. Aiken C. Pseudotyping human immunodeficiency virus type 1 (HIV-1) by the glycoprotein of vesicular stomatitis virus targets HIV-1 entry to an endocytic pathway and suppresses both the requirement for Nef and the sensitivity to cyclosporin A. J Virol. 1997;71(8):5871–7.
18. Yu D, Wang W, Yoder A, Spear M, Wu Y. The HIV envelope but not VSV glycoprotein is capable of mediating HIV latent infection of resting CD4 T cells. PLoS Pathog. 2009;5(10):e1000633.
19. Finkelshtein D, Werman A, Novick D, Barak S, Rubinstein M. LDL receptor and its family members serve as the cellular receptors for vesicular stomatitis virus. Proc Natl Acad Sci USA. 2013;110(18):7306–11.
20. McClure MO, Sommerfelt MA, Marsh M, Weiss RA. The pH independence of mammalian retrovirus infection. J Gen Virol. 1990;71(4):767–73.
21. Rasmussen I, Vilhardt F. Macropinocytosis is the entry mechanism of amphotropic murine leukemia virus. J Virol. 2015;89:1851–66.
22. Lu X, Silver J. Ecotropic murine leukemia virus receptor is physically associated with caveolin and membrane rafts. Virology. 2000;276(2):251–8.
23. Mothes W, Boerger AL, Narayan S, Cunningham JM, Young JA. Retroviral entry mediated by receptor priming and low pH triggering of an envelope glycoprotein. Cell. 2000;103(4):679–89.
24. Nussbaum O, Roop A, Anderson WF. Sequences determining the pH dependence of viral entry are distinct from the host range-determining region of the murine ecotropic and amphotropic retrovirus envelope proteins. J Virol. 1993;67(12):7402–5.
25. Besnier C, Takeuchi Y, Towers G. Restriction of lentivirus in monkeys. Proc Natl Acad Sci USA. 2002;99(18):11920–5.
26. Cowan S, Hatziioannou T, Cunningham T, Muesing MA, Gottlinger HG, Bieniasz PD. Cellular inhibitors with Fv1-like activity restrict human and simian immunodeficiency virus tropism. Proc Natl Acad Sci USA. 2002;99(18):11914–9.
27. Towers G, Bock M, Martin S, Takeuchi Y, Stoye JP, Danos O. A conserved mechanism of retrovirus restriction in mammals. Proc Natl Acad Sci USA. 2000;97(22):12295–9.
28. Towers G, Collins M, Takeuchi Y. Abrogation of Ref1 retrovirus restriction in human cells. J Virol. 2002;76(5):2548–50.
29. Münk C, Brandt SM, Lucero G, Landau NR. A dominant block to HIV-1 replication at reverse transcription in simian cells. Proc Natl Acad Sci USA. 2002;99(21):13843–8.
30. Noser JA, Towers GJ, Sakuma R, Dumont J-M, Collins MKL, Ikeda Y. Cyclosporine increases human immunodeficiency virus type 1 vector transduction of primary mouse cells. J Virol. 2006;80(15):7769–74.
31. Petrillo C, Cesana D, Piras F, Bartolaccini S, Naldini L, Montini E, et al. Cyclosporin A and rapamycin relieve distinct lentiviral restriction blocks in hematopoietic stem and progenitor cells. Mol Ther. 2014;23(2):352–62.
32. Santoni de Sio FR, Gritti A, Cascio P, Neri M, Sampaolesi M, Galli C, et al. Lentiviral vector gene transfer is limited by the proteasome at postentry steps in various types of stem cells. Stem Cells. 2008;26(8):2142–52.
33. Anderson JL, Campbell EM, Wu X, Vandegraaff N, Engelman A, Hope TJ. Proteasome inhibition reveals that a functional preintegration complex intermediate can be generated during restriction by diverse TRIM5 proteins. J Virol. 2006;80(19):9754–60.
34. Goujon C, Moncorgé O, Bauby H, Doyle T, Ward CC, Schaller T, et al. Human MX2 is an interferon-induced post-entry inhibitor of HIV-1 infection. Nature. 2013;502(7472):559–62.

35. Kane M, Yadav SS, Bitzegeio J, Kutluay SB, Zang T, Wilson SJ, et al. MX2 is an interferon-induced inhibitor of HIV-1 infection. Nature. 2013;502(7472):563–6.
36. Liu Z, Pan Q, Ding S, Qian J, Xu F, Zhou J, et al. The interferon-inducible MxB protein inhibits HIV-1 infection. Cell Host Microbe. 2013;14(4):398–410.
37. Berthoux L, Sebastian S, Sokolskaja E, Luban J. Cyclophilin A is required for TRIM5alpha-mediated resistance to HIV-1 in old world monkey cells. Proc Natl Acad Sci USA. 2005;102(41):14849–53.
38. Yoo S, Myszka DG, Yeh C, McMurray M, Hill CP, Sundquist WI. Molecular recognition in the HIV-1 capsid/cyclophilin A complex. J Mol Biol. 1997;269(5):780–95.
39. Rits MAN, van Dort KA, Münk C, Meijer AB, Kootstra NA. Efficient transduction of simian cells by HIV-1-based lentiviral vectors that contain mutations in the capsid protein. Mol Ther. 2007;15(5):930–7.
40. Lee K, Ambrose Z, Martin TD, Oztop I, Mulky A, Julias JG, et al. Flexible use of nuclear import pathways by HIV-1. Cell Host Microbe. 2010;7(3):221–33.
41. Price AJ, Fletcher AJ, Schaller T, Elliott T, Lee KE, KewalRamani VN, et al. CPSF6 defines a conserved capsid interface that modulates HIV-1 replication. PLoS Pathog. 2012;8(8):e1002896.
42. Ambrose Z, Lee K, Ndjomou J, Xu H, Oztop I, Matous J, et al. Human immunodeficiency virus type 1 capsid mutation N74D alters cyclophilin A dependence and impairs macrophage infection. J Virol. 2012;86:4708–14.
43. Sloan RD, Wainberg MA. The role of unintegrated DNA in HIV infection. Retrovirology. 2011;8(1):52.
44. Bukrinsky MI, Sharova N, Dempsey MP, Stanwick TL, Bukrinskaya AG, Haggerty S, et al. Active nuclear import of human immunodeficiency virus type 1 preintegration complexes. Proc Natl Acad Sci USA. 1992;89(14):6580–4.
45. De Iaco A, Luban J. Cyclophilin A promotes HIV-1 reverse transcription but its effect on transduction correlates best with its effect on nuclear entry of viral cDNA. Retrovirology. 2014;11:11.
46. Cabezas-Wallscheid N, Klimmeck D, Hansson J, Lipka DB, Reyes A, Wang Q, et al. Identification of regulatory networks in HSCs and their immediate progeny via integrated proteome, transcriptome, and DNA methylome analysis. Cell Stem Cell. 2014;15(4):507–22.
47. Griffin DO, Goff SP. HIV-1 is restricted prior to integration of viral DNA in primary cord-derived human CD34 + cells. J Virol. 2015;89(15):8096–100.
48. Griffin DO, Goff SP. Restriction of HIV-1-based lentiviral vectors in adult primary marrow-derived and peripheral mobilized human CD34 + hematopoietic stem and progenitor cells occurs prior to viral DNA integration. Retrovirology. 2016;13(1):14.
49. Goldstone DC, Yap MW, Robertson LE, Haire LF, Taylor WR, Katzourakis A, et al. Structural and functional analysis of prehistoric lentiviruses uncovers an ancient molecular interface. Cell Host Microbe. 2010;8(3):248–59.
50. Dull T, Zufferey R, Kelly M, Mandel RJ, Nguyen M, Trono D, et al. A third-generation lentivirus vector with a conditional packaging system. J Virol. 1998;72(11):8463–71.
51. Zufferey R, Dull T, Mandel RJ, Bukovsky A, Quiroz D, Naldini L, et al. Self-inactivating lentivirus vector for safe and efficient in vivo gene delivery. J Virol. 1998;72(12):9873–80.
52. Schambach A, Mueller D, Galla M, Verstegen MMA, Wagemaker G, Loew R, et al. Overcoming promoter competition in packaging cells improves production of self-inactivating retroviral vectors. Gene Ther. 2006;13(21):1524–33.
53. Voelkel C, Galla M, Dannhauser PN, Maetzig T, Sodeik B, Schambach A, et al. Pseudotype-independent nonspecific uptake of gammaretroviral and lentiviral particles in human cells. Hum Gene Ther. 2012;23:274–86.
54. Warlich E, Kuehle J, Cantz T, Brugman MH, Maetzig T, Galla M, et al. Lentiviral vector design and imaging approaches to visualize the early stages of cellular reprogramming. Mol Ther. 2011;19(4):782–9.
55. Turan S, Kuehle J, Schambach A, Baum C, Bode J. Multiplexing RMCE: versatile extensions of the Flp-recombinase-mediated cassette-exchange technology. J Mol Biol. 2010;402(1):52–69.
56. Pear WS, Nolan GP, Scott ML, Baltimore D. Production of high-titer helper-free retroviruses by transient transfection. Proc Natl Acad Sci USA. 1993;90(18):8392–6.
57. Rasheed S, Nelson-Rees WA, Toth EM, Arnstein P, Gardner MB. Characterization of a newly derived human sarcoma cell line (HT-1080). Cancer. 1974;33(4):1027–33.
58. Yee JK, Miyanohara A, LaPorte P, Bouic K, Burns JC, Friedmann T. A general method for the generation of high-titer, pantropic retroviral vectors: highly efficient infection of primary hepatocytes. Proc Natl Acad Sci USA. 1994;91(20):9564–8.
59. Schambach A, Bohne J, Chandra S, Will E, Margison GP, Williams DA, et al. Equal potency of gammaretroviral and lentiviral SIN vectors for expression of O6-methylguanine-DNA methyltransferase in hematopoietic cells. Mol Ther. 2006;13(2):391–400.
60. Morita S, Kojima T, Kitamura T. Plat-E: an efficient and stable system for transient packaging of retroviruses. Gene Ther. 2000;7(12):1063–6.
61. Beyer WR, Westphal M, Ostertag W, von Laer D. Oncoretrovirus and lentivirus vectors pseudotyped with lymphocytic choriomeningitis virus glycoprotein: generation, concentration, and broad host range. J Virol. 2002;76(3):1488–95.
62. Pfaffl MW. A new mathematical model for relative quantification in real-time RT-PCR. Nucleic Acids Res. 2001;29(9):e45.
63. Schmittgen TD, Livak KJ. Analyzing real-time PCR data by the comparative C(T) method. Nat Protoc. 2008;3(6):1101–8.
64. Maetzig T, Kuehle J, Schwarzer A, Turan S, Rothe M, Chaturvedi A, et al. All-in-one inducible lentiviral vector systems based on drug controlled Flp recombinase. Biomaterials. 2014;35(14):4345–56.
65. Tervo H-M, Goffinet C, Keppler OT. Mouse T-cells restrict replication of human immunodeficiency virus at the level of integration. Retrovirology. 2008;5:58.
66. Voelkel C, Galla M, Maetzig T, Warlich E, Kuehle J, Zychlinski D, et al. Protein transduction from retroviral Gag precursors. Proc Natl Acad Sci USA. 2010;107(17):7805–10.

2

Distinct gene expression signatures induced by viral transactivators of different HTLV-1 subgroups that confer a different risk of HAM/TSP

Tadasuke Naito[1], Jun-ichirou Yasunaga[2], Yuichi Mitobe[2,8], Kazumasa Shirai[3], Hiroe Sejima[1], Hiroshi Ushirogawa[1], Yuetsu Tanaka[4], Tatsufumi Nakamura[5], Kousuke Hanada[3], Masahiro Fujii[6], Masao Matsuoka[2,7] and Mineki Saito[1*]

Abstract

Background: Among human T cell leukemia virus type 1 (HTLV-1)-infected individuals, there is an association between HTLV-1 *tax* subgroups (subgroup-A or subgroup-B) and the risk of HAM/TSP in the Japanese population. To investigate the role of HTLV-1 subgroups in viral pathogenesis, we studied the functional difference in the subgroup-specific viral transcriptional regulators Tax and HBZ using microarray analysis, reporter gene assays, and evaluation of viral-host protein–protein interaction.

Results: (1) Transcriptional changes in Jurkat Tet-On human T-cells that express each subgroup of Tax or HBZ protein under the control of an inducible promoter revealed different target gene profiles; (2) the number of differentially regulated genes induced by HBZ was 2–3 times higher than that induced by Tax; (3) Tax and HBZ induced the expression of different classes of non-coding RNAs (ncRNAs); (4) the chemokine CXCL10, which has been proposed as a prognostic biomarker for HAM/TSP, was more efficiently induced by subgroup-A Tax (Tax-A) than subgroup-B Tax (Tax-B), in vitro as well as in unmanipulated (ex vivo) PBMCs obtained from HAM/TSP patients; (5) reporter gene assays indicated that although transient Tax expression in an HTLV-1-negative human T-cell line activated the CXCL10 gene promoter through the NF-κB pathway, there was no difference in the ability of each subgroup of Tax to activate the CXCL10 promoter; however, (6) chromatin immunoprecipitation assays showed that the ternary complex containing Tax-A is more efficiently recruited onto the promoter region of CXCL10, which contains two NF-κB binding sites, than that containing Tax-B.

Conclusions: Our results indicate that different HTLV-1 subgroups are characterized by different patterns of host gene expression. Differential expression of pathogenesis-related genes by subgroup-specific Tax or HBZ may be associated with the onset of HAM/TSP.

Keywords: HTLV-1, Virus subgroup, HAM/TSP, Tax, HBZ, Microarray

*Correspondence: mineki@med.kawasaki-m.ac.jp
[1] Department of Microbiology, Kawasaki Medical School, 577 Matsushima, Kurashiki, Okayama 701-0192, Japan
Full list of author information is available at the end of the article

Introduction

Human T-cell leukemia virus type 1 (HTLV-1) was the first human oncogenic retrovirus to be identified [1] and associated with distinct human diseases such as adult T-cell leukemia (ATL) [2, 3] and HTLV-1-associated myelopathy/tropical spastic paraparesis (HAM/TSP) [4, 5]. HTLV-1 can be divided into 7 "subtypes" (subtypes 1a–1g), including 5 African subtypes (subtypes 1b, 1d, 1e, 1f, and 1g), a Melanesian/Australian subtype (subtype 1c), and a cosmopolitan subtype (subtype 1a) based on a phylogenetic analysis of its long terminal repeat (LTR) region. The cosmopolitan subtype has spread worldwide and is further divided into 5 "subgroups" (subgroup A–E of subtype 1a) [6–8]. In the Japanese population, nucleotide sequence variation in the viral transactivator *tax* also determines the HTLV-1 subgroups—namely, "*tax* subgroup-A" and "*tax* subgroup-B" correspond to LTR-based "cosmopolitan subtype 1a subgroup A" and "cosmopolitan subtype 1a subgroup B," respectively [9]. We therefore refer to "*tax* subgroup-A" and "*tax* subgroup-B" as "subgroup-A" and "subgroup-B" hereafter.

It is well established that both the Tax and HBZ proteins of HTLV-1 transactivate viral and cellular genes and play a key role in HTLV-1 replication and pathogenesis [10–16]. A difference of four nucleotides exists in *tax* and *HBZ* coding regions (i.e., nucleotides 7897, 7959, 8208 and 8344) between subgroup-A Tax (Tax-A) and subgroup-B Tax (Tax-B), which result in two and one amino acid coding changes, respectively, in Tax and HBZ [9]. The most important observation concerning these virus subgroups is that the incidence of HAM/TSP in asymptomatic healthy carriers (HCs) infected with subgroup-A is 2.5 times higher than that in individuals infected with subgroup-B in southern Japan, where both subgroups co-exist [9]. Recently, we reported that this is also the case for inhabitants of Okinawa Prefecture, Japan, which consists of 160 islands and is located in the subtropical southernmost point of Japan [17]. We have also reported that although different HTLV-1 subgroups are characterized by different patterns of *HBZ* and *FoxP3* gene expression in HAM/TSP patients via independent mechanisms of direct transcriptional regulation, these differences do not significantly affect the clinical and laboratory characteristics of HAM/TSP patients [18]. Thus, the mechanism by which HTLV-1 subgroups differ in the risk for HAM/TSP is still largely unknown.

The rationale of this study is that a microarray-based study of subgroup-specific Tax- or HBZ-induced changes of cellular genes would reveal the downstream targets and effectors of these viral transcriptional factors and identify which targets differ between the viral strains. The results will cast light on the causes of HAM/TSP and identify attractive targets for novel therapeutics.

Methods

Patients and preparation of clinical samples

This study was approved by the Research Ethics Committee of Kawasaki Medical School (approval number: 1422-3). Written informed consent was obtained from all individuals. Clinical samples from 37 patients with HAM/TSP (19 subgroup-A and 18 subgroup-B infected patients), 20 HCs, and 20 HTLV-1-uninfected normal control subjects (NCs) were analyzed. The diagnosis of HAM/TSP was made according to the World Health Organization diagnostic criteria [19]. The detail information of the patients' characteristics including proviral load (PVL) was presented in Table 1. Fresh peripheral blood mononuclear cells (PBMCs) were isolated using Histopaque-1077 (Sigma, St. Louis, MO, USA) density gradient centrifugation, washed twice in RPMI medium, and stored in liquid nitrogen as stocked lymphocytes until use.

Cells

Twelve HTLV-1-infected human T-cell lines (MT2, MT4, C5MJ, MT1, ATL43Tb, ATL55T, ED, TLOm1, ILT-M1, HCT1, HCT4, and HCT5) and three HTLV-1-uninfected T-cell lines (Jurkat, CEM, and Molt4) were used in this study. MT2, MT4, and C5MJ are chronically HTLV-1-infected cell lines derived from cord blood mononuclear cells exposed to HTLV-1 from patients with ATL, i.e., HTLV-1-transformed T-cell lines. MT1, ATL43Tb, ATL55T, ED, and TLOm1 are HTLV-1-infected cell lines derived from patients with ATL. Among these ATL cell lines, only ATL55T is IL-2-dependent. ILT-M1, HCT1, HCT4, and HCT5 are IL-2-dependent HTLV-1-infected T-cell lines derived from patients with HAM/TSP. The Tax-inducible JPX-9 cell line was derived from the Jurkat HTLV-1-negative human T-cell leukemia cell line and expresses biologically active Tax protein under the

Table 1 Clinical profiles of HTLV-1-associated myelopathy/tropical spastic paraparesis (HAM/TSP) patients

	Tax A (n = 19)	Tax B (n = 18)	*p* value
Age at onset	47.0 ± 16.7	44.2 ± 15.9	0.3467
Sex (Male: Female)	9: 19	14: 70	0.1374
Serum anti-HTLV-1 antibody titer[a]	283.4 ± 461.3	441.4 ± 1049.0	0.6889
HTLV-1 provirus load in PBMCs[b]	873.2 ± 1046.9	745.3 ± 609.2	0.7394
OMDS[c]	5.9 ± 2.1	5.8 ± 2.3	0.6908
Duration of illness	15.1 ± 9.2	17.1 ± 9.8	0.3621

The results represent the mean ± SD

[a] Anti-HTLV-1 antibodies were titrated using the particle agglutination method

[b] HTLV-1 Tax copy number per 10^4 PBMCs

[c] OMDS: Osame Motor Disability Score (see Additional file 1: Table S1)

control of the metallothionein promoter [20]. These cells were cultured in RPMI 1640 medium supplemented with 10% heat-inactivated fetal calf serum (FCS), 50 U/ml penicillin, and 50 μg/ml streptomycin (Wako Pure Chemical, Osaka, Japan) at 37 °C and 5% CO_2. For IL-2-dependent cell lines, 10 U/ml (for ATL55T, HCT1, HCT4, and HCT5) or 30 U/ml (for ILT-M1) of recombinant human IL-2 (Wako) were added to the culture. Jurkat Tet-On 3G (Takara Bio USA, Mountain View, CA) is a Jurkat-derived T-cell line made for use in the Tet-On 3G Inducible Gene Expression System (Takara Bio USA). Jurkat Tet-On 3G cells were cultured in RPMI1640 supplemented with 10% FCS, 100 units/ml penicillin, 100 μg/ml streptomycin (Wako), and 200 μg/ml G418 (SIGMA-ALDRICH Japan, Tokyo, Japan) at 37 °C in 5% CO_2. Human embryonic kidney 293T cells were grown in Dulbecco's modified Eagle's medium (DMEM) supplemented with 4 mM glutamine, 10% FCS, 100 units/ml penicillin, and 100 μg/ml streptomycin at 37 °C and 5% CO_2.

Plasmid construction

All primers used for the plasmid construction are listed in Additional file 2: Table S2. The following pTRE3G-based plasmids were constructed for inducible expression in the presence of doxycycline (Dox) (Takara Bio USA) in Jurkat Tet-On 3G cells. To construct pTRE3G-Tax-A or pTRE3G-Tax-B, DNA fragments corresponding to the Tax coding sequence were amplified via PCR, using Tax-FOR and Tax-REV as primers and pCG-Tax-A or pCG-Tax-B [18] as a template. PCR products were digested with SalI and BglII, then cloned into SalI- and BamHI-digested pTRE3G-IRES (Takara Bio USA). To construct pTRE3G-HBZ-A or pTRE3G-HBZ-B, DNA fragments corresponding to the HBZ coding sequence were amplified via PCR, using HBZ-FOR and HBZ-REV as primers and pCMV-HA-HBZ-A or pCMV-HA-HBZ-B [18] as the template. PCR products were digested with SalI and BglII, then cloned into SalI- and BglII-digested pTRE3G-IRES.

The following pCAGGS-P7- and pGL3-based plasmids were constructed for the immunoprecipitation assay and luciferase assay. To construct pCAGGS-P7-Tax-A and pCAGGS-P7-Tax-B, DNA fragments corresponding to the Tax coding sequence were amplified by PCR using Tax-FOR and Tax-Stop-REV as primers and pCG-Tax-A or pCG-Tax-B [18] as a template. PCR products were phosphorylated with T4 polynucleotide kinase and digested with SalI and cloned into SalI- and EcoRV-digested pCAGGS-P7.

To construct pCAGGS-P7-Tax-A-FLAG, pCAGGS-P7-Tax-B-FLAG, and pCAGGS-P7-Tax-1 (225-232)-FLAG, DNA fragments corresponding to the Tax coding sequence were amplified via 1st-PCR using Tax-FOR and Tax-FLAG-REV as primers and pCAGGS-P7-Tax-A, pCAGGS-P7-Tax-B, or CSII-EF-Tax-1(225-232)-FLAG [21] as the templates. To add a tandem FLAG-tag sequence to the C-terminus of Tax proteins, DNA fragments were amplified via 2nd-PCR using Tax-FOR and FLAG-Tandem-REV as primers and the 1st-PCR products as templates. Next, to fuse the NotI-recognized DNA sequence at the 3′ site of the FLAG-tag, DNA fragments were amplified via 3rd-PCR using Tax-FOR and FLAG-Tandem-NotI-REV as primers and the 2nd-PCR products as templates. The 3rd-PCR products were digested with SalI and NotI, and cloned into SalI- and NotI-digested pCAGGS-P7. To construct pCAGGS-P7-Tax-2B-FLAG, DNA fragments were amplified via PCR using Tax-2-FOR and Tax-2-FLAG-Tandem-REV as primers and CSII-EF-Tax-2-FLAG [21] as the template. To add a tandem FLAG-tag sequence to the C-terminus of Tax proteins, DNA fragments were amplified via 2nd-PCR using Tax-2-FOR and FLAG-Tandem-REV as primers and the 1st-PCR products as templates. Next, to fuse the NotI-recognized DNA sequence to the 3′ site of the FLAG-tag, DNA fragments were amplified via 3rd-PCR using Tax-2-FOR and FLAG-Tandem-*NotI*-REV as primers and the 2nd-PCR products as templates. The 3rd-PCR products were digested with SalI and NotI, and cloned into SalI- and NotI-digested pCAGGS-P7.

To amplify the human C-X-C motif chemokine 10 (CXCL10) promoter region, harboring a sequence spanning nucleotide positions − 875 to + 97 (where the transcription start site is set to be + 1) or − 279 to + 97 [22], PCR was carried out using CXCL10-875-FOR or CXCL10-279-FOR and CXCL10-REV as primer sets and genomic DNA derived from Jurkat cells as templates. PCR products were digested with KpnI and XhoI and cloned into KpnI- and XhoI-digested pGL3-Basic (Promega Corporation, Madison, WI). The resultant plasmid was designated pGL3-CXCL10-Long-Luc or pGL3-CXCL10-Short-Luc. Mutations in the NF-κB1 (5′-GGGACTTCCC-3′), NF-κB2 (5′-GGGAAATTCC-3′), and AP-1 (5′-TAAGTCA-3′) binding sites in the CXCL10 promoter were generated using overlapping-PCR using the primers listed in Additional file 2: Table S2. To construct pGL3-CXCL10 (κB1-mut)-Luc (κB1 mutant sequence: 5′-GTGACTTCAC-3′), two DNA fragments corresponding to the CXCL10 promoter sequence were amplified via PCR using CXCL10-279-FOR and κB1-mut-REV or CXCL10-REV and κB1-mut-FOR as primers and pGL3-CXCL10-Short-Luc as the PCR template. Full-length CXCL10 (κB1-mut) DNA fragment was amplified via PCR using CXCL10-279-FOR and CXCL10-REV as primers. PCR products were digested with KpnI and XhoI and cloned into the KpnI- and XhoI-digested pGL3-Basic. Likewise, pGL3-CXCL10 (κB2-mut)-Luc (κB2 mutant sequence: 5′-GTGAAATTAC-3′), pGL3-CXCL10 (κB1+κB2-mut)-Luc (mutation of both the κB1 and κB2), and pGL3-CXCL10 (AP-1-mut)-Luc (AP-1 mutant sequence: 5′-TAAGAGA-3′) were constructed using primers listed in Additional file 2: Table S2.

The sequences of all recombinant plasmids were confirmed using Sanger sequencing.

Transfection by electroporation

DNA electroporation was performed with the Neon™ Transfection System (Invitrogen, Carlsbad, CA). For the electroporation, Jurkat and Jurkat Tet-On 3G cells were grown in non-treated cell culture dishes and transferred into a 15 ml polypropylene tube. Cells were centrifuged at 700×*g* for 3 min. The pellet was re-suspended in 10 ml of PBS, and cells were counted. Cells were pelleted again and re-suspended in Buffer R (included with Neon™ Kits) to a final concentration of 2.0×10^7/ml. 100 μl or 10 μl of cell suspension containing 2.0×10^6 cells or 2.0×10^5 cells, respectively, and 10 μg or 3 μg of DNA plasmid were transferred into the Neon™ tip (100 μl tip or 10 μl tip). The electroporation was carried out under appropriate conditions as per the manufacturer's instructions. After 24 h transfection, Tax-A, Tax-B, HBZ-A, or HBZ-B proteins were induced in Jurkat Tet-On 3G cells by adding Dox (final concentration: 3 ng/ml) for 24 h. Uninduced Jurkat Tet-On 3G cells were used as a baseline reference.

Western blotting

Whole-cell lysates were subjected to SDS–polyacrylamide gel electrophoresis (SDS-PAGE) and transferred to polyvinylidene difluoride (PVDF) membranes (pore size 0.45 μm, Merck Millipore, MA, USA). PVDF membranes were blocked with 5% skim milk or Blocking One (Nacalai tesque, Kyoto Japan) in Tris-buffered saline containing 0.1% Tween 20 (TBS-T) and probed with anti-Tax mouse monoclonal (Lt-4), anti-HBZ mouse monoclonal (#7-1) [23], anti-HBZ rat monoclonal (#4B12), anti-β-actin rabbit polyclonal (PM053, MEDICAL & BIOLOGICAL LABORATORIES, Nagoya, Japan), or anti-α-tubulin rabbit polyclonal (PM054, MEDICAL & BIOLOGICAL LABORATORIES) antibodies. PVDF membranes were washed with TBS-T and incubated with IRDye 680RD Goat anti-Mouse IgG, IRDye 800CW goat anti-rat IgG, or IRDye 800CW goat anti-rabbit IgG (LI-COR Biosciences, Lincoln, NE). After washing with TBS-T, the proteins were detected using Odyssey CLx Infrared Imaging System (LI-COR Biosciences).

Microarray analysis

The microarray experiments were performed in triplicate, and data are shown as mean values.

RNA processing and labeling

After 24 h of cultivation, the transfected cells were harvested and processed for subsequent analysis. RNA was prepared using the RNAeasy Mini Kit column purification (QIAGEN, Tokyo, Japan) following the manufacturer's instructions. RNA was quantified using a NanoVue™ spectrophotometer (GE Healthcare Japan, Tokyo, Japan) and quality was monitored using 1% Agarose/Formaldehyde gel electrophoresis. Cyanine-3 (Cy3) labeled cRNA was prepared from 0.1 μg RNA using the One-Color Low Input Quick Amp Labeling Kit (Agilent Technologies, Tokyo, Japan) according to the manufacturer's instructions, followed by RNAeasy Mini Kit column purification (QIAGEN). Dye incorporation and cRNA yield were assessed using the NanoDrop™ spectrophotometer (Thermo Fisher Scientific) and the Agilent 2100 Bioanalyzer (Agilent).

cRNA hybridization and scanning

600 ng of Cy3-labelled cRNA (specific activity > 6.0 pmol Cy3/μg cRNA) was fragmented at 60 °C for 30 min in a reaction volume of 25 μl containing 1 × Agilent fragmentation buffer and 2 × Agilent blocking agent following the manufacturer's instructions. On completion of the fragmentation reaction, 25 μl of 2 × GE Agilent hybridization buffer, HI-RPM was added to the fragmentation mixture and hybridized to Agilent SurePrint G3 Human GE 8 × 60 K V2 Microarrays (Catalog Code: G4851B) and V3 Microarrays (G4851C) (Agilent) for 17 h at 65 °C in a rotating Agilent hybridization oven. After hybridization, microarrays were washed for 1 min at room temperature with GE Wash Buffer 1 (Agilent) and 1 min with 37 °C

GE Wash buffer 2 (Agilent), then dried immediately by brief centrifugation. Slides were scanned immediately after washing on the Agilent DNA Microarray Scanner (G2505C) (Agilent) using one color scan setting for 8 × 60 k array slides (Scan Area 61 × 21.6 mm, Scan resolution 3 μm, dye channel set to Green).

The scanned images were analyzed using Feature Extraction Software 11.0.1.1 (Agilent) at default parameters (protocol GE1-1100_Jul11 and Grid: 039494_D_F_20140813) to obtain background subtracted and spatially detrended Processed Signal intensities. Features flagged in Feature Extraction as Feature Non-uniform outliers were excluded. Data were analyzed using GeneSpring GX (Agilent). Microarray data have been deposited in the NCBI Gene Expression Omnibus (www.ncbi.nlm.nih.gov/geo/) (GEO ID: GSE103323 and GSE121201). To detect differentially expressed genes, we applied Significance Analysis of Microarrays (SAM) with R package 'samr'. In the SAM analysis, we used two-class unpaired comparison with a threshold of q-value < 0.05 between control and Tax-A, control and Tax-B, control and HBZ-A, control and HBZ-B. When a gene has more than two probes, fold-change in the gene is defined to be the average fold-changes in multiple probes assigned to be the gene.

Genomic DNA and RNA extraction and cDNA synthesis

Genomic DNA was extracted from PBMCs using the QIAamp Blood Kit (Qiagen, Tokyo, Japan). RNA was extracted from PBMCs using the RNeasy Mini Kit with on-column DNase digestion (Qiagen). cDNA was synthesized using the PrimeScript® RT Reagent Kit (Takara, Kyoto, Japan). All reaction procedures were performed as suggested by the manufacturer.

Quantification of HTLV-1 proviral load and anti-HTLV-1 antibody titers

To examine the HTLV-1 proviral load (PVL), quantitative PCR was performed using 100 ng of genomic DNA (roughly equivalent to 10^4 cells) from PBMCs as previously reported [24]. The amount of HTLV-1 proviral DNA was determined using the following formula: copy number of HTLV-1 *tax* per 1×10^4 PBMCs = [(copy number of tax)/(copy number of β − actin/2)] × 10^4. All samples were examined in triplicate. Antibody titers to HTLV-1 were determined using a particle agglutination method (Serodia-HTLV-1®; Fujirebio, Tokyo, Japan), according to the manufacturer's instructions.

Real-time quantitative reverse transcription PCR analysis

To estimate *tax* and *HBZ* mRNA expression levels, a real-time quantitative reverse transcription PCR (real-time qRT-PCR) was performed as previously reported [14]. Human *CXCL10* (Applied Biosystems Hs01085834_m1) gene specific assays were used for *CXCL10* quantifications. The expression levels of these genes were normalized to the expression levels of human hypoxanthine phosphoribosyltransferase 1 (*HPRT1*) (Human HPRT1 Endogenous Control 4333768; Applied Biosystems, Foster City, CA). All assays were performed in triplicate.

ELISA

Human CXCL10 levels were determined using ELISA, employing the Quantikine® ELISA Human CXCL10/IP-10 Immunoassay kit (R&D Systems, Minneapolis, MN) according to the manufacturer's instruction.

Luciferase assay

To compare CXCL10 promoter activity, Jurkat cells (2.0×10^5 cells) were transfected with 1.5 μg of firefly luciferase reporter plasmid, i.e., pGL3-CXCL10-Long-Luc, pGL3-CXCL10-Short-Luc, pGL3-CXCL10 (κB1-mut)-Luc, pGL3-CXCL10 (κB2-mut)-Luc, pGL3-CXCL10 (κB1 + κB2-mut)-Luc, or pGL3-CXCL10 (AP-1-mut)-Luc, 1.5 μg of Tax protein expression plasmid pCAGGS-P7-Tax-A-FLAG or pCAGGS-P7-Tax-B-FLAG), and 250 ng of a *Renilla* reporter plasmid (pRL-RSV) per well by electroporation. The reporter construct containing a luciferase gene fused to five repeats of the NF-κB site of the IL-2Ra gene [25] was tested with same procedure. We also tested CXCL10 promoter activity using 293T cells. The reporter plasmid pGL3-CXCL10-Long-Luc or pGL3-CXCL10-Short-Luc, 1.5 μg of Tax protein expression plasmid (pCG-Tax-A or pCG-Tax-B), and 250 ng of a *Renilla* reporter plasmid (pRL-RSV) were transfected into 293T cells (2.0×10^5 cells) per well by using Lipofectamine LTX with PLUS reagent (Invitrogen, Carlsbad, CA, USA) according to the manufacturer's instructions. After 24 h transfection, cells were harvested and lysed, and reporter activities were measured using the Dual-Luciferase Reporter Assay System (Promega, Tokyo, Japan). Each experiment was performed in triplicate, and the data represent the mean ± SD of three independent experiments. Firefly Luciferase values were normalized to values of the *Renilla* reporter.

Immunoprecipitation assay

Jurkat cells (2.0 × 10^6 cells) were transfected with 10 μg of each indicated plasmid, i.e., pCAGGS-P7-empty, pCAGGS-P7-Tax-A-FLAG, pCAGGS-P7-Tax-B-FLAG, pCAGGS-P7-Tax-1(225-232)-FLAG, and pCAGGS-P7-Tax-2B-FLAG, by electroporation and then cultured for 24 h at 37 °C. The harvested cells were solubilized using a lysis buffer (20 mM Tris–HCl [pH 7.9], 100 mM NaCl, 0.1% Triton X-100). After sonication, homogenates were centrifuged at 14,000×*g* at 4 °C for 5 min, and the supernatant fraction was used as extracts for immunoprecipitation. Then, Tax protein was immunoprecipitated with anti-FLAG gels. The cell extracts were incubated with ANTI-FLAG M2 Affinity Gel (SIGMA-ALDRICH Japan, Tokyo, Japan) at 4 °C for 1 h. After incubation, the resins were collected via brief centrifugation and washed two times with the lysis buffer. The resin-bound proteins were eluted by boiling in the SDS-PAGE Sample Loading Buffer (Takara Bio USA) and subjected to 10% SDS-PAGE, followed by western blotting using anti-FLAG M2 (SIGMA-ALDRICH), anti-p105/p50 (#3035: Cell Signaling Technology Japan, Tokyo, Japan), anti-p100/p52 (#4882: Cell Signaling Technology), anti-p65 (#4764: Cell Signaling Technology), anti-RelB (#4922: Cell Signaling Technology), and anti-c-Rel (#4727: Cell Signaling Technology) antibodies. An aliquot of the cell lysates removed before immunoprecipitation was also characterized as an input (Input). The amounts of p100, p52, RelB, p65, p105, p50, and c-Rel proteins in immunoprecipitated samples were quantified by the densitometry scanning of corresponding bands of the western blot using Image J software.

Chromatin immunoprecipitation (ChIP) assay

Jurkat cells (2.0 × 10^6 cells) were co-transfected with 9 μg of each indicated plasmid, i.e., pCAGGS-P7-empty, pCAGGS-P7-Tax-A-FLAG, and pCAGGS-P7-Tax-B-FLAG, and 1 μg of pEF-321T (plasmid encoding SV40 T antigen for amplification of transfected pCAGGS-P7) by electroporation and then cultured for 24 h at 37 °C. Transfected cells were treated with 1% formaldehyde at room temperature for 15 min. Fixation was quenched by the addition of glycine at the final concentration of 125 mM, then cells were washed twice with PBS and collected by centrifugation. Cell pellets were resuspended in ChIP lysis buffer (50 mM Tris–HCl [pH 7.9], 1% Triton X-100, 0.1% SDS, and 10 mM EDTA). Cell lysates were sonicated to shear the chromatin DNA to ~ 500 base-pairs in size and then diluted with four volumes of ChIP dilution buffer (12.5 mM Tris–HCl [pH 7.9], 187.5 mM NaCl, and 1% Triton X-100). Lysates clarified by centrifugation were incubated with anti-Lt-4 antibody at 4 °C overnight. Antibody-protein-DNA complexes were incubated with Salmon Sperm DNA/Protein A Agarose (MILLIPORE, Temecula CA), and immunoprecipitates collected onto Protein A Agarose were washed with a high salt wash buffer (20 mM Tris–HCl [pH 7.9], 500 mM NaCl, 2 mM EDTA, 0.1% SDS, and 1% Triton X-100), a low salt wash buffer (20 mM Tris–HCl[pH 7.9], 100 mM NaCl, 2 mM EDTA, 0.1% SDS, and 1% Triton X-100), LiCl buffer (10 mM Tris–HCl[pH 7.9], 250 mM LiCl, 1% NP-40, and 1 mM EDTA), and twice with TE (10 mM Tris–HCl[pH 7.9] and 1 mM EDTA) successively. Then, bound proteins were eluted from the agarose beads in elution buffer (1% SDS and 100 mM $NaHCO_3$) by incubation at room temperature for 15 min. Crosslinking was reversed by incubation at 65 °C overnight. All samples were treated with 45 μg/ml of proteinase K (Nacalai tesque, Kyoto, Japan) at 50 °C for 2 h, then 40 μg of glycogen (Nacalai tesque) was added, followed by extraction with phenol:chloroform:isoamyl alcohol and precipitation with ethanol. DNA fragments were subjected to PCR or quantitative real-time PCR using EmeraldAmp Master Mix (TaKaRa, Shiga, Japan) or SYBR Premix EX Taq II (TaKaRa, Shiga, Japan) with specific primer sets: 5′-GAA AGT GAA ACC TAA TTC ACT ATT ACC AA-3′ and 5′-ACT TAG CAA AAC CTG CTG GCT GTT-3′ for genomic DNA fragment containing both NF-κB and AP-1 binding sites. Three independent experiments were performed, and the data represent the mean ± SD.

Statistical analysis

To test for significant differences among four different groups of subjects (HAM/TSP patients with subgroup-A, HAM/TSP patients with subgroup-B, HCs and NCs), the data were statistically analyzed using one-way analysis of variance (ANOVA). Inter-group comparisons were done by Scheffe's post hoc multiple comparisons test. The Mann–Whitney U test was used to compare data between two groups (i.e., difference between Tax-A and Tax-B with respect to transcriptional activity for both the short and long CXCL10 promoters, see Fig. 6b). Correlations between variables were examined using Spearman rank correlation analyses. The results shown are the mean ± SD where applicable. Results were considered statistically significant at p values < 0.05.

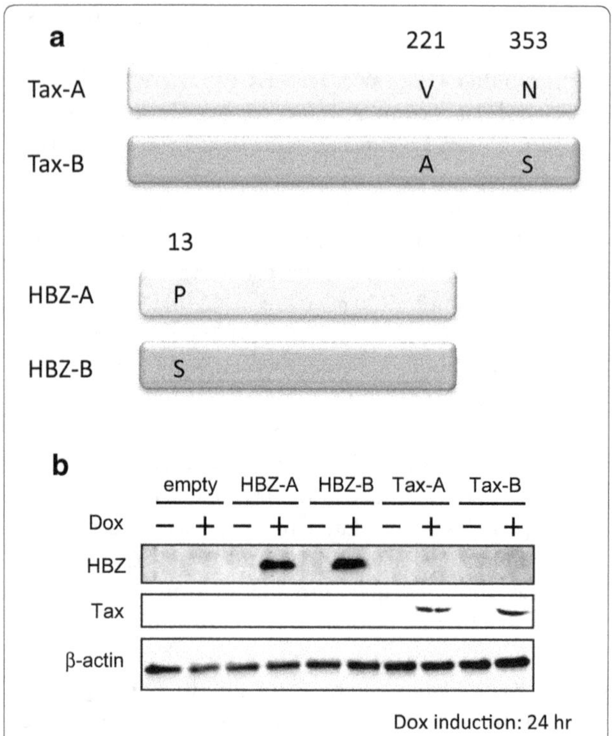

Fig. 1 Expression of each subgroup Tax or HBZ protein under transcriptional control of a doxycycline (Dox)-inducible promoter. **a** Schematic representation of the subgroup-specific Tax and HBZ proteins. Proteins are shown as grey bars, and differences in amino acid sequences between subgroup-specific Tax, i.e., Tax-A and Tax-B, or HBZ, i.e., HBZ-A and HBZ-B, proteins are indicated by a single letter amino acid abbreviation. **b** The Dox-induced expression of each Tax or HBZ protein is shown. Western blot analysis confirmed that each subgroup-specific Tax and HBZ protein was appropriately induced following Dox treatment

Results

Expression of each subgroup Tax or HBZ protein under the control of an inducible promoter revealed different target gene profiles

The differences in amino acid sequences between subgroup-specific Tax, i.e., Tax-A and Tax-B, or HBZ, i.e., HBZ-A and HBZ-B, proteins are depicted in Fig. 1a. Protein expression in Jurkat Tet-On 3G cells was induced by adding Dox to culture media at a final concentration of 3 ng/ml for 24 h at 37 °C. Western blot analysis confirmed that each subgroup-specific Tax and HBZ protein was appropriately induced following Dox treatment (Fig. 1b). Microarray was performed on RNAs derived from these samples.

The results of microarray analysis are shown in Fig. 2. In our microarray analysis, genes were considered differentially expressed on the significant level using SAM analysis (q-value < 0.05). The numbers of genes that occurred in each cluster are shown in Fig. 2a; we identified 231, 676, and 1712 genes specifically regulated by Tax-A, by both Tax-A and Tax-B, and by Tax-B, respectively. In HBZ, 500, 731, and 6843 genes were specifically regulated by HBZ-A, by both HBZ-A and HBZ-B, and by HBZ-B respectively. These data indicated that the number of genes regulated by HBZ was 2–3 times higher than that regulated by Tax.

Next, to define the gene expression profiles of each subgroup-specific Tax or HBZ, we manually created a list of the top 50 genes up-regulated by each subgroup of Tax, i.e., Tax-A and -B) (Table 2), or HBZ, i.e., HBZ-A and -B (Table 3). The data show the average fold-change expression in each subgroup-specific Tax (Table 2) or HBZ (Table 3) in Dox-treated Jurkat Tet-On 3G cells, compared to untreated cells. Gene expression levels were represented with red (high expression) or blue (low expression) in each block. The genes are ranked in order of their average fold-change expression of Tax-A, and the next column represents the average fold-change expression of Tax-B. Interestingly, as shown in Table 2, all of the top 50 genes regulated by Tax were more strongly upregulated by Tax-A than Tax-B (Table 2). In contrast, as shown in Table 3, some target genes were preferentially induced by HBZ-A, rather than by HBZ-B, whereas other target genes were preferentially induced by HBZ-B, rather than by HBZ-A. There were also some target genes regulated by both subtypes HBZ at similar efficiency. These results indicate that the Tax and HBZ molecule of each respective subgroup has a different target gene profile. Importantly, some of the genes listed in Table 2 (shown in red letters), such as EBI3 (= Epstein-Barr virus-induced gene 3, a component of IL-27) [26], VCAM-1 [27], IL-13 [28], and CCL1 [29] have already been identified as Tax target genes in previous studies, validating our gene induction system. It is also noteworthy that among the Tax- or HBZ-regulated genes, there were a set of ncRNAs, including miRNAs and long non-coding RNAs (lncRNAs) (Fig. 2b and Tables 2, 3).

Validation of differentially expressed genes identified by microarray using qRT-PCR in PBMCs of HAM/TSP patients with HTLV-1 subgroup A or B

To confirm whether the obtained in vitro gene expression profile had clinical significance, we performed real-time qRT-PCR on the PBMCs of 37 patients with HAM/TSP (19 subgroup-A and 18 subgroup-B), 20 HCs, and 20 NCs (Fig. 3). Among the identified Tax-responsive genes,

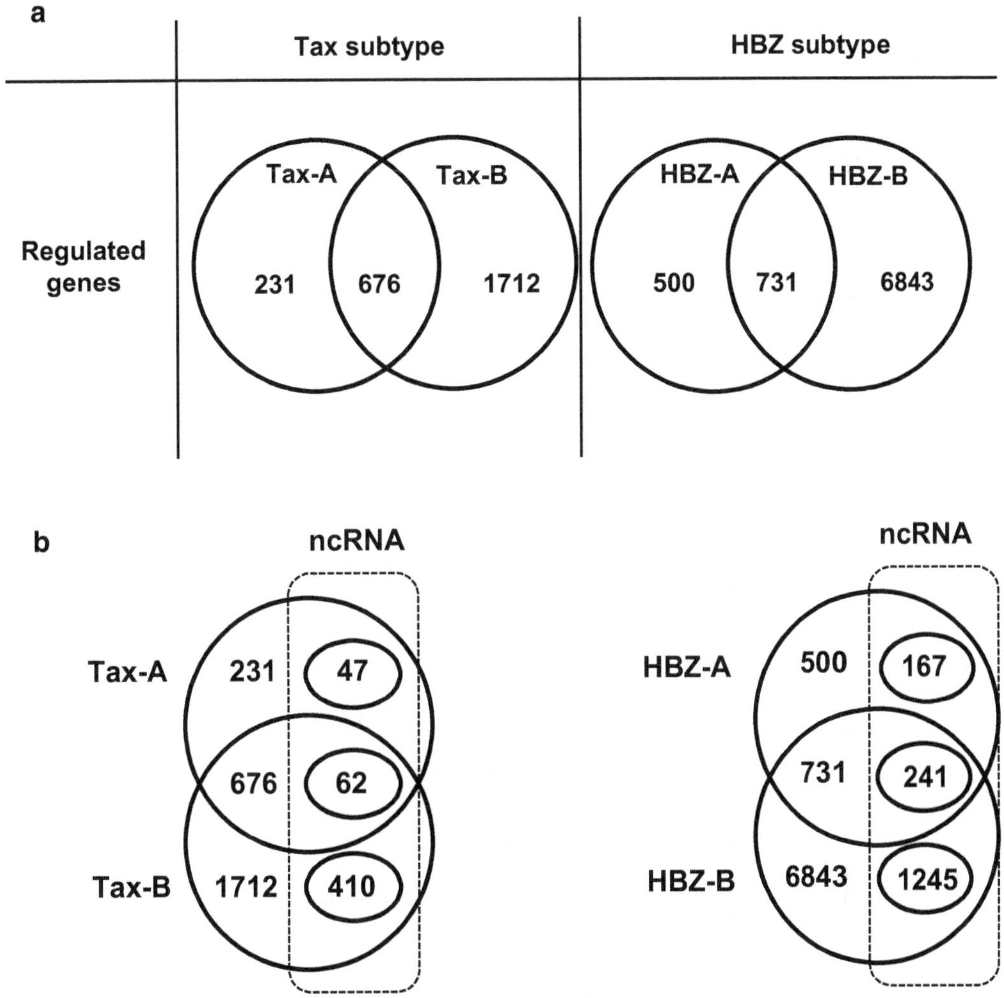

Fig. 2 Results of cDNA microarrays to examine subgroup-specific Tax or HBZ-mediated transcriptional changes in Jurkat Tet-On human T-cells. After transfection, Tax-A, Tax-B, HBZ-A or HBZ-B proteins were induced in Jurkat Tet-On 3G cells by adding Dox (final concentration: 3 ng/ml) for 24 h. Uninduced Jurkat Tet-On 3G cells were used as a baseline reference. Microarray was performed on RNAs derived from these samples. **a** The numbers of genes that occurred in each cluster are shown. We identified 231, 676, and 1712 genes that were specifically up- or down-regulated by Tax-A, by both Tax-A and Tax-B, and by Tax-B, respectively. In HBZ, 500, 731, and 6843 genes were specifically up- or down-regulated by HBZ-A, by both HBZ-A and HBZ-B, and by HBZ-B respectively. **b** Among 231, 676, and 1712 genes were specifically upregulated by HBZ-A, by both HBZ-A and HBZ-B, and by HBZ-B, respectively, 47, 62, and 410 genes were ncRNAs. Similarly, among 500, 731, and 6843 genes that were specifically upregulated by Tax-A, by both Tax-A and Tax-B, and by Tax-B, respectively, 167, 241, and 1245 genes were ncRNAs

we measured the expression levels of *CXCL10* mRNA, since *CXCL10* was preferentially induced by Tax-A rather than by Tax-B (Table 2). In accordance with the microarray data, the results showed that the expression levels of *CXCL10* in HAM/TSP patients with subgroup-A were significantly higher than in HAM/TSP patients with subgroup-B ($p=0.0334$ by Scheffe's post hoc multiple comparisons test.) (Fig. 3a). Meanwhile, there was no significant difference in HTLV-1 PVL between HAM/TSP patients with subgroup-A or -B (Fig. 3b).

More efficient induction of chemokine CXCL10 by Tax-A than by Tax-B

Among the genes differentially induced by Tax-A and Tax-B (Table 2), we focused our attention especially on the chemokine genes (Table 4), particularly because chemokines have been considered to play relevant roles in the pathogenesis of HAM/TSP [30, 31], and both CXCL9 [32] and CXCL10 [33] (listed in Table 2) have already been identified as Tax target genes in previous studies, providing further validation of our gene induction system. In particular, CXCL10 has been proposed as the most viable, prognostic biomarker for HAM/TSP, as the cerebrospinal

Table 2 Top 50 up-regulated genes in Tax-induced Jurkat Tet-On 3G cells

Gene symbol[a]	Tax-A[b]	Tax-B[c]	Gene name
KIF25	[illegible]	567.2	kinesin family member 25
EBI3	304.0	140.5	Epstein-Barr virus induced 3
EGR2	195.1	106.2	early growth response 2
UNQ6494	178.3	106.6	uncharacterized LOC100129066
FAM129A	142.1	71.5	family with sequence similarity 129 member A
GEM	121.4	88.3	GTP binding protein overexpressed in skeletal muscle
WNT10A	103.9	69.5	Wnt family member 10A
CDH16	91.4	36.7	cadherin 16
CXCL10	90.9	62.4	C-X-C motif chemokine ligand 10
ANXA8L1	88.0	33.7	annexin A8 like 1
EGR1	82.8	59.4	early growth response 1
LYZL4	80.4	52.9	lysozyme like 4
FRG2	78.9	57.5	FSHD region gene 2
ENTPD2	72.8	48.5	ectonucleoside triphosphate diphosphohydrolase 2
IL3	70.5	42.8	interleukin 3
GADD45G	67.6	35.2	growth arrest and DNA damage inducible gamma
CRTAM	64.3	36.9	cytotoxic and regulatory T cell molecule
SDC4	58.4	27.7	syndecan 4
FSCN1	58.2	33.4	fascin actin-bundling protein 1
VCAM1	48.8	32.9	vascular cell adhesion molecule 1
FRG2C	47.5	39.7	FSHD region gene 2 family member C
LRG1	44.5	23.8	leucine rich alpha-2-glycoprotein 1
BCL3	40.7	26.2	B cell CLL/lymphoma 3
BCL2A1	39.2	23.7	BCL2 related protein A1
JPH4	38.8	16.2	junctophilin 4
SGPP2	38.2	24.0	sphingosine-1-phosphate phosphatase 2
RGS1	37.9	26.9	regulator of G protein signaling 1
IKZF4	35.6	22.5	IKAROS family zinc finger 4
IL13	33.2	19.1	interleukin 13
TSPO	32.2	15.2	translocator protein
PTGIR	29.7	12.4	prostaglandin I2 receptor
IL36G	29.5	20.8	interleukin 36 gamma
KCNJ18	28.5	20.1	potassium voltage-gated channel subfamily J member 18
BDKRB2	27.8	14.9	bradykinin receptor B2
DEPP1	26.5	21.6	DEPP1, autophagy regulator
CSF2	26.1	10.9	colony stimulating factor 2
UBD	24.9	13.8	ubiquitin D
TNFRSF4	24.8	15.0	TNF receptor superfamily member 4
LINC01589	23.7	17.8	long intergenic non-protein coding RNA 1589
HGC6.3	23.1	15.7	uncharacterized LOC100128124
TNFAIP2	23.0	13.7	TNF alpha induced protein 2
LTB	22.6	19.1	lymphotoxin beta
LAIR2	22.0	16.4	leukocyte associated immunoglobulin like receptor 2
CXCL9	21.8	11.4	C-X-C motif chemokine ligand 9
GPR183	21.4	10.1	G protein-coupled receptor 183
CCL1	21.2	10.6	C-C motif chemokine ligand 1
KIF25-AS1	21.1	16.1	KIF25 antisense RNA 1
AHSP	20.7	11.9	alpha hemoglobin stabilizing protein
LOXL4	20.6	11.8	lysyl oxidase like 4
LHX2	20.2	10.6	LIM homeobox 2

[a] Genes were ranked in order of fold change by Tax-A induction

[b,c] In order to identify up-regulated genes in Tax-A or Tax-B induction, cells transfected with pTRE3G-empty were used as normalization samples. The microarray experiments were performed in triplicate, and data were shown as mean values

fluid (CSF) levels of CXCL10 were well correlated with disease progression of HAM/TSP, better even than HTLV-1 PVL in PBMCs, i.e., the number of HTLV-1-infected cells [31, 32]. Our data demonstrated that CXCL10 was more efficiently upregulated by Tax-A, which is associated with an increased risk of developing HAM/TSP, than by Tax-B (Table 2), suggesting that the HTLV-1 subgroups are associated with changes in host gene expression closely associated with HAM/TSP pathogenesis. To further validate the results using a different inducible protein expression

Table 3 Top 50 up-regulated genes in HBZ-induced Jurkat Tet-On 3G cells

Gene symbol[a]	HBZ-A[b]	HBZ-B[c]	Gene name
lnc-GADD45G-5	23.0	[illegible]	lnc-GADD45G-5:2
RNU6ATAC	20.7	21.8	RNA, U6atac small nuclear (U12-dependent splicing)
NRARP	19.2	14.1	NOTCH regulated ankyrin repeat protein
OR52E8	16.4	18.3	olfactory receptor family 52 subfamily E member 8
CCL21	16.0	0.1	C-C motif chemokine ligand 21
LINC00353	15.9	22.1	long intergenic non-protein coding RNA 353
RPPH1	15.8	14.3	ribonuclease P RNA component H1
LINC00399	15.4	0.1	long intergenic non-protein coding RNA 399
CLDN6	15.2	0.2	claudin 6
CPXM2	14.6	0.2	carboxypeptidase X, M14 family member 2
STARD13	14.5	20.1	StAR related lipid transfer domain containing 13
LOC107986042	14.1	18.7	uncharacterized LOC107986042
lnc-C17orf48-4	14.1	0.1	lnc-C17orf48-4:1
lnc-NDUFV2-1	13.9	9.2	lnc-NDUFV2-1:1
TRHDE-AS1	13.5	19.4	TRHDE antisense RNA 1
LOC101927439	13.4	15.7	uncharacterized LOC101927439
OR10J1	12.6	0.2	olfactory receptor family 10 subfamily J member 1
LOC101927859	12.3	12.8	uncharacterized LOC101927859
TM4SF5	12.1	0.2	transmembrane 4 L six family member 5
LOC389199	11.8	12.2	uncharacterized LOC389199
THBS4	11.7	15.5	thrombospondin 4
lnc-TTLL2-1	11.6	14.9	lnc-TTLL2-1:1
RALYL	11.5	13.8	RALY RNA binding protein like
GCSAML-AS1	11.4	0.2	GCSAML antisense RNA 1
lnc-MAP2K6-2	11.3	10.7	lnc-MAP2K6-2:1
LINC02218	11.3	0.2	long intergenic non-protein coding RNA 2218
LOC105376650	11.2	0.2	uncharacterized LOC105376650
LOC101927882	11.2	14.4	uncharacterized LOC101927882
LOC105373067	11.2	0.1	uncharacterized LOC105373067
CNTNAP3B	11.2	0.2	contactin associated protein like 3B
LOC105370672	11.1	16.9	uncharacterized LOC105370672
lnc-RAD23B-1	11.0	0.2	lnc-RAD23B-1:1
XAGE-4	11.0	10.9	XAGE-4 protein
lnc-TNP1-2	10.9	9.7	lnc-TNP1-2:1
LINC00944	10.7	11.9	long intergenic non-protein coding RNA 944
lnc-CDH3-2	10.6	0.2	lnc-CDH3-2:1
LINC02219	10.6	5.0	long intergenic non-protein coding RNA 2219
CDC14C	10.5	0.3	cell division cycle 14C
lnc-REEP5-1	10.5	12.1	lnc-REEP5-1:1
KRT79	10.3	10.9	keratin 79
TMEM238L	10.3	0.2	transmembrane protein 238 like
LOC285626	10.3	0.3	uncharacterized LOC285626
RTL1	10.2	10.5	retrotransposon Gag like 1
lnc-STXBP6-1	10.0	0.2	lnc-STXBP6-1:2
POLR2M	9.8	11.6	RNA polymerase II subunit M
FAM216B	9.7	12.1	family with sequence similarity 216 member B
PPP1R1A	9.7	6.6	protein phosphatase 1 regulatory inhibitor subunit 1A
LOC107984697	9.7	12.5	uncharacterized LOC107984697
CA6	9.7	0.2	carbonic anhydrase 6
LOC105372061	9.7	6.7	uncharacterized LOC105372061

[a] Genes were ranked in order of fold change by HBZ-A induction

[b,c] In order to identify up-regulated genes in HBZ-A or HBZ-B induction, cells transfected with pTRE3G-empty were used as normalization samples. The microarray experiments were performed in triplicate, and data were shown as mean values

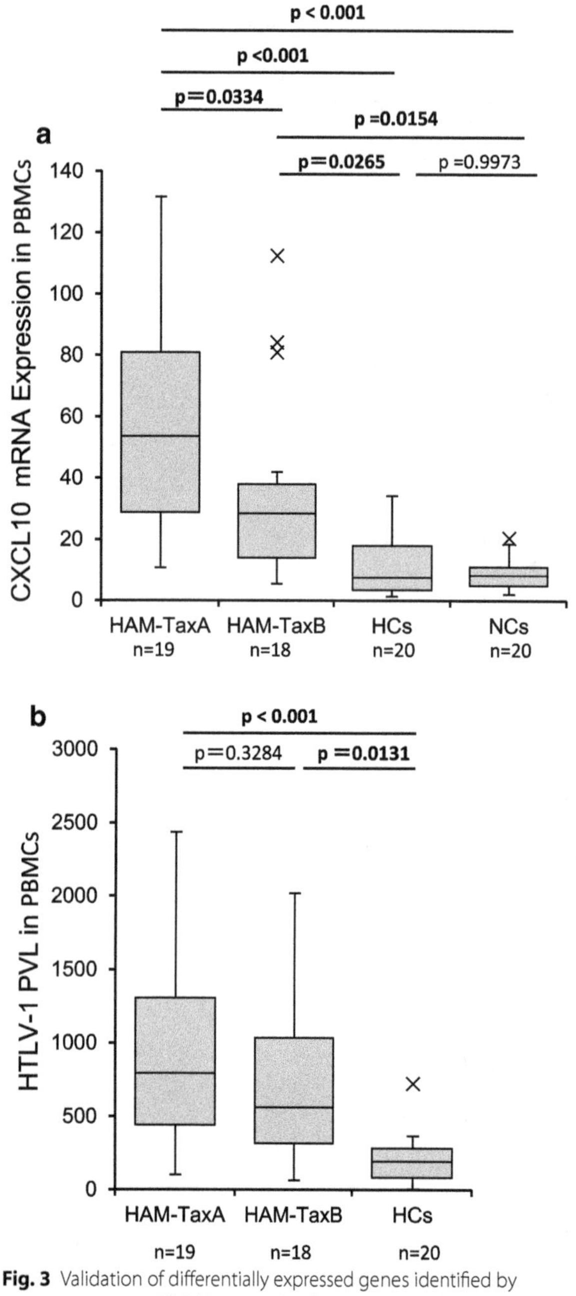

Fig. 3 Validation of differentially expressed genes identified by microarray using qRT-PCR in PBMCs of HAM/TSP patients with HTLV-1 subgroup (**a**) or (**b**). To confirm whether the obtained in vitro gene expression profiles have clinical significance, we performed real-time qRT-PCR on PBMCs of 37 patients with HAM/TSP (19 subgroup-A and 18 subgroup-B), 20 HCs, and 20 NCs. Among identified Tax-responsive genes, we measured the expression levels of *CXCL10* mRNA levels. **a** The expression levels of *CXCL10* in HAM/TSP patients with subgroup-A was significantly higher than in HAM/TSP patients with subgroup-B ($p = 0.0334$ by Scheffe's post hoc multiple comparisons test). **b** There was no significant difference in HTLV-1 PVL between HAM/TSP patients with subgroup-A or -B

system, we used JPX-9 cells [20], a subclone of Jurkat generated by stable transfection of a functional Tax expression-plasmid vector, and induced Tax expression by adding cadmium chloride ($CdCl_2$) to the culture medium (10 μM final concentration). Tax protein was almost undetectable in JPX-9 before the induction but became detectable 6 h after the addition of $CdCl_2$ to the culture medium (Fig. 4a). Quantitative real-time PCR results indicate that CXCL10 mRNA expression was induced along with Tax protein expression in JPX-9 cells (Fig. 4b, white bars). ELISA data indicated that CXCL10 protein became detectable 12 h after the addition of $CdCl_2$ to the culture medium, i.e. 6 h after detection of Tax protein and 3 h after detection of CXCL10 mRNA (Fig. 4b, gray bars), and the CXCL10 protein levels were still increasing even 120 h after the addition of $CdCl_2$.

Preferential expression of CXCL10 in HTLV-1-infected T-cell lines derived from patients with ATL and HAM/TSP

We measured the levels of CXCL10 mRNA and protein expression in HTLV-1-infected and -uninfected human T-cell lines. As shown in Fig. 5a, CXCL10 mRNA was preferentially expressed in HTLV-1-infected human T-cell lines derived from patients with ATL (3 out of 5 tested) and HAM/TSP (4 out of 4 tested), compared with HTLV-1-transformed T-cell lines (0 out of 3) and HTLV-1 negative human T-cell lines (0 out of 3). Quantitative real-time PCR and ELISA analysis revealed that, although the expression levels of the viral RNAs *tax* and *HBZ* in ATL-derived cell lines were low (Fig. 5b), high levels of CXCL10 mRNA (Fig. 5a) and protein expression (Fig. 5c) were observed in ATL cell lines, suggesting that over-expression of CXCL10 in ATL-derived cell lines was independent of *tax* gene expression. Western blot also revealed that over-expression of CXCL10 in ATL-derived cell lines was independent of Tax protein expression (Fig. 5d). Importantly, consistent with the cell line data, the expression levels of *CXCL10* mRNA in PBMCs of HAM/TSP patients were also not significantly correlated with the *tax* expression ($p = 0.4247$, $r = -0.406$ by Spearman rank correlation analysis), indicating that increased *CXCL10* expression is independent of *tax* expression. Thus, our data indicated that increased expression of *CXCL10* mRNA is associated with "having HTLV-1 subgroup-A or subgroup-B", not "tax expression in infected cells" both in vitro and ex vivo (i.e., in unmanipulated PBMCs obtained from HAM/TSP patients).

No difference in subgroup-specific Tax molecules with respect to transcriptional activation of the CXCL10 promoter via NF-κB

We further examined whether there were any differences in the ability of subgroup-specific Tax to activate

Table 4 Chemokine genes differentially induced by Tax-A and Tax-B

	Fold change					
	Tax-A	**Tax-B**	**Tax A/B ratio**	**Gene name**	**Partner (receptor or ligand)**	**References**
Chemokine	90.9	62.4	1.5	CXCL10	CXCR3	[32, 45, 46]
	21.8	11.4	1.9	CXCL9	CXCR3	[32, 46]
	N.D.[a]	2.4		CCL22	CCR4	[47]
	21.2	10.6	2.0	CCL1	CCR8	[29, 48]
	N.D.	2.9		CCL5	CCR1, CCR3, CCR5	[49, 50]
	5.8	2.3	2.5	CCL20	CCR6	[51]
	9.0	6.0	1.5	XCL1	XCR1	[52]
Chemokine receptor	5.6	3.8	1.5	CXCR6	CXCL16	None reported
	2.2	2.2	1.0	CCR7	CCL19, CCL21	[53]
	1.5	N.D.		CCRL2/CCR11	CCL2, CCL5, CCL7, CCL8	None reported

[a] Not detected

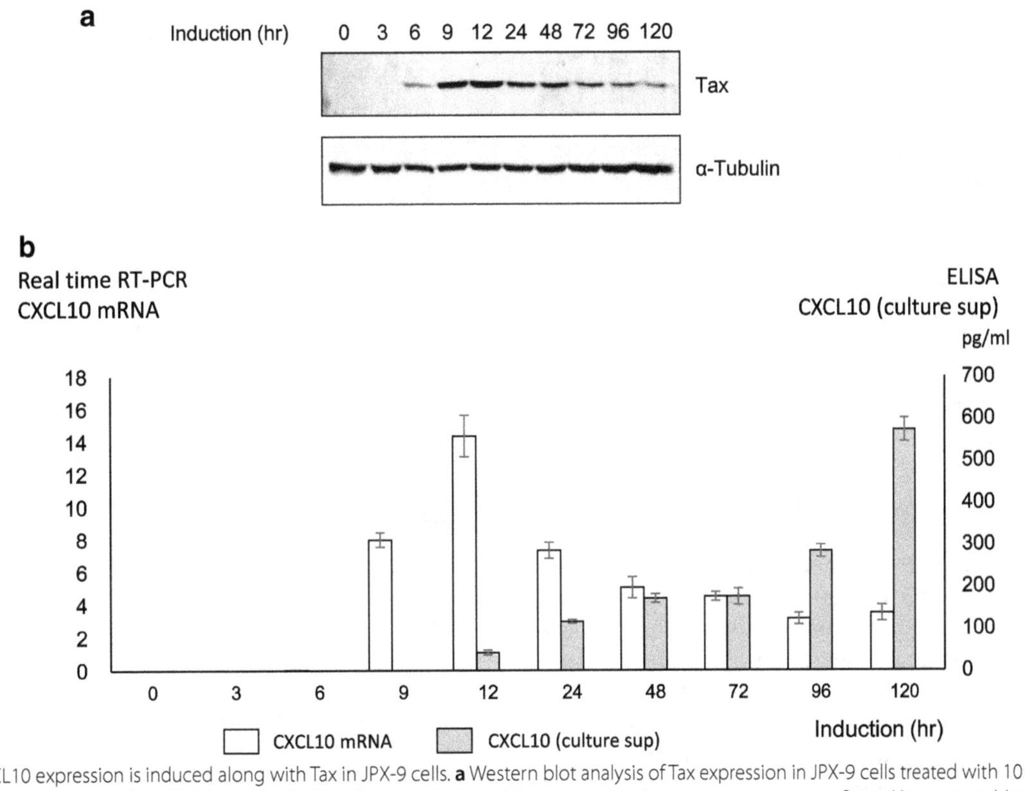

Fig. 4 CXCL10 expression is induced along with Tax in JPX-9 cells. **a** Western blot analysis of Tax expression in JPX-9 cells treated with 10 μM $CdCl_2$. Cell lysates were prepared from $CdCl_2$-treated JPX-9 cells at the indicated time points, and Tax expression was confirmed by western blotting with Lt-4 anti-Tax monoclonal antibody. Equal sample loading was verified with anti-α-Tubulin (bottom). Tax protein was almost undetectable in JPX-9 before the induction, but became detectable 6 h after the addition of $CdCl_2$ to the culture medium. **b** Induction of CXCL10 mRNA and protein expression in JPX-9 cells treated with $CdCl_2$. CXCL10 mRNA and protein levels after Tax induction were detected by quantitative real-time PCR and ELISA, respectively. The copy number of CXCL10 mRNA was normalized by the copy number of hypoxanthine guanine phosphoribosyl transferase 1 (HPRT1) mRNA. Quantitative real-time PCR results indicate that CXCL10 mRNA expression was induced along with Tax protein expression in JPX-9 cells (white bars). ELISA data indicated that CXCL10 protein became detectable 12 h after the addition of $CdCl_2$ to the culture medium, i.e. 6 h after detection of Tax protein and 3 h after detection of CXCL10 mRNA, and the CXCL10 protein levels were still increasing even at 120 h after the addition of $CdCl_2$ (gray bars). The data are representative of 3 independent experiments. Data shown as mean ± SD, n = 3

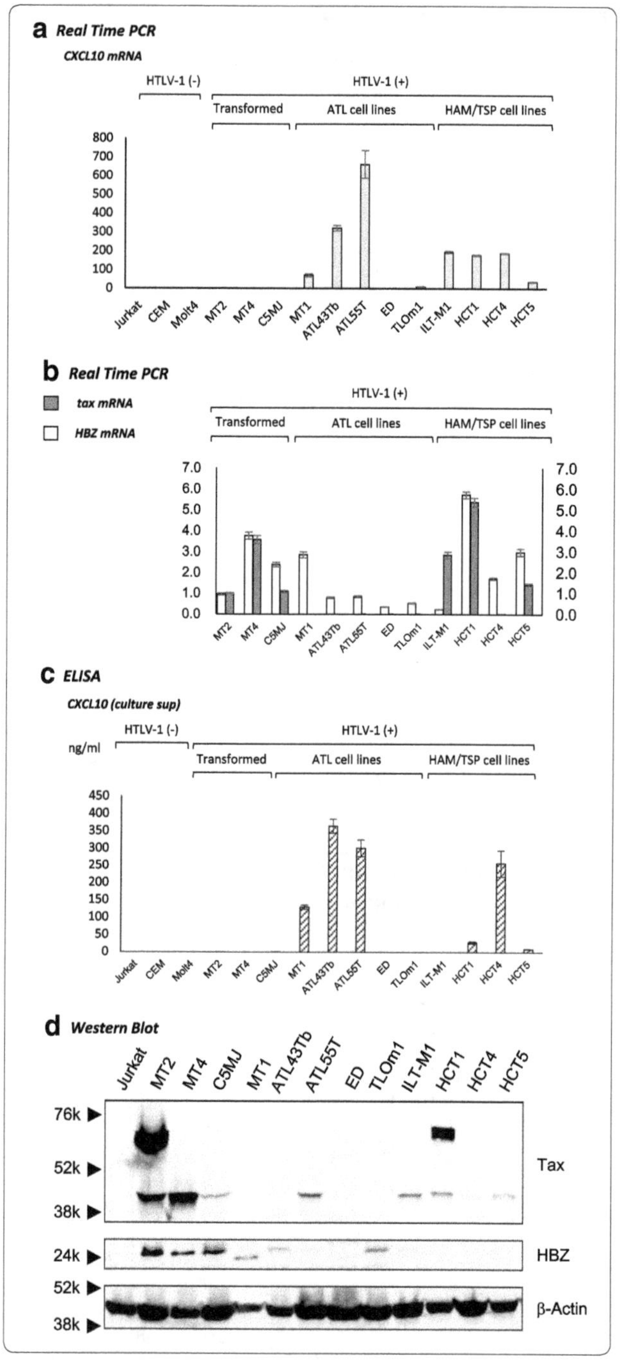

Fig. 5 Preferential expression of CXCL10 in Human T-cell leukemia virus type-1 (HTLV-1)-infected T-cell lines derived from patients with adult T-cell leukemia (ATL) and HTLV-1-associated myelopathy/tropical spastic paraparesis (HAM/TSP). **a** Expression of CXCL10 mRNA was examined by real-time PCR in HTLV-1-infected and -uninfected T-cell lines. CXCL10 mRNA was preferentially expressed in HTLV-1-infected human T-cell lines derived from patients with ATL (3 out of 5 tested) and HAM/TSP (4 out of 4 tested), compared with HTLV-1-transformed T-cell lines (0 out of 3) and HTLV-1 negative human T-cell lines (0 out of 3). **b** The expressions of *tax* and *HBZ* mRNA were examined by quantitative real-time PCR in HTLV-1-infected T-cell lines. The expression levels of the viral RNAs *tax* and *HBZ* were relatively high in T-cell lines derived from patients with HAM/TSP and HTLV-1-transformed T-cell lines when compared with those in ATL cell lines. **c** CXCL10 levels in culture supernatants from HTLV-1-infected and -uninfected human T-cell lines were assessed by ELISA. Significant levels of CXCL10 was observed in the cultured supernatants from HTLV-1-infected human T-cell lines derived from patients with ATL and HAM/TSP, whereas it was not detectable in any of the other cell lines tested. **d** Expression of Tax and HBZ was examined by Western blot in HTLV-1-infected and -uninfected T-cell lines

the human CXCL10 promoter [22]. Jurkat cells were co-transfected with Tax effector plasmids and a reporter gene construct containing the region between − 875 and +97 nucleotides (pGL3-CXCL10-Long-Luc) and − 279 and + 97 nucleotides (pGL3-CXCL10-Short-Luc) of the CXCL10 upstream regulatory sequences (Fig. 6a). After 24 h transfection, Tax-A or Tax-B protein was induced by adding Dox for 24 h. Then, the reporter gene assay was carried out for Tax-mediated transcriptional activation. As shown in Fig. 6b, there was no difference between Tax-A and Tax-B with respect to transcriptional activity for both the short and long CXCL10 promoters in the epithelial cell line 293T and the T cell line, Jurkat. The data also showed that the Tax mutant M22 [34], which is defective in NF-κB activation, failed to activate the CXCL10 promoters (both short and long) (Fig. 6b). In contrast, the Tax 703 mutant [35], which can activate NF-κB but not CREB, efficiently activated the CXCL10 promoters (both short and long) (Fig. 6b). To further determine whether either the NF-κB or the AP-1 sequence was required for Tax-mediated activation of the CXCL10 promoter, we constructed reporter plasmids by site-directed mutagenesis. Then mutant reporter constructs were co-transfected along with the Tax expression plasmid, and luciferase activity was determined for each of the four mutants (Fig. 6c). As a result, Tax-induced luciferase activity was significantly reduced by mutation in one of the two NF-κB binding site sequence, but not reduced by mutation in the AP-1 site, indicating that Tax transactivation of the CXCL10 involves both κB binding sites. Namely, Tax-induced activation of NF-κB pathway is responsible for the upregulation of CXCL10 expression, but subgroup-specific Tax molecules have similar effects on the transcriptional activity of the CXCL10 promoter. This was also the case for another reporter construct containing NF-κB binding site sequence. As shown in Additional file 3: Fig. S1, subgroup-specific Tax molecules do not differ in the transcriptional activation of the reporter construct containing a luciferase gene under control of the NF-κB binding sequence of the IL-2Ra gene [25].

(See figure on next page.)
Fig. 6 Subgroup-specific Tax molecules do not differ in the transcriptional activation of the CXCL10 promoter via NF-κB. **a** Schematic representation of the CXCL10 promoter sequence cloned into pGL3-Basic. The firefly luciferase gene was used to monitor the activity of the CXCL10 gene promoter. **b** These reporter constructs were independently transfected into Jurkat human T-cells and 293T cells with or without the Tax expression plasmid. Luciferase assays were performed 24 h after transfection. There was no difference between Tax-A and Tax-B with respect to transcriptional activity for both the short and long CXCL10 promoters. The data also showed that the Tax mutant M22, which is defective in NF-κB activation, failed to activate the CXCL10 promoters (both short and long). In contrast, the Tax 703 mutant, which can activate NF-κB but not CREB, efficiently activated the CXCL10 promoters (both short and long). **c** To determine whether either the NF-κB or the AP-1 sequence was required for Tax-mediated activation of the CXCL10 promoter, mutant reporter constructs were co-transfected along with the Tax expression plasmid, and luciferase activity was determined for each of the four mutants. Tax-induced luciferase activity was significantly reduced by mutation in one of the two NF-κB binding site sequence, but not reduced by mutation in the AP-1 site, indicating that Tax transactivation of the CXCL10 involves both κB binding sites. Three independent experiments were performed. Data shown as mean ± SD, n = 3

No difference in Tax-A and Tax-B with respect to binding to NF-κB proteins

While HTLV-1 is the causative agent of ATL and HAM/TSP, the closely related virus HTLV-2 is not clearly associated with a known clinical disease [36]. It has been reported that HTLV-1 Tax (i.e., Tax1) but not HTLV-2 Tax (Tax2) interacts with NF-κB2/p100/p52 and RelB, and this interaction was mediated by the Tax1 amino acid region 225-232 [37]. As one of the two sites of amino acid differences between Tax-A and Tax-B is located just beside this region (i.e., at position 221), we tested whether this amino acid difference affects binding affinity to NF-κB protein, thereby altering the ability to activate the NF-κB pathway and its downstream target genes. Jurkat cells were transfected with each Tax expression plasmid and cells were harvested after 24 h, then western blot and immunoprecipitation analyses were performed. The amounts of p100, p52, RelB, p65, p105, p50, and c-Rel proteins in immunoprecipitated samples were quantified by densitometric scanning of corresponding bands of the western blot. As shown in Fig. 7a, b, western blot analysis performed on immunoprecipitated Tax protein revealed that each subgroup of Tax (i.e., Tax-A and -B) interacted with each NF-κB component (i.e., c-Rel, RelA, RelB, p50/p105, and p52/p100) with similar efficiency.

Higher abundance of the DNA fragment bound by ternary complex including Tax-A than that including Tax-B in the CXCL10 promoter

To investigate whether HTLV-1 subgroups affect the binding of Tax to the endogenous CXCL10 promoter, we performed ChIP assays (Fig. 8). Chromatin fragments were prepared from Jurkat T-cells transfected with plasmid expressing Tax and immunoprecipitated with specific anti-Tax monoclonal antibodies (Lt-4). The immunoprecipitated DNA was then amplified via PCR using specific primer pairs that amplify the CXCL10 proximal promoter sequence including the two NF-κB sites and an AP-1 site (Fig. 8a, "Target region"). As shown in Fig. 8a, the ternary protein complex including Tax was found to be associated with the CXCL10 proximal promoter sequence including the two NF-κB sites and an AP-1 site (Fig. 8a). Moreover, the abundance of a specific DNA fragment bound by ternary complex including Tax is higher for Tax-A than for Tax-B (Fig. 8b, c).

Discussion

In this study, to investigate the role of HTLV-1 subgroups in viral pathogenesis, we first performed a microarray-based study to define the gene expression profiles of each subgroup-specific Tax or HBZ. Our results showed that some Tax-responsive genes identified by microarray, such as EBI3 [26], VCAM-1 [27], IL-13 [28], and CCL1 [29] have already been identified as Tax target genes in previous studies, validating our gene induction system. Furthermore, Tax-induced expression of CXCL10 gene identified by microarray was confirmed by qRT-PCR.

Our microarray analysis showed that out of a total of 50,599 genes including Entrez genes and lincRNA genes screened, approximately 2619 and 8074 were differentially expressed upon Tax or HBZ induction, respectively. The list of the top 50 genes up-regulated by each subgroup of Tax (i.e., Tax-A and -B) or HBZ (i.e., HBZ-A and -B) showed that the genes that were differentially expressed in the presence of Tax included those related to apoptosis (BCL2-related protein A1: BCL2A1), the cell cycle and DNA repair (growth arrest and DNA-damage-inducible, gamma: GADD45G), cytokines (IL3, IL13, IL36G), chemokines (CXCL9, CXCL10), and adhesion molecules (vascular cell adhesion molecule 1: VCAM1). In contrast, the genes that were differentially expressed in the presence of HBZ include those related to RNA-splicing (RNA, U6atac small nuclear: RNU6ATAC), the Notch signaling pathway (NOTCH-regulated ankyrin repeat protein: NRARP) and chemokine (CCL21). These results suggest that Tax mainly induces expression of genes related to the activation and transformation of CD4+ T cells, whereas HBZ modulates a variety of cellular signaling pathways which are related to the immune response, differentiation, and growth of T-cells.

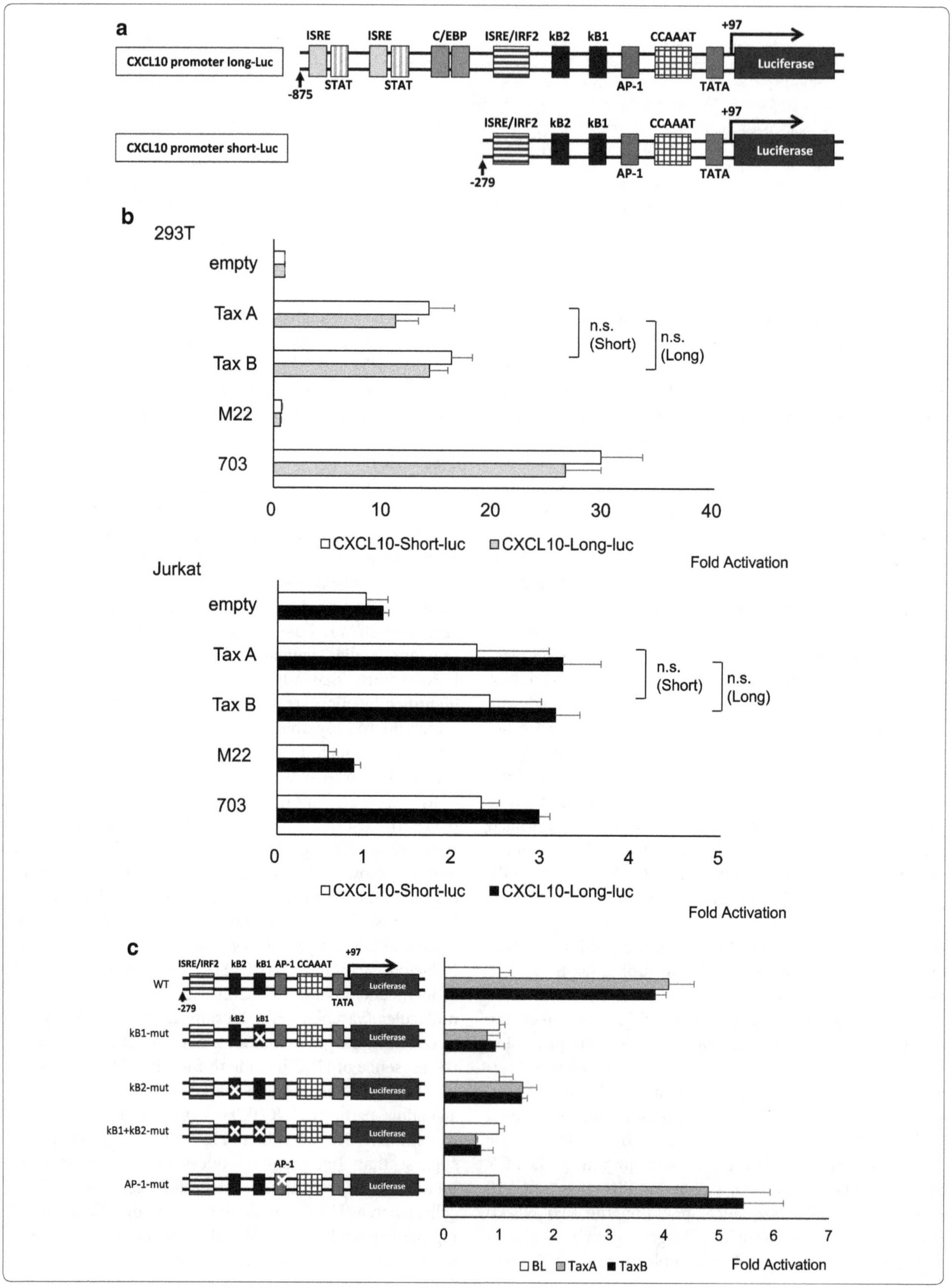
a
CXCL10 promoter long-Luc
ISRE
ISRE
C/EBP
ISRE/IRF2
kB2
kB1
CCAAAT
+97
Luciferase
-875
STAT
STAT
AP-1
TATA
CXCL10 promoter short-Luc
-279
b
293T
empty
Tax A
Tax B
M22
703
n.s. (Short)
n.s. (Long)
0
10
20
30
40
CXCL10-Short-luc
CXCL10-Long-luc
Fold Activation
Jurkat
0
1
2
3
4
5
c
WT
kB1-mut
kB2-mut
kB1+kB2-mut
AP-1-mut
6
7
BL
TaxA
TaxB

(See figure on next page.)
Fig. 7 Tax-A and Tax-B do not differ in their binding to NF-κB proteins. To test whether one of the two amino acid differences between Tax-A and Tax-B, i.e., at position 221, which are located just beside the Tax1 (225-232) motif is involved in the p100 processing, we tested whether this amino acid difference affects the binding affinity to the NF-κB protein, thereby altering the ability to activate the NF-κB pathway and their downstream target genes. Jurkat cells were transfected with each expression plasmid and cells were harvested after 24 h, then immunoprecipitation and western blot analyses were performed. **a** Western blot analysis performed on immunoprecipitated Tax protein revealed that each subgroup of Tax, i.e., Tax-A and -B, interacted with each NF-κB components, i.e., c-Rel, RelA, RelB, p50/p105, and p52/p100, with similar efficiencies. **b** The amounts of p100, p52, RelB, p65, p105, p50, and c-Rel proteins in immunoprecipitated samples were quantified by densitometric scanning of corresponding bands of the western blot using Image J software. Vertical bars indicate mean ± SD of the densitometric analysis from four independent experiments

Our microarray results also showed that various kinds of ncRNAs frequently appeared among both Tax- and HBZ-regulated genes. It is now well established that the majority of transcribed genomic sequences are associated with ncRNAs rather than protein-coding RNAs, and such ncRNAs, including regulatory miRNAs and lncRNAs that are functionally involved in a variety of host immune responses and in the pathogenesis of human diseases [38]. It is, therefore, plausible to propose that deregulation of the ncRNA signature caused by a virus infection will strongly affect the phenotype and function of the infected cell. Indeed, in HTLV-1 infection, it has been reported that the expression of miR-31, which is a negative regulator of the noncanonical NF-κB pathway, was genetically and epigenetically silenced in ATL cells, resulting in constitutive NF-κB activation [39]. In addition, lncRNAs, a heterogeneous class of RNAs defined as non-protein-coding transcripts longer than 200 nucleotides, are thought to play a role in proteasomal and ubiquitination pathways, apoptosis, DNA damage responses, and cell cycle regulation [40]. lncRNAs have been directly linked to human diseases such as certain cancers and autoimmune and neurodegenerative diseases [40]. However, to date, the role of lncRNAs in viral infections remains largely unknown. Our results provide new insights into the hitherto unknown functions of lncRNAs in infection by viruses including HTLV-1.

A previous report suggested that serum CXCL10 is significantly higher in HAM/TSP patients than in HCs, and the CSF level of CXCL10 was strongly correlated with disease severity [32]. Thus, CXCL10 concentration was proposed as a potential prognostic biomarker for HAM/TSP. Our data showed that among the Tax-regulated target genes, CXCL10 was approximately 1.5 times more strongly induced by Tax-A than Tax-B. More importantly, real-time qRT-PCR on PBMCs obtained from HAM/TSP patients indicated that the expression levels of CXCL10 in HAM/TSP patients with subgroup-A were significantly higher than those in HAM/TSP patients with subgroup-B. The difference in transcription is likely to be due to a difference in the action of the NF-κB/Rel family of transcription factors. In HTLV-1 infection, Tax-mediated NF-κB activation is recognized as a crucial factor associated with the development of HTLV-1-associated diseases [41], since NF-κB, which consists of five molecules (RelA (p65), RelB, c-Rel, p50, and p52) that form transcriptionally active complexes in various combinations, has an essential role in inflammation, innate immunity, and many steps of cancer initiation and progression [42]. Indeed, ATL cells and their derivative cell lines carry constitutively active NF-κB regardless of their Tax expression, and NF-κB is required for immortalization and also the survival of HTLV-1 transformed cells [43]. We therefore determined whether the HTLV-1 viral protein Tax activates the expression of CXCL10 at the transcriptional level and whether there were any differences in the ability of subgroup-specific Tax molecules to activate the CXCL10 promoter. However, contrary to our expectations, there was no difference in the ability of each subgroup Tax to activate the CXCL10 promoter, although transient Tax expression in an HTLV-1-negative human T-cell line activated the CXCL10 gene promoter through the NF-κB pathway.

In clear contrast to HTLV-1, HTLV-2 has not been associated with ATL or other types of malignancies [36]. As HTLV-1 Tax (Tax1) and HTLV-2 Tax (Tax2) have many shared activities but also certain significantly distinct activities [36], the difference between those two Tax proteins may reveal the key roles in HTLV-1-induced malignant transformation. Most importantly, although Tax2 activates the classical pathway of NF-κB, similar to Tax1, malignant transformation by Tax2 has rarely been reported [44]. Thus, one significant difference between Tax1 and Tax2 is the activation of transcription factor NF-κB2/p100/p52, which is a key player in the alternative, non-classical NF-κB pathway [44]. Interestingly, Tax1 but not Tax2 was reported to interact with NF-κB2/p100/p52 and RelB, and the distinct interaction activity was mediated by the Tax1 amino acid region 225-232, and one of the two sites of amino acid differences between Tax-A and Tax-B is located just beside this region (i.e., at position 221). We therefore tested whether this amino acid difference affected the binding affinity to NF-κB protein, thereby altering the ability to activate the NF-κB pathway and their downstream target genes. The results showed that both Tax-A and Tax-B showed

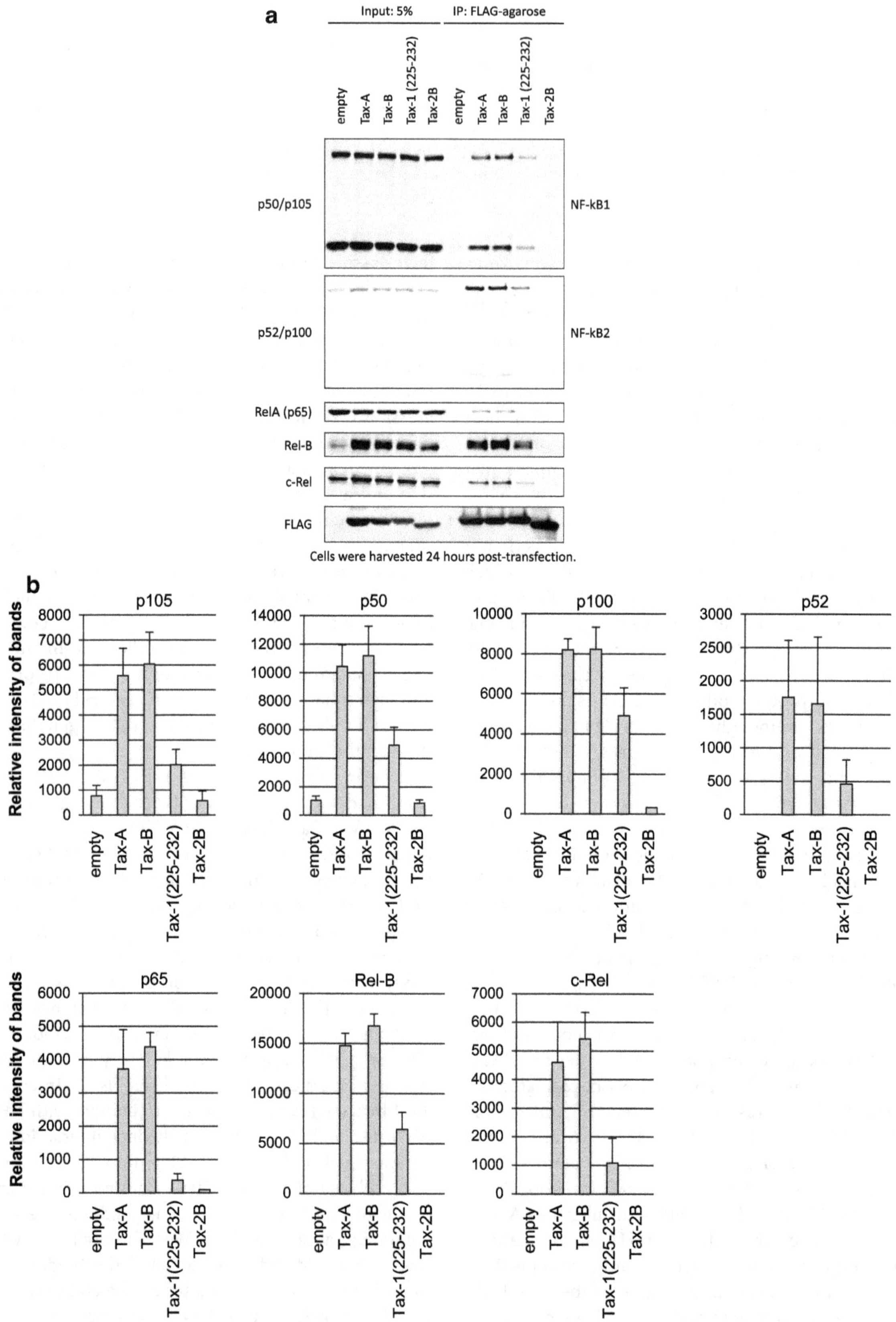
a
Input: 5%
IP: FLAG-agarose
empty
Tax-A
Tax-B
Tax-1 (225-232)
Tax-2B
p50/p105
NF-kB1
p52/p100
NF-kB2
RelA (p65)
Rel-B
c-Rel
FLAG
Cells were harvested 24 hours post-transfection.
b
Relative intensity of bands
p105
p50
p100
p52
p65
Rel-B
c-Rel
empty
Tax-A
Tax-B
Tax-1(225-232)
Tax-2B

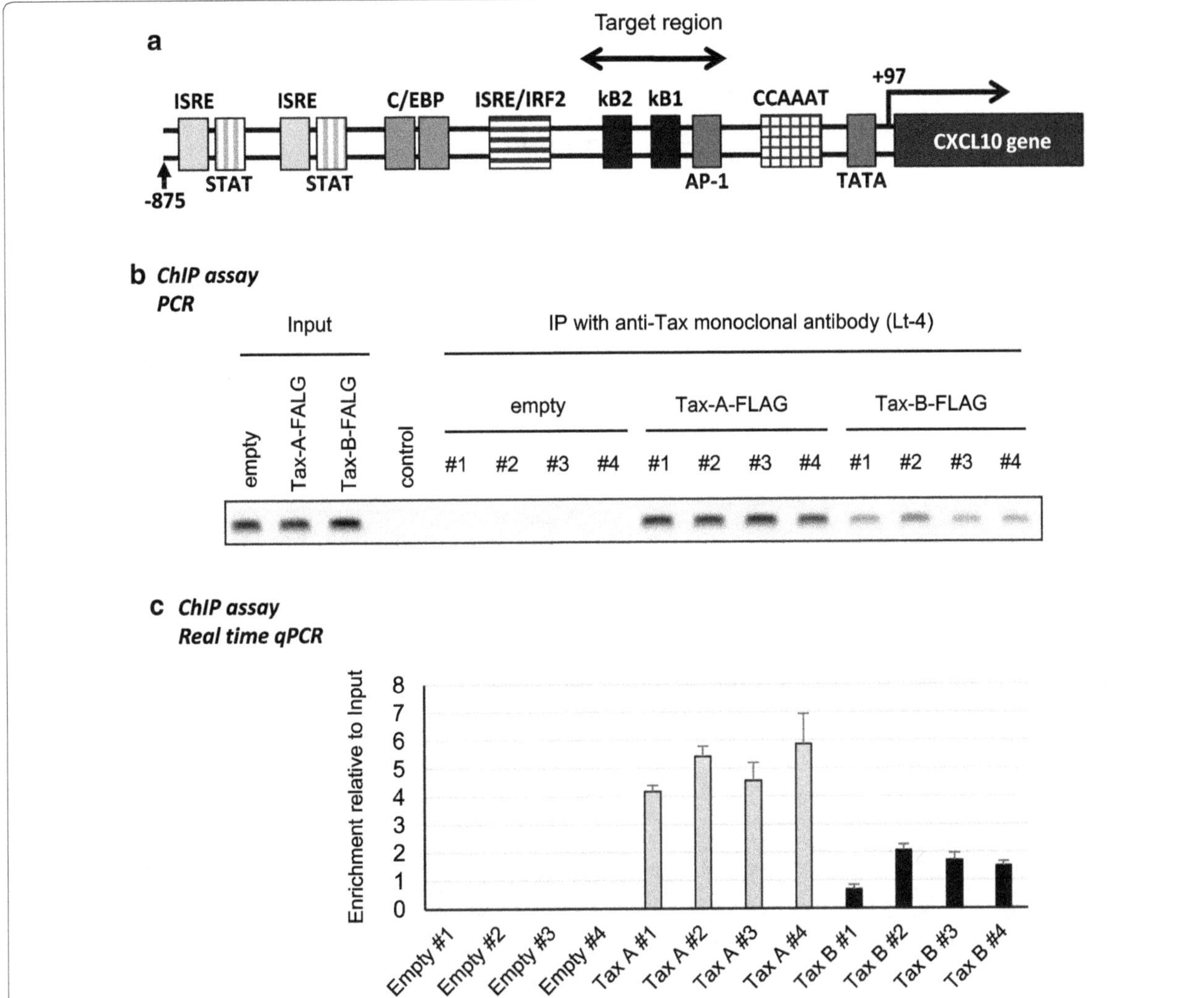

Fig. 8 DNA fragment bound by the ternary complex including Tax-A is more abundant than that including Tax-B in the CXCL10 promoter. To investigate whether HTLV-1 subgroups affects the binding of Tax to the endogenous CXCL10 promoter, we performed chromatin immunoprecipitation (ChIP) assays. Chromatin fragments were prepared from Jurkat T-cells transfected with plasmid expressing Tax and immunoprecipitated with specific anti-Tax monoclonal antibody (Lt-4). The immunoprecipitated DNA was then amplified by PCR. **a** Schematic representation of the CXCL10 promoter sequence. "Target region", which was amplified by PCR, includes two NF-κB sites and an AP-1 site. **b, c** Chromatin immunoprecipitation (ChIP) assays. Representative examples of PCR products obtained from chromatin material that was directly immunoprecipitated with anti-Tax monoclonal antibody (Lt-4) (**b**). The ternary protein complex including Tax was found to be associated with the CXCL10 proximal promoter sequence including the two NF-κB sites and an AP-1 site. The abundance of a specific DNA fragment bound by the ternary complex, including Tax, is higher for Tax-A than for Tax-B. Cell lysates were subjected to ChIP assays using Lt-4 antibody followed by qRT-PCR using the specific primer sets (**c**). Three independent experiments were performed. Data shown as mean ± SD

similar levels of respective binding to each of the NF-κB proteins, i.e., NF-κB1 (p105/p50), NF-κB2 (p100/p52), RelA (p65), RelB, and c-Rel. This is consistent with the CXCL10 reporter gene assay, which revealed no difference in the ability of each subgroup Tax to activate the CXCL10 promoter via NF-κB pathway as a transcriptional regulator.

To further investigate how the subgroup-specific Tax regulates the CXCL10 gene expression in T-cells, we performed the ChIP assay using Jurkat T-cells. The results showed that DNA fragments bound by ternary protein complexes including Tax-A are more abundant than those including Tax-B in the CXCL10 promoter, suggesting that Tax A is more frequently recruited at this genomic region than Tax-B.

In conclusion, we demonstrate that different HTLV-1 subgroups are characterized by different patterns of host gene expression. Differential expression of pathogenesis-related genes regulated directly or indirectly by subgroup-specific Tax or HBZ may be associated with the onset of HAM/TSP. To better understand the pathogenesis of HAM/TSP, further studies are needed to elucidate the role of subgroup-specific viral transcription factors and its target genes in more detail.

Additional files

Additional file 1: Table S1. Osame Motor Disability Score (OMDS).

Additional file 2: Table S2. Primer sequences for plasmid construction.

Additional file 3: Fig. S1. Subgroup-specific Tax molecules do not differ in the transcriptional activation of the reporter construct containing a luciferase gene under control of the NF-κB binding sequence of the IL-2Ra gene. The reporter construct containing a luciferase gene fused to five repeats of the NF-κB site of the IL-2Ra gene was independently transfected into Jurkat human T-cells with or without the Tax expression plasmid. Luciferase assays were performed 24 h after transfection. There was no difference between Tax-A and Tax-B with respect to transcriptional activity. Three independent experiments were performed. Data shown as mean ± SD, n = 3.

Abbreviations

HTLV-1: human T-cell leukemia virus type-1; ATL: adult T-cell leukemia; HAM/TSP: HTLV-1-associated myelopathy/tropical spastic paraparesis; HCs: asymptomatic carriers; NCs: normal uninfected healthy controls; CSF: cerebrospinal fluid; PVL: proviral load; CNS: central nervous system.

Authors' contributions

TN designed and performed the experiments, analyzed the data, and wrote the paper; JY, YM, HS and HU performed experiments and analyzed and interpreted the data; KS and KH analyzed and interpreted the microarray data; YT, TN, and MF provided essential reagents and helped with writing the paper; MM contributed to the conception and design of the study. MS designed and supervised the research, performed experiments, and wrote the paper. All authors read and approved the final manuscript.

Author details

[1] Department of Microbiology, Kawasaki Medical School, 577 Matsushima, Kurashiki, Okayama 701-0192, Japan. [2] Laboratory of Virus Control, Institute for Frontier Life and Medical Sciences, Kyoto University, Kyoto, Japan. [3] Department of Bioscience and Bioinformatics, Kyushu Institute of Technology, Fukuoka, Japan. [4] Department of Immunology, Graduate School of Medicine, University of the Ryukyus, Okinawa, Japan. [5] Department of Social Work, Faculty of Human and Social Studies, Nagasaki International University, 2825-7 Huis Ten Bosch Machi, Sasebo, Nagasaki 859-3298, Japan. [6] Division of Virology, Niigata University Graduate School of Medical and Dental Sciences, Niigata, Japan. [7] Department of Hematology, Rheumatology and Infectious Disease, Faculty of Life Sciences, Kumamoto University, Kumamoto, Japan. [8] Present Address: Division of Gene Regulation and Signal Transduction, Research Center for Genomic Medicine, Saitama Medical University, Saitama, Japan.

Acknowledgements

The ILT-M1 cell line was kindly provided by Professor Mari Kannagi, from Tokyo Medical and Dental University. The expression vector pCG-Tax and the control vector pCG-BL were kindly provided by Professor Jun-ichi Fujisawa, from Kansai Medical University. This work was supported by the Joint Usage/Research Center Program of the Institute for Frontier Life and Medical Sciences, Kyoto University. This work was financially supported by KAKENHI, from the Japan Society for the Promotion of Science (JSPS) (Grant Number JP15K09345 to M.S.), the Japan Agency for Medical Research and Development (AMED) (Grant Numbers 16ek0109026h0003 and 16ek0109010h0003 to M.S.). The funding agency had no role in the study design, data collection, and analysis, decision to publish, or preparation of the manuscript.

Competing interests

The authors declare that they have no competing interests.

References

1. Poiesz BJ, Ruscetti FW, Gazdar AF, Bunn PA, Minna JD, Gallo RC. Detection and isolation of type C retrovirus particles from fresh and cultured lymphocytes of a patient with cutaneous T-cell lymphoma. Proc Natl Acad Sci U S A. 1980;77:7415–9.
2. Yoshida M, Seiki M, Yamaguchi K, Takatsuki K. Monoclonal integration of human T-cell leukemia provirus in all primary tumors of adult T-cell leukemia suggests causative role of human T-cell leukemia virus in the disease. Proc Natl Acad Sci U S A. 1984;81:2534–7.
3. Hinuma Y, Nagata K, Hanaoka M, Nakai M, Matsumoto T, Kinoshita KI, Shirakawa S, Miyoshi I. Adult T-cell leukemia: antigen in an ATL cell line and detection of antibodies to the antigen in human sera. Proc Natl Acad Sci USA. 1981;78:6476–80.
4. Osame M, Usuku K, Izumo S, Ijichi N, Amitani H, Igata A, Matsumoto M, Tara M. HTLV-I associated myelopathy, a new clinical entity. Lancet. 1986;1:1031–2.
5. Gessain A, Barin F, Vernant JC, Gout O, Maurs L, Calender A, de The G. Antibodies to human T-lymphotropic virus type-I in patients with tropical spastic paraparesis. Lancet. 1985;2:407–10.
6. Proietti FA, Carneiro-Proietti AB, Catalan-Soares BC, Murphy EL. Global epidemiology of HTLV-I infection and associated diseases. Oncogene. 2005;24:6058–68.
7. van Tienen C, de Silva TI, Alcantara LC, Onyango CO, Jarju S, Goncalves N, Vincent T, Aaby P, Whittle H, Schim van der Loeff M, Cotten M. Molecular epidemiology of endemic human T-lymphotropic virus type 1 in a rural community in Guinea-Bissau. PLoS Negl Trop Dis. 2012;6:e1690.
8. Wolfe ND, Heneine W, Carr JK, Garcia AD, Shanmugam V, Tamoufe U, Torimiro JN, Prosser AT, Lebreton M, Mpoudi-Ngole E, et al. Emergence of unique primate T-lymphotropic viruses among central African bushmeat hunters. Proc Natl Acad Sci USA. 2005;102:7994–9.
9. Furukawa Y, Yamashita M, Usuku K, Izumo S, Nakagawa M, Osame M. Phylogenetic subgroups of human T cell lymphotropic virus (HTLV) type I in the tax gene and their association with different risks for HTLV-I-associated myelopathy/tropical spastic paraparesis. J Infect Dis. 2000;182:1343–9.
10. Yoshida M. Molecular approach to human leukemia: isolation and characterization of the first human retrovirus HTLV-1 and its impact on tumorigenesis in adult T-cell leukemia. Proc Jpn Acad Ser B Phys Biol Sci. 2010;86:117–30.
11. Gaudray G, Gachon F, Basbous J, Biard-Piechaczyk M, Devaux C, Mesnard JM. The complementary strand of the human T-cell leukemia virus type 1 RNA genome encodes a bZIP transcription factor that down-regulates viral transcription. J Virol. 2002;76:12813–22.
12. Fan J, Ma G, Nosaka K, Tanabe J, Satou Y, Koito A, Wain-Hobson S, Vartanian JP, Matsuoka M. APOBEC3G generates nonsense mutations in human T-cell leukemia virus type 1 proviral genomes in vivo. J Virol. 2010;84:7278–87.
13. Satou Y, Yasunaga J, Yoshida M, Matsuoka M. HTLV-I basic leucine zipper factor gene mRNA supports proliferation of adult T cell leukemia cells. Proc Natl Acad Sci U S A. 2006;103:720–5.
14. Saito M, Matsuzaki T, Satou Y, Yasunaga J, Saito K, Arimura K, Matsuoka M, Ohara Y. In vivo expression of the HBZ gene of HTLV-1 correlates with proviral load, inflammatory markers and disease severity in HTLV-1 associated myelopathy/tropical spastic paraparesis (HAM/TSP). Retrovirology. 2009;6:19.
15. Arnold J, Zimmerman B, Li M, Lairmore MD, Green PL. Human T-cell leukemia virus type-1 antisense-encoded gene, Hbz, promotes T-lymphocyte proliferation. Blood. 2008;112:3788–97.
16. Satou Y, Yasunaga J, Zhao T, Yoshida M, Miyazato P, Takai K, Shimizu K, Ohshima K, Green PL, Ohkura N, et al. HTLV-1 bZIP factor induces T-cell lymphoma and systemic inflammation in vivo. PLoS Pathog. 2011;7:e1001274.

17. Sakihama S, Saito M, Kuba-Miyara M, Tomoyose T, Taira N, Miyagi T, Hayashi M, Kinjo S, Nakachi S, Tedokon I, et al. Human T-cell leukemia virus type I Tax genotype analysis in Okinawa, the southernmost and remotest islands of Japan: different distributions compared with mainland Japan and the potential value for the prognosis of aggressive adult T-cell leukemia/lymphoma. Leuk Res. 2017;61:18–24.
18. Yasuma K, Matsuzaki T, Yamano Y, Takashima H, Matsuoka M, Saito M. HTLV-1 subgroups associated with the risk of HAM/TSP are related to viral and host gene expression in peripheral blood mononuclear cells, independent of the transactivation functions of the viral factors. J Neurovirol. 2016;22:416–30.
19. Osame M. Review of WHO Kagoshima meeting and diagnostic guidelines for HAM/TSP. New York: Raven Press; 1990.
20. Nagata K, Ohtani K, Nakamura M, Sugamura K. Activation of endogenous c-fos proto-oncogene expression by human T-cell leukemia virus type I-encoded p40tax protein in the human T-cell line. Jurkat. J Virol. 1989;63:3220–6.
21. Motai Y, Takahashi M, Takachi T, Higuchi M, Hara T, Mizuguchi M, Aoyagi Y, Terai S, Tanaka Y, Fujii M. Human T-cell leukemia virus type 1 (HTLV-1) Tax1 oncoprotein but not HTLV-2 Tax2 induces the expression of OX40 ligand by interacting with p52/p100 and RelB. Virus Genes. 2016;52:4–13.
22. Spurrell JC, Wiehler S, Zaheer RS, Sanders SP, Proud D. Human airway epithelial cells produce IP-10 (CXCL10) in vitro and in vivo upon rhinovirus infection. Am J Physiol Lung Cell Mol Physiol. 2005;289:L85–95.
23. Shiohama Y, Naito T, Matsuzaki T, Tanaka R, Tomoyose T, Takashima H, Fukushima T, Tanaka Y, Saito M. Absolute quantification of HTLV-1 basic leucine zipper factor (HBZ) protein and its plasma antibody in HTLV-1 infected individuals with different clinical status. Retrovirology. 2016;13:29.
24. Nagai M, Usuku K, Matsumoto W, Kodama D, Takenouchi N, Moritoyo T, Hashiguchi S, Ichinose M, Bangham CR, Izumo S, Osame M. Analysis of HTLV-I proviral load in 202 HAM/TSP patients and 243 asymptomatic HTLV-I carriers: high proviral load strongly predisposes to HAM/TSP. J Neurovirol. 1998;4:586–93.
25. Hirai H, Suzuki T, Fujisawa J, Inoue J, Yoshida M. Tax protein of human T-cell leukemia virus type I binds to the ankyrin motifs of inhibitory factor kappa B and induces nuclear translocation of transcription factor NF-kappa B proteins for transcriptional activation. Proc Natl Acad Sci U S A. 1994;91:3584–8.
26. Larousserie F, Bardel E, Pflanz S, Arnulf B, Lome-Maldonado C, Hermine O, Bregeaud L, Perennec M, Brousse N, Kastelein R, Devergne O. Analysis of interleukin-27 (EBI3/p28) expression in Epstein-Barr virus- and human T-cell leukemia virus type 1-associated lymphomas: heterogeneous expression of EBI3 subunit by tumoral cells. Am J Pathol. 2005;166:1217–28.
27. Valentin H, Lemasson I, Hamaia S, Casse H, Konig S, Devaux C, Gazzolo L. Transcriptional activation of the vascular cell adhesion molecule-1 gene in T lymphocytes expressing human T-cell leukemia virus type 1 Tax protein. J Virol. 1997;71:8522–30.
28. Silbermann K, Schneider G, Grassmann R. Stimulation of interleukin-13 expression by human T-cell leukemia virus type 1 oncoprotein Tax via a dually active promoter element responsive to NF-kappaB and NFAT. J Gen Virol. 2008;89:2788–98.
29. Saito M, Sejima H, Naito T, Ushirogawa H, Matsuzaki T, Matsuura E, Tanaka Y, Nakamura T, Takashima H. The CC chemokine ligand (CCL) 1, upregulated by the viral transactivator Tax, can be downregulated by minocycline: possible implications for long-term treatment of HTLV-1-associated myelopathy/tropical spastic paraparesis. Virol J. 2017;14:234.
30. Yamano Y, Sato T. Clinical pathophysiology of human T-lymphotropic virus-type 1-associated myelopathy/tropical spastic paraparesis. Front Microbiol. 2012;3:389.
31. Yamano Y, Coler-Reilly A. HTLV-1 induces a Th1-like state in CD4+ CCR4+ T cells that produces an inflammatory positive feedback loop via astrocytes in HAM/TSP. J Neuroimmunol. 2017;304:51–5.
32. Sato T, Coler-Reilly A, Utsunomiya A, Araya N, Yagishita N, Ando H, Yamauchi J, Inoue E, Ueno T, Hasegawa Y, et al. CSF CXCL10, CXCL9, and neopterin as candidate prognostic biomarkers for HTLV-1-associated myelopathy/tropical spastic paraparesis. PLoS Negl Trop Dis. 2013;7:e2479.
33. Ando H, Sato T, Tomaru U, Yoshida M, Utsunomiya A, Yamauchi J, Araya N, Yagishita N, Coler-Reilly A, Shimizu Y, et al. Positive feedback loop via astrocytes causes chronic inflammation in virus-associated myelopathy. Brain. 2013;136:2876–87.
34. Smith MR, Greene WC. Identification of HTLV-I tax trans-activator mutants exhibiting novel transcriptional phenotypes. Genes Dev. 1990;4:1875–85.
35. Matsumoto K, Shibata H, Fujisawa JI, Inoue H, Hakura A, Tsukahara T, Fujii M. Human T-cell leukemia virus type 1 Tax protein transforms rat fibroblasts via two distinct pathways. J Virol. 1997;71:4445–51.
36. Ciminale V, Rende F, Bertazzoni U, Romanelli MG. HTLV-1 and HTLV-2: highly similar viruses with distinct oncogenic properties. Front Microbiol. 2014;5:398.
37. Shoji T, Higuchi M, Kondo R, Takahashi M, Oie M, Tanaka Y, Aoyagi Y, Fujii M. Identification of a novel motif responsible for the distinctive transforming activity of human T-cell leukemia virus (HTLV) type 1 Tax1 protein from HTLV-2 Tax2. Retrovirology. 2009;6:83.
38. Cech TR, Steitz JA. The noncoding RNA revolution-trashing old rules to forge new ones. Cell. 2014;157:77–94.
39. Yamagishi M, Nakano K, Miyake A, Yamochi T, Kagami Y, Tsutsumi A, Matsuda Y, Sato-Otsubo A, Muto S, Utsunomiya A, et al. Polycomb-mediated loss of miR-31 activates NIK-dependent NF-kappaB pathway in adult T cell leukemia and other cancers. Cancer Cell. 2012;21:121–35.
40. Wu H, Yang L, Chen LL. The Diversity of Long Noncoding RNAs and Their Generation. Trends Genet. 2017;33:540–52.
41. Grassmann R, Aboud M, Jeang KT. Molecular mechanisms of cellular transformation by HTLV-1 Tax. Oncogene. 2005;24:5976–85.
42. Hoesel B, Schmid JA. The complexity of NF-kappaB signaling in inflammation and cancer. Mol Cancer. 2013;12:86.
43. Matsuoka M, Jeang KT. Human T-cell leukemia virus type 1 (HTLV-1) and leukemic transformation: viral infectivity, Tax. HBZ and therapy. Oncogene. 2011;30:1379–89.
44. Higuchi M, Fujii M. Distinct functions of HTLV-1 Tax1 from HTLV-2 Tax2 contribute key roles to viral pathogenesis. Retrovirology. 2009;6:117.
45. Narikawa K, Fujihara K, Misu T, Feng J, Fujimori J, Nakashima I, Miyazawa I, Saito H, Sato S, Itoyama Y. CSF-chemokines in HTLV-I-associated myelopathy: CXCL10 up-regulation and therapeutic effect of interferon-alpha. J Neuroimmunol. 2005;159:177–82.
46. Guerreiro JB, Santos SB, Morgan DJ, Porto AF, Muniz AL, Ho JL, Teixeira AL Jr, Teixeira MM, Carvalho EM. Levels of serum chemokines discriminate clinical myelopathy associated with human T lymphotropic virus type 1 (HTLV-1)/tropical spastic paraparesis (HAM/TSP) disease from HTLV-1 carrier state. Clin Exp Immunol. 2006;145:296–301.
47. Hieshima K, Nagakubo D, Nakayama T, Shirakawa AK, Jin Z, Yoshie O. Tax-inducible production of CC chemokine ligand 22 by human T cell leukemia virus type 1 (HTLV-1)-infected T cells promotes preferential transmission of HTLV-1 to CCR4-expressing CD4+ T cells. J Immunol. 2008;180:931–9.
48. Ruckes T, Saul D, Van Snick J, Hermine O, Grassmann R. Autocrine antiapoptotic stimulation of cultured adult T-cell leukemia cells by overexpression of the chemokine I-309. Blood. 2001;98:1150–9.
49. Mori N, Krensky AM, Ohshima K, Tomita M, Matsuda T, Ohta T, Yamada Y, Tomonaga M, Ikeda S, Yamamoto N. Elevated expression of CCL5/RANTES in adult T-cell leukemia cells: possible transactivation of the CCL5 gene by human T-cell leukemia virus type I tax. Int J Cancer. 2004;111:548–57.
50. Montanheiro P, Vergara MP, Smid J, da Silva Duarte AJ, de Oliveira AC, Casseb J. High production of RANTES and MIP-1alpha in the tropical spastic paraparesis/HTLV-1-associated myelopathy (TSP/HAM). J Neuroimmunol. 2007;188:138–42.
51. Rafatpanah H, Felegari M, Azarpazhooh MR, Vakili R, Rajaei T, Hampson I, Hassanshahi G, Valizadeh N, Gerayli S, Hamid F, et al. Altered expression of CXCR3 and CCR6 and their ligands in HTLV-1 carriers and HAM/TSP patients. J Med Virol. 2017;89:1461–8.
52. Baba M, Imai T, Yoshida T, Yoshie O. Constitutive expression of various chemokine genes in human T-cell lines infected with human T-cell leukemia virus type 1: role of the viral transactivator Tax. Int J Cancer 1996;66:124–9.
53. Kataoka K, Nagata Y, Kitanaka A, Shiraishi Y, Shimamura T, Yasunaga J, Totoki Y, Chiba K, Sato-Otsubo A, Nagae G, et al. Integrated molecular analysis of adult T cell leukemia/lymphoma. Nat Genet. 2015;47:1304–15.

3

Inhibitors of the integrase–transportin-SR2 interaction block HIV nuclear import

Jonas Demeulemeester[1,4†], Jolien Blokken[1†], Stéphanie De Houwer[1], Lieve Dirix[1], Hugo Klaassen[2], Arnaud Marchand[2], Patrick Chaltin[2,3], Frauke Christ[1] and Zeger Debyser[1*]

Abstract

Background: Combination antiretroviral therapy efficiently suppresses HIV replication in infected patients, transforming HIV/AIDS into a chronic disease. Viral resistance does develop however, especially under suboptimal treatment conditions such as poor adherence. As a consequence, continued exploration of novel targets is paramount to identify novel antivirals that do not suffer from cross-resistance with existing drugs. One new promising class of targets are HIV protein–cofactor interactions. Transportin-SR2 (TRN-SR2) is a β-karyopherin that was recently identified as an HIV-1 cofactor. It has been implicated in nuclear import of the viral pre-integration complex and was confirmed as a direct binding partner of HIV-1 integrase (IN). Nevertheless, consensus on its mechanism of action is yet to be reached.

Results: Here we describe the development and use of an AlphaScreen-based high-throughput screening cascade for small molecule inhibitors of the HIV-1 IN–TRN-SR2 interaction. False positives and nonspecific protein–protein interaction inhibitors were eliminated through different counterscreens. We identified and confirmed 2 active compound series from an initial screen of 25,608 small molecules. These compounds significantly reduced nuclear import of fluorescently labeled HIV particles.

Conclusions: Alphascreen-based high-throughput screening can allow the identification of compounds representing a novel class of HIV inhibitors. These results corroborate the role of the IN–TRN-SR2 interaction in nuclear import. These compounds represent the first in class small molecule inhibitors of HIV-1 nuclear import.

Keywords: Drug discovery, Integrase, Transportin-SR2, HIV, Nuclear import

Background

Control of human immunodeficiency virus type 1 (HIV-1) infection still poses considerable challenges. Viral resistance selection under suboptimal treatment conditions together with long-term adverse effects of chronic combination antiretroviral therapy (cART), highlight the need for novel antivirals with (1) a higher barrier towards resistance development, (2) pharmacokinetic properties allowing once-daily dosing and (3) less adverse effects. Novel targets, both viral and cellular, therefore need to be evaluated for their potential to be targeted with a small molecule in order to develop new classes of antivirals. HIV is characterized by its ability to stably integrate into the host cell genome and to infect both dividing and non-dividing cells [1]. As a result, the pre-integration complex (PIC) needs to be actively transported across the nuclear envelope. Integration into host cell chromatin is a complex process catalyzed by the viral integrase (IN). Once integration has been achieved, the fate of virus and host cell are irreversibly linked, highlighting this step as the point-of-no-return in the viral replication cycle.

*Correspondence: zeger.debyser@kuleuven.be
†Jonas Demeulemeester and Jolien Blokken shared first author
[1] Laboratory for Molecular Virology and Gene Therapy, Department of Pharmaceutical and Pharmacological Sciences, KU Leuven, Kapucijnenvoer 33, VCTB +5, Bus 7001, 3000 Leuven, Flanders, Belgium
Full list of author information is available at the end of the article

Due to technical issues such as poor in vitro solubility, IN was last of the three HIV enzymes to be targeted by cART. To date, three IN inhibitors have been approved for clinical use (raltegravir, elvitegravir and dolutegravir), all belonging to a class of catalytic site inhibitors known as integrase strand transfer inhibitors (INSTIs, see [2] for a recent review). As a consequence of their common mechanism of action these inhibitors suffer a significant degree of cross-resistance. Allosteric inhibitors potently blocking IN indirectly can circumvent cross-resistance in analogy to non-nucleoside reverse transcriptase inhibitors (NNRTIs) [3]. Of note, successful integration and completion of the viral replication cycle in general, are dependent on more than just the catalytic activity of IN but also on its multimerization dynamics and interplay with cellular cofactors such as LEDGF/p75 [4–8]. Screening for inhibitors of either the LEDGF/p75-IN interaction or the integrase 3′-processing reaction independently led to the discovery of a second class of IN inhibitors, LEDGINs (also referred to as non-catalytic site integrase inhibitors, NCINIs) [9, 10]. By binding the LEDGF/p75 binding site across the IN dimerization interface, LEDGINs not only block LEDGF/p75 binding but also perturb IN multimerization, which in turn leads to inhibited catalytic activity (both 3′processing and strand transfer) and aberrant viral maturation [4–7, 11, 12].

In 2008, another IN cofactor was identified, Transportin-SR2 (TRN-SR2, Tnpo3), encoded by the *TNPO3* gene. TRN-SR2 was picked up as a cellular cofactor of HIV-1 in two genome-wide siRNA screens [13, 14] and as a binding partner of HIV IN in a yeast two-hybrid screen [15]. Through q-PCR analysis and the use of a cellular nuclear import assay [16], Christ et al. [15] showed a clear reduction in HIV nuclear import after depletion of TRN-SR2, supporting a role of TRN-SR2 in this process. Transportin-SR2 belongs to the β-karyopherin family. It has been shown to import splicing factors to the nucleus, most of which contain an RS (arginine–serine) repeat region and/or an RNA recognition motif (RRM) domain [17–19]. Its overall toroid structure, composed of stacked HEAT repeats, provides flexibility to accommodate a variety of cellular cargoes [19–21]. Charged residues on and around an Arg-rich helix in TRN-SR2 are critical for recognition of the phosphorylated RS region of cargo and hence its nuclear import [19]. Until now, crystal structures of TRN-SR2 alone [19], in complex with RanGTP [21] and in complex with the cellular cargo ASF/SF2 [19] have been described. A crystal structure of TRN-SR2 in complex with IN is not available.

A variety of viral components have been linked to nuclear import of the HIV pre-integration complex (PIC): capsid (CA), the central polypurine tract (cPPT), IN, matrix and viral protein R [22–24]. Also for the host cell, a plethora of import factors have been implicated, most notably importin-α/β [25, 26], importin-α3 [27] and importin-7 [28]. Despite the general agreement on the importance of TRN-SR2 for HIV nuclear import, the exact mechanism of action remains a matter of debate. The TRN-SR2–CA interaction has been reported to play a role in nuclear import by some groups [29, 30], while others published evidence for a direct interaction with HIV IN [15, 31–33]. Moreover, IN was shown to be displaced from TRN-SR2 upon addition of RanGTP, as is the case with normal cargoes [20]. An IN R263A/K264A mutant is partially deficient for the interaction with TRN-SR2 [33, 34] and the corresponding virus was affected at the nuclear import step, supporting the notion that the IN–TRN-SR2 interaction is responsible for this process [34].

As evidenced by the discovery and development of LEDGINs, targeting protein–protein interactions between IN and cellular cofactors can yield new classes of viral replication inhibitors [35, 36]. Since nuclear import represents a bottleneck during HIV replication [15] we reasoned that inhibitors of this interaction might have the potential to become potent antivirals and we embarked on a drug discovery campaign targeting the interaction between HIV-1 IN and TRN-SR2. Small molecules disrupting the interaction and blocking nuclear import would additionally be valuable to study HIV nuclear import and therefore increase our understanding of this crucial step in its replication cycle.

At the time this study was initiated, the interface between TRN-SR2 and IN had not been defined and no crystal structure of TRN-SR2 was available. Therefore, we opted for a high-throughput screening (HTS) approach. Here, we describe the development and use of an amplified luminescent proximity homogenous assay (AlphaScreen)-based screening cascade to identify small-molecule inhibitors of the HIV-1 IN–TRN-SR2 interaction from a library of 25,608 compounds. We eliminated false positives and nonspecific protein–protein interaction inhibitors through the implementation of appropriate counterscreens. Five compound classes provided modest protection against HIV-1 during multiple round replication. Finally, four representative compounds were tested in a cellular fluorescent HIV nuclear import assay. Two compounds significantly reduced the number of nuclear PICs, suggesting these molecules represent a novel class of inhibitors targeting HIV nuclear import. These novel inhibitors validate the IN–TRN-SR2 interaction as an antiviral target and warrant further exploration.

Methods

Recombinant protein purification

Recombinant proteins were expressed in *E. coli* strain BL21-CodonPlus2 (DE3). N-terminally His_6-tagged and untagged IN, His_6-tagged glutathione-S-transferase (GST-His_6), maltose-binding protein (MBP)-tagged JPO2 (Cell division cycle-associated 7-like protein, CDCA7L), 3xflag-tagged LEDGF/p75 and GST-TRN-SR2 were purified as described previously [20, 37–39].

High-throughput screening

Compound stocks were dissolved at approximately 5 mM in dimethylsulfoxide (DMSO) in 96-well plates. Stocks were first diluted to 125 μM in assay buffer (150 mM NaCl, 25 mM Tris–HCl pH 7.4, 1 mM $MgCl_2$, 0.1% (v/v) Tween-20 and 0.1% (w/v) bovine serum albumin (BSA)) supplemented with 2.5% (v/v) DMSO before 10 μl was transferred to a dry 384-well PS 384-OptiPlate (Perkin Elmer) on a Freedom EVO200 liquid handling robot (Tecan). Untagged TRN-SR2 at 2.5 μM and assay buffer containing 5% DMSO were transferred from a separate 96-well plate to the assay plates and represented the positive and negative controls, respectively. Next, 5 μl each of 5× working dilutions of GST-TRN-SR2 and His_6-IN were added using an XRD-384 automated reagent dispenser (FluidX). The plate was left to incubate for 1 h at 7 °C before 5 μl of a glutathione donor and Ni^{2+}-chelate acceptor AlphaScreen bead mixture (Perkin Elmer) was added. This brought the final assay volume to 25 μl and established final concentrations of 50 μM for the compounds, 10 nM GST-TRN-SR2, 40 nM His_6-IN, 10 μg/ml donor and acceptor beads and 2% DMSO.

Hit validation and counterscreens

Both the hit validation assay and the two counterscreens (GST-His_6 and LEDGF/p75–JPO2) were performed in duplicate. Compounds were first diluted to 125 μM as during the main screen but now two doses were transferred to a dry assay plate. From this point on, the hit validation assay was performed identically to the main screen.

For the GST-His_6 counterscreen, 10 μl of a 2.5× working dilution of the recombinant GST-His_6 protein (10 nM final assay concentration) was added to the plate instead of the GST-TRN-SR2 and His_6-IN. In this case, 50 μM bromophenol blue and buffer containing the appropriate amount of DMSO represented the positive and negative controls, respectively.

The LEDGF/p75–JPO2 specificity counterscreen was performed similarly, but instead of GST-TRN-SR2 and His_6-IN, maltose-binding protein (MBP)-tagged JPO2 and 3xflag-tagged LEDGF/p75 were employed at final concentrations of 5 nM. The AlphaScreen bead mixture was modified accordingly to anti-flag acceptor beads and streptavidin donor beads coated with anti-MBP antibody according to the manufacturer's protocol (Perkin Elmer). Untagged LEDGF/p75 at a final concentration of 1 μM and buffer containing the appropriate amount of DMSO were used as positive and negative controls, respectively.

Multiple round antiviral activity assay

The inhibitory effect of antiviral compounds on the HIV-induced cytopathic effect (CPE) in MT-4 cell culture was determined by the MTT-assay [40]. The assay is based on the reduction of the yellow colored 3-(4,5-dimethylthiazol-2-yl)-2,5- diphenyltetrazolium bromide (MTT) by mitochondrial dehydrogenase of metabolically active cells to a blue formazan derivative, which can be measured spectrophotometrically. The 50% cell culture infective dose of the HIV strain was determined by titration of the virus stock on MT-4 cells. For the antiviral activity assays, MT-4 cells were infected with 100–300× 50% cell culture infective doses in the presence of five-fold serial dilutions of the compounds. The concentration achieving 50% protection against the HIV CPE (the 50% effective concentration, EC_{50}), was determined as well as the concentration killing 50% of the MT-4 cells (the 50% cytotoxic concentration, CC_{50}).

Single round antiviral activity assay

20,000 HeLaP4 cells were seeded into 96-well plates on the day prior to infection. Cells were infected in triplicate with 3 dilutions (typically 1×10^5, 3.3×10^4 and 1.1×10^4 pg p24) of VSV-G pseudotyped single-round HIV-1 supplemented with one of the compounds in a total volume of 200 μl per well. The virus was produced as described previously [31]. The final compound concentration in the assay was 100 μM. p24 measurements were performed with the Innotest HIV Antigen mAb kit (Fujirebio). 24 h after infection the supernatant was replaced by fresh medium. 72 h post infection cells were lysed in 50 μl of lysis buffer (50 mM Tris/HCl, pH 7.3, 200 mM NaCl, 0.2% NP40, 5% glycerol) and analyzed for firefly luciferase activity (ONE-Glo™, Promega, Belgium) according to the manufacturer's protocol. Chemiluminescence was measured with a Glomax luminometer (Promega, Belgium). The signals were normalized for protein content as determined by a BCA protein assay (Thermo Scientific Pierce).

PIC nuclear import assay

The PIC nuclear import assay was performed as previously described [34]. Briefly, to produce the fluorescent HIV particles, 293T cells were transfected with 15 μg pVpr-IN-eGFP, 15 μg pNL4-3.Luc.R-.E- (obtained from the NIH AIDS Reference and Reagent Program), and 5 μg of the pMD.G plasmid encoding the vesicular stomatitis virus glycoprotein (VSV-G) [41]. Supernatant was collected after 48 h, filtered and concentrated by ultracentrifugation. p24 was measured with the Innotest HIV Antigen mAb kit (Fujirebio). A viral inoculum of 3.10^6 pg p24 was used to infect 30,000 HeLaP4 cells in the presence of 100 μM of hit compound. PF-3450074 (PF74, Sigma-Aldrich) was tested at 2 μM and 10 μM. Five hours after infection, cells were fixed and the nuclear lamina was visualized with a monoclonal anti-lamin A/C antibody (Santa Cruz, sc-7292) followed by a secondary goat anti-mouse IgG Alexa-Fluor 633. Three-dimensional image stacks were acquired on a Zeiss LSM510 multiphoton confocal microscope (Cell Imaging Core CIC, University of Leuven) equipped with a Plan-Apochromat 63×/1.4 Oil DIC objective. The Z-step size was 0.3 μm. The quantification of PICs was performed using a homemade MatLab routine (The MathWorks, Inc.). A fluorescent spot was assigned as a PIC if at least two adjacent pixels were above the threshold and if the signal was present in at least two consecutive Z-planes. PICs were classified as cytoplasmic or nuclear based on the nuclear lamin staining.

Results

Optimization of the IN–TRN-SR2 AlphaScreen

Development of robust high-throughput assays requires optimal buffer conditions in which both proteins and the detection method are stable and reproducible. Initial conditions for an AlphaScreen assay measuring the direct protein–protein interaction between HIV-1 IN and TRN-SR2 were previously reported by our group [31]. This experience allowed us to readily adapt an assay buffer to meet the requirements for a high-throughput screen (HTS) for inhibitors of the interaction. Two modifications were made to the previously reported assay buffer: First, measurements were done in the presence of 1 mM dithiothreitol (DTT) to maintain a sufficiently reducing environment and prevent aggregation of glutathione-S-transferase (GST)-tagged protein on the surface of the glutathione donor beads [39]. Second, since protein–protein interactions can be difficult to disrupt by small molecules, we aimed to screen at a relatively high compound concentration of 50 μM. To aid compound solubility at these concentrations, aside from the already present 0.1% (v/v) Tween-20 and 0.1% (w/v) bovine serum albumin (BSA), we increased the final concentration of dimethylsulfoxide (DMSO). To assess the tolerance of the assay for DMSO, we titrated it out in a two-fold dilution series between 20 and 0.16% (v/v). The AlphaScreen signal was normalized to the 0% DMSO condition and the background signal (100 and 0%, respectively). Figure 1 shows that despite a high tolerance of the assay for DMSO the variability increases to unacceptable levels when more

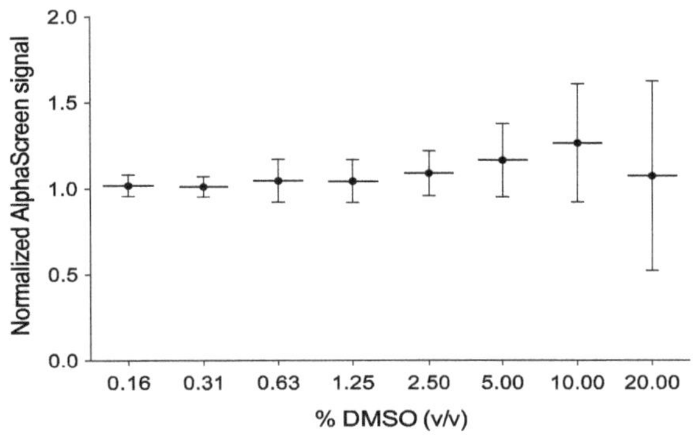

Fig. 1 Tolerance for dimethylsulfoxide (DMSO) in the AlphaScreen assay. DMSO was titrated out starting from 20% (v/v) in the IN-TRN-SR2 interaction assay and the resulting AlphaScreen signal was measured. Results are normalized to the 0% DMSO condition

than 2.5% DMSO is present. Based on these results, we opted for a final DMSO concentration in the assay buffer of 2%.

HTS assay optimization is a fine balance between identifying robust assay conditions with minimal variability and keeping costs acceptable. In the present case, costs were mainly driven by the amount of AlphaScreen beads used. However, reducing the bead concentration leads to a roughly linear decrease of the signal-to-background (S/B) ratio, and hence of the assay quality as well [39]. As a compromise, we decided to lower the final bead concentrations only two-fold to 10 μg/ml.

Next, optimal protein concentrations were determined through cross-titration of the binding partners (Fig. 2a). Concentrations of 10 nM GST-TRN-SR2 and 40 nM His_6-IN provided a good S/B ratio (> 25) with minimal protein consumption and remained well below the AlphaScreen hooking range. Hooking can be seen to occur at concentrations of 300 nM His_6-IN. Under these conditions, the Ni^{2+}-chelate acceptor beads are expected to be fully saturated and the excess of free IN protein will compete with IN on the bead surface for binding to TRN-SR2, effectively inhibiting the signal.

To monitor per plate quality, each 384-well plate contained a total of 64 control wells arranged in 2 columns of 8 positive and 8 negative wells on either side. As a positive control, we evaluated both untagged TRN-SR2 and IN to corroborate the ability of our assay to pick up inhibitors. However, untagged IN proved unable to inhibit the AlphaScreen signal to background levels (Fig. 2b), most likely due to the proneness of the enzyme to form multimers at higher concentrations [42, 43] which could still bring both beads together. Untagged TRN-SR2 did inhibit the signal (Fig. 2b) and was chosen as the positive control at a final concentration of 1 μM.

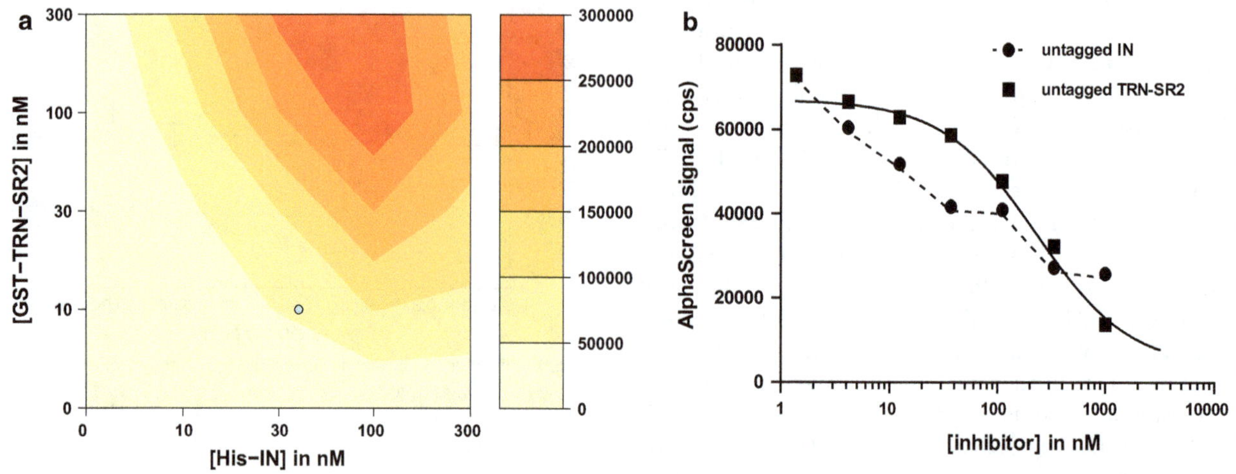

Fig. 2 Cross-titration of His_6-IN and GST-TRN-SR2 and evaluation of positive controls. **a** Both proteins were titrated against one another and the AlphaScreen signal measured (from yellow/low to red/high signal). The optimal condition containing 10 nM GST-TRN-SR2 and 40 nM His_6-IN is indicated with a blue circle. GST, glutathione S-transferase; IN, integrase. **b** Untagged TRN-SR2 and IN were titrated out in the IN–TRN-SR2 interaction assay. Untagged IN was unable to reduce the signal to background levels most likely due to its ability to multimerize. Untagged TRN-SR2 could block the interaction completely and was selected as positive control

High-throughput screening, false positives and negatives

The optimized assay was used to screen a part of the CD3 (Center for Drug Design and Discovery) library. The set of test compounds consisted of 25,608 small molecules which were selected based on (1) chemical diversity, (2) druglike properties (Lipinski rule of five compliant), (3) exclusion of known toxicophores and purchased from multiple commercial suppliers. Assay performance was evaluated on a plate-by-plate basis and remained robust throughout the entire screening campaign (Fig. 3). A median Z′-factor of 0.76 was obtained and no plates failed during screening. Considering a cut-off of 50% inhibition (percentage of inhibition (PIN) of 50%), we identified a total of 409 initial hits from the collection.

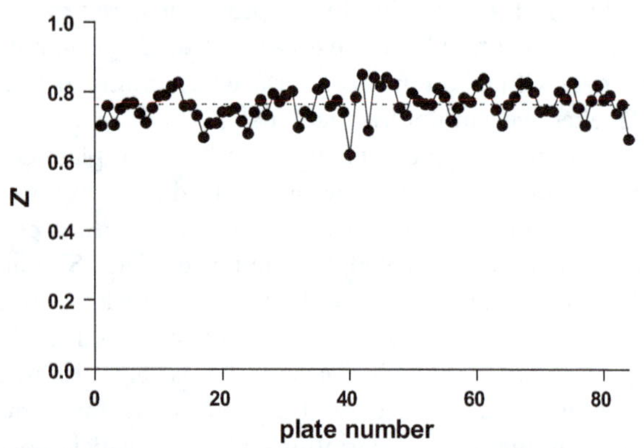

Fig. 3 Assay performance throughout the screening campaign. Assay performance per plate was monitored throughout the screening campaign. A median Z′-factor of 0.76 was determined. No plates had to be repeated

False positives often dominate initial hit lists obtained from HTS campaigns. True false positives were weeded out from the cherry-picked hit compounds by a confirmation screen in which the 409 hits were retested in duplicate in the IN–TRN-SR2 interaction assay. Figure 4a shows percentages of inhibition (PIN) of the duplicates of each compound normalized to the positive (100%) and negative (0%) controls. Linear regression results in a slope of 1.000 ± 0.004 and an R^2 of 0.9675, indicating the robustness of the platform.

AlphaScreen technology is not insensitive to interference [44]. To rid our hit lists of AlphaScreen-specific false positives including inner filter effect, $^1\Delta O_2$ quenchers, and Ni^{2+} chelators, we included in our cascade a GST-His_6 counterscreen described previously [39]. Figure 4b shows PIN values for the duplicates of each compound in this counterscreen normalized to the positive (bromophenol blue, inner filter effect, 100%) and negative (0%) controls. Here as well, the slope of 0.9892 ± 0.006 and R^2 of 0.9640 confirm the platform's robustness.

Combining both results by plotting the average PIN in the GST-His_6 counterscreen against that from the confirmation screen allowed us to confidently select positive hits (Fig. 4c). First, we opted for compounds that met the cut-off of 50% inhibition of the IN–TRN-SR2 interaction, eliminating 126 false positives. Second, compounds should preferentially be devoid of quenching at the active concentrations. Specifically, the percentage of inhibition in the GST-His_6 counterscreen should be below 20% or the activity observed against the IN–TRN-SR2 interaction should be > 50% stronger than the quenching. These

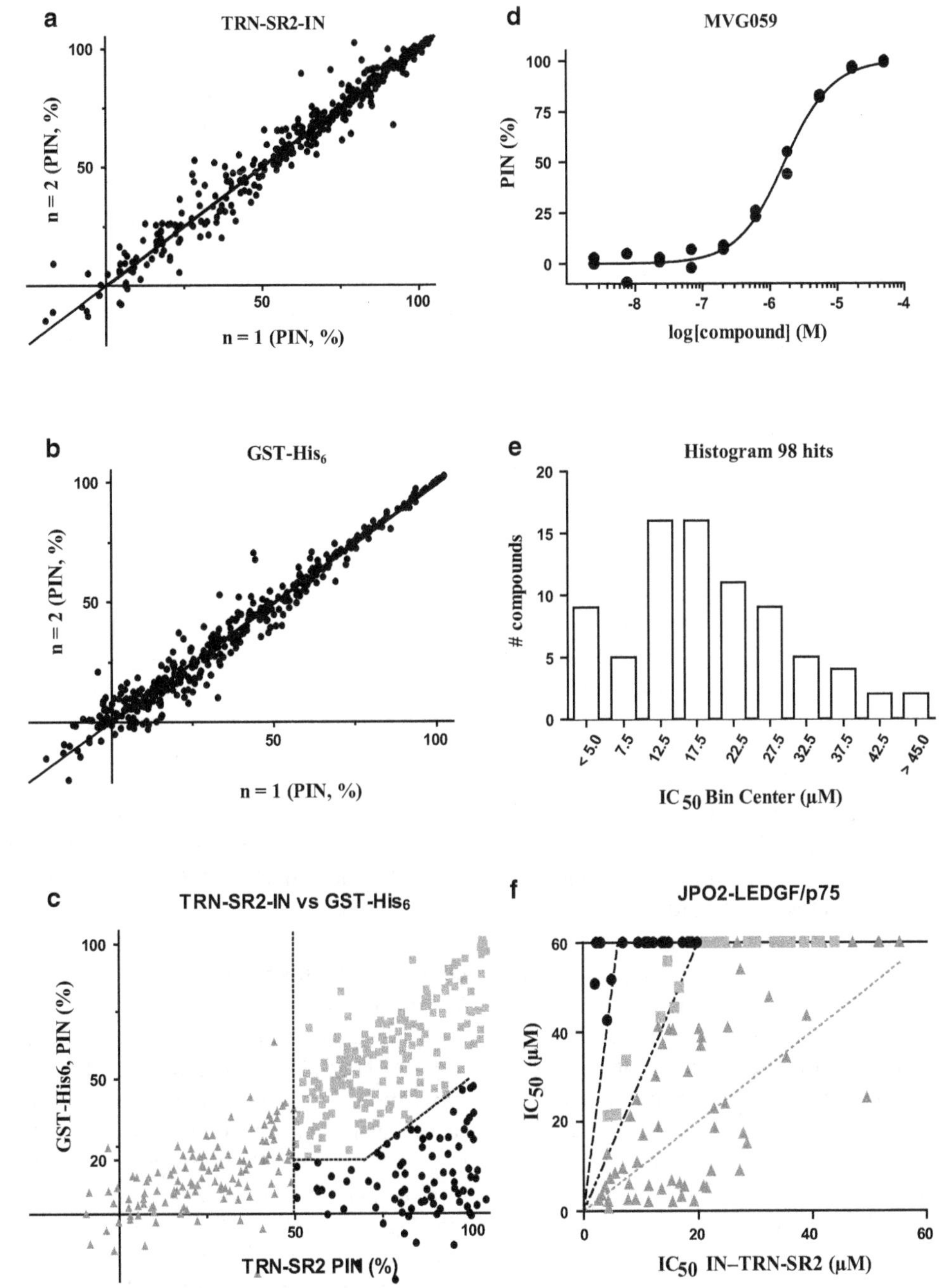

Fig. 4 AlphaScreen screening strategy (**a, b**). Normalized percentage inhibition (PIN) of duplicates of each of the 409 primary hits in (**a**) a confirmatory repeat of the IN–TRN-SR2 AlphaScreen or (**b**) the GST-His_6 false positive counterscreen. In both cases, the linear regression fit to the data is shown: slopes are not significantly different from 1 and R^2 values are 0.9675 and 0.9640 for (**a**) and (**b**), respectively. **c** Average percentage inhibition in the GST-His_6 counterscreen plotted against that observed in the IN–TRN-SR2 assay. Hits of interest show > 50% inhibition of the IN–TRN-SR2 interaction and are devoid of quenching at the active concentration (< 20% inhibition of the GST-His_6 signal or 50% stronger inhibition of the IN–TRN-SR2 interaction). Primary hits occupying the different areas of the result space are plotted as dark grey triangles (false positives), light grey squares (quenchers) and black circles (hits of interest). **d** Titration results and dose–response curve fit obtained for compound MVG059. Duplicates are plotted. **e** Histogram of the obtained IC_{50} values for all 98 hits sorted into 5 µM bins. **f** Hit compound specificity. IC_{50} values obtained against the IN–TRN-SR2 and LEDGF/p75–JPO2 interactions were plotted against one another. Compounds having an IC_{50} < 20 µM against the IN–TRN-SR2 interaction and being tenfold less active or having an IC_{50} > 60 µM against LEDGF/p75–JPO2 are preferred (black circles). Hits that were only between 10 and 3 times more active in the IN–TRN-SR2 assay are less promising (light grey squares). Those that were less potent ($IC_{50} \geq 20$ µM) but did not inhibit the LEDGF/p75–JPO2 interaction (IC_{50} > 60 µM) were looked at case-by-case. Compounds falling outside of these regions were considered non-specific and not investigated further (dark grey triangles)

requirements delineated an area of the results space containing 98 clean and confirmed hit compounds (Fig. 4c, black circles).

Determination of potency and specificity

The 98 confirmed hits were next subjected to IC_{50} determination in the IN–TRN-SR2 AlphaScreen interaction assay. Compounds were titrated in duplicate in a threefold dilution series starting at 50 μM and dose–response curves were fit to the resulting data. Figure 4d shows a representative titration result while Fig. 4e provides a histogram of the obtained IC_{50} data. In total, 14 out of the 98 hits had an IC_{50} value in the single digit micromolar range.

Since the IN–TRN-SR2 protein–protein interface is believed to be relatively flat and featureless, we decided to perform a specificity counterscreen. Compounds were assayed for their inhibition of an unrelated protein–protein interaction (LEDGF/p75–JPO2) [45, 46] and their IC_{50} values were determined. Compounds were titrated as before and the inhibition of the interaction between full-length 3xflag-tagged LEDGF/p75 and MBP-tagged JPO2 was measured by AlphaScreen. Dose–response curves were fit and the resulting IC_{50} values compared to those obtained for the IN–TRN-SR2 interaction. Compounds that showed high activity (IC_{50} < 20 μM) in the IN–TRN-SR2 interaction assay and were tenfold less active or completely inactive (IC_{50} > 60 μM) against the LEDGF/p75–JPO2 interaction were prioritized (Fig. 4f). Compounds that were between 10 and 3 times more active in the IN–TRN-SR2 assay were kept as back-up compounds. Those that were less potent ($IC_{50} \geq$ 20 μM) but did not inhibit the LEDGF/p75–JPO2 interaction (IC_{50} > 60 μM) were looked at case-by-case and evaluated as potential back-up compounds based on their PIN at the highest concentration used. All compounds falling outside these regions of the results space were considered nonspecific protein–protein interaction inhibitors and discarded from the hit list. In the end, this resulted in 23 first priority and 25 fallback compounds.

Analogues

The 23 first priority compounds were clustered into 12 classes based on their structure and the hit compounds of each cluster were repurchased. Unfortunately, the IN–TRN-SR2 activity could not be confirmed for 5 hits which left us with 7 hit compounds for which commercially available analogues were selected and ordered. These analogues were funneled through the same screening cascade consisting of the IN–TRN-SR2 interaction assay and both counterscreens (GST-His_6 for quenching and LEDGF/p75–JPO2 for specificity) leading to a selection of compounds which were further evaluated for inhibition of HIV-1 replication and cellular toxicity in the MTT/MT-4 viral replication assay (see Table 1).

Antiviral activity in cell culture

A selection of active compounds was then evaluated for inhibition of HIV-1 replication and cellular toxicity in the MTT/MT-4 viral replication assay. The percentage protection against the HIV CPE was measured. Modest (up to 46% at 100 μM) protection was observed for a limited number of compounds in classes 3, 7 and 9 but no 50% effective concentration (EC_{50}) was reached at concentrations lower than the observed 50% cytotoxic concentration (CC_{50}) (Table 1).

Rapid growth (and frequent cell divisions) of the MT-4 cells during the viral replication assay however, may mask viral nuclear import defects, in particular with low potency compounds. This could imply that the MTT/MT-4 assay is not well suited to detect a block in viral nuclear import. We hence decided to evaluate four representative compounds from the most promising classes 6, 7, 9 and 10 in a single round antiviral activity assay on HeLaP4 cells. Compounds were tested at a final concentration of 100 μM and cells were infected with a threefold dilution series of a single-round virus expressing the firefly luciferase reporter gene (Fluc). The percentage inhibition of HIV replication compared to the DMSO control was determined. For compound MVG036 no inhibition could be detected while MVG010, MVG044 and MVG030 showed 34.2, 16.3 and 23.3% inhibition, respectively (Fig. 5).

Effect on HIV nuclear import

We next analyzed the selected compounds, again at 100 μM concentration, in a low throughput PIC nuclear import assay [34] to test whether the decrease in viral replication is due to inhibition of PIC nuclear import. In this assay fluorescently labeled viral PICs are visualized using confocal microscopy and their subcellular localization is evaluated based on a nuclear lamina staining (Fig. 6a, b). The ratio of nuclear PICs over the total number of PICs (percentage nuclear PICs) was then calculated as a measure of nuclear import. DMSO and Raltegravir (RAL) were used as negative controls in two independent experiments as they should not affect HIV nuclear import. The capsid-binder PF74 was used as a positive control. This compound is known to inhibit nuclear import at low concentrations (2 μM) whereas it also inhibits reverse transcription at higher concentrations (10 μM) (Fig. 6c) [47, 48]. MVG044 and MVG030 significantly reduced the number of nuclear PICs compared to both DMSO and RAL ($p < 0.05$) while MVG036 only showed significance compared to DMSO (Fig. 6d; Table 2).

Table 1 Antiviral activity and toxicity of selected hits in multiple round MTT assay

Class	Compound ID	IN–TRN-SR2 Pin at 50 μM (%)[a]	EC_{50} (μM)[b]		CC_{50} (μM)[c]		Protection (%)[d]
3	MVG001	65.3	<	8	<	8	2
	MVG002	43.4	>	13	=	13	5
	MVG003	47.2	>	80	=	80	13
	MVG004	82.6	>	21	=	21	4
	MVG005	39.7	>	57	=	57	6
	MVG006	50.5	>	63	=	63	6
	MVG007	64.8	>	74	=	74	6
	MVG008	30.1	>	99	=	99	23
	MVG009	32.0	>	51	=	51	4
6	MVG010	37.1	>	100	=	100	6
	MVG011	41.7	>	102	=	102	1
	MVG012	107.1	>	74	=	74	10
	MVG013	51.9	>	95	=	95	3
	MVG014	63.8	>	87	=	87	4
	MVG015	48.1	>	87	=	87	5
	MVG016	46.9	>	244	>	244	20
	MVG017	< 10	=	195	=	229	52
	MVG018	< 10	>	137	>	137	7
	MVG019	< 10	>	250	>	250	6
	MVG020	< 10	>	250	>	250	18
	MVG021	< 10	>	250	>	250	34
	MVG022	44.9	=	67	>	136	53
	MVG023	< 10	>	110	>	110	0
	MVG024	< 10	>	36	>	45	49
7	MVG025	72.9	>	34	=	34	4
	MVG026	45.7	>	35	=	48	4
	MVG027	43.2	>	57	=	57	6
	MVG028	58.7	>	56	=	80	13
	MVG029	48.8	>	57	=	57	3
	MVG030	71.6	>	97	=	97	25
	MVG031	62.3	>	53	=	53	4
	MVG032	46.5	>	48	=	48	5
	MVG033	49.5	>	59	=	59	20
	MVG034	43.0	>	73	=	73	16
9	MVG035	43.8	>	36	=	36	1
	MVG036	105.6	>	55	=	65	46
	MVG037	43.1	>	85	=	85	22
	MVG038	47.2	>	59	=	59	0
	MVG039	40.1	>	17	=	17	1
	MVG040	43.6	>	25	=	25	25
	MVG041	54.0	>	35	=	35	2
10	MVG042	89.6	>	55	=	55	10
	MVG043	102.0	>	67	=	67	4
	MVG044	75.9	>	250	>	250	22
	MVG045	11.4	>	203	=	203	21
	MVG046	10.3	>	207	=	207	32
	MVG047	< 10	>	250	>	250	7
	MVG048	< 10	>	87	=	87	0
	MVG049	< 10	=	165	>	231	55
	MVG050	44.1	>	250	>	250	0

Table 1 continued

Class	Compound ID	IN–TRN-SR2 Pin at 50 μM (%)[a]	EC_{50} (μM)[b]		CC_{50} (μM)[c]		Protection (%)[d]
	MVG051	< 10	>	250	>	250	22
	MVG052	20.1	>	66	=	66	8
	MVG053	< 10	>	167	>	167	39
	MVG054	18.0	>	225	=	225	0
	MVG055	< 10	>	167	>	167	21
	MVG056	71.1	>	250	>	250	24
	MVG057	23.3	>	250	>	250	16
	MVG058	53.7	=	170	>	250	24

[a] Compounds were tested in AlphaScreen at 50 μM for their ability to inhibit the IN-TRN-SR2 interaction. Percentage inhibition (PIN), relative to the DMSO control

[b,c] MT-4 cells were infected with HIV at 100 to 300× the 50% cell culture infective doses in the presence of 100 μM of the antiviral drugs. The compound concentration achieving 50% protection against the cytopathic effect of HIV, the 50% effective concentration (EC_{50}), was determined. The concentration of the compound killing 50% of the MT-4 cells, the 50% cytotoxic concentration (CC_{50}), was determined as well

[d] The percentage of protection against the cytopathic effect of HIV was measured in MT-4 cells as well. Values below 20% were not considered as real activities. Data are averages of triplicate measurements

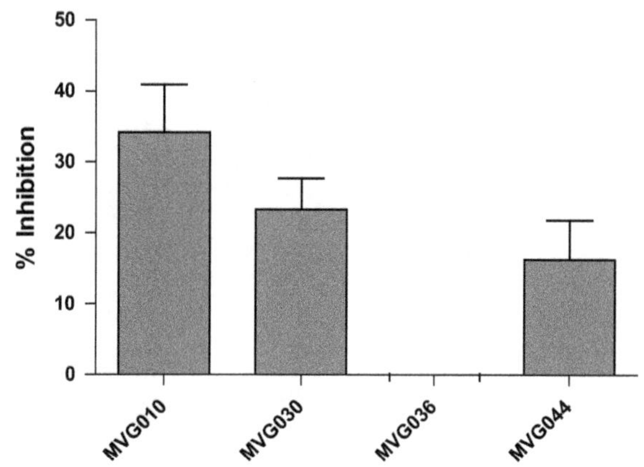

Fig. 5 Representative hit compounds inhibit single round viral replication. HeLaP4 cells were infected with 3.3 × 10^4 pg p24 of a single-round virus expressing the firefly luciferase reporter gene (Fluc) in the presence of 100 μM of one of the compounds. After 72 h, cells were lysed and the percentage inhibition of HIV replication compared to the DMSO control was measured. Mean and standard deviation of two independent experiments, each performed in triplicate, are presented

MVG044 from compound family 10 induced the most pronounced block to nuclear import: the mean % nuclear PICs was 4.8% for DMSO (95% confidence interval [3.8; 5.9%]), 4.5% for Ral [3.4; 5.6%] and 2.1% [0.9; 3.2%] and 2.0% [1.5; 2.7%] for the two MVG044 conditions (DMSO, n = 49 cells; MVG044, n = 50 cells; $p = 0.00001$, Mann–Whitney *U* test and Ral, n = 35 cells; MVG044, n = 31 cells; $p = 0.0002$, Mann–Whitney *U* test) (Table 2).

In Table 3 we present an overview of the data for the four representative compounds. At first sight, there does not seem to be a strong correlation between the different read outs. This is not entirely unexpected, mainly due to the modest activity of our compounds. In addition, these compounds are products of commercially available libraries and have not been optimized yet for activity. The experiments on the potency determination in MT4 cells as well as the inhibition of viral infectivity in a single round viral replication assay in HeLaP4 cells, were experiments lasting for 5 and 3 days, respectively. Chemical (in)stability of the compounds might lead to a more pronounced effect when early time points are analyzed (PIC assay) in contrast to analysis after incubation over several days.

Discussion

Here we present the discovery of small HIV-1 nuclear import inhibitors targeting the interaction of HIV-1 IN and its cellular co-factor, the karyopherin TRN-SR2. We developed an AlphaScreen-based HTS assay to screen for inhibitors of the HIV IN–TRN-SR2 interaction (Figs. 1, 2, 3). We screened a diverse library of 25,608 compounds, yielding a total of 409 hits (Fig. 4). A confirmation assay on the cherry-picked hits in combination with a GST-His_6 counterscreen to remove AlphaScreen technology-interfering false positives narrowed down the hit list to 98 inhibitors (Fig. 4c). Of these 98 hits, 14 were found to have an IC_{50} value in the single-digit micromolar range as determined in the TRN-SR2–HIV IN AlphaScreen assay (Fig. 4e). Because we anticipated the IN–TRN-SR2 interface to be relatively flat and featureless, we performed an additional specificity counterscreen for an unrelated protein–protein interaction (LEDGF/p75–JPO2). Selecting compounds that did not inhibit the LEDGF/p75–JPO2 interaction or were 10 times more potent against the IN–TRN-SR2 interaction resulted in a list of 23 first priority compounds (Fig. 4f). After clustering, hit confirmation with a new "fresh" sample, and analogs selection, the

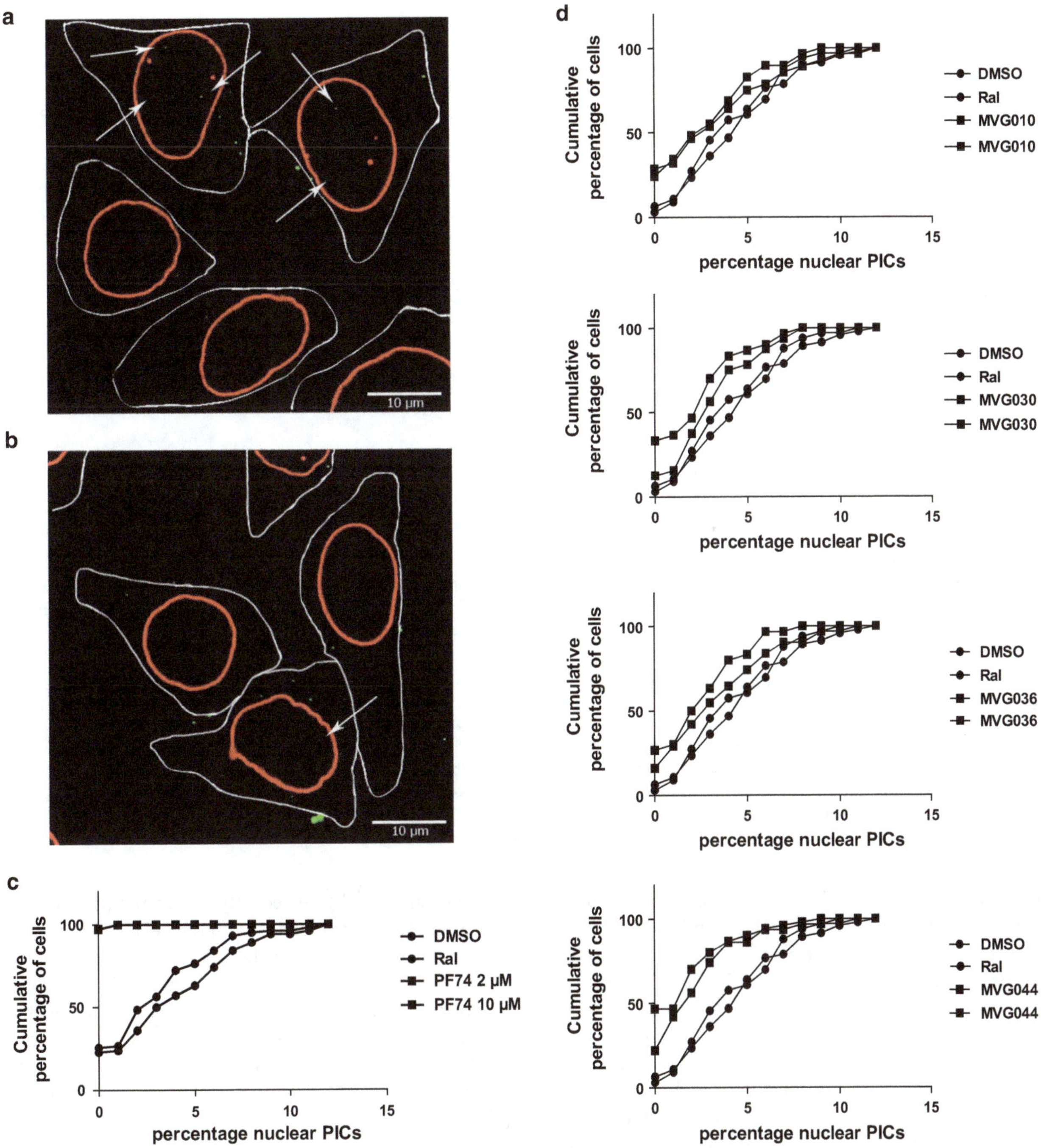

Fig. 6 Representative hit compounds reduce nuclear import of HIV. Five hours after infection with IN-eGFP labeled virus in the presence of one of four representative hit compounds (MVG010, MVG044, MVG030 or MVG036), PF74 and DMSO or RAL as controls, HeLaP4 cells were fixed and analyzed by laser-scanning confocal microscopy. The ratio of nuclear/total amount of PICs was quantified (percentage nuclear PICs). **a, b** Representative slice of a stack of cells infected with eGFP-IN labeled virus in the presence of DMSO (**a**) or MVG044 (b). The nuclear lamina was immunostained with anti-lamin a/c (red). PICs are identified as green dots and nuclear PICs are highlighted by white arrows. **c, d** Presented are the cumulative distributions of the percentage of cells containing the indicated percentage nuclear PICs for the positive control PF74 (**c**) and four representative hit compounds (MVG010, MVG044, MVG030 or MVG036) or DMSO and Ral as controls (**d**). Two independent experiments, with distinct virus productions, are presented for each compound

Table 2 PIC assay statistics

	Compared to DMSO				Compared to Raltegravir			
	n^a	p value[b]	% nuclear PICs[c]	95% CI[d]	n^a	p value[b]	% nuclear PICs[c]	95% CI[d]
DMSO		–	–	–	49	0.8099	4.8	[3.8; 5.9]
Ral	35	0.8099	4.5	[3.4; 5.6]		–	–	–
MVG010	32	0.1209	3.8	[2.3; 5.4]	29	0.1152	4.2	[2.2; 6.3]
MVG044	31	0.00001	2.1	[1.5; 2.7]	50	0.0002	2.0	[0.9; 3.2]
MVG030	30	0.0411	3.3	[2.3; 4.2]	33	0.0019	2.2	[1.4; 3.0]
MVG036	31	0.0011	2.3	[1.6; 3.1]	30	0.0700	3.2	[2.2; 4.2]

Summary of two independent PIC assays comparing the four representative hit compounds to the DMSO and Ral control. HeLaP4 cells were fixed and analyzed by confocal microscopy 5 h after infection with an eGFP-labeled IN virus. PICs were assigned as cytoplasmic or nuclear based on nuclear lamina staining

[a] Number of HeLaP4 cells counted in each condition

[b] p values (Mann–Whitney U test) compared to DMSO or RAL control

[c] Mean percentage of nuclear PICs in each cell calculated using a homemade MatLab routine (The MathWorks, Inc.)

[d] 95% confidence interval (CI) for the percentage of nuclear PICs

Table 3 Overview of the data for four representative compounds

	In vitro		MT4 cells			HeLaP4 cells	
	AlphaScreen[a]		MTT[b]			Fluc[c]	PIC assay[d]
	IC_{50} (μM)	PIN (%)	EC_{50} (μM)	CC_{50} (μM)	% Protection	PIN (%)	p value
MVG010	54.6	37.1	> 100	100	6	34.2	0.1209
MVG030	33.6	71.6	> 97	97	25	23.3	0.0411
MVG036	37.6	105.6	> 55	65	46	–	0.0011
MVG044	17.1	75.9	> 250	> 250	22	16.3	0.00001

[a] HIV-1 IN–TRN-SR2 AlphaScreen IC_{50} values and percentage inhibition (PIN) are presented. All compounds were tested at 50 μM

[b] EC_{50} and CC_{50} values in MT4 cells are given as well as the percentage of protection against the cytopathic effect of HIV

[c] HeLaP4 cells were infected with a single-round virus expressing the firefly luciferase reporter gene (Fluc) in the presence of one of the compounds (100 μM) and the results are expressed as a percentage inhibition relative to the DMSO control

[d] For the PIC assay in HeLaP4 cells, p values (Mann–Whitney U test), compared to the DMSO control, are given

best compounds selected from 6 series were evaluated for their ability to block HIV replication in infected cells. For five of the classes we were able to detect modest protection against the HIV CPE in an MTT/MT-4 antiviral activity assay (Table 1). In a next step, compounds from the most promising classes 6, 7, 9 and 10 were tested in a single-round antiviral activity assay and a HIV nuclear import assay on HeLaP4 cells. Although only modest inhibition of viral replication could be detected (Fig. 5), two of the compounds significantly reduced the number of nuclear PICs while the other two showed a trend towards less nuclear import (Fig. 6; Table 3).

As previously mentioned, no information on the structure and interface of the TRN-SR2–HIV IN complex was available at the start of our screening campaign. In 2012, De Houwer et al. identified the interaction hot spots in IN for the IN–TRN-SR2 interaction [32]. Amino acids R262/R263/K264 and K266/R269 in the C-terminal domain of IN were found to be the main determinants of the interaction. These results were independently confirmed by Larue et al. [33]. More recently, the structures of TRN-SR2 alone and its complexes with Ran-GTP and ASF/SF2 were published [19, 21]. Although the structure of TRN-SR2–IN has not yet been solved, structural information together with the mutagenesis data point towards a large, charged and relatively flat interaction interface. Targeting the TRN-SR2–IN interaction with high affinity may therefore require design of completely new chemotypes [49].

Conclusions

Our efforts evidenced that an Alphascreen-based HTS can allow the identification of compounds representing a novel class of HIV inhibitors. Their activity in the PIC nuclear import assay confirms that nuclear import of HIV can be targeted by small molecules. While the effects are clear-cut in the PIC assay, the activities of the compounds in the MTT/MT-4 and the single round antiviral activity assay are only modest (Table 3). Several reasons may be conceived to explain this discrepancy. First, rapid growth (and frequent cell divisions) of the cell culture-adapted MT-4 cells during the viral replication assay may relieve the bottleneck of nuclear import. Disassembly and reassembly of the nuclear membrane during the frequent mitoses could give incoming viral PICs periodic access to the condensed, LEDGF/p75-decorated chromatin, masking nuclear import defects. Second, while the single round antiviral activity assay gives a global view of viral replication after 72 h, the PIC assay is a kinetic assay, providing a snapshot of the population of cytoplasmic and nuclear PICs at 5 h after infection. As nuclear import is a bottleneck process [15], PICs accumulating at the nuclear periphery may remain there for some time before being bound and imported by TRN-SR2 (or being degraded). Depending on its affinity and kinetics, an inhibitor of the IN–TRN-SR2 interaction could delay the import process and shift PICs towards degradation in the cytoplasm. Our compounds may not induce sufficient delay in import and a few PICs may make it into the cell's nucleus masking the inhibition.

One of the caveats of targeting pathogen-host protein–protein interactions is the risk of inducing cellular toxicity due to inhibition of the host protein function. Notably, none of the active classes exhibited prominent toxicity in the MTT/MT-4 assay.

Recently, De Houwer et al. [34] reported an IN mutant virus that is partially defective for interaction with TRN-SR2 and nuclear import. Together with these findings, the identification of molecules inhibiting HIV-1 PIC nuclear import from a HTS campaign against the IN–TRN-SR2 interaction underscores the importance of the interaction for HIV-1 nuclear import.

Authors' contributions

JD, JB, SDH and LD performed the experiments; JD, JB, SDH and LD analysed the data; JD, JB and SDH have written the manuscript with help from FC, AM and ZD. HK, AM and PC contributed with compounds, reagents, materials and analysis tools. All authors read and approved the final manuscript.

Author details

[1] Laboratory for Molecular Virology and Gene Therapy, Department of Pharmaceutical and Pharmacological Sciences, KU Leuven, Kapucijnenvoer 33, VCTB +5, Bus 7001, 3000 Leuven, Flanders, Belgium. [2] Center for Innovation and Stimulation of Drug Discovery (CISTIM), Leuven, Belgium. [3] Center for Drug Design and Development (CD3), KU Leuven R&D, Leuven, Belgium. [4] Present Address: The Francis Crick Institute, London, UK.

Acknowledgements

We thank Barbara Van Remoortel and Nam Joo Van der Veken for excellent technical support and Dr. Woan-Yuh Tarn (Inst. of Biomedical Sciences, Taiwan) for the pGEX-TRN-SR2 expression plasmid.

Competing interests

The authors declare that they have no competing interests.

Consent for publication

Not applicable.

Funding

This work was supported by grants from the FWO (G.0487.10 N), (G.0A53.16 N), the IAP BelVir, FP7 CHAARM, hiveranet EURECA and the KU Leuven BOF. JD and LD were doctoral fellows of the Research Foundation Flanders (FWO) and SDH was a doctoral fellow of the agency for Innovation by Science and Technology (IWT). JB is part of the academic and teaching staff of KU Leuven. FC is an Industrial Research Fund (IOF) fellow.

References

1. Weinberg JB, Matthews TJ, Cullen BR, Malim MH. Productive human immunodeficiency virus type 1 (HIV-1) infection of nonproliferating human monocytes. J Exp Med. 1991;174(6):1477–82 **(Epub 1991/12/01)**.
2. Podany AT, Scarsi KK, Fletcher CV. Comparative clinical pharmacokinetics and pharmacodynamics of HIV-1 integrase strand transfer inhibitors. Clin Pharmacokinet. 2017;56(1):25–40 **(Epub 2016/06/19)**.
3. Christ F, Debyser Z. The LEDGF/p75 integrase interaction, a novel target for anti-HIV therapy. Virology. 2013;435(1):102–9 **(Epub 2012/12/12)**.
4. Desimmie BA, Schrijvers R, Demeulemeester J, Borrenberghs D, Weydert C, Thys W, et al. LEDGINs inhibit late stage HIV-1 replication by modulating integrase multimerization in the virions. Retrovirology. 2013;10:57 **(Epub 2013/06/01)**.
5. Christ F, Shaw S, Demeulemeester J, Desimmie BA, Marchand A, Butler S, et al. Small-molecule inhibitors of the LEDGF/p75 binding site of integrase block HIV replication and modulate integrase multimerization. Antimicrob Agents Chemother. 2012;56(8):4365–74 **(Epub 2012/06/06)**.
6. Balakrishnan M, Yant SR, Tsai L, O'Sullivan C, Bam RA, Tsai A, et al. Non-catalytic site HIV-1 integrase inhibitors disrupt core maturation and induce a reverse transcription block in target cells. PLoS ONE. 2013;8(9):e74163 **(Epub 2013/09/17)**.
7. Jurado KA, Engelman A. Multimodal mechanism of action of allosteric HIV-1 integrase inhibitors. Expert Rev Mol Med. 2013;15:e14 **(Epub 2013/11/28)**.
8. Cherepanov P, Maertens G, Proost P, Devreese B, Van Beeumen J, Engelborghs Y, et al. HIV-1 integrase forms stable tetramers and associates with LEDGF/p75 protein in human cells. J Biol Chem. 2003;278(1):372–81 **(Epub 2002/10/31)**.
9. Fader LD, Malenfant E, Parisien M, Carson R, Bilodeau F, Landry S, et al. Discovery of BI 224436, a noncatalytic site integrase inhibitor (NCINI) of HIV-1. ACS Med Chem Lett. 2014;5(4):422–7 **(Epub 2014/06/06)**.
10. Christ F, Voet A, Marchand A, Nicolet S, Desimmie BA, Marchand D, et al. Rational design of small-molecule inhibitors of the LEDGF/p75-integrase interaction and HIV replication. Nat Chem Biol. 2010;6(6):442–8 **(Epub 2010/05/18)**.
11. Kessl JJ, Jena N, Koh Y, Taskent-Sezgin H, Slaughter A, Feng L, et al. Multimode, cooperative mechanism of action of allosteric HIV-1 integrase inhibitors. J Biol Chem. 2012;287(20):16801–11 **(Epub 2012/03/23)**.
12. Tsiang M, Jones GS, Niedziela-Majka A, Kan E, Lansdon EB, Huang W, et al. New class of HIV-1 integrase (IN) inhibitors with a dual mode of action. J Biol Chem. 2012;287(25):21189–203 **(Epub 2012/04/27)**.
13. Brass AL, Dykxhoorn DM, Benita Y, Yan N, Engelman A, Xavier RJ, et al. Identification of host proteins required for HIV infection through a functional genomic screen. Science. 2008;319(5865):921–6 **(Epub 2008/01/12)**.
14. Konig R, Zhou Y, Elleder D, Diamond TL, Bonamy GM, Irelan JT, et al. Global analysis of host-pathogen interactions that regulate early-stage HIV-1 replication. Cell. 2008;135(1):49–60 **(Epub 2008/10/16)**.
15. Christ F, Thys W, De Rijck J, Gijsbers R, Albanese A, Arosio D, et al. Transportin-SR2 imports HIV into the nucleus. Curr Biol CB. 2008;18(16):1192–202 **(Epub 2008/08/30)**.
16. Albanese A, Arosio D, Terreni M, Cereseto A. HIV-1 pre-integration complexes selectively target decondensed chromatin in the nuclear periphery. PLoS ONE. 2008;3(6):e2413 **(Epub 2008/06/12)**.

17. Lai MC, Lin RI, Huang SY, Tsai CW, Tarn WY. A human importin-beta family protein, transportin-SR2, interacts with the phosphorylated RS domain of SR proteins. J Biol Chem. 2000;275(11):7950–7 **(Epub 2000/03/14)**.
18. Lai MC, Kuo HW, Chang WC, Tarn WY. A novel splicing regulator shares a nuclear import pathway with SR proteins. EMBO J. 2003;22(6):1359–69 **(Epub 2003/03/12)**.
19. Maertens GN, Cook NJ, Wang W, Hare S, Gupta SS, Oztop I, et al. Structural basis for nuclear import of splicing factors by human transportin 3. Proc Natl Acad Sci USA. 2014;111(7):2728–33 **(Epub 2014/01/23)**.
20. Taltynov O, Demeulemeester J, Christ F, De Houwer S, Tsirkone VG, Gerard M, et al. Interaction of transportin-SR2 with Ras-related nuclear protein (Ran) GTPase. J Biol Chem. 2013;288(35):25603–13 **(Epub 2013/07/24)**.
21. Tsirkone VG, Beutels KG, Demeulemeester J, Debyser Z, Christ F, Strelkov SV. Structure of transportin SR2, a karyopherin involved in human disease, in complex with Ran. Acta Crystallogr Sect F Struct Biol Commun. 2014;70(Pt 6):723–9 **(Epub 2014/06/11)**.
22. Yamashita M, Emerman M. The cell cycle independence of HIV infections is not determined by known karyophilic viral elements. PLoS Pathog. 2005;1(3):e18 **(Epub 2005/11/18)**.
23. De Rijck J, Vandekerckhove L, Christ F, Debyser Z. Lentiviral nuclear import: a complex interplay between virus and host. BioEssays News Rev Mol Cell Dev Biol. 2007;29(5):441–51 **(Epub 2007/04/24)**.
24. Suzuki Y, Craigie R. The road to chromatin—nuclear entry of retroviruses. Nat Rev Microbiol. 2007;5(3):187–96 **(Epub 2007/02/17)**.
25. Gallay P, Hope T, Chin D, Trono D. HIV-1 infection of nondividing cells through the recognition of integrase by the importin/karyopherin pathway. Proc Natl Acad Sci USA. 1997;94(18):9825–30 **(Epub 1997/09/02)**.
26. Kamata M, Nitahara-Kasahara Y, Miyamoto Y, Yoneda Y, Aida Y. Importin-alpha promotes passage through the nuclear pore complex of human immunodeficiency virus type 1 Vpr. J Virol. 2005;79(6):3557–64 **(Epub 2005/02/26)**.
27. Ao Z, Danappa Jayappa K, Wang B, Zheng Y, Kung S, Rassart E, et al. Importin alpha3 interacts with HIV-1 integrase and contributes to HIV-1 nuclear import and replication. J Virol. 2010;84(17):8650–63 **(Epub 2010/06/18)**.
28. Fassati A, Gorlich D, Harrison I, Zaytseva L, Mingot JM. Nuclear import of HIV-1 intracellular reverse transcription complexes is mediated by importin 7. EMBO J. 2003;22(14):3675–85 **(Epub 2003/07/11)**.
29. Krishnan L, Matreyek KA, Oztop I, Lee K, Tipper CH, Li X, et al. The requirement for cellular transportin 3 (TNPO3 or TRN-SR2) during infection maps to human immunodeficiency virus type 1 capsid and not integrase. J Virol. 2010;84(1):397–406 **(Epub 2009/10/23)**.
30. De Iaco A, Luban J. Inhibition of HIV-1 infection by TNPO3 depletion is determined by capsid and detectable after viral cDNA enters the nucleus. Retrovirology. 2011;8:98 **(Epub 2011/12/08)**.
31. Thys W, De Houwer S, Demeulemeester J, Taltynov O, Vancraenenbroeck R, Gerard M, et al. Interplay between HIV entry and transportin-SR2 dependency. Retrovirology. 2011;8:7 **(Epub 2011/02/01)**.
32. De Houwer S, Demeulemeester J, Thys W, Taltynov O, Zmajkovicova K, Christ F, et al. Identification of residues in the C-terminal domain of HIV-1 integrase that mediate binding to the transportin-SR2 protein. J Biol Chem. 2012;287(41):34059–68 **(Epub 2012/08/09)**.
33. Larue R, Gupta K, Wuensch C, Shkriabai N, Kessl JJ, Danhart E, et al. Interaction of the HIV-1 intasome with transportin 3 protein (TNPO3 or TRN-SR2). J Biol Chem. 2012;287(41):34044–58 **(Epub 2012/08/09)**.
34. De Houwer S, Demeulemeester J, Thys W, Rocha S, Dirix L, Gijsbers R, et al. The HIV-1 integrase mutant R263A/K264A is 2-fold defective for TRN-SR2 binding and viral nuclear import. J Biol Chem. 2014;289(36):25351–61 **(Epub 2014/07/27)**.
35. Debyser Z, Christ F, De Rijck J, Gijsbers R. Host factors for retroviral integration site selection. Trends Biochem Sci. 2015;40(2):108–16 **(Epub 2015/01/04)**.
36. Demeulemeester J, Chaltin P, Marchand A, De Maeyer M, Debyser Z, Christ F. LEDGINs, non-catalytic site inhibitors of HIV-1 integrase: a patent review (2006–2014). Expert Opin Ther Pat. 2014;24(6):609–32 **(Epub 2014/03/29)**.
37. Maertens G, Cherepanov P, Pluymers W, Busschots K, De Clercq E, Debyser Z, et al. LEDGF/p75 is essential for nuclear and chromosomal targeting of HIV-1 integrase in human cells. J Biol Chem. 2003;278(35):33528–39 **(Epub 2003/06/11)**.
38. Busschots K, Vercammen J, Emiliani S, Benarous R, Engelborghs Y, Christ F, et al. The interaction of LEDGF/p75 with integrase is lentivirus-specific and promotes DNA binding. J Biol Chem. 2005;280(18):17841–7 **(Epub 2005/03/08)**.
39. Demeulemeester J, Tintori C, Botta M, Debyser Z, Christ F. Development of an AlphaScreen-based HIV-1 integrase dimerization assay for discovery of novel allosteric inhibitors. J Biomol Screen. 2012;17(5):618–28 **(Epub 2012/02/18)**.
40. Pauwels R, Balzarini J, Baba M, Snoeck R, Schols D, Herdewijn P, et al. Rapid and automated tetrazolium-based colorimetric assay for the detection of anti-HIV compounds. J Virol Methods. 1988;20(4):309–21 **(Epub 1988/08/01)**.
41. Naldini L, Blomer U, Gallay P, Ory D, Mulligan R, Gage FH, et al. In vivo gene delivery and stable transduction of nondividing cells by a lentiviral vector. Science. 1996;272(5259):263–7 **(Epub 1996/04/12)**.
42. Guiot E, Carayon K, Delelis O, Simon F, Tauc P, Zubin E, et al. Relationship between the oligomeric status of HIV-1 integrase on DNA and enzymatic activity. J Biol Chem. 2006;281(32):22707–19 **(Epub 2006/06/16)**.
43. Deprez E, Tauc P, Leh H, Mouscadet JF, Auclair C, Brochon JC. Oligomeric states of the HIV-1 integrase as measured by time-resolved fluorescence anisotropy. Biochemistry. 2000;39(31):9275–84 **(Epub 2000/08/05)**.
44. Baell JB, Holloway GA. New substructure filters for removal of pan assay interference compounds (PAINS) from screening libraries and for their exclusion in bioassays. J Med Chem. 2010;53(7):2719–40 **(Epub 2010/02/06)**.
45. Maertens GN, Cherepanov P, Engelman A. Transcriptional co-activator p75 binds and tethers the Myc-interacting protein JPO2 to chromatin. J Cell Sci. 2006;119(Pt 12):2563–71 **(Epub 2006/06/01)**.
46. Bartholomeeusen K, Christ F, Hendrix J, Rain JC, Emiliani S, Benarous R, et al. Lens epithelium-derived growth factor/p75 interacts with the transposase-derived DDE domain of PogZ. J Biol Chem. 2009;284(17):11467–77 **(Epub 2009/02/27)**.
47. Shi J, Zhou J, Shah VB, Aiken C, Whitby K. Small-molecule inhibition of human immunodeficiency virus type 1 infection by virus capsid destabilization. J Virol. 2011;85(1):542–9 **(Epub 2010/10/22)**.
48. Peng K, Muranyi W, Glass B, Laketa V, Yant SR, Tsai L, et al. Quantitative microscopy of functional HIV post-entry complexes reveals association of replication with the viral capsid. ELife. 2014;3:e04114 **(Epub 2014/12/18)**.
49. Wells JA, McClendon CL. Reaching for high-hanging fruit in drug discovery at protein–protein interfaces. Nature. 2007;450(7172):1001–9 **(Epub 2007/12/14)**.

4

Quantitation of the latent HIV-1 reservoir from the sequence diversity in viral outgrowth assays

Art F. Y. Poon[1*], Jessica L. Prodger[2,4], Briana A. Lynch[3], Jun Lai[4], Steven J. Reynolds[4,5], Jingo Kasule[5], Adam A. Capoferri[4], Susanna L. Lamers[6], Christopher W. Rodriguez[6], Daniel Bruno[7], Stephen F. Porcella[7], Craig Martens[7], Thomas C. Quinn[3,4] and Andrew D. Redd[3,4]

Abstract

Background: The ability of HIV-1 to integrate into the genomes of quiescent host immune cells, establishing a long-lived latent viral reservoir (LVR), is the primary obstacle to curing these infections. Quantitative viral outgrowth assays (QVOAs) are the gold standard for estimating the size of the replication-competent HIV-1 LVR, measured by the number of infectious units per million (IUPM) cells. QVOAs are time-consuming because they rely on culturing replicate wells to amplify the production of virus antigen or nucleic acid to reproducibly detectable levels. Sequence analysis can reduce the required number of culture wells because the virus genetic diversity within the LVR provides an internal replication and dilution series. Here we develop a Bayesian method to jointly estimate the IUPM and variant frequencies (a measure of clonality) from the sequence diversity of QVOAs.

Results: Using simulation experiments, we find our Bayesian approach confers significantly greater accuracy over current methods to estimate the IUPM, particularly for reduced numbers of QVOA replicates and/or increasing actual IUPM. Furthermore, we determine that the improvement in accuracy is greater with increasing genetic diversity in the sample population. We contrast results of these different methods applied to new HIV-1 sequence data derived from QVOAs from two individuals with suppressed viral loads from the Rakai Health Sciences Program in Uganda.

Conclusions: Utilizing sequence variation has the additional benefit of providing information on the contribution of clonality of the LVR, where high clonality (the predominance of a single genetic variant) suggests a role for cell division in the long-term persistence of the reservoir. In addition, our Bayesian approach can be adapted to other limiting dilution assays where positive outcomes can be partitioned by their genetic heterogeneity, such as immune cell populations and other viruses.

Keywords: Latent viral reservoir, HIV-1, Viral outgrowth assay, Limiting dilution assay, Next-generation sequencing, Bayesian inference

*Correspondence: apoon42@uwo.ca
[1] Department of Pathology and Laboratory Medicine, Western University, London, ON, Canada
Full list of author information is available at the end of the article

Background

HIV-1 persists in individuals, despite fully suppressive anti-retroviral therapy (ART), due to the presence of resting CD4+ (rCD4) T cells that are latently infected with replication competent, integrated copies of the virus [1]. These resting cells can reactivate through natural immunological challenge within the body, and upon reactivation will produce progeny virus. In the case of discontinuation of ART, this reactivation will lead to full viral rebound within a matter of weeks in virtually all individuals [2]. The size of this pool of latently infected cells, referred to as the latent viral reservoir (LVR), has proven difficult to measure as the vast majority of integrated proviral DNA is defective [3, 4]. Therefore current methods for quantifying replication-competent provirus rely on the production of infectious virus in vitro; the quantitative viral outgrowth assay (QVOA) is presently the gold standard for LVR quantification. The QVOA is based on the in vitro reactivation of rCD4 T cells isolated from an infected individual in a limiting dilution of replicate wells, with a readout of HIV-1 p24 antigen production to identify wells containing at least one latently infected cell. However, the QVOA is labor-intensive and difficult to scale up to larger numbers of wells. Each well is cultured for several weeks with the repeated addition of uninfected CD8-depleted lymphoblasts, or the single addition of MOLT-4/CCR5 cells, to amplify any viruses released so that HIV p24 antigen can be produced at detectable levels [5, 6] (Fig. 1). Consequently, it would be beneficial to minimize the number of replicate wells that need to be cultivated for every patient sample. Sequencing individual viruses from the positive wells internally dilutes each virus population by partitioning the binary outcome (p24 antigen positive) into multinomial outcomes (presence or absence of sequence variants) [7]. In other words, the genetic diversity of the virus population provides an intrinsic replication and dilution series that can enable the investigator to culture a smaller number of wells, and has therefore been proposed as a tool for refining the estimates of infectious units per million (IUPM) cells [7], the standard LVR measure.

Additionally, sequencing of outgrowth virus can provide information on the contribution of rCD4 T cell division to persistence of the LVR. The long half-life of

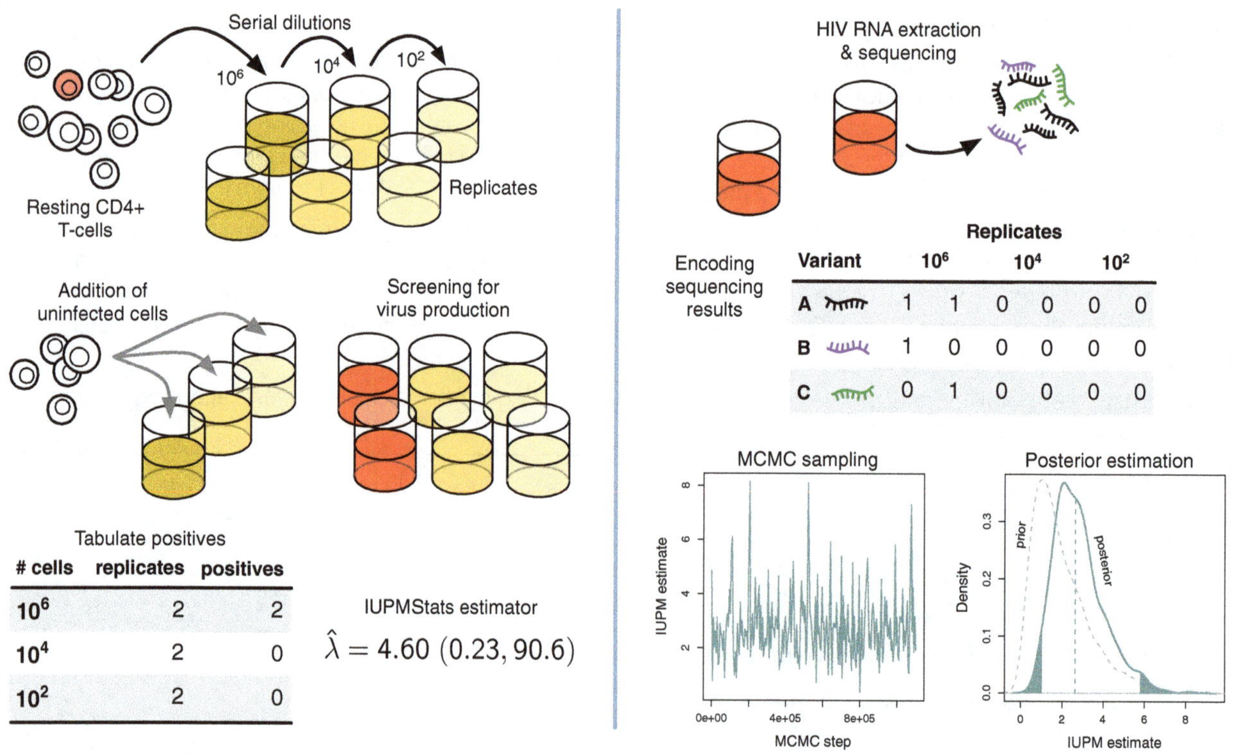

# cells	replicates	positives
10^6	2	2
10^4	2	0
10^2	2	0

Variant	Replicates 10^6		10^4		10^2	
A	1	1	0	0	0	0
B	1	0	0	0	0	0
C	0	1	0	0	0	0

Fig. 1 Schematic diagrams of experimental and data analysis procedures. (Left panel) Resting CD4+ T cells sampled from an HIV+ patient are serially diluted in replicate culture wells. Uninfected cells are added to the culture wells to amplify viral outgrowth (red). IUPMStats estimates the rate parameter $\hat{\lambda}$ of the single-hit Poisson model from the numbers of positive wells at varying dilutions. (Right panel) HIV-1 RNA is extracted from each positive well and amplified for library construction and sequencing. The presence/absence of different sequence variants are tabulated and used to fit a multi-target Poisson model by Markov chain Monte Carlo (MCMC) sampling, in which a prior distribution on the IUPM (lower right, grey dashed curve) is updated by the data to estimate the posterior distribution (solid curve)

the HIV-1 LVR that averages several years in duration [8, 9], in combination with the presence of clonal populations of rCD4 T-cells that contain identical HIV-1 proviral sequences [10–12], suggests that rCD4 T-cell proliferation may play a significant role in maintaining the LVR. The role that this phenomenon plays in maintaining the LVR is of critical importance for HIV-1 cure efforts, as it will help direct the ideal path for targeting and eventually destroying these latently infected cells. To this end, multiple groups have begun using viral sequencing of both latent HIV-1 proviral DNA, as well as populations derived from viral outgrowth assays (such as the quantitative viral outgrowth assay [QVOA]), to characterize the contribution of this clonal expansion to the maintenance of the LVR [10, 13–15].

The single-hit Poisson model

The current standard approach for estimating the IUPM from QVOA results is based on the single-hit Poisson model, which has long been used for the analysis of limiting dilution assays [16]. The expected IUPM is estimated by maximizing the likelihood for a binomial distribution:

$$L(Y|\lambda,N) \propto p_\lambda^Y(1-p_\lambda)^{N-Y} \quad (1)$$

where N is the number of wells in the QVOA; Y is the number of p24-positive wells; and $p_\lambda = 1 - \exp(-n\lambda)$ is the probability that a well is positive because one or more of the cells within it were infected (hence 'single-hit'), given that each well contained n cells. The rate parameter λ is the proportion of cells that are infected, from which we derive our estimate of the IUPM as $\lambda \times 10^6$. This model assumes that all cells in the sample population have a uniform probability of being infected. It is conventionally assumed that n (the number of cells per well) is known without error, so that the maximum likelihood estimate (MLE) of p_λ is a direct estimate of λ. Serial dilution is typically used to vary n across sets of wells, which reduces the chance that the wells are either all positive or all negative; neither of these extreme outcomes can be used effectively to estimate λ. Thus, a more general formulation of equation (1) is:

$$L(Y|\lambda,N) = \prod_{i=1}^{D}(1-\exp(n_i\lambda))^{Y_i}\exp(n_i\lambda)^{N_i-Y_i} \quad (2)$$

where D is the number of dilution factors, N_i is the number of wells with the i-th dilution factor, Y_i is the number of these wells that test positive, and n_i is the number of cells per well [17].

Incorporating sequence variation

The defining feature of the single-hit Poisson (SHP) model is that each well has a binary all-or-nothing outcome, in which the well either tests positive or negative for viral outgrowth (Fig. 1). Estimating IUPM is contingent on having a mix of positive and negative outcomes, where the probability of each positive outcome (that a well contains at least one infected cell, and therefore p24 is detected) is modulated by dilution. However, if we obtain genetic sequences for individual viruses within a given well, then we have additional information about the number of infected cells per well [7]. If two distinct sequence variants are observed from a well, for instance, then we can assume that the well initially contained at least two infected cells. Additionally, we can track the occurrence of specific variants across wells to simultaneously estimate the relative frequency of variants in the sample population of infected cells, which also provides information about the clonal expansion of infected cells. For example, consider a QVOA where four wells are plated with 10^6 cells per well, and p24 ELISA shows that all four wells are positive for HIV. From these data, there is no sensible maximum likelihood estimate for λ under the SHP model because the likelihood surface asymptotes as λ approaches infinity. However, suppose there are five genetic variants (labeled *A* through *E*) with a uniform frequency of 0.2 in the sample population of viruses, and the true IUPM is 5. Then we can simulate a random outcome represented by the following binary (presence/absence) matrix:

Well	p24	*A*	*B*	*C*	*D*	*E*
1	+	0	1	0	1	0
2	+	1	0	0	0	0
3	+	1	1	0	0	0
4	+	0	1	1	1	0

Under these conditions, the probability that all four wells are positive is $(1-e^{-5})^4 \approx 0.973$. In other words, there is a good chance that we will be unable to estimate the IUPM with the SHP model. On the other hand, we expect that each variant will be present in a given well only $1-e^{-0.2\times5} \approx 0.63$ of the time, and all four wells only about $0.63^4 \approx 16\%$ of the time; the chance that all variants appear in all wells is only about one-hundredth of a percent. Therefore, if we use sequencing to determine which variants appear in each well, then it is very likely that we can regain the ability to estimate the IUPM in an 'all positive' scenario. Incorporating sequence information constitutes a more difficult problem because we need to simultaneously estimate the variant frequencies along with λ.

Recently, Lee et al. [7] described using a primer ID-based next-generation sequencing assay (NGSA) to sequence the virus populations in positive QVOA wells. This QVOA-NGSA method employs a maximum likelihood estimator for IUPM that is the sum of variant-specific quantities, $\hat{\lambda} = \sum_i -\log(1 - y_i/K)$, where y_i is the number of K wells containing the i-th variant. This estimator stipulates that every well contains the same number of cells, such that the estimate $\hat{\lambda}$ can be rescaled by a constant factor to estimate the IUPM; for instance, if each well contained 2.5×10^6 cells then IUPM $= \hat{\lambda}/2.5$. Furthermore, this estimator has the same problem as the SHP model in that $\hat{\lambda}$ goes to ∞ if any variant is observed in every well ($y_i = K$). This potentially limits the utility of the estimator when the number of wells is small, the number of cells per well is large, or the actual IUPM is large and/or predominated by a single variant (high clonality) [14].

In this study, we developed a Bayesian method (IUPMBayes; Fig. 1) to jointly estimate the IUPM and the frequency distribution of variants from next-generation sequence data obtained from the positive wells of viral outgrowth assays. We perform an extensive simulation study to compare our Bayesian method to the maximum likelihood estimators developed for the SHP model [IUPMStats; 17] and the QVOA-NGSA method [7]. Lastly, we compare the performance of these methods using next-generation sequencing of two genetic regions (*pol* and *gp41*) of viral outgrowth populations from positive wells, derived from peripheral blood samples from two ART-suppressed individuals with markedly different levels of HIV sequence diversity in the LVR.

Results

Simulation results

We simulated data sets under an experimental design in which 12 wells were inoculated with serial dilutions of 10^6, 2×10^5, 4×10^4, 8000, 1600 and 320 cells per well in duplicate. This design was configured to be representative of viral outgrowth assays used in previous empirical studies [6, 8]. Next, we used these simulations to evaluate the accuracy of IUPMStats, which combines a maximum likelihood estimator on the single-hit Poisson model with a Bayesian estimator for cases where no positive wells were observed, and our Bayesian method (IUPMBayes) that utilizes sequence variation in addition to the number of positive wells. For these experiments, we used the uninformative prior distribution on $\lambda \sim U(0, \infty)$. Our results are summarized in Fig. 2. First, we observed that the relative errors associated with estimates from IUPMStats were relatively consistent across the IUPM values used to simulate the data sets, with an overall median relative error of about 0.49. When the true IUPM was low (0.2 per million cells), then similar estimates were obtained using either method. However, the overall relative error decreased significantly with increasing IUPM (log-link gamma generalized linear model, GLM; $t = -9.0$, $P < 10^{-15}$). If we assumed that the sample population contained only two variants at equal frequencies (1:1), then we observed no significant improvement in IUPMBayes above IUPMStats (paired t-test, $t = -1.05$, $P = 0.29$). However, IUPMBayes gained a significant advantage with increasing virus sequence diversity and IUPM (interaction term $t = -4.5$, $P = 6.0 \times 10^{-6}$; Fig. 2), where we converted the variant frequency distributions into Shannon entropy values to quantify diversity [18]. Adding this interaction term to the gamma GLM conferred a significant improvement in fit (likelihood ratio test, LRT; df $= 2$, $P = 1.8 \times 10^{-5}$).

Why do we observe significant differences in accuracy between IUPMStats and our Bayesian method? First, the maximum likelihood estimator (MLE) used in the IUPMStats can only assume a finite number of values because the outcome space is discrete. When there are a large number of wells in which a positive outcome is likely to occur, then there is a high probability that the true value is close to the MLE. As the expected number of positive wells declines, however, the distribution of potential MLEs becomes increasingly sparse. For example, the most frequent IUPMStats estimates when the true IUPM was set to 1 were 0.51 and 1.61 (Fig. 3), which correspond respectively to the outcomes where one or both

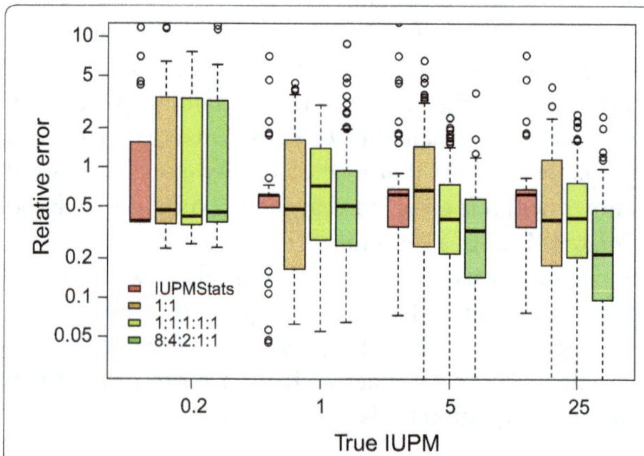

Fig. 2 Relative error in estimating IUPM for data simulated under the standard experimental design. We calculated the relative error of an estimate $\hat{x}$ given true value x as $|(\hat{x} - x)/x|$. Each set of box-and-whisker plots summarizes the relative errors for estimates obtained by IUPMStats (red) and IUPMBayes under three different sets of variant frequencies (see inset legend) for a given true value of IUPM (0.2, 1, 5 and 25 per million cells). We used a log-transformation of relative errors and rescaled the y-axis to clarify differences between methods and simulation conditions; 25 outliers with relative errors below .03 were excluded from this plot region

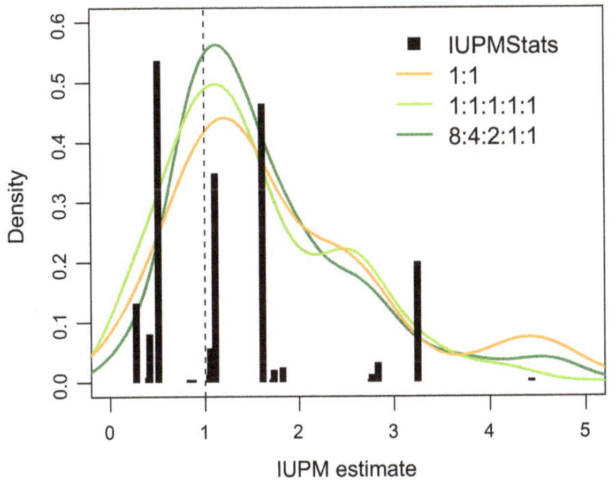

Fig. 3 Distribution of estimates from IUPMStats and IUPMBayes for simulations given IUPM = 1. We used a barplot to summarize the distribution of IUPMStats estimates, which makes clear that these estimates are limited to a relatively small number of values given the dilution series and number of replicate wells in our simulation experiments. The leftmost bar (red) corresponds to the median Bayesian posterior estimate employed by IUPMStats when all wells are negative. The distributions of estimates from IUPMBayes under two sets of variant frequencies (1:1 and 1:1:1:1:1) are summarized with Gaussian kernel densities (curves). Unlike the IUPMStats estimates, these distributions were unimodal and centred near the true value. We obtained similar results under varying conditions and IUPM values

of the wells with 10^6 cells plated were positive. The next two most frequent MLEs corresponded to the outcomes where both 10^6 wells were positive and either one or both of the wells with 2×10^5 cells were positive. In contrast, a Bayesian approach attempts to generate a random sample from a continuous posterior distribution over the IUPM parameter.

Second, the wells with high cell counts become less informative with increasing IUPM, since they will nearly always be positive when the IUPM is sufficiently high. For example, a well with 10^6 cells has a probability $1 - \exp(-5) = 0.993$ of being positive when the IUPM is 5. Sequencing the outgrowth variants in positive wells can restore the information content of these wells, because these additional data partition the outcomes into variant-defined types. Thus, the accuracy of IUPM-Bayes improves with increasing IUPM but only when the sequence variants are present at informative frequencies. For instance, the advantage of utilizing sequence information deteriorates as the frequency distribution is skewed toward a single predominant variant (Additional file 1: Fig. S1), which also reduces the Shannon entropy. In this extreme case, most of the variants are so rare that they are seldom sampled in the positive wells.

Reducing the number of wells

We conducted an additional set of simulation experiments in which the same number of cells (10^6) were plated in replicate wells. This configuration facilitated the comparison of our method to the QVOA-NGSA maximum likelihood estimator proposed by Lee et al. [7], which requires that the number of cells is constant across wells. Moreover, we used these experiments to evaluate the sensitivity of the methods to reducing the total number of wells to 8, 4 and 2 per experiment. Given the sparseness of these data, we employed an informative prior distribution on $\lambda_m \sim \Gamma(\alpha = 2, \beta = 1)$. Our results are summarized in Fig. 4. First, we observed that IUPM-Bayes tended to overestimate the IUPM relative to the other methods when the true value was 0.2, where full sets of negative wells were common. The median relative error was 1.09, which corresponds to an overestimate of about 0.37 cells per million. When the IUPM is set this low, many replicate experiments will result in all-negative outcomes with probability $\exp(-0.2 \times n)$, where n is the total number of replicates. This lack of information in turn results in greater uncertainty in estimating IUPM. For this situation, IUPMStats uses the posterior median estimate assuming a uniform prior on IUPM from 0 to 1 cells per million, whereas we used the same prior distribution across the same set of simulations irrespective of outcomes.

When the true IUPM was increased to 1.0, we observed that both QVOA-NGSA and IUPMBayes tended to have lower relative errors than IUPMStats; the difference was highly significant for IUPMBayes (Wilcoxon signed rank test, $P = 2.84 \times 10^{-7}$; Fig. 4). Furthermore, the number of 'all-positive' cases that resulted in $\hat{\lambda} = \infty$ estimates from either IUPMStats or QVOA-NGSA became more frequent with decreasing sample size. For instance, IUPMStats was unable to generate a meaningful estimate of IUPM for 46% of replicate simulations with 2 wells. Even though QVOA-NGSA utilized sequence variation to ameliorate this effect, if any of the variants appeared in all replicate wells, then its maximum likelihood estimator similarly resulted in an infinite value. This effect became severe when we increased the true IUPM to 5.0; IUPMStats yielded valid estimates of IUPM in only 2% of replicates irrespective of sample size, and QVOA-NGSA produced estimates for no more than 12% (Fig. 4). In contrast, IUPMBayes was surprisingly robust to reductions in sample size, with relative errors increasingly only slightly with decreasing numbers of wells when the true IUPM was 1.0 or greater.

We also observed that the 95% confidence intervals generated by the IUPMStats method tend to be slightly broader than the analogous Bayesian credibility intervals we obtained from the empirical 95% intervals. For instance, when the true IUPM was 1 and we simulated

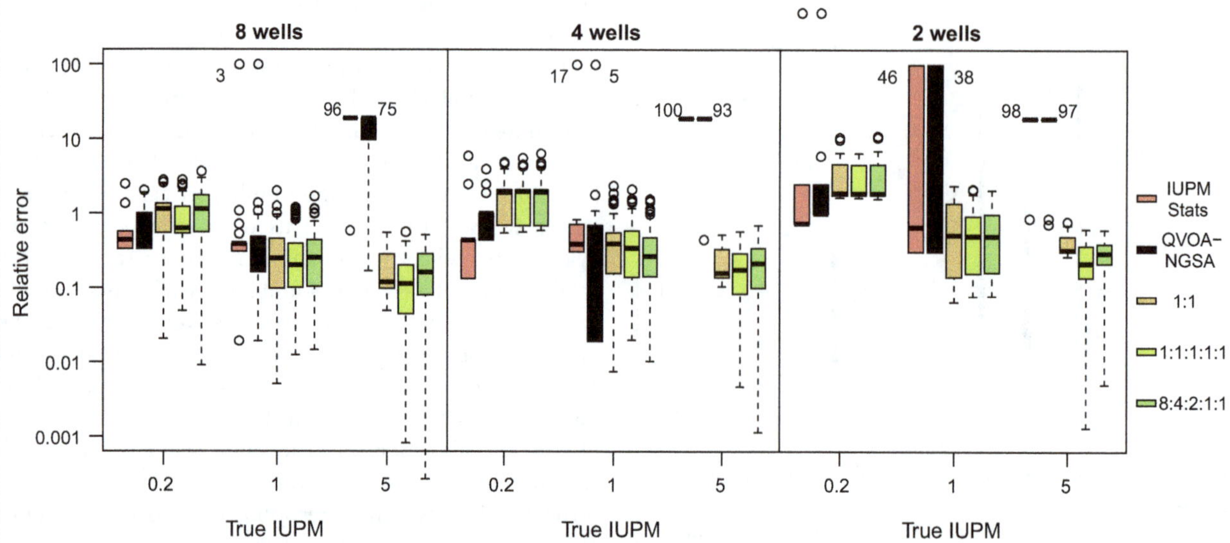

Fig. 4 Relative errors in IUPM estimates for simulated data with uniform cell counts. Each box plot summarizes the distribution of relative error for 100 replicate simulations for the respective methods: IUPMStats, QVOA-NGSA, and IUPMBayes (three sets of variant frequencies, see legend). The plots are drawn on a log-transformed *y*-axis to facilitate comparison between methods under varying conditions. When the maximum likelihood estimators employed by IUPMStats or QVOA-NGSA were unable to produce a finite estimate of IUPM, we assigned an arbitrary value of 100 cells per million. The respective box plots are directly labeled with the numbers of non-finite estimates at the outliers or median bands

four replicate wells, there were five possible outcomes ranging from zero to 4 positives. The IUPM estimates and confidence/credibility intervals obtained by IUPMStats and IUPMBayes for the respective outcomes are summarized in Table 1. The most frequent outcome was that 3 of 4 wells were positive: for this outcome, the IUPMStats estimate was 1.39 with a 95% CI from 0.41 to 4.72, while the median IUPMBayes estimate was 1.36 with a narrower 95% CI from 0.52 to 2.87. Note that these intervals describe the confidence within a single experiment, as opposed to the variance in estimates across replicate experiments. Some of this difference could be attributed to using a prior distribution on λ. However, we obtained similar results in the simulation experiments with duplicate serial fivefold dilutions and a uniform prior on λ (Fig. 2). For example, when the IUPM was set to 1, the most frequent outcome was a single positive well with 10^6 cells; the 95% C.I. around the IUPMStats estimate of 0.51 was 0.07 to 3.70, while the median IUPMBayes estimate (assuming five variants with frequency ratio 8:4:2:1:1) was 0.91 with a 95% C.I. = (0.14, 3.06). The next most frequent outcome was two positive wells with 10^6 cells, and the respective results for IUPMStats and IUPMBayes were 1.61 (0.34, 7.51) and 1.91 (0.53, 4.63). Thus, the incorporation of additional information from the presence or absence of specific variants in the QVOA augments our precision in estimating λ.

Computing time

Since we employ Bayesian sampling to estimate the model parameters, including IUPM, our method is more time-consuming than methods based on maximum likelihood estimators such as IUPMStats. We evaluated the time required to generate IUPM estimates by measuring run times for IUPMBayes on 10 replicate data sets simulated under the first set of conditions (previous section). On a single core of an Intel E5-1620v2 processor, it required an average of 358.8 seconds (roughly six minutes) to run a chain sample for one million steps, which we have found to be adequate for convergence under a variety of simulated conditions. In contrast, our implementation of IUPMStats in *R* used only 208.3 microseconds on average to process the same data in the same computing environment.

Empirical data

We applied both the IUPMStats and IUPMBayes methods to actual experimental results derived from two patients (106 and 111). These patients had been selected from a larger study in progress because the overall sequence diversity in viral outgrowth wells derived from the latent reservoirs of the respective individuals were characterized by either very high (106) or very low (111) levels of clonality. We use the term 'clonality' to refer to the scenario where the virus population is dominated by a variant or variants that are repeated, such that most sequences in a random sample will be identical or have another well containing that same variant. Based on this prior information, we adjusted the hyperparameters of

Table 1 Comparison of estimates and 95% confidence/credibility intervals for IUPMStats and IUPMBayes

Positives	n	IUPMStats			IUPMBayes		
		Estimate	Lower 95%	Upper 95%	Median	Lower 95%	Upper 95%
0	3	0.17*	0	0.75	0.34	0.05	1.10
1	16	0.29	0.04	2.06	0.61	0.15	1.68
2	28	0.69	0.17	2.85	1.04	0.36	2.41
3	37	1.39	0.41	4.72	1.36	0.52	2.87
4	16	∞	Undefined		1.85	0.80	3.63

We summarized the results of each method on 100 simulations of 4 replicate wells with 10^6 cells each, where the true IUPM was set to 1.0 with five sequence variants of equal frequency (1:1:1:1:1). We obtained results for the IUPMStats method directly from the online calculator. Entries for IUPMBayes were averaged across replicate simulations for each number of positives. *IUPMStats uses a median posterior estimate instead of the maximum likelihood estimate (0) when none of the wells are positive

Table 2 Summary of outgrowth assay results for two patients

Patient	Positive wells (total wells)						IUPMStats	# variants
	10^6	2×10^5	4×10^4	8000	1600	320 cells	(95% CI)	(*gp41*, *pol*)
106	8 (8)	2 (2)	0 (2)	0 (2)	0 (2)	0 (2)	8.148 (1.863, 35.635)	10, 4
111	13 (16)	0 (2)	0 (2)	0 (2)	0 (2)	0 (2)	1.551 (0.851, 2.825)	13, 20

For each dilution (number of cells per well), we report the number of positive wells (total number of wells). IUPM estimates and 95% confidence intervals (CIs) under the single-hit Poisson model were generated using the JavaScript calculator at http://silicianolab.johnshopkins.edu (last access date: June 23, 2017). The final column reports the number of sequence variants that were observed in regions within HIV *gp41* and *pol*, respectively, when HIV RNA was extracted, amplified and sequenced from the positive wells

the Dirichlet prior distribution to $\alpha_1 = 10, \alpha_{i \neq 1} = 1$ for patient 106 and $\alpha_i = 1$ for patient 111. We assessed the sensitivity of our results to these prior distributions in Additional file 2: Fig. S2. The dilution series for allocating numbers of cells per well was the same as experimental design 1 (10^6, 2×10^5, 4×10^4, 8000, 1600, 320). However, the number of wells carrying 10^6 cells was varied with 8 wells for patient 106 and 18 wells for patient 111.

We generated ML estimates under the SHP model using both our implementation of this method in *R* and the online calculator at http://silicianolab.johnshopkins.edu. These experimental data and results are summarized in Table 2. Our implementation of SHP model estimation in *R* produced numerically identical estimates to the online calculator for both patients. The IUPMStats estimate was significantly higher for patient 106 (8.2 per million) than 111 (1.6 per million). Additionally, we observed fewer distinct sequence variants in well samples from patient 106 than 111 for both the *gp41* and *pol* regions.

Results from our Bayesian analysis of the same data sets under the MTP model are summarized in Fig. 5. To facilitate evaluation of convergence, we ran three replicate Markov chain samples per data set with different initial parameters. We found that for both patients and genes, the median posterior estimates of IUPM were substantially lower than the ML estimates obtained under the SHP model. For patient 106, the median estimates were 5.28 (95% C.I., 3.19, 9.26) per million for *gp41* and 3.03 (1.60, 5.54) for *pol*. This difference was driven in part by the occurrence of the predominant *gp41* variant in all eight wells carrying 10^6 cells, whereas the predominant *pol* variant occurred in only seven of these eight. Combining these estimates, we predict that the IUPM in patient 106 was about 3.8 infected cells per million—substantially lower than the IUPMStats estimate, but within its 95% confidence interval (Table 2). The discrepancies between these estimates was consistent with the levels of absolute error we observed with simulated data. In contrast, we obtained more concordant IUPMBayes estimates between gene regions for patient 111; the median posterior estimates were 1.13 (95% C.I., 0.66, 1.74) per million for *gp41* and 1.36 (0.85, 2.00) for *pol*, respectively.

Discussion

Maximum likelihood (ML) is a powerful technique for estimating the rate parameter of the single-hit Poisson model [19, 20]. However, there can be issues with ML estimation that are exacerbated with small sample sizes. First, the single-hit Poisson estimator becomes infinite when all the wells are positive [21]. We have

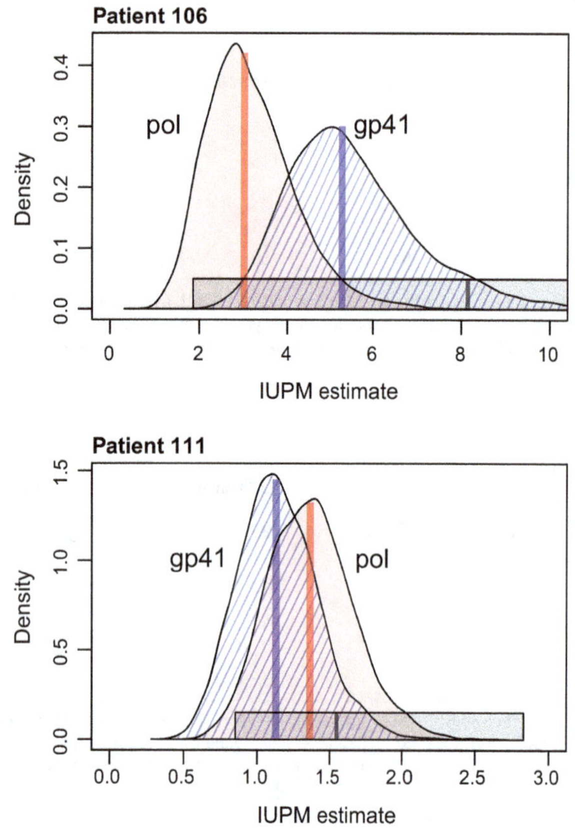

Fig. 5 Summary of IUPM estimates obtained by a Bayesian analysis of experimental data from subjects 106 (high clonality) and 111 (low clonality). IUPM was estimated separately using sequence data obtained for HIV regions *pol* (red) and *gp41* (blue, hatched). Three replicate chains were combined for each patient and gene after assessing that the chains had converged to the posterior distribution over the IUPM parameter. Median estimates are indicated by vertical line segments within each density plot. Grey bars represent the maximum likelihood estimate and 95% confidence interval obtained by IUPMStats [17]. In the case of patient 106, the upper confidence limit from IUPMStats extends to 35.6 cells per million

demonstrated here that this issue can even affect estimators that partition the outcomes by sequence variants, i.e., QVOA-NGSA [7]. Methods that reject data sets where all wells are positive may systematically underestimate the IUPM when the actual value is high, because the investigator is required to perform additional experiments until one or more negative outcomes are obtained. However this scenario is unlikely when the experimental design utilizes a wide-ranging dilution series and can also be mitigated by increasing the number of replicate wells. Second, ML estimators based on discrete outcomes can take only a finite number of values, which we have shown can limit the estimator's accuracy when the number of informative observations is small (Fig. 3). Both issues are exacerbated by reducing the number of wells used in the experiment. On the other hand, there are significant practical benefits to reducing the number of replicate wells required to estimate the IUPM, not only because the assay requires weeks of culturing cells, but also because scaling up these experiments to large numbers of samples is necessary to detect statistical associations between the IUPM and clinical variables [9, 22].

Our results support the concept of applying Bayesian methods to sequence diversity to overcome the limitations of maximum likelihood for small samples (fewer wells). Although previous studies have implemented Bayesian methods to parameterize the single-hit Poisson model [23], we are not aware of another study taking this approach to incorporate sequence information for estimating the number of infectious units. Additionally, the Bayesian approach designed here could be adapted to more properly incorporate viral sequence data in other LVR assays that estimate the size of inducible or intact proviral HIV infected cell populations, and thereby improve their accuracy and usefulness in HIV cure studies [24, 25]. Using simulations, we have demonstrated that our Bayesian method becomes more accurate than current methods as the true IUPM increases above 1, and that this advantage is greater with increasing sequence diversity in the latent HIV reservoir (Fig. 2). The emerging consensus from empirical evidence is that, on average, slightly less than one in a million resting CD4+ T cells contain replication competent latent HIV. If actual IUPM values tended to be tightly clustered around this expectation, then there would be limited use in incorporating sequence variation for quantifying the latent reservoir. However, empirical evidence indicates that there is substantial variability in IUPM around the mean. For example, Crooks et al. [9] recently reported estimates from 37 patients that ranged by more than two orders of magnitude around a mean of 0.42 per million (about 0.02 to 20 IUPM). Similarly, Eriksson et al. [22] reported a geometric mean of 0.64 IUPM among 30 patients on suppressive therapy with a range also spanning about two orders of magnitude. Therefore it is reasonable to expect a substantial fraction (~10%) of individuals to have an IUPM that is measurably greater than 1.

A core assumption of our method is that detecting more than one sequence variant in a culture well indicates the presence of at least that number of latently infected cells. It is possible that a cell is multiply infected with two or more integrated HIV DNA variants [26]. This scenario would cause our method to overestimate the IUPM. However, recent work determined that over 85% of infected CD4+ T cells in peripheral blood carried only a single integrated HIV DNA molecule [27], and the majority of these proviruses are defective [4]—thus, the generally low prevalence of replication competent virus in the LVR (see above) supports this simplifying assumption of the model. In addition, our method assumes that any variant that is present in a particular well will be detected by next-generation sequencing, irrespective of its frequency in the initial sample population. In other words, we assume that viral outgrowth is completely efficient; the other methods for analyzing QVOA data require this same assumption irrespective of genetic variation.

The considerable depth at which next-generation sequencing samples templates from the outgrowth population makes it unlikely that variants present in the well are unobserved. However, the rapid evolution of HIV-1 means there may be an unknown number of rare genetic variants in the LVR that fail to become sampled from the source population into any replicate wells of the QVOA experiment, immediately precluding these variants from outgrowth and sequencing. In fitting the multi-hit Poisson model, we have out of necessity restricted the model to the observed variants in the experiment by marginalizing out the number of unobserved variants, M^*, which is exceedingly difficult to estimate from these data. This raises a potential issue in that selecting the prior distribution on variant frequencies, $P(f|M)$, becomes directly influenced by the number of observed variants in the data. Strict adherence to Bayesian principles discourages this repeated use of the data, but there is growing recognition that such 'data-dependent priors' are routinely used for empirical Bayesian inference and can retain statistically desirable properties, e.g., [28]. For example, IUPMStats employs a data-dependent prior that is a uniform distribution over the interval $\lambda_m = (0, 1)$ if all wells in the QVOA experiment are negative, and $\lambda_m = (0, \infty)$ otherwise. There are fully hierarchical Bayesian methods available to accommodate the uncertainty in M^*, such as the Dirichlet process prior, which describes a distribution over the different Dirichlet prior distributions for varying numbers of variants. However, MCMC sampling under the Dirichlet process prior is not trivial to implement and can be slow to converge, in part because the number of prior distributions is infinite [29]. Furthermore, we are pessimistic that the presence-absence matrix derived from sequencing QVOA experiments can sustain such an expansion of the model parameter space.

Although our method was developed for the specific purpose of measuring the latent HIV reservoir, it can be adapted to other applications of limiting dilution assays where the target has genetic variation. For example, limiting dilution has been used to measure the frequency of reactivation in cells latently infected with human cytomegalovirus [30]. It could also be adapted for the quantitation of nucleic acids by dilution and amplification-based assays, which have been developed for viruses such as hepatitis C virus [31], and modernized with high-throughput technologies that rely on the same underlying statistical model, such as digital PCR [32]. Finally, limiting dilution and the single-hit Poisson model have historically been deployed in the study of immunocompetent cell populations [19] and continues to play a fundamental role in ongoing studies of the diverse cellular subsets of the immune response [33].

Conclusions

The existence of the latent viral reservoir is a key barrier to curing HIV-1 despite highly effective drug treatments. In this study, we demonstrate that sequencing individual viruses from the positive wells of a viral outgrowth assay can provide a more robust and accurate measure of the latent reservoir. Furthermore, this approach can facilitate scaling up the assay to large numbers of samples by reducing the number of wells necessary to measure the latent viral reservoir.

Methods

Models

To facilitate batch processing on simulated and experimental data sets, we re-implemented the SHP model in the *R* programming language. For brevity, we will use the notation $\lambda_m = \lambda \times 10^6$ for IUPM. We obtained a maximum likelihood estimate ($\hat{\lambda}_m$) given equation (2) using Brent's root-finding method [34] with a lower bound of $\lambda_m > 0$ and an arbitrary upper bound (default $\lambda_m < 10^3$). Further, we confirmed that our implementation yielded results consistent with the web-based IUPMStats calculator (http://silicianolab.johnshopkins.edu/, version 1.0). We estimated the 95% confidence interval from the likelihood function using numerical root-finding [34] to solve the equality $L(\lambda_m)/L(\hat{\lambda}_m) = 0.147$ on the intervals $\lambda_m = (10^{-3}, \hat{\lambda}_m)$ and $\lambda_m = (\hat{\lambda}_m, 10^3)$; this approach assumes the sample size is sufficiently large that the log-likelihood ratio can be approximated by the χ^2-distribution. When the data comprise entirely of negative wells, the MLE for λ in the SHP model is zero. In this case, the IUPMStats calculator substitutes the posterior median estimate given a uniform prior distribution over the interval $\lambda_m = (0, 1)$ [17].

Suppose that an experiment has K wells and M sequence variants. Let $v_{k,i} = \{0, 1\}$ be the binomial outcome for the presence or absence of the i-th sequence variant in well k. Let f_i be the frequency of the i-th variant in the population of infected cells in the undiluted QVOA wells. The likelihood that at least one cell in a particular well was infected with the i-th variant is described by combining the Bernoulli and Poisson distributions:

$$L(v_{k,i}|\lambda,f_i) = \left(1-\exp(-f_i n_k \lambda)\right)^{v_{k,i}} \exp(-f_i n_k \lambda)^{1-v_{k,i}} \quad (3)$$

where $f_i n_k \lambda$ is the expected number of cells infected by variant i in well k. Following standard practice such as the single-hit Poisson model [17], we assume that n_k are known without error. Equation (3) is similar to equation (2), except that (3) accommodates variation in the relative frequencies of sequence variants in the infected cell population. The likelihood for the data set is calculated by taking the product across all K wells and M variants:

$$\mathbf{L}(\mathbf{v}|\lambda,f,M) = \prod_{k=1}^{K}\prod_{i=1}^{M} L(v_{k,i}|\lambda,f_i) \quad (4)$$

where $\mathbf{v}$ is a binary $K \times M$ matrix with entries $\{v_{k,i}\}$. This model is similar to the two-target Poisson models used in immunology to accommodate subsets of cells with different frequencies [35]—accordingly, we will refer to equation (4) as the multi-target Poisson (MTP) model.

HIV-1 can become highly diverse within hosts [36] and not all variants in the population are necessarily observed in the experiment, especially those at very low frequencies. The number of unobserved variants, which we denote by M^*, is an uncertain quantity and not feasible to estimate from the QVOA data. We will express this by expanding equation (4) as follows:

$$\mathbf{L}(\mathbf{v}|\lambda,f,M,M^*) = \prod_{k=1}^{K}\left(\prod_{i=1}^{M} L(v_{k,i}|\lambda,f_i)\prod_{i=M+1}^{M+M^*} L(0|\lambda,f_i)\right) \quad (5)$$

where M now refers to the observed number of variants instead of the total number. Our objective is use this likelihood to estimate λ where f, M and M^* are nuisance parameters. The viral outgrowth intrinsic to this assay precludes the use of next-generation sequencing to directly estimate variant frequencies f, which become obscured by the highly stochastic nature of viral amplification. Thus, we used Bayesian inference to estimate the posterior distribution of λ while accommodating the uncertainty in the other model parameters. Let the total number of variants be represented by an unspecified prior distribution $P(M,M^*)$ that ranges from 1 to ∞. We used the standard Dirichlet prior distribution for the variant frequencies f given $M+M^*$ variants:

$$P(f|M,M^*) = \Gamma\left(\sum_{i=1}^{M+M^*}\alpha_i\right)\prod_{i=1}^{M+M^*}\frac{f_i^{\alpha_i-1}}{\Gamma(\alpha_i)} \quad (6)$$

where the hyperparameters α_i can be interpreted as the respective counts of the variants in a hypothetical sample a priori. By default, we set $\alpha_i = 1$ for all i to obtain a flat uninformative prior distribution over all possible values of f. We evaluated different prior distributions for λ: first, we omitted λ from our calculation of prior probability, which was equivalent to an unbounded uniform prior; second, we used a gamma distribution with shape parameter $\alpha = 2$ and rate parameter $\beta = 1$, which yields a median of 1.68 and 95% interval of (0.24, 5.6). The posterior probability of the multi-hit Poisson model can thereby be written:

$$P(\lambda,f,M,M^*|\mathbf{v}) \propto \mathbf{L}(\mathbf{v}|\lambda,f,M,M^*)P(\lambda) P(f|M,M^*)P(M,M^*) \quad (7)$$

Rather than sample over possible values of M^* for which we have no information, we exclude the unobserved variants from the data matrix and marginalize out this parameter:

$$P(\lambda,f|\mathbf{v}^+) \propto \sum_{M^*=0}^{\infty} \mathbf{L}(\mathbf{v}^+|\lambda,f,M=M_{\mathbf{v}^+}) P(\lambda)P(f|M=M_{\mathbf{v}^+})P(M^*|M=M_{\mathbf{v}^+}) \propto \left(\prod_{k=1}^{K}\prod_{i=1}^{M_{\mathbf{v}^+}} L(v_{k,i}|\lambda,f_i)\right)P(\lambda)P(f|M=M_{\mathbf{v}^+}) \quad (8)$$

where $\mathbf{v}^+$ excludes all M^* columns in $\mathbf{v}$ where $\sum_{k=0}^{K} v_{k,\cdot} = 0$, f is renormalized so that $\sum_{i=1}^{M} f_i = 1$, and $\mathbf{L}(\mathbf{v}^+|\lambda,f,M=M_{\mathbf{v}^+})$ is the likelihood conditioned on observing $M_{\mathbf{v}^+}$ variants in the data. Since this conditional likelihood and the priors $P(\lambda)$ and $P(f|M)$ are assumed to be independent of M^*, they can be moved outside of the sum, which becomes a constant term that drops out of the proportionality relation. Note that in equation (8), M is set to the observed number of variants in the data ($M_{\mathbf{v}^+}$) instead of being inferred as a model parameter. We will revisit these assumptions in the "Discussion" section.

Markov chain Monte Carlo (MCMC) samples were propagated over the model parameters λ and f by Metropolis–Hastings sampling. The chains were arbitrarily initialized at the parameter values $\lambda_m = 5$ and $f_i = 1/M$ for all i. We used a Gaussian proposal function for λ with a mean centred at the current value of λ and standard deviation $\sigma = 0.1$, i.e., $\lambda' \sim \mathcal{N}(\lambda, 0.1)$. Because λ has a lower bound of zero, we took the absolute value of the proposed λ; the resulting proposal distributions retain the important property that $q(x|x') = q(x'|x)$. We proposed new variant frequencies from the uniform distributions $(f_i - \delta, f_i + \delta)$ where δ was set to 0.005, and normalized the results so that $\sum_{i=1}^{M} f_i = 1$. Again, these proposal distributions were reflected at zero so that all $f_i > 0$. The MCMC sampler was implemented in *R*.

Method validation

To evaluate the performance of different methods to estimate IUPM, we simulated virus outgrowth data sets in *R* by first sampling the number of infected cells per well, $Y \sim \text{Poisson}(n\lambda)$. Next, the infected cells were partitioned among *N* sequence variants by drawing from the multinomial distribution defined by the frequency vector $f = (f_1, \ldots, f_N)$ (note we changed notation to avoid confusing this parameter with M and M^*). If a sequence variant was completely absent from the simulated data, then we omitted the corresponding column of zero counts from the data set (subsetting $\mathbf{v}$ to $\mathbf{v}^+$ as above) as a necessary part of calculating the conditional likelihood (Eq. 8). Put another way, if a particular variant was not observed any wells, then we assumed that there was no way of inferring that variant's presence in the source population.

To systematically evaluate the impact of different simulation conditions, we conducted simulations for all combinations of rate parameters $\lambda_m = \{0.2, 1, 5, 25 \text{ per million cells}\}$ and variant frequencies $f = \{(0.5, 0.5), (0.2, 0.2, 0.2, 0.2, 0.2),)$ $(0.5, 0.25, 0.125, 0.0625, 0.0625)\}$, which we denote more concisely as ratios 1:1, 1:1:1:1:1 and 8:4:2:1:1. For each combination of model parameters, we generated 100 replicate data sets and applied the maximum likelihood and Bayesian methods to estimate the IUPM. Each MCMC sample was propagated for 10^6 steps with the first 10^5 steps discarded as burn-in; the remaining chain sample was thinned at regular intervals of 1000 steps. We chose these settings based on our analysis of the convergence behaviour and autocorrelation of preliminary runs (Additional file 3: Fig. S3). To reduce computing time, all simulations and model analyses were run on an Intel Xeon-based computing cluster using the *R* module *parallel*. The concordance of model estimates and actual values was visually assessed with density plots and quantified by the relative error, which was calculated as the absolute difference between the actual and true values, normalized by the true value.

Data collection and processing

Actual HIV-1 sequence data was obtained from QVOA previously performed on a population of HIV-individuals recruited through the Rakai Health Science Program in Uganda, originally enrolled to examine the size of the HIV-1 LVR [37]. Participants were HIV-1-infected adults ($\geq$18 years) on ART who had been virally suppressed for > 1 year at the time of enrollment (two historical plasma HIV-1 RNA measurements < 40 copies/mL, obtained 10–18 months apart with no intervening detectable viral load result).

p24 positive viral outgrowth wells derived from QVOA of rCD4 T cells from these Ugandan subjects were collected and stored at − 80°C. HIV viral RNA was extracted from 140 µL of outgrowth media, amplified with a directed nested RT-PCR assay for both *gp41* (~ 350 bp) and *pol* (~530 bp). Paired-end libraries (2 × 300 bases) were generated from the amplicons and barcoded as previously described [38] for sequencing on an Illumina MiSeq system. Adapter sequences were trimmed from the short read data using *cutadapt* (version 1.8) [39] and mate pairs were merged using *AdapterRemoval* [40] with a minimum read overlap of 15 nucleotides. Merged sequences containing ambiguous nucleotide calls were filtered out and clustered using USEARCH (ver.10) with an identity threshold of 0.98 to generate a consensus sequence. The number of times each sequence variant was observed in the data was recorded in the sequence label. Any sequence variant representing < 2.5% of the total number of sequences was excluded from further analysis [7]. The remaining sequence variants were manually screened for large internal deletions.

Wells containing three or more sequence variants (that were greater than 2.5% of the total number of sequences) were manually examined for evidence of recombination as a result of ex vivo recombination in the well or crossover events during library preparation and sequencing, by visually inspecting a Highlighter plot [41] generated from all variant consensus sequences. Any potential recombinant was removed from the data unless the same sequence appeared in another well from the same patient, which implied that the sequence represented a true recombination event within that patient's virus population (Additional file 4: Fig. S4). A multiple sequence alignment (*MAFFT* ver. 7) was generated from the resulting set of sequence variants and used to reconstruct a phylogenetic tree using the implementation of the neighbor-joining method [42] in *R* for visual inspection. Each unique sequence variant was assigned a unique number and the presence/absence of the variant was recorded for each well.

Empirical data analysis

We used the SHP (IUPMStats) and Bayesian MTP (IUPMBayes) methods to analyze sequence data from viral outgrowth assays for samples derived from two individuals (denoted 106 and 111) from the Ugandan study cohort. There were two sequence data sets per patient corresponding to the *pol* and *gp41* amplicons, for a total of four data sets. The clonal prediction scores, which quantify the proportions (maximum 100) of unique sequence variants that would be correctly identified by a particular subsequence [43], were 90 ($\pm$14) and 83 ($\pm$26) for these regions due in part to the relatively small amplicon size required for targeted NGS of the Illumina system. Since we were analyzing a much smaller number of data sets than the simulation study, we ran three replicate chain samples per data set to evaluate convergence. Each chain was initialized with random parameter values; specifically, λ_m was drawn from a uniform distribution spanning 1 to 10, and f was drawn from a flat Dirichlet distribution ($\alpha_i = 1 \; \forall \; i$). We ran each chain sample for 10^7 steps with the first 5×10^5 steps discarded as the burn-in interval. Chain sample states (log posterior probability and model parameter values) were written to log files at regular intervals of 5000 steps. Convergence was assessed using the Gelman-Rubin diagnostic implemented in the *R* package *coda*.

Additional files

Additional file 1. Fig. S1. Relative errors in IUPM estimates for simulated data with highly skewed variant frequencies. The composition of this figure is similar to Fig. 2.

Additional file 2. Fig. S2. Effect of misspecified prior distributions on posterior estimates of IUPM from real data sets. The filled curves represent the same posterior distributions as in Fig. 5, where the prior distributions on variant frequencies were set to $a = \{10, 1, \ldots, 1\}$ for patient 106 and $a = \{1, \ldots, 1\}$ for patient 111. The dashed curves represent the posterior distributions obtained when these priors were swapped between the patient data sets. These results illustrate that misspecification of the prior distribution on variant frequencies can have a measurable effect on posterior estimates of IUPM where the underlying variant frequencies are skewed toward a single common variant (patient 106). However when the virus population has low clonality and there is a mixture of positive and negative wells at the lowest dilution of the QVOA (patient 111, see Table 2), the posterior estimates are more robust to the prior settings.

Additional file 3. Fig. S3. Summary of convergence properties for MCMC sampling. The plots display results for three replicate Markov chain samples (black, red, blue) on the same simulated data set, where there were eight wells with 106 cells, four wells with 4 × 104 cells, and four wells with 320 cells; the true IUPM was set to $\lambda = 1$; and the variant frequencies were set to $\mathbf{f} = \{0.5, 0.25, 0.125, 0.0625, 0.0625\}$. (left) Decay of autocorrelation with increasing lag between samples in the Markov chain. Throughout the study, we thinned chain samples at a lag of 1000 steps. (right) Traces of posterior probability for the first 10,000 steps of the three replicate Markov chains, which corresponds to the length of the burn-in period used in this study. Based on these results, the rate of approach to the posterior distribution was fairly rapid: on the order of 1000 steps.

Additional file 4. Fig. S4. Prominent species identification and elimination of potential outgrowth derived recombinant sequences. **a** All Pol derived consensus sequences with amplicon totals greater than 0.02% of the total amplicon number for well 5M4 (1,000,000 rCD4+ cells plated) from patient 111 were aligned and viewed in a Neighbor-Joining (NJ) phylogenetic tree. Prominent species were defined as those with amplicon totals > 2.5% of the total amplicon read number for well 5M4 and are indicated with red arrows. **b** Since the number of prominent species in well 5M4 were ≥ 3 the prominent sequences were aligned and viewed in a highlighter plot. The probable outgrowth derived recombinant sequence is indicated with a red X and was removed from the analysis (probable recombination area highlighted in red box). **c** All Pol prominent outgrowth sequences for patient 111 were viewed in a NJ tree and variants assigned based on clonality (well 5M4 indicated with red arrows).

Authors' contributions

AFYP, TCQ and ADR conceived the study. AFYP designed and implemented the Bayesian method, analyzed the data, generated all figures and wrote the manuscript. JLP coordinated sample processing and data analysis. BAL and DB performed next-generation sequencing. SFP directed the next-generation sequencing work. JL and AAC performed viral outgrowth assays. SJR coordinated the clinical latent reservoir study. JK coordinated the field team for the clinical study. SL and CWR designed the sequence curation program. CM performed the next-generation sequencing curation steps. TCQ coordinated the laboratory latent reservoir study and data analysis. ADR coordinated next-generation sequencing and sequence analysis. All authors reviewed and contributed revisions to the manuscript. All authors read and approved the final manuscript.

Author details

[1] Department of Pathology and Laboratory Medicine, Western University, London, ON, Canada. [2] Department of Microbiology and Immunology, Western University, London, ON, Canada. [3] Laboratory of Immunoregulation, National Institute of Allergy and Infectious Diseases, National Institutes of Health, Baltimore, MD, USA. [4] Department of Medicine, Johns Hopkins School of Medicine, Baltimore, MD, USA. [5] Rakai Health Sciences Program, Kalisizo, Uganda. [6] BioInfoExperts, LLC, Thibodaux, LA, USA. [7] Genomics Unit, Research Technologies Branch, Rocky Mountain Laboratories, National Institute of Allergy and Infectious Diseases, National Institutes of Health, Hamilton, MT, USA.

Acknowledgements

The authors would like to thank the participants and staff of the Rakai Health Sciences Program, Dr. Greg Gloor for helpful discussions on the Bayesian analysis of compositional data, and Drs. Janet and Robert Siliciano and two anonymous reviewers for their feedback on earlier versions of this manuscript.

Competing interests

The authors declare that they have no competing interests.

Funding

This work was supported in part by the Government of Canada through Genome Canada and the Ontario Genomics Institute (OGI-131 to AFYP); by the Division of Intramural Research, National Institute of Allergy and Infectious Diseases, National Institutes of Health; the Canadian Institutes of Health Research (CIHR Grants 201302MFE-300992 to JLP and PJT-153391, PJT-156178, PJT-155990, FRN-130609 and BOP-149562 to AFYP); and the Gilead HIV Cure Grants Program (90072171). The funders had no role in study design, data collection and interpretation, or the decision to submit the work for publication.

References

1. Churchill MJ, Deeks SG, Margolis DM, Siliciano RF, Swanstrom R. HIV reservoirs: what, where and how to target them. Nat Rev Microbiol. 2016;14(1):55–60.
2. Harrigan PR, Whaley M, Montaner JS. Rate of HIV-1 RNA rebound upon stopping antiretroviral therapy. AIDS. 1999;13(8):F59–62.
3. Bruner KM, Hosmane NN, Siliciano RF. Towards an HIV-1 cure: measuring the latent reservoir. Trends Microbiol. 2015;23(4):192–203.
4. Ho YC, Shan L, Hosmane NN, Wang J, Laskey SB, Rosenbloom DI, et al. Replication-competent noninduced proviruses in the latent reservoir increase barrier to HIV-1 cure. Cell. 2013;155(3):540–51.
5. Finzi D, Hermankova M, Pierson T, Carruth LM, Buck C, Chaisson RE, et al. Identification of a reservoir for HIV-1 in patients on highly active antiretroviral therapy. Science. 1997;278(5341):1295–300.
6. Laird GM, Rosenbloom DI, Lai J, Siliciano RF, Siliciano JD. Measuring the frequency of latent HIV-1 in resting CD4+ T cells using a limiting dilution coculture assay. In: Prasad VR, Kalpana GV, editors. HIV Protocols. New York, NY: Humana Press; 2016. p. 239–53.
7. Lee SK, Zhou S, Baldoni PL, Spielvogel E, Archin NM, Hudgens MG, et al. Quantification of the latent HIV-1 reservoir using ultra deep sequencing and primer ID in a viral outgrowth assay. J Acquir Immune Defic Syndr. 2017;74(2):221–8.
8. Finzi D, Blankson J, Siliciano JD, Margolick JB, Chadwick K, Pierson T, et al. Latent infection of CD4+ T cells provides a mechanism for lifelong persistence of HIV-1, even in patients on effective combination therapy. Nat Med. 1999;5(5):512–7.
9. Crooks AM, Bateson R, Cope AB, Dahl NP, Griggs MK, Kuruc JD, et al. Precise quantitation of the latent HIV-1 reservoir: implications for eradication strategies. J Infect Dis. 2015;212(9):1361–5.

10. Hosmane NN, Kwon KJ, Bruner KM, Capoferri AA, Beg S, Rosenbloom DIS, et al. Proliferation of latently infected CD4(+) T cells carrying replication-competent HIV-1: potential role in latent reservoir dynamics. J Exp Med. 2017;214(4):959–72.
11. Maldarelli F, Wu X, Su L, Simonetti FR, Shao W, Hill S, et al. HIV latency. Specific HIV integration sites are linked to clonal expansion and persistence of infected cells. Science. 2014;345(6193):179–83.
12. Simonetti FR, Sobolewski MD, Fyne E, Shao W, Spindler J, Hattori J, et al. Clonally expanded CD4+ T cells can produce infectious HIV-1 in vivo. Proc Natl Acad Sci USA. 2016;113(7):1883–8.
13. Chun TW, Carruth L, Finzi D, Shen X, DiGiuseppe JA, Taylor H, et al. Quantification of latent tissue reservoirs and total body viral load in HIV-1 infection. Nature. 1997;387(6629):183–8.
14. Bui JK, Sobolewski MD, Keele BF, Spindler J, Musick A, Wiegand A, et al. Proviruses with identical sequences comprise a large fraction of the replication-competent HIV reservoir. PLoS Pathog. 2017;13(3):e1006283.
15. Lorenzi JCC, Cohen YZ, Cohn LB, Kreider EF, Barton JP, Learn GH, et al. Paired quantitative and qualitative assessment of the replication-competent HIV-1 reservoir and comparison with integrated proviral DNA. Proc Natl Acad Sci USA. 2016;113(49):E7908–16.
16. Miller RG, Teh HS, Harley E, Phillips RA. Quantitative studies of the activation of cytotoxic lymphocyte precursor cells. Immunol Rev. 1977;35:38–58.
17. Rosenbloom DI, Elliott O, Hill AL, Henrich TJ, Siliciano JM, Siliciano RF. Designing and interpreting limiting dilution assays: general principles and applications to the latent reservoir for human immunodeficiency virus-1. Open Forum. Infect Dis. 2015;2(4):ofv123.
18. Wineberg M, Oppacher F. The underlying similarity of diversity measures used in evolutionary computation. In: Genetic and evolutionary computation—GECCO 2003. Springer; 2003. p. 206–206.
19. Taswell C. Limiting dilution assays for the determination of immunocompetent cell frequencies. I. Data analysis. J Immunol. 1981;126(4):1614–9.
20. Stallard N, Gravenor MB, Curnow RN. Estimating numbers of infectious units from serial dilution assays. J R Stat Soc C. 2006;55(1):15–30.
21. Mehrabi Y, Matthews JNS. Likelihood-based methods for bias reduction in limiting dilution assays. Biometrics. 1995;51(4):1543–9.
22. Eriksson S, Graf EH, Dahl V, Strain MC, Yukl SA, Lysenko ES, et al. Comparative analysis of measures of viral reservoirs in HIV-1 eradication studies. PLoS Pathog. 2013;9(2):e1003174.
23. Mehrabi Y, Matthews JNS. Implementable Bayesian designs for limiting dilution assays. Biometrics. 1998;54(4):1398–406.
24. Yucha RW, Hobbs KS, Hanhauser E, Hogan LE, Nieves W, Ozen MO, et al. High-throughput characterization of HIV-1 reservoir reactivation using a single-cell-in-droplet PCR assay. EBioMedicine. 2017;20:217–29.
25. Sanyal A, Mailliard RB, Rinaldo CR, Ratner D, Ding M, Chen Y, et al. Novel assay reveals a large, inducible, replication-competent HIV-1 reservoir in resting CD4(+) T cells. Nat Med. 2017;23(7):885–9.
26. Jung A, Maier R, Vartanian JP, Bocharov G, Jung V, Fischer U, et al. Recombination: multiply infected spleen cells in HIV patients. Nature. 2002;418(6894):144.
27. Josefsson L, King MS, Makitalo B, Brännström J, Shao W, Maldarelli F, et al. Majority of CD4+ T cells from peripheral blood of HIV-1-infected individuals contain only one HIV DNA molecule. Proc Natl Acad Sci. 2011;108(27):11199–204.
28. Wasserman L. Asymptotic inference for mixture models by using data-dependent priors. J R Stat Soc Ser B (Stat Methodol). 2000;62(1):159–80.
29. Ghosal S. The Dirichlet process, related priors and posterior asymptotics. Bayesian Nonparametr. 2010;28:35.
30. Goodrum F, Jordan CT, Terhune SS, High K, Shenk T. Differential outcomes of human cytomegalovirus infection in primitive hematopoietic cell subpopulations. Blood. 2004;104(3):687–95.
31. Hawkins A, Davidson F, Simmonds P. Comparison of plasma virus loads among individuals infected with hepatitis C virus (HCV) genotypes 1, 2, and 3 by quantiplex HCV RNA assay versions 1 and 2, Roche Monitor assay, and an in-house limiting dilution method. J Clin Microbiol. 1997;35(1):187–92.
32. McCaughan F, Dear PH. Single-molecule genomics. J Pathol. 2010;220(2):297–306.
33. Tubo NJ, Pagán AJ, Taylor JJ, Nelson RW, Linehan JL, Ertelt JM, et al. Single naive CD4+ T cells from a diverse repertoire produce different effector cell types during infection. Cell. 2013;153(4):785–96.
34. Brent RP. Algorithms for minimization without derivatives. North Chelmsford: Courier Corporation; 1973.
35. Bonnefoix T, Bonnefoix P, Verdiel P, Sotto JJ. Fitting limiting dilution experiments with generalized linear models results in a test of the single-hit Poisson assumption. J Immunol Methods. 1996;194(2):113–9.
36. Shankarappa R, Margolick JB, Gange SJ, Rodrigo AG, Upchurch D, Farzadegan H, et al. Consistent viral evolutionary changes associated with the progression of human immunodeficiency virus type 1 infection. J Virol. 1999;73(12):10489–502.
37. Prodger JL, Lai J, Reynolds SJ, Keruly JC, Moore RD, Kasule J, et al. Reduced frequency of cells latently infected with replication-competent HIV-1 in virally suppressed individuals living in Rakai, Uganda. Clin Infect Dis. 2017;65(8):1308–15.
38. Courtney CR, Mayr L, Nanfack AJ, Banin AN, Tuen M, Pan R, et al. Contrasting antibody responses to intrasubtype superinfection with CRF02_AG. PLoS ONE. 2017;12(3):e0173705.
39. Martin M. Cutadapt removes adapter sequences from high-throughput sequencing reads. EMBnet.journal. 2011;17(1):10–12.
40. Lindgreen S. AdapterRemoval: easy cleaning of next-generation sequencing reads. BMC Res Not. 2012;5:337.
41. Keele BF, Giorgi EE, Salazar-Gonzalez JF, Decker JM, Pham KT, Salazar MG, et al. Identification and characterization of transmitted and early founder virus envelopes in primary HIV-1 infection. Proc Natl Acad Sci. 2008;105(21):7552–7.
42. Thompson JD, Higgins DG, Gibson TJ. CLUSTAL W: improving the sensitivity of progressive multiple sequence alignment through sequence weighting, position-specific gap penalties and weight matrix choice. Nucleic Acids Res. 1994;22(22):4673–80.
43. Laskey SB, Pohlmeyer CW, Bruner KM, Siliciano RF. Evaluating clonal expansion of HIV-infected cells: optimization of PCR strategies to predict clonality. PLoS Pathog. 2016;12(8):e1005689.

5

Myricetin antagonizes semen-derived enhancer of viral infection (SEVI) formation and influences its infection-enhancing activity

Ruxia Ren[1,2†], Shuwen Yin[1†], Baolong Lai[2], Lingzhen Ma[1], Jiayong Wen[1], Xuanxuan Zhang[1], Fangyuan Lai[1], Shuwen Liu[1*] and Lin Li[1*]

Abstract

Background: Semen is a critical vector for human immunodeficiency virus (HIV) sexual transmission and harbors seminal amyloid fibrils that can markedly enhance HIV infection. Semen-derived enhancer of viral infection (SEVI) is one of the best-characterized seminal amyloid fibrils. Due to their highly cationic properties, SEVI fibrils can capture HIV virions, increase viral attachment to target cells, and augment viral fusion. Some studies have reported that myricetin antagonizes amyloid β-protein (Aβ) formation; myricetin also displays strong anti-HIV activity in vitro.

Results: Here, we report that myricetin inhibits the formation of SEVI fibrils by binding to the amyloidogenic region of the SEVI precursor peptide (PAP248–286) and disrupting PAP248–286 oligomerization. In addition, myricetin was found to remodel preformed SEVI fibrils and to influence the activity of SEVI in promoting HIV-1 infection. Moreover, myricetin showed synergistic effects against HIV-1 infection in combination with other antiretroviral drugs in semen.

Conclusions: Incorporation of myricetin into a combination bifunctional microbicide with both anti-SEVI and anti-HIV activities is a highly promising approach to preventing sexual transmission of HIV.

Keywords: HIV, Myricetin, Amyloid fibrils, SEVI, Synergistic antiviral effects

Background

Since the first cases of acquired immune deficiency syndrome (AIDS) were reported in 1981, more than 70 million people have been infected by human immunodeficiency virus (HIV), and approximately 1 million die of the disease annually [1]. Currently, unprotected sex remains the major route of HIV transmission, accounting for more than 80% of new HIV infections worldwide [2].

It has been shown that semen functions as a critical carrier of virus during sexual transmission [3]. Notably, amyloid fibrils in semen are considered to be responsible for the reduced efficacy of antiretroviral therapies (cART) in vivo because they facilitate virus attachment and internalization into cells [4]. A naturally occurring C-proximal proteolytic fragment of prostatic acid phosphatase (PAP) in semen, PAP248–286, has been reported to form aggregated amyloid fibrils (known as semen-derived enhancer of virus infection or SEVI) and to increase by several orders of magnitude the rate of HIV infection in vitro [5]. Because SEVI is highly cationic and the fibrils it forms are positively charged, it not only effectively captures HIV virions but also promotes viral attachment to target cells by neutralizing the inherent electrostatic repulsion between the negative charges on the surfaces of the virions and target cells [6]. Some other endogenous amyloid aggregates that increase HIV-1 infectivity, including the PAP N-proximal fragment (PAP85–120) and semenogelins (SEM1 and SEM2), have been identified in human semen [7, 8], which may explain the discrepancy between the low infectiousness of HIV in vitro and its observed efficient sexual transmission.

*Correspondence: liusw@smu.edu.cn; li75lin@126.com
†Ruxia Ren and Shuwen Yin contributed equally to this work
[1] Guangdong Provincial Key Laboratory of New Drug Screening, Guangzhou Key Laboratory of Drug Research for Emerging Virus Prevention and Treatment, School of Pharmaceutical Sciences, Southern Medical University, 1838 Guangzhou Avenue North, Guangzhou 510515, Guangdong, China
Full list of author information is available at the end of the article

As amyloid species that are naturally abundant in semen, including SEVI, play a critical role in the spread of HIV, they represent a particularly attractive target for the development of molecules that can reduce spread of the virus. Theoretically, eliminating seminal amyloid fibrils by antagonizing fibril formation or enhancing the degradation of mature endogenous fibrils can massively reduce any viral infection enhancement [9–11]. This strategy might be advantageous because it targets host factors rather than the virus itself.

It is well known that the aggregation of proteins into amyloid fibrils is associated with fatal diseases, including Alzheimer's disease, Parkinson's disease and diabetes, and many molecules that inhibit fibrillization in vitro have been reported. For instance, epigallocatechin-3-gallate (EGCG), the major catechin present in green tea, not only potently inhibits the amyloidogenesis of various polypeptides but is also able to disassemble a wide range of preformed amyloid fibrils [12, 13].

Myricetin, a common dietary flavonoid found in foods such as walnuts, onions, berries, herbs and red grapes, exerts a wide variety of biological and nutraceutical effects, including antioxidative, anti-inflammatory, antidiabetic, antitumor and free radical-scavenging activities [14–16]. Moreover, myricetin shows activity against HIV-1 infection by inhibiting HIV-1 integrase [17, 18]. It is noteworthy that myricetin inhibits amyloid fibrillization by a variety of disease-associated amyloidogenic proteins, including amyloid β-protein (Aβ), tau protein, islet amyloid polypeptide and other amyloidogenic peptides [19–21]. As described above, SEVI is a possible drug target for preventing HIV sexual transmission due to its stable structure and high levels in semen [22]. The concentration of endogenous SEVI in pooled semen is approximately 35 μg/ml, which considerably exceeds the level (≥ 2 μg/ml) required to enhance infection [7, 23]. However, it is not known whether myricetin affects the formation of SEVI amyloid fibrils in semen. In the current study, we sought to elucidate the effects of myricetin on the formation of SEVI amyloid fibrils and on the enhancement of HIV-1 infection by SEVI. Strikingly, the results showed that myricetin influences the activity of SEVI in enhancing HIV-1 infection and that it exhibits synergistic effects in semen against HIV-1 infection when applied to cells in combination with other antiretroviral (ARV) agents. These findings will be helpful in the development of dual-function antiviral drugs or microbicides that possess both anti-HIV and anti-SEVI-enhancing activities.

Results

Myricetin inhibits the formation of SEVI fibrils and other seminal fibrils

Thioflavin T (ThT), a dye that is commonly used in the detection of amyloid fibrils, can intercalate into the β-sheet structure of amyloid fibrils, resulting in enhanced fluorescence and a characteristic redshift in the emission spectrum. Congo red, another amyloid-binding dye, exhibits apple-green birefringence under polarized light as well as increased fluorescence [7, 24]. To evaluate the effects of myricetin on the formation of SEVI fibrils by PAP248–286, we collected PAP248–286 aggregates formed in the presence or absence of various concentrations of myricetin at different time points and monitored them using both ThT fluorescence and Congo red staining assays. The results showed that myricetin antagonized the assembly of PAP248–286 monomers into SEVI fibrils in a dose-dependent manner. The addition of myricetin at 75 μg/ml increased the lag time of SEVI amyloid fibril formation by PAP248–286 to more than 48 h, as measured by ThT fluorescence (Fig. 1a), whereas the lag time of amyloid fibril formation by PAP248–286 in the absence of myricetin was less than 6 h. Congo red assays showed that the lag time of SEVI amyloid fibril formation in the presence of various concentrations of myricetin was extended to approximately 18 h (Fig. 1b). The morphology of SEVI fibrils in the presence and absence of myricetin was imaged by transmission electron microscopy (TEM). As shown in Fig. 1c, mature SEVI fibrils formed by PAP248–286 were identified after shaking for 24 h, whereas no amyloid fibrils were observed after adding 16 μg/ml of myricetin, even after shaking for 48 h.

Far-UV (190–260 nm) circular dichroism (CD) spectrum measurement is commonly used to detect the presence of the typical secondary β-sheet structure of fibrils; such fibrils display a spectrum with a minimum at approximately 218 nm. A characteristic β-sheet secondary structure undergoes higher-level association to form protein aggregates and mature fibrils. Thus, we further investigated the secondary structure of SEVI fibrils formed in the presence or absence of myricetin by measuring their CD spectra. As illustrated in Fig. 1e, the spectrum of PAP248–286 after agitation for 24 h displayed a single minimum at approximately 230 nm, indicating the presence of a typical β-sheet component. However, the spectrum of PAP248–286 after agitation for 24 h in the presence of 5 μg/ml myricetin (final concentration) indicated the presence of a characteristic random coil structure, suggesting that PAP248–286 did not adopt any ordered or stable conformation and indicating that β-sheet aggregation had not occurred.

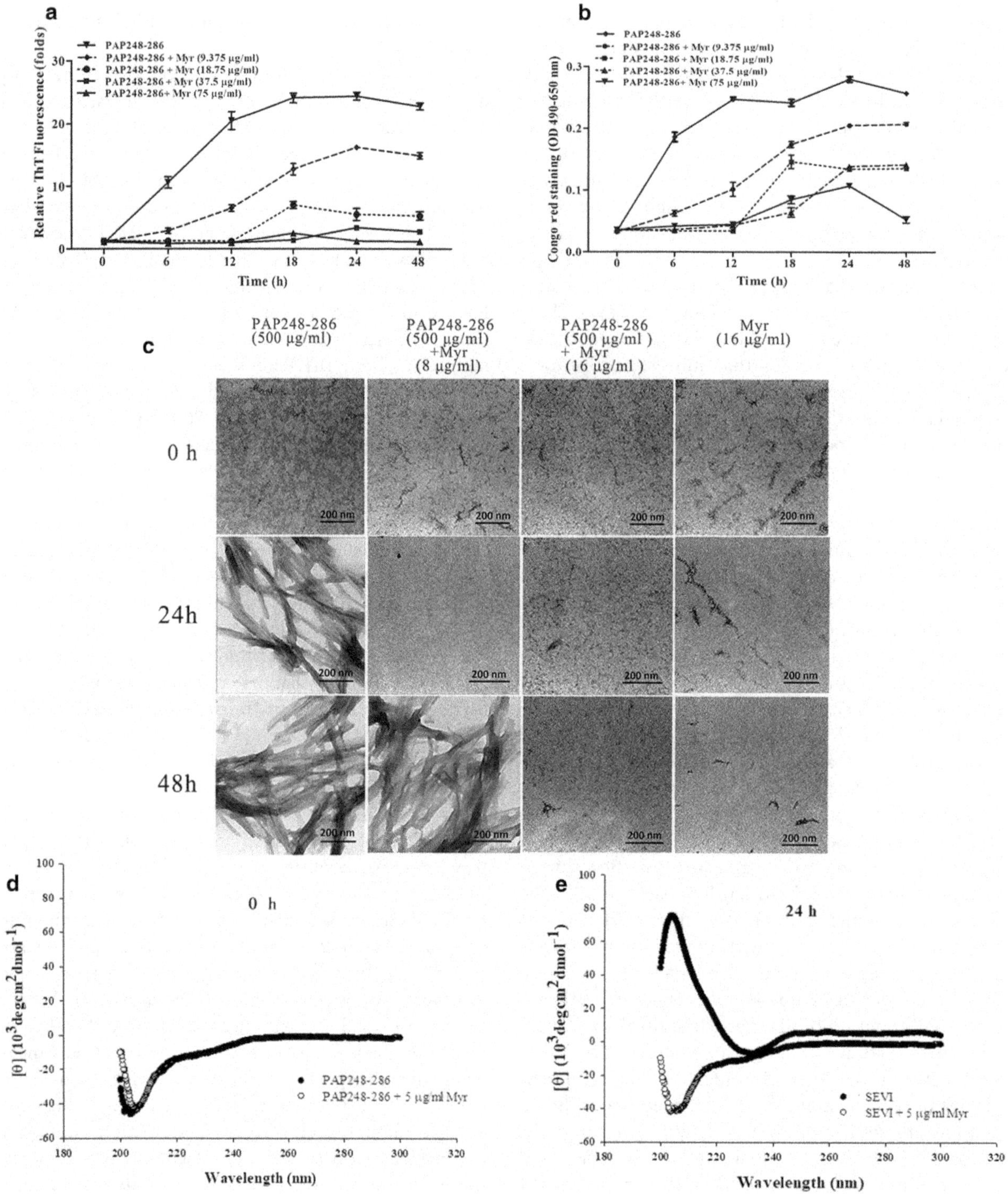

Fig. 1 Myricetin inhibits the assembly of PAP248–286. PAP248–286 (3 mg/ml) was incubated with myricetin (75, 37.5, 18.75, 9.375 and 0 μg/ml), and the mixture was agitated at 1400 rpm at 37 °C. Samples were collected at various time points (0, 6, 12, 18, 24 and 48 h) and monitored by ThT (**a**) or Congo red staining (**b**). Average values (± SD) were calculated from triplicate measurements; the data shown represent one representative trial of three independent experiments. **c** Amyloid fibril samples (500 μg/ml) in the presence or absence of myricetin (8 and 16 μg/ml) at various time points (0, 24 and 48 h), as visualized by TEM. The scale bar is 200 nm. **d, e** The secondary structure of PAP248–286 or SEVI fibrils (300 μg/ml) in the presence or absence of myricetin (5 μg/ml) at various time points (0 and 24 h) was measured using CD spectroscopy

Several studies have reported that the PAP248–286 monomer lacks the ability to promote HIV-1 infection [25]. To determine whether enhancement of HIV-1 infection by PAP248–286 is lost after the addition of myricetin, we characterized the HIV-1 infection-enhancing activity of PAP248–286 samples after agitation in the presence or absence of myricetin for various times (samples in Fig. 1a, b); the final concentrations of PAP and myricetin used in the cell-based HIV infection assay are only one-sixtieth of their initial concentrations. To eliminate the anti-HIV activity of myricetin itself, the above samples were first centrifuged at 12,000 rpm for 5 min to remove free myricetin. As shown in Fig. 2 and Additional file 1: Figure S2, PAP248–286 that had been agitated for 6–48 h effectively enhanced HIV-1 infection by two HIV-1 clones in a time-dependent manner. However, the ability of PAP248–286 to enhance HIV-1 infection decreased after agitation in the presence of myricetin in a dose-dependent manner. The highest final concentration of myricetin tested (1.25 μg/ml) significantly neutralized SEVI-mediated enhancing activity by inhibiting the formation of SEVI fibrils (Fig. 2, Additional file 1: Figure S2). It is worth mentioning that inhibition of SEVI fibril formation was not observed after agitation of PAP248–286 in the presence of lower concentrations of myricetin for 24 h but was observed after agitation of PAP248–286 in

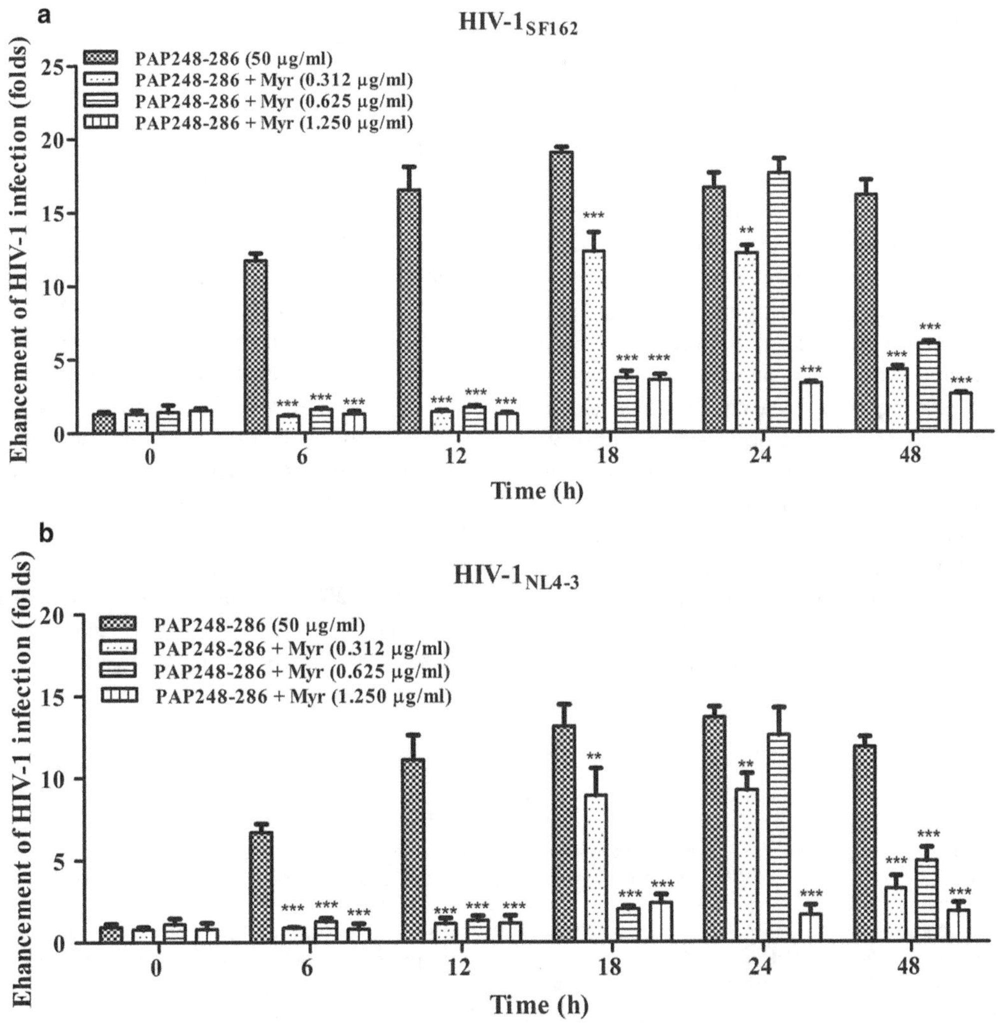

Fig. 2 Amyloid fibril samples display loss of enhancement of HIV-1 infection in the presence of myricetin. Mixed SEVI fibril samples prepared as described above in the presence or absence of myricetin were diluted to a final concentration of 50 μg/ml (PAP248–286). The samples were then incubated with HIV-1_{SF162} (**a**) and HIV-1_{NL4-3} (**b**) infectious clones. The mixtures were added to prepared TZM-b1 cells. Luciferase activities were measured at 72 h post-infection. The values shown represent the mean ± SD (n = 3). One-way ANOVA with Dunnett's post hoc multiple comparisons test was used to statistically analyze the differences between samples containing PAP248–286 alone and samples containing PAP248–286 and myricetin (*$p < 0.05$; **$p < 0.01$, ***$p < 0.001$)

the presence of the same concentrations of myricetin for 48 h (Fig. 2). The most likely reason for this is that the ability of myricetin to inhibit fibril formation may require long exposure times and full interaction. The results also indicated that myricetin may play a role in the degradation of mature SEVI fibrils. Taken together, these findings show that myricetin abrogates the formation of SEVI fibrils by PAP248–286, thereby influencing the activity of SEVI in enhancing HIV-1 infection.

The effects of myricetin on the formation of two other common seminal amyloid fibrils, SEM1 and SEM2, were also assessed. According to the results, myricetin also inhibited the formation of fibrils by both $SEM1_{86-107}$ and $SEM2_{86-107}$ in a dose-dependent manner (Additional file 1: Figure S1). The addition of myricetin at a concentration of 200 µg/ml completely blocked amyloid fibril formation, as measured by ThT fluorescence.

Myricetin shows binding affinity to functional regions of the PAP248–286 monomer

To clarify the molecular mechanism by which myricetin inhibits assembly of PAP248–286 into fibrils, we first analyzed potential interaction between myricetin and the PAP248–286 monomer by western blotting (WB). Myricetin is a negatively charged compound, whereas PAP248–286 is positively charged peptide. The inherent electrostatic repulsion that exists between two oppositely charged molecules might account for deposition of the PAP248–286 peptide via a change in isoelectric point. Using native acid gel electrophoresis, we observed a marked decrease in the level of PAP248–286 monomer in the supernatant after adding myricetin, a decrease that was dose dependent (Fig. 3a). The results indicate that myricetin binds to PAP248–286 monomers and forms insoluble aggregates via natural electrostatic interaction. The affinity between myricetin and the PAP248–286 monomer was further investigated using surface plasmon resonance (SPR). As shown in Fig. 3b, myricetin bound to the PAP248–286 monomer with high affinity ($K_D = 3.02 \times 10^{-7}$ M). To confirm the specificity of this binding, an irrelevant peptide (N36) and an irrelevant flavonoid (phlorizin) were used as negative controls; N36 is a 36-mer synthetic peptide derived from the N-terminal heptad repeat region of the HIV gp41 envelope protein [26], and phlorizin is a natural phenol that is found in apple and pear tree leaves. Our results confirmed that myricetin does not bind to N36 and that phlorizin does not bind to the PAP248–286 monomer (Fig. 3c, d).

To further substantiate the interaction of myricetin with PAP248–286 at the atomic level, computational molecular docking was conducted to identify the potential amino acids involved in the binding of myricetin to PAP248–286 monomers. Our analysis, which was based on the 3D molecular structure of myricetin and the monocrystal PAP248–286 protein structure reported in Protein Data Bank (PDB; code No. 2L3H), showed that myricetin is able to bind to PAP248–286 and form hydrogen bonds via four potential residues: Leu258, Gln259, Met271 and Arg273. Myricetin is also able to bind to PAP248–286 via hydrophobic interactions with Val264 and Leu268 (Fig. 3e).

Some studies have reported that the central region of the SEVI precursor peptide PAP248–286, namely, PAP257–267, is an amyloidogenic region with high fibril-forming propensity [27]. Coincidentally, PAP268–271 plays a vital role in promoting SEVI fibrillation [27]. According to our computational molecular docking results, six potential residues of PAP248–286 are associated with myricetin binding (Fig. 3e). To replace PAP248–286, we synthesized a 15-amino acid PAP257–271 fragment that includes both of the amyloidogenic regions (PAP257–267 and PAP268–271) and all six of the amino acids (Leu258, Gln259, Val264, Leu268, Met271 and Arg273) shown to be important in the molecular

(See figure on next page.)
Fig. 3 Myricetin exhibits high-affinity binding to the functional region of the PAP248–286 monomer. **a** The electrostatic interaction of myricetin with PAP248–286. PAP248–286 (100 µg/ml) was incubated with serially diluted myricetin (12.5, 6.25, 3.13 and 1.56 µg/ml) at 37 °C for 30 min. The samples were then centrifuged at 5000 rpm for 15 min; the peptides remaining in the supernatant were electrophoresed through 10% acidic native polyacrylamide gels and detected by WB using a polyclonal antibody against PAP248–286. **b** Binding affinity of myricetin to PAP248–286. PAP248–286 (50 µg/ml) was immobilized on a sensor chip. Subsequently, twofold serial dilutions of recombinant myricetin were injected as the analyte. The affinity constant K_D is the ratio of the dissociation constant K_d to the association constant K_a ($K_D = K_d/K_a$). **c** Binding affinity of myricetin to an irrelevant N36 peptide (50 µg/ml). **d** Binding affinity of an irrelevant flavonoid (phlorizin) to PAP248–286 (50 µg/ml). **e** Presumed binding sites of myricetin to PAP248–286. According to the computational docking results, myricetin formed hydrogen bonds with Leu258, Gln259, Met271 and Arg273, and it bound to Val264 and Leu268 by hydrophobic interactions. **f** Amyloid fibril formation by wild-type PAP257–273 and six mutants. Wild-type and mutant peptides (3 mg/ml) were agitated at 1400 rpm at 37 °C for 48 h; the presence of fibrils was then monitored using ThT assays. **g** Myricetin inhibition of fibril formation by wild-type and mutants PAP257–273. Wild-type and mutants PAP257–273 were mixed with myricetin (200, 100, 50 and 25 µg/ml), and the mixtures were agitated at 1400 rpm at 37 °C. The samples were then collected and monitored using ThT assays. Percent inhibition was calculated from the following equation: (1-ThT fluorescence of fibrils with Myr/ThT fluorescence of fibrils without Myr) × 100. Average values (± SD) were calculated from triplicate measurements; the data shown represent one representative trial of three independent experiments

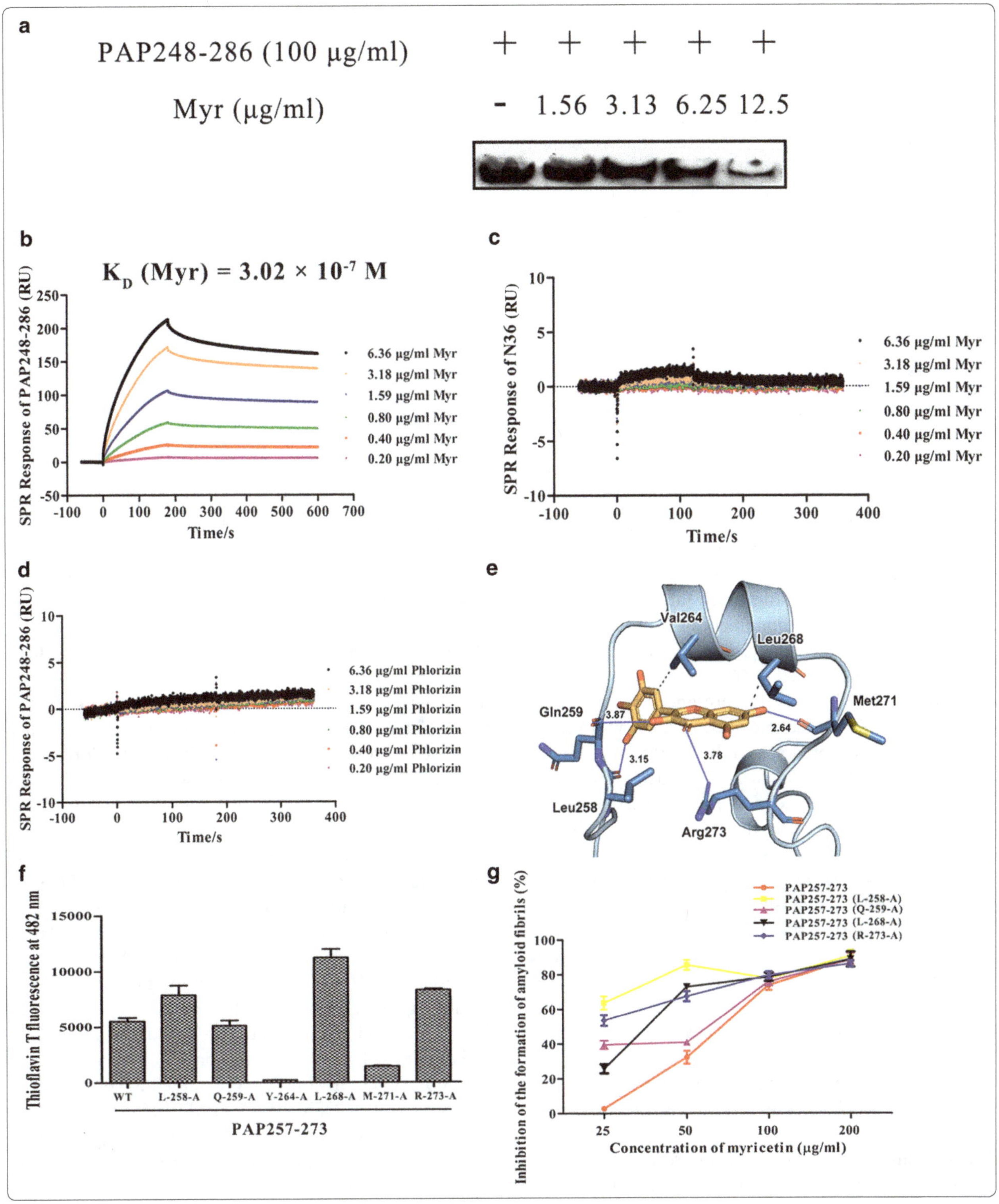

docking analysis. Importantly, due to its short length, the PAP257–271 peptide is simpler to synthesize on a large scale than is the SEVI peptide. In this study, we confirmed that the wild-type PAP257–273 fragment readily forms amyloid fibrils (Fig. 3f). To investigate whether the above-named six residues are essential for the formation of SEVI fibrils, we assessed the aggregation of PAP257–273 peptides containing site-directed mutations of these residues. As shown in Fig. 3f, four mutants of PAP257–273, i.e., Leu258A, Gln259A, Leu268A and Arg273A, exhibited fibril-forming propensity (Fig. 3f). Notably, PAP257–273 peptides with Val264A and Met271A mutations displayed attenuated aggregation during fibril formation, indicating that Val264 and Met271 of PAP248–286 might be critical for the assembly of PAP248–286 into fibrils. The inhibitory effects of myricetin on fibril formation by the four mutants of PAP257–273 were similar to or even stronger than their effects on wild-type PAP257–273, as measured by ThT assays (Fig. 3g).

Myricetin inhibits oligomerization of PAP248–286 monomers

Considerable evidence shows that the presence of low-*n*-order oligomers during the early stages of fibril formation indicates a lag phase in fibril assembly. It has been reported that myricetin inhibits Aβ aggregation by preventing amyloid β-protein oligomerization [28]. We therefore used photo-induced cross-linking of unmodified proteins (PICUP) to determine whether myricetin inhibits PAP248–286 monomer oligomerization and found that myricetin significantly inhibited oligomerization of PAP248–286 monomers in a dose-dependent manner (Fig. 4a). When 20 μg/ml of myricetin was mixed with 500 μg/ml of PAP248–286, oligomerization did not proceed to completion. In the presence of cross-linking agents, the PAP248–286 monomer, which was used as a positive control, predominantly existed as a mixture of oligomers on the order of 2–7. An irrelevant protein (glutathione S-transferase, GST) and an irrelevant flavonoid (phlorizin) were used as negative controls; GST normally exists as a mixture of homodimers and higher-order cross-linked species, thus providing an appropriate positive control for cross-linking chemistry. The results showed that phlorizin did not inhibit early oligomerization of PAP248–286 and that myricetin did not block oligomerization of GST (Fig. 4b, c).

Myricetin antagonizes assembly of SEVI seeded by preformed fibrils

It has been reported that the addition of a small amount of preformed SEVI fibrils as seeds to a suspension of soluble PAP248–286 monomers promotes fibril polymerization and eliminates the lag phase for assembly [23, 29]. In the present study, myricetin-mediated inhibition of SEVI aggregation seeded by preformed fibrils was determined by the ThT assay. In contrast to unseeded PAP248–286 solutions, PAP248–286 solutions seeded with preformed SEVI fibrils showed rapid assembly into fibrils (Fig. 4d). In addition to antagonizing unseeded PAP248–286 assembly (Fig. 1a, b), myricetin also completely obstructed SEVI-seeded fibrillization (Fig. 4d), indicating that myricetin antagonizes the growth and aggregation of SEVI fibrils after nucleation. ThT assay results were utilized to generate a dose–response curve for myricetin inhibition of PAP248–286 (2 mg/ml) seeded or not with 10% wt/wt mature SEVI fibrils agitated for 48 h (Fig. 4e, f), and we obtained a half maximal inhibitory concentration (IC_{50}) value of 41.51 μg/ml for myricetin inhibition of seeded PAP248–286 assembly compared to an IC_{50} value of 22.59 μg/ml for myricetin inhibition of spontaneous PAP248–286 assembly. Thus, higher concentrations of myricetin (200 μg/ml) were required to inhibit PAP248–286 assembly after the addition of SEVI seeds (Fig. 4d–f).

Myricetin remodels mature SEVI fibrils

ThT and Congo red assays were employed to determine whether myricetin can disassemble preformed fibrils, and the results showed that myricetin remodeled mature SEVI fibrils immediately in a dose-dependent but not in a time-dependent manner (Fig. 5a, b). Similar results showing that myricetin leads to a decrease in the β-sheet minimum of SEVI fibrils were obtained by CD, both immediately at 1 min and up to 48 h (Fig. 5c–e).

Myricetin influences the activity of SEVI in enhancing HIV-1 infection

The effects of myricetin on the SEVI-mediated enhancement of HIV-1 infection were assessed by fluorescence microscopy, flow cytometry and HIV-1 infection assays. CEMx174 5.25 M7 cells expressing a green fluorescent protein (GFP) reporter gene under the control of the HIV-1 promoter were chosen as target cells. Fluorescence microscopy images revealed faint background GFP^+ fluorescence expression in uninfected CEMx174 5.25 M7 cells and that SEVI fibril formation increased the fluorescence intensity of HIV-1-infected cells. However, myricetin notably decreased the fluorescence intensity of HIV-1-infected cells in the presence of SEVI fibrils (Fig. 6a). Subsequent flow cytometry assays further confirmed that myricetin reduced the number of cells infected by HIV-1 in the presence of SEVI fibrils (Fig. 6b). When SEVI fibrils were added to cells, the percentage of GFP^+ CEMx174 5.25 M7 cells was increased to 15.5%, whereas the initial percentage of GFP^+ cells of the control group in the absence of SEVI fibrils and HIV-1_{SF162} was 2.37%. The percentage of GFP^+ cells of the HIV-1_{SF162} group without SEVI fibrils was 2.57%, though the percentage of GFP^+ cells decreased to 1.97% upon incubation with myricetin (Fig. 6b). Similar to the HIV-1_{SF162} control, the initial percentage of GFP^+ cells after incubation with HIV-1_{SF162} and myricetin in the absence of SEVI fibrils was approximately 2.69%. To demonstrate

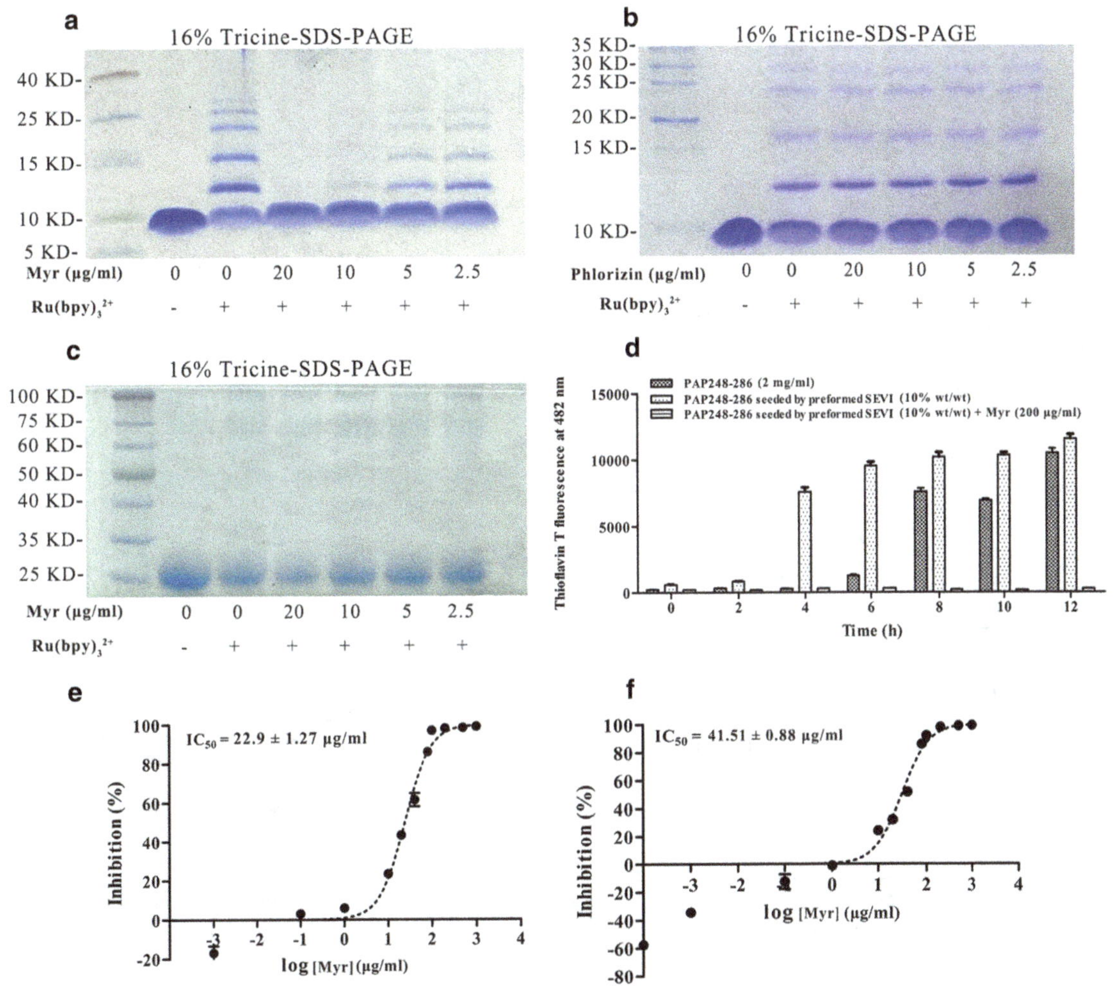

Fig. 4 Myricetin inhibits early oligomerization, fibril extension and degradation of PAP248–286. **a** The effects of myricetin on the oligomerization of PAP248–286, as assessed using 16% Tricine-SDS-PAGE and Coomassie blue staining. PAP248–286 monomer at 500 μg/ml with or without cross-linking served as the control. **b** The effects of an irrelevant flavonoid (phlorizin) on the oligomerization of PAP248–286. **c** The effects of myricetin on the oligomerization of GST. **d** The effects of myricetin on fibril extension by PAP248–286. Mature SEVI fibril seeds (1% wt/wt) were mixed with PAP248–286 peptide (2 mg/ml) in the presence or absence of myricetin (200 μg/ml), and the mixture was agitated at 1400 rpm at 37 °C. Aliquots were collected at various time points and assessed by ThT fluorescence. **e** Dose–response curve for myricetin inhibition of PAP248–286 (2 mg/ml) fibrillization after 48 h of agitation by ThT assay. **f** Dose–response curve for myricetin inhibition of PAP248–286 (2 mg/ml) fibrillization seeded by preformed SEVI fibrils (10% wt/wt) after 48 h of agitation by ThT assay. IC_{50} values were calculated. The values shown represent the mean ± SD (n = 3)

the reproducibility of the results, additional bar graph showing the averaged data of the percentages of GFP^+ CEMx174 5.25 M7 cells from multiple experiments is presented in Fig. 6c.

To directly measure the effects of myricetin on SEVI-mediated enhancement of HIV-1 infection, we performed HIV-1 infection assays in the presence of SEVI fibrils. To eliminate the anti-HIV activity of myricetin itself as much as possible, we removed free myricetin by washing myricetin-treated SEVI samples one to five times with phosphate-buffered saline (PBS) followed by centrifugation prior to the assays. In contrast to samples containing only SEVI fibrils, myricetin-treated SEVI fibrils after one wash resulted in significant attenuation in infectivity of CCR5-tropic HIV-1_{SF162} and CXCR4-tropic HIV-1_{NL4-3} infectious clones in a dose-dependent manner (Fig. 6e, f) [7]. Our results further showed that myricetin-mediated inhibition of SEVI-mediated enhancement of HIV infection was still observed after five washes (Additional file 1: Figure S3).

We further assessed the viability of on both CEMx174 5.25 M7 and TZM-bl cells after exposure to myricetin

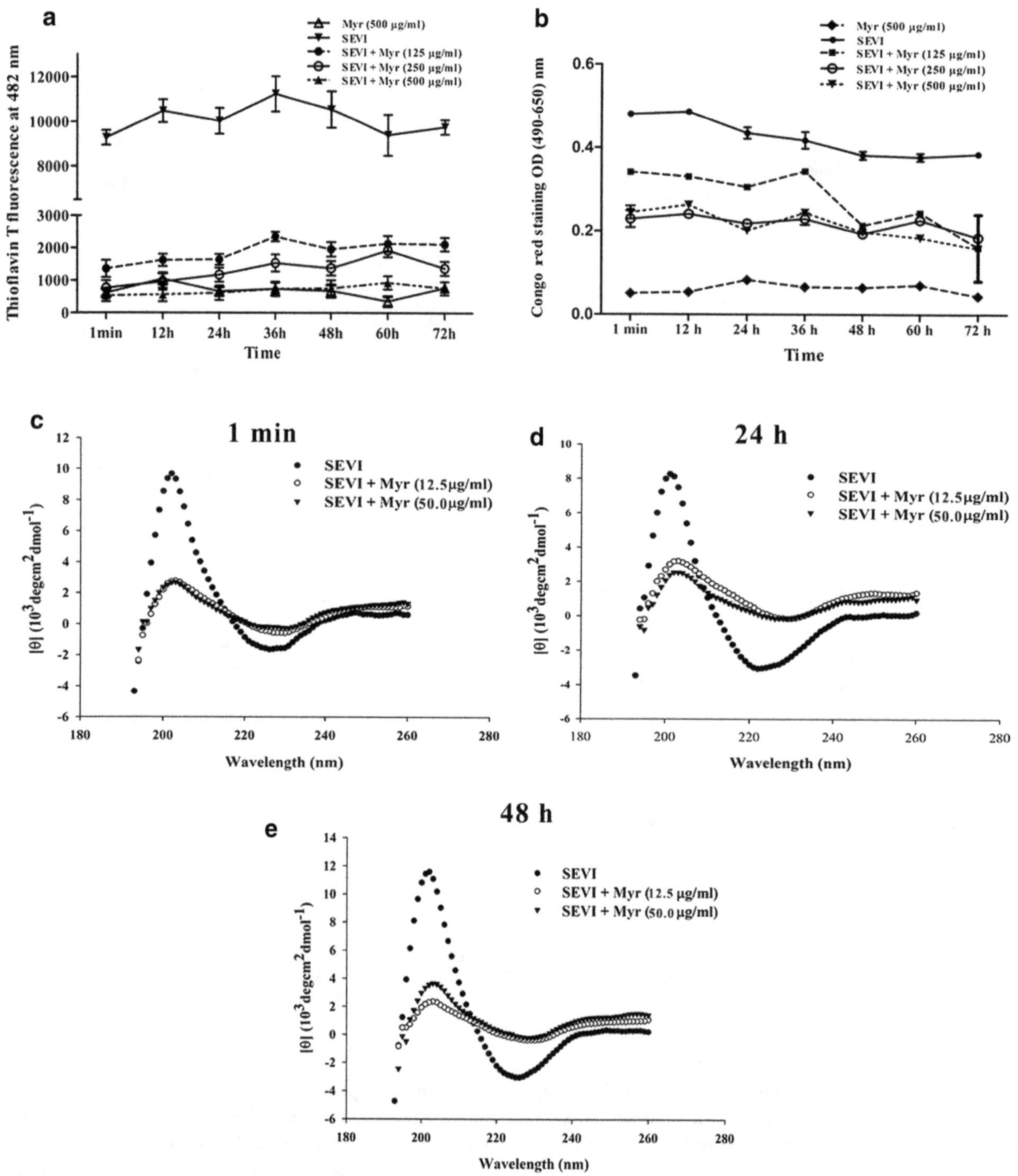

Fig. 5 Remodeling of fibrils was monitored using ThT (**a**) or Congo red (**b**) assays. Mature SEVI fibrils were incubated with various concentrations of myricetin (500, 250, 125 μg/ml) or PBS and agitated as described above for 72 h. Aliquots were removed at the indicated time points. **c–e** At various time points (1 min, 24 and 48 h), the samples collected in the above-described remodeling experiment were measured by CD. Average values (± SD) were calculated from triplicate measurements; the data shown represent one representative trial of three independent experiments

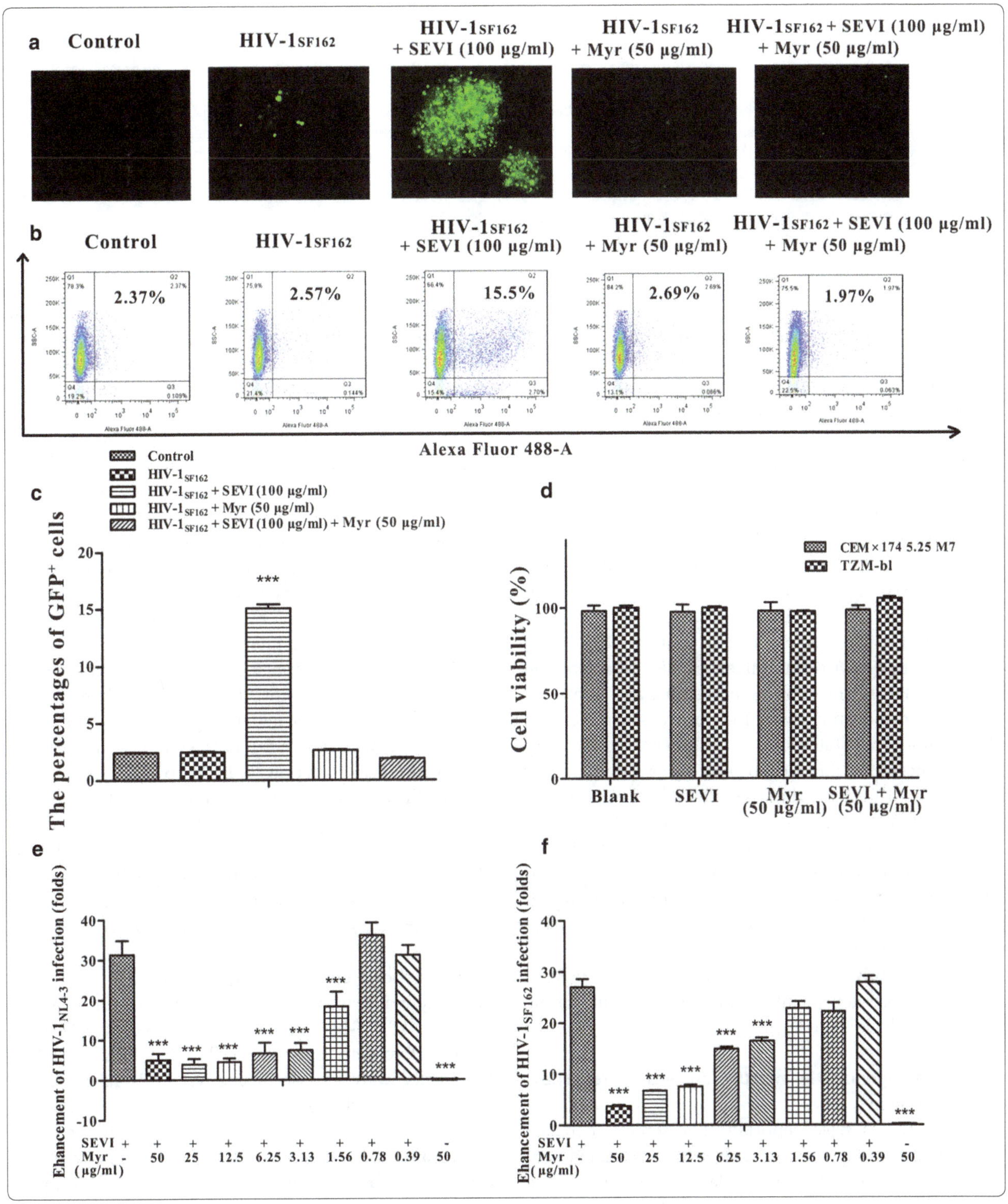
a
Control
HIV-1SF162
HIV-1SF162 + SEVI (100 μg/ml)
HIV-1SF162 + Myr (50 μg/ml)
HIV-1SF162 + SEVI (100 μg/ml) + Myr (50 μg/ml)
b
Control
HIV-1SF162
HIV-1SF162 + SEVI (100 μg/ml)
HIV-1SF162 + Myr (50 μg/ml)
HIV-1SF162 + SEVI (100 μg/ml) + Myr (50 μg/ml)
2.37%
2.57%
15.5%
2.69%
1.97%
Alexa Fluor 488-A
c
Control
HIV-1SF162
HIV-1SF162 + SEVI (100 μg/ml)
HIV-1SF162 + Myr (50 μg/ml)
HIV-1SF162 + SEVI (100 μg/ml) + Myr (50 μg/ml)
The percentages of GFP+ cells
***
d
CEM×174 5.25 M7
TZM-bl
Cell viability (%)
Blank
SEVI
Myr (50 μg/ml)
SEVI + Myr (50 μg/ml)
e
Ehancement of HIV-1NL4-3 infection (folds)
SEVI + + + + + + + + + -
Myr - 50 25 12.5 6.25 3.13 1.56 0.78 0.39 50
(μg/ml)
f
Ehancement of HIV-1SF162 infection (folds)
SEVI + + + + + + + + + -
Myr - 50 25 12.5 6.25 3.13 1.56 0.78 0.39 50
(μg/ml)

(See figure on previous page.)
Fig. 6 Myricetin reduces the activity of SEVI in enhancing HIV-1 infection. CEMx174 5.25M7 cells infected by HIV-1 $_{SF162}$ in the presence or absence of SEVI (100 μg/ml) and myricetin (50 μg/ml) were imaged by fluorescence microscopy (**a**) and flow cytometric analysis (**b**). **c** The percentages of GFP^{+} CEMx174 5.25 M7 cells from multiple experiments. **d** Cell viability assay of CEMx174 5.25 M7 and TZM-bl cells with or without SEVI (100 or 50 μg/ml) or/and myricetin (50 μg/ml). SEVI (50 μg/ml) was incubated with myricetin at various concentrations (50, 25, 12.5, 6.25, 3.13, 1.56, 0.78 and 0.39 μg/ml). The mixtures were centrifuged to remove soluble myricetin. The pellets were resuspended in the original volume of medium and mixed with CCR5-tropic HIV-1_{SF162} (**e**) or CXCR4-tropic HIV-1_{NL4-3} (**f**). The luciferase activities of the cultures were measured at 72 h post-infection. Average values (± SD) were calculated from triplicate measurements; the data shown represent one representative trial of three independent experiments. One-way ANOVA with Dunnett's post hoc multiple comparisons test was used to statistically analyze differences between samples containing SEVI alone and samples containing SEVI and myricetin (*$p<0.05$; **$p<0.01$, ***$p<0.001$)

at the highest concentration in parallel with all infection assays under the exact same conditions. As shown in Fig. 6d, myricetin at 50 μg/ml exhibited no cytotoxicity toward CEMx174 5.25 M7 and TZM-b1 cells, with almost 100% cell viability being observed.

Interaction of SEVI with virions is affected by myricetin [30]

The effects of myricetin on the virus-binding ability of SEVI were examined using a virus pull-down assay. The results showed that SEVI fibrils alone bound to more than 51.6% of the input HIV-1 virions, including HIV-1_{SF162} and HIV-1_{NL4-3} infectious clones. Myricetin significantly abrogated the binding of all tested HIV-1 virions to SEVI fibrils in a dose-dependent manner (Fig. 7a–c), indicating that myricetin influences SEVI's activity in enhancing HIV-1 infection by preventing the formation of virion-amyloid complexes. The positive surface charge of SEVI fibrils facilitates virion attachment to target cells and enhances HIV infection by serving as a polycationic bridge [7]. Thus, we determined whether negatively charged myricetin attenuates the interaction of SEVI with HIV-1 virions by affecting the cationic properties of SEVI fibrils. Zeta potential is commonly employed to quantitate the magnitude of charge of materials; such charge is an important indicator of the stability and degree of electrostatic repulsion of colloidal dispersions [31, 32]. Myricetin significantly decreased the zeta potential of SEVI fibrils in a dose-dependent manner (Fig. 7d), indicating that it neutralized their surface positive charge.

Myricetin antagonizes the infection-enhancing properties of human semen

Multiple studies have reported that semen increases HIV-1 infection in vitro and that semen-mediated promotion of HIV-1 infection correlates with the levels of amyloidogenic fragments in semen. The combined effects of myricetin on fibril architecture and the formation of fibril–virion complexes led us to investigate whether myricetin possibly diminishes the infection-enhancing properties of the amyloid fibrils present in human semen. Figure 8a shows that after agitation in the absence of myricetin, seminal fluid (SE-F) samples displayed a slight increase in fluorescence intensity, indicating the formation of new seminal fibrils. Remarkably, myricetin slightly abrogated the new formation of seminal fibrils in SE-F. To determine whether myricetin depletes or affects semen of endogenous amyloids, we further measured the effects of myricetin using a sensitive TEM assay, the results of which showed weak inhibition of seminal fibril formation by myricetin at a final concentration of 40 μg/ml after agitation of 8 h (Fig. 8b). We further investigated whether myricetin affects the role of endogenous seminal fibrils in enriching virions. Based on fluorescence confocal microscopy, myricetin markedly inhibited the capture of HIV-1 virions by endogenous seminal fibrils (Fig. 8c). In our HIV-1 infection assay, semen was first incubated with myricetin for 1 h, and the sample was then centrifuged to remove the remaining free myricetin to avoid any direct activity against HIV. The treatment of cells with 2.5% SE-F in the absence of myricetin enhanced HIV-1 infection by 15- to 16-fold. In contrast, seminal fibrils alone promoted aggregation of virions and fibril–virion complex formation. However, the addition of myricetin decreased the enhancement of infection by SE-F in a dose-dependent manner, suggesting that myricetin antagonizes semen-mediated infection enhancement by HIV-1_{SF162} (Fig. 8d) and HIV-1_{NL4-3} (Fig. 8e).

Myricetin shows low cytotoxicity in vitro

The potential cytotoxic effects of myricetin on HIV target cells (TZM-bl and CEMx174 5.25 M7 cells) and reproductive tract epithelial cells (VK2/E6E7 and Ect/E6E7 cells) were assessed using MTT (3-[4,5-dimethyl-2-thiazolyl]-2,5-diphenyl-2*H*-tetrazolium bromide) or XTT [2,3-bis(2-methoxy-4-nitro-5-sulfophenyl)-5-(phenylamino)carbonyl-2*H*-tetrazolium hydroxide] and tris (2,2-bipyridyl)dichlororuthenium(II)($Ru(bpy)_3^{2+}$) assays. Myricetin displayed low cytotoxicity in vitro toward all cell lines tested, with 50% cytotoxic concentration (CC_{50}) values ranging from 95.6 to 107.2 μg/ml (Fig. 9a–d). It should be noted that the final concentrations of myricetin used in these experiments were far below the CC_{50} values. At a concentration of 50 μg/ml, myricetin displayed no obvious cytotoxicity toward

Fig. 7 Myricetin blocks the interaction of SEVI with HIV-1 virions. SEVI samples (200 μg/ml) were first incubated with graded concentrations of myricetin for 30 min. The samples were then centrifuged at 12,000 rpm for 5 min; the supernatant was removed, and the pellets were resuspended and mixed with HIV-1_{SF162} (**a**) and HIV-1_{NL4-3} (**b**). These mixtures were centrifuged, and the p24 antigens present in the pellet were evaluated using a p24-antigen ELISA. **c** The zeta potential of SEVI fibrils in the presence or absence of myricetin was measured using Zeta Nanosizer. Average values (± SD) were calculated from triplicate measurements; the data shown represent one representative trial of three independent experiments. One-way ANOVA with Dunnett's post hoc multiple comparisons test was used to statistically analyze the differences between samples containing SEVI alone and samples containing SEVI and myricetin (*$p<0.05$; **$p<0.01$, ***$p<0.001$)

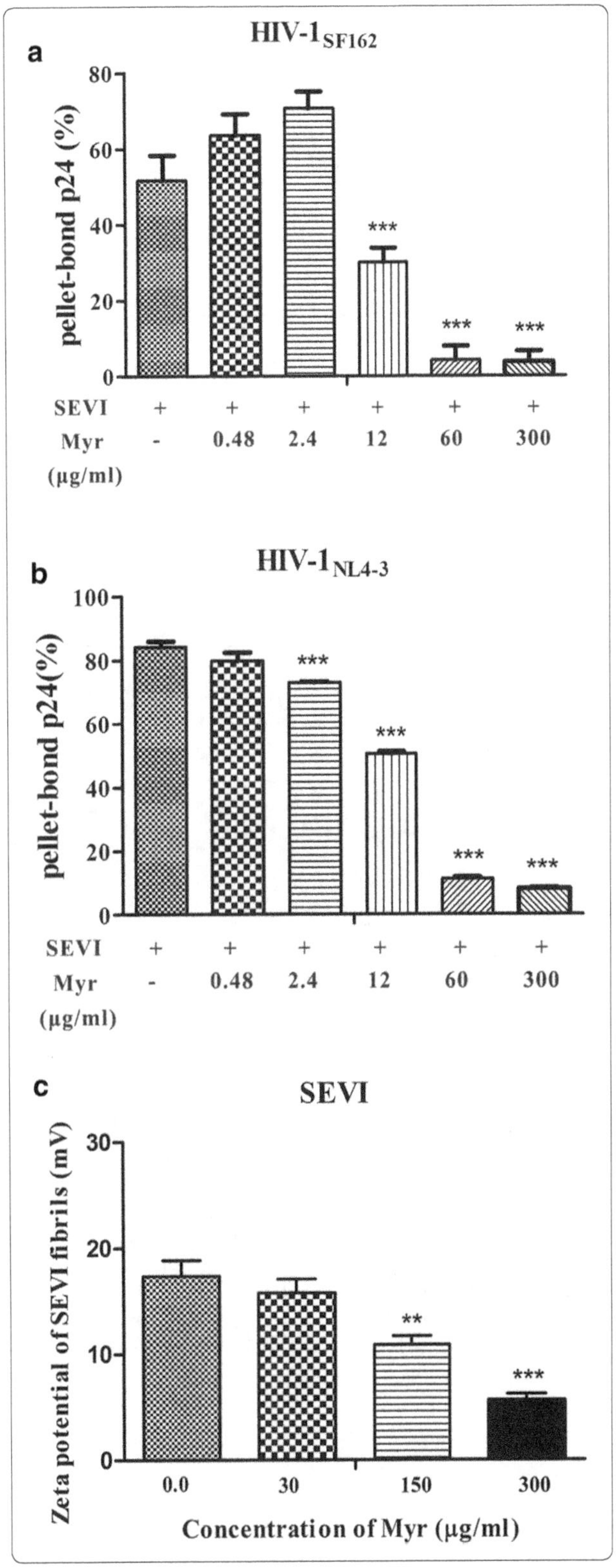

CEMx174 5.25 M7 or TZM-b1 cells (Fig. 6d). We also assessed direct inhibition of HIV-1_{SF162} infection by myricetin alone and obtained an IC_{50} of 1.952 μg/ml (Fig. 9e). The selectivity index ($SI = CC_{50}/IC_{50}$) ranged from 49 to 55, indicating that myricetin might be safe for use in patients (Fig. 9f).

Combinations of myricetin and ARV drugs in 1% SE-F display synergistic effects against HIV-1 infection

In addition to binding to seminal amyloid fibrils, myricetin also appears to directly inhibit HIV-1 infection by targeting HIV integrase [18]. Accordingly, we determined the overall complementary effects of myricetin combined with various anti-retroviral (ARV) drugs on transmission of HIV-1 through semen, which is the most common route. The capacity of myricetin and ARV drugs alone or in combination to prevent infection of TZM-bl cells by 100 $TCID_{50}$ of HIV-1_{SF162} in 1% SE-F was determined using luciferase assays. The ARV drugs used included an HIV entry inhibitor (maraviroc, MAR), nucleoside reverse transcriptase inhibitors (NRTIs) (zidovudine (AZT) and tenofovir (TNF)), non-nucleoside reverse transcriptase inhibitors (NNRTIs) (nevirapine (NVP) and efavirenz (EFV)) and an integrase inhibitor (raltegravir, RAL). The data in Table 1 show that combining myricetin with any of the above ARV drugs produced strong synergistic and complementary effects against infection by HIV-1_{SF162} in 1% SE-F. The observed 50% competitive index (CI_{50}) values ranged from 0.051 to 0.450, and the IC_{50} values of the individual drugs in each combination were reduced by approximately 2.91- to 154.87-fold. The synergistic effects of these combinations were also assessed intuitively, as shown in Fig. 10. The strongest

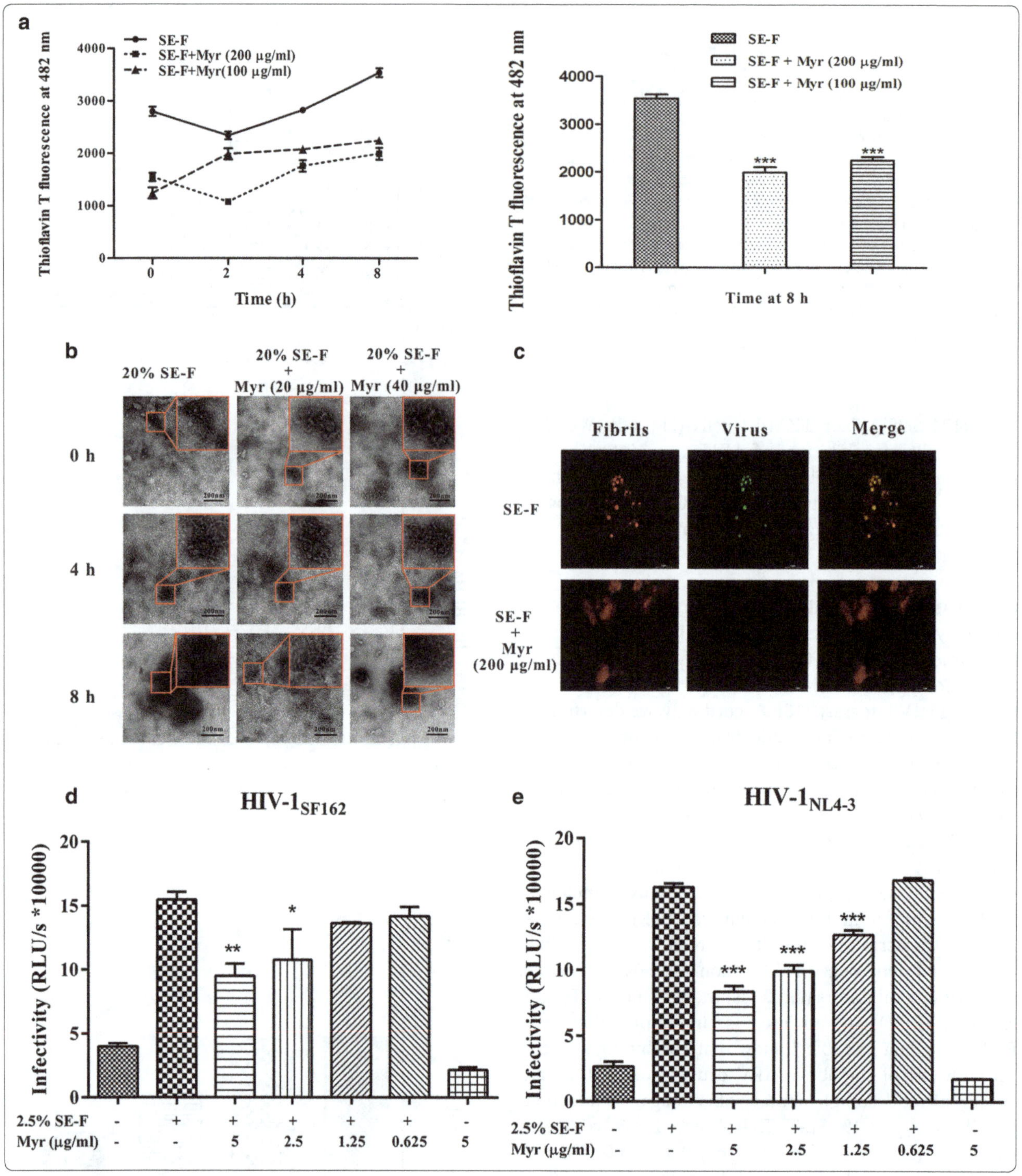

synergism was observed when myricetin was combined with the NRTI TNF: $CI_{50} = 0.051$, and dose reductions in the IC_{50} values of myricetin and TNF of approximately 24- and 107-fold, respectively, were observed (Table 1, Fig. 10). This combination is expected to exert synergistic antiviral effects and to help prevent the development of drug resistance during the prevention and treatment of HIV infection.

(See figure on previous page.)
Fig. 8 Myricetin antagonizes the infection-enhancing properties of human seminal fluid (SE-F). **a** Seminal amyloid fibril formation was antagonized by myricetin in a dose-dependent manner. SE-F samples containing myricetin (200, 100 or 0 µg/ml) were agitated at 1400 rpm at 37 °C for 8 h. Fibril integrity at the indicated time points (0, 4 and 8 h) was assessed using ThT fluorescence. **b** Seminal fibril samples (1:5 dilution) in the presence or absence of myricetin (20 and 40 µg/ml) at various time points (0, 4 and 8 h), as visualized by TEM. The scale bar is 200 nm. **c** Myricetin inhibited the assembly and attachment of HIV-1 virions to seminal fibrils by fluorescence confocal microscopy. The scale bar is 5 µm. The infection-enhancing properties of SE-F on HIV-1_{SF162} (**d**) and HIV-1_{NL4-3} (**e**) infection were attenuated by myricetin. SE-F samples were incubated with myricetin at various concentrations (200, 100, 50, 25 and 0 µg/ml) for 1 h. After centrifugation, the pellets were dissolved in fresh medium at a final dilution of 1:40 and incubated with various HIV-1 infectious clones. Luciferase activities were measured at 72 h post-infection. Average values (±SD) were calculated from triplicate measurements; the data shown represent one representative trial of three independent experiments. One-way ANOVA with Dunnett's post hoc multiple comparisons test was used to statistically analyze differences between samples containing SE-F alone and samples containing SE-F and myricetin (*$p < 0.05$; **$p < 0.01$, ***$p < 0.001$)

Discussion

There is evidence to support that human semen within the context of HIV-1 sexual transmission may have contributed to the failure of microbicides in previous clinical trials [33, 34]. Indeed, the antiviral efficacy of many candidate microbicides is greatly diminished in the presence of semen, and seminal amyloid fibrils have been found to be responsible for the semen-mediated enhancement of HIV infection [4, 9]. Due to their powerful positive charge, amyloid fibrils in semen can effectively capture virus and promote attachment to target cells by neutralizing the inherent electrostatic repulsion between the negative charges on the surfaces of HIV virions and target cells. Thus, a strategy of neutralizing the enhancing activity of seminal amyloid fibrils might be an attractive option for the development of ideal microbicides.

Several agents that block the enhancing activity of seminal amyloid fibrils in HIV infection have been reported. For example, based on its ability to abrogate SEVI-mediated enhancement of HIV-1 infection, the natural ingredient of green tea EGCG appears to be a promising supplement to antiretroviral microbicides [12, 13]. A component of EGCG, gallic acid, is able to directly abrogate the viral enhancement of seminal fibrils and might be able to prevent HIV infection via sexual transmission [35]. Unfortunately, EGCG has poor solubility and poor oral bioavailability and easily undergoes oxidation in vivo, complicating its use in clinical trials [36, 37]. Furthermore, EGCG was recently reported to promote the formation of toxic tau oligomers [38]. The small "molecular tweezer" CLR01 interacts with lysine and arginine residues on SEVI, leading to inhibition of the formation of infectivity-enhancing seminal amyloids and preformed fibril remodeling [29]. Of note, CLR01 interacts promiscuously and nonspecifically with the virus envelope; thus, it is also a broad-spectrum inhibitor of infection by enveloped viruses, including HIV-1, Herpes, Ebola and Zika [39]. The simplest and most logical explanation for the broad range of activity of CLR01 is that the inherent "stickiness" of the molecule confers its promiscuous nature. However, the nonspecific binding of CLR01 to alkaline amino acids in the human body decreases the suitability of CLR01 as a therapy. Therefore, the development of effective, affordable and specific semen fibril antagonists is urgently needed.

The available literature verifies that myricetin possesses the potential to inhibit aggregation of the amyloid fibrils that play an important role in disease. In this work, we explored the potential impact of myricetin on the aggregation of seminal amyloid fibrils. First, myricetin at a concentration of 75 µg/ml was found to completely inhibit amyloidogenesis by peptide PAP248–286 in vitro (Fig. 1a, b); myricetin at a concentration of 200 or 40 µg/ml partially inhibited the formation of new amyloid fibrils in semen, as based on ThT and TEM assays (Fig. 8a, b). The ability of seminal fibrils to cause virion enrichment was inhibited by myricetin at a final concentration of 200 µg/ml (Fig. 8c). Second, myricetin was shown to be capable of blocking the enhancement of viral infection by abrogating interaction between SEVI and HIV-1 virions (Figs. 6, 7). It was clearly demonstrated that an SE-F sample agitated at 37 °C for 8 h significantly lost HIV-enhancing activity compared to the SE-F sample with no agitation [11, 40]. Correspondingly, the ability of semen to enhance HIV-1 infection after the addition of myricetin for 1 h incubation was decreased in a dose-dependent manner (Fig. 8d, e).

It is noteworthy that myricetin is known as a strong inhibitor of HIV reverse transcriptase and integrase, and it is possible that all of the inhibition described above is simply due to the known ability of myricetin to inhibit HIV infection, as opposed to its effects on SEVI. Indeed, we plan to compare the inhibitory activities of myricetin on HIV infection in the absence and presence of SEVI and semen. Regardless, one issue with such a comparison is that SEVI- or semen-mediated enhancement of HIV-1 infection occurs only at low infectivity [5], with myricetin completely preventing HIV infection, even very low concentrations. To avoid the direct activity of myricetin

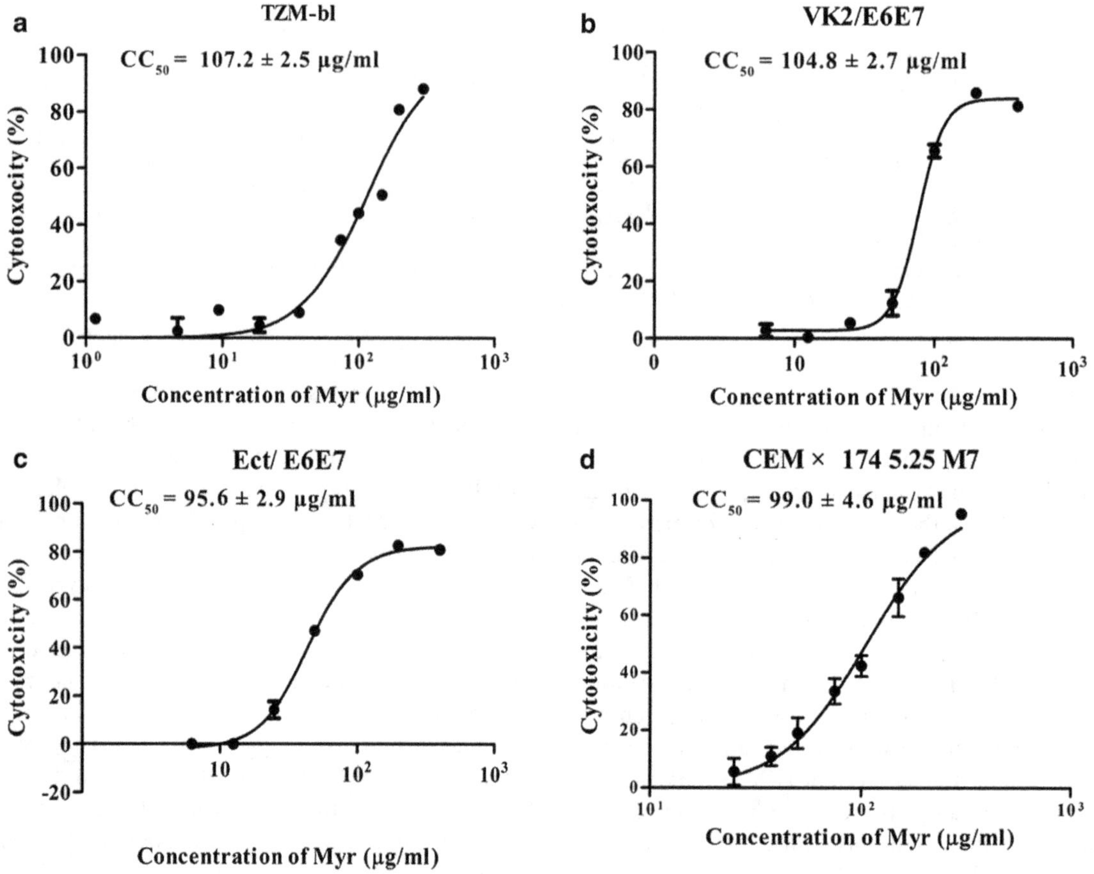

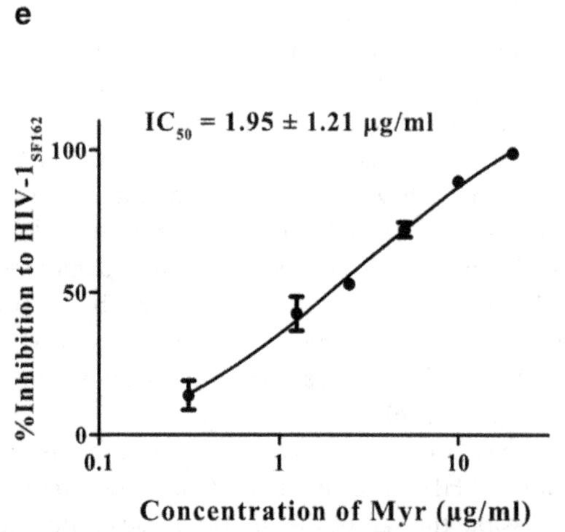

f

Cells	CC_{50} (µg/ml)	CC_{90} (µg/ml)	SI
TZM-b1	107.2±2.5	353.1±30.4	55.0
VK2/E6E7	104.8 ± 2.7	177.6 ± 2.5	53.7
Ect/E6E7	95.6 ± 2.9	170.3 ± 6.6	49.0
CEMx174 5.25 M7	99.0±4.6	256.9±20.7	50.7

Fig. 9 Cytotoxicity of myricetin in vitro. **a** TZM-b1, **b** VK2/E6E7, **c** Ect/E6E7 and **d** CEMx174 5.25 M7 cells. The concentrations of myricetin for (**a**) and (**d**) were 300, 200, 150, 100, 75, 50, 37.5, and 25 µg/ml, for (**b, c**) were double dilution from 400 to 6.25 µg/ml. The MTT assay was used to evaluate adherent cells, and the XTT assay was used for cells in suspension. **e** %Inhibition of HIV-1_{SF162} infection by myricetin alone without SEVI. The indicated concentrations of myricetin were incubated with the HIV-1_{SF162} infectious clone for 30 min. The mixtures were then added to prepared TZM-b1 cells. Luciferase activities were measured at 72 h post-infection. Each sample was tested in triplicate, and the data are presented as the mean ± SD. **f** CC_{50}, CC_{90} and SI values of myricetin against all tested cells. The assay was performed in triplicate; the data are presented as the mean ± SD

Table 1 Combination index (CI) and dose-reduction values for inhibition of HIV-1$_{SF162}$ infection determined by combining myricetin with ARVs in semen[a]

Drug combination, % inhibition	CI[b]	Mean value for[a]:					
		Myr			ARVs		
		Conc. (ng/ml)		Dose reduction	Conc. (ng/ml)		Dose reduction
Conc. (molar ratio)		Alone	Mixture		Alone	Mixture	
Myr:AZT (1571:1)							
50	0.331	600.52	105.97	5.67	0.37	0.06	6.17
90	0.145	16,962.76	1626.10	10.43	17.81	0.87	20.47
Myr:EFV (3142:1)							
50	0.450	2040.81	654.58	3.12	0.80	0.10	8.00
90	0.466	9407.22	2299.93	4.09	1.64	0.36	4.56
Myr:MAR (2095:1)							
50	0.275	1916.80	298.25	6.43	1.92	0.23	8.35
90	0.234	14,712.25	2335.84	6.30	24.03	1.80	13.35
Myr:RAL (3142:1)							
50	0.413	1613.13	372.91	4.33	0.46	0.08	5.75
90	0.514	7554.10	2592.53	2.91	3.37	0.58	5.81
Myr:TNF (62.8:1)							
50	0.051	1976.91	82.31	24.02	113.10	1.06	106.70
90	0.151	9440.37	1366.11	6.91	2721.14	17.57	154.87
Myr:NVP (628:1)							
50	0.262	2199.95	183.01	12.02	2.15	0.38	5.66
90	0.437	15,403.41	1709.88	9.01	11.02	3.59	3.07

[a] The data shown are the means of three independent assays performed in triplicate

[b] The CI value reflects the nature of the interaction between compounds

against HIV as much as possible, the mixture of SEVI/semen and myricetin was centrifuged at 12,000 rpm for 5 min to remove free myricetin from the samples. Thus, most of the inhibition caused by myricetin in our study is likely to have resulted from its effects on SEVI. Our findings indicate that myricetin is worth investigating as a possible candidate for the development of an ideal bifunctional microbicide with both anti-HIV and anti-SEVI activities.

According to native acid gel electrophoresis and SPR analyses, myricetin at 12.5 and 6.36 μg/ml was found to be able to bind to the PAP248–286 monomer (Fig. 3a, b). By using a computational molecular docking assay, we found that myricetin binds to specific amino acids in the PAP248–286 monomer, namely, Leu-258, Gln-259, Val-264, Leu-268, Met-271 and Arg-273, through salt bridges, hydrophobic interactions, and hydrogen bonds (Fig. 3e). Van der Waals forces may contribute to the interaction between myricetin and PAP248–286. Interestingly, the six residues identified as being involved in myricetin and PAP248–286 interaction are contained within the reported functional regions of PAP248–286. PAP257–267 is considered to be the central region of PAP248–286, and this amyloidogenic region of the SEVI precursor peptide PAP248–286 has high fibril-forming propensity. Coincidentally, PAP268–271 has been reported to play an important role in promoting SEVI fibrillation [23]. Residues G260–N265 are thought to be involved in the initiation of fibrillation [40], and residues V262–H270 were shown to be the amyloidogenic region for SEVI fibrillation [41]. Therefore, one or several of the residues present at those six sites might be essential for the formation of SEVI fibrils [42]. The results of our study confirmed that wild-type PAP257–273, as well as PAP257–273 mutants containing Leu258A, Gln259A, Leu268A and Arg273A, readily formed amyloid fibrils (Fig. 3f). Notably, Val264A and Met271A mutations of PAP257–273 exhibited attenuated aggregation during fibril formation, indicating that the Val264 and Met271 residues of PAP248–286 may play a critical role in the assembly of PAP248–286 into fibrils. Myricetin clearly inhibited fibril formation by wild-type PAP257–273 and by four of the PAP257–273 mutants in a dose-dependent manner (Fig. 3g). These results suggest that inhibition of SEVI fibril formation by myricetin might be caused by the binding of myricetin to Val264 and Met271 of PAP248–286.

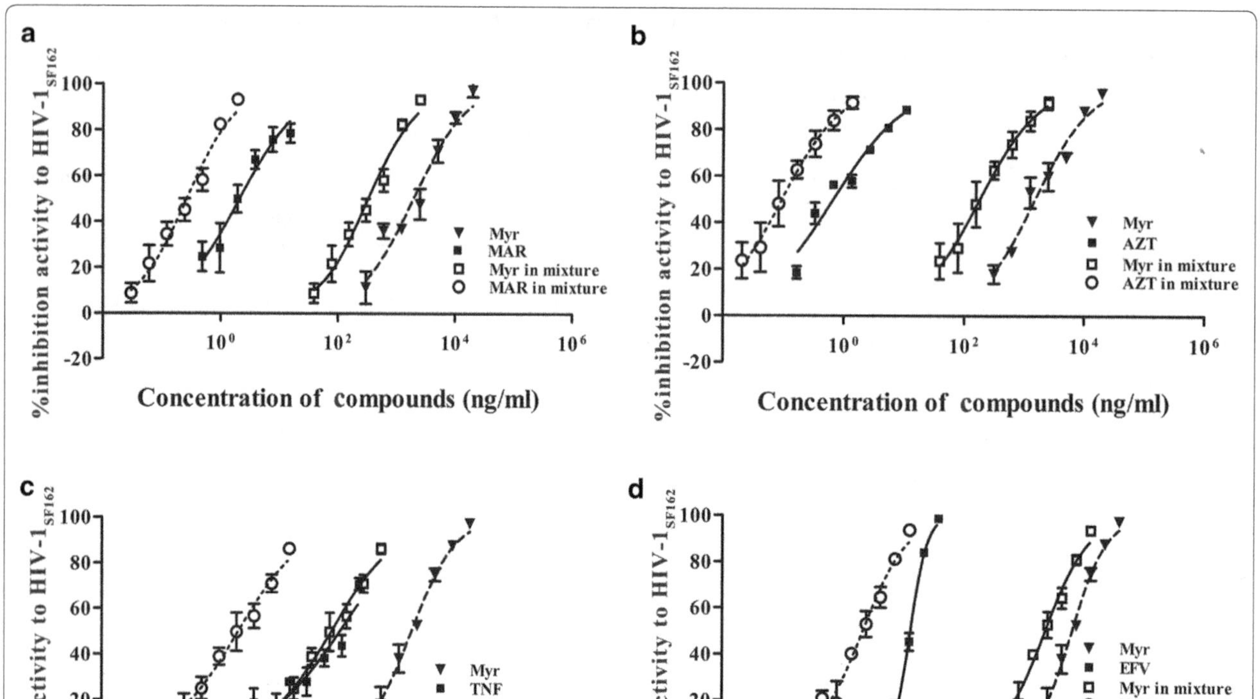

Fig. 10 Synergism achieved by combining myricetin with ARVs for inhibition of infection by HIV-1_{SF162} in semen. Compounds were examined at fixed molar ratios individually and in combination. **a** Myr and MAR (2095:1); **b** Myr and AZT (1571:1); **c** Myr and TNF (62.8:1); **d** Myr and EFV (3142:1); **e** Myr and NVP (628:1); **f** Myr and RAL (3142:1). Luciferase activities were measured at 72 h post-infection. Each sample was tested in triplicate, and the data are presented as the mean ± SD

It is generally recognized that the formation of mature amyloid fibrils involves three stages: (i) monomers are first assembled into oligomers; (ii) fibrils evolve from protofibrils after nucleation; and (iii) the fibrils rapidly aggregate until equilibrium is reached [43]. Continuing investigations suggest that an oligomeric form of amyloid proteins plays a key role in disease causation and that low-*n*-order oligomers are especially important [28]. It would therefore be expected that the most efficacious therapeutic agents would be those that block the early assembly processes associated with amyloid protein oligomerization. Using a PICUP assay, we found that myricetin efficiently abrogated oligomerization of PAP248–286 in a dose-dependent manner (Fig. 4a). CD data also confirmed that myricetin at 5 μg/ml successfully inhibited the formation of β-sheet oligomers by PAP248–286 monomers (Fig. 1e).

Notably, myricetin decreased the zeta potential of SEVI in a dose-dependent manner (Fig. 7d). This might reflect a completely independent mechanism of inhibition of SEVI activity that is distinct from fibrillation inhibition. In general, the molecular electron-donating ability of a compound is determined by the highest occupied molecular orbital (HOMO) eigenvalue. Myricetin is a highly negatively charged compound with a HOMO eigenvalue of −5.7 eV [44]. The electrostatic interaction between myricetin and SEVI fibrils might cause myricetin to shield the surface cationic property of SEVI fibrils, leading to competitive inhibition of the binding of SEVI to the virus. In a similar manner, virus pull-down assays confirmed that myricetin inhibits the binding of HIV to SEVI fibrils in a dose-dependent manner (Fig. 7a–c). As shown in Additional file 1: Figure S4, myricetin did not affect the total amount of viral core protein p24, an observation that tends to exclude an antiviral effect of myricetin on the virus particles themselves. The observations are consistent with the idea that myricetin is an integrase inhibitor that affects the stage of viral growth but has no effect on the viral envelope.

Although anionic polymers are effective inhibitors of SEVI activity and HIV infection, they are not ideal candidates as anti-HIV vaginal microbicides because they are highly pro-inflammatory, disrupting the vaginal epithelial barrier [45]. One question that arises is does intravaginal application of anionic myricetin cause inflammation in female genital tract? There is a growing body of data indicating that myricetin displays strong anti-inflammatory activity in a variety of in vitro assays, as well as in both acute and chronic in vivo animal models, through regulation of multiple signaling pathways [15, 46, 47]. Therefore, myricetin with anti-inflammatory, anti-HIV and anti-SEVI activities might be developed as a promising microbicide to reduce the risk of HIV infection.

Previous studies have reported that fresh ejaculate contains semen fibrils; thus, assembly blockers might be inactive [10]. However, it is rare that agents exhibit the ability to degrade highly stable amyloid fibrils. Fortunately, we found that myricetin degraded SEVI fibrils in a dose-dependent manner, resulting in a lack of ability to seed the assembly of soluble PAP248–286 by SEVI (Fig. 5a, b). In addition, CD results showed that myricetin could immediately remodel SEVI into non-templating conformers (Fig. 5c). One of the advantages of using myricetin is that degradation occurs immediately, whereas other reported remodeling inhibitors, including EGCG and CLR01, must be applied after several hours [12, 29]. Another advantage of using myricetin is its broad inhibition of the formation of seminal fibrils including SEVI, SEM1 and SEM2. Consequently, it is preferable to develop semen fibril antagonists that not only antagonize fibrillation but also remodel preformed fibrils. Future studies designed to determine the exact mechanism of action of myricetin with regard to the degradation of fibrils should be conducted.

It has previously been reported that brief exposure of HIV to semen reduced the antiviral efficacy of most microbicides [48]. Notably, myricetin is known to be an inhibitor of HIV-1 integrase, and we found that myricetin can partially weaken the semen-mediated enhancement of viral infection. Moreover, myricetin displays synergistic effects against HIV-1 infection in combination with other ARV agents in semen (Table 1; Fig. 10). As a key ingredient of various human foods and beverages, myricetin is relatively safe when used in patients. Our results also confirmed that myricetin has low cytotoxicity toward all tested cell lines in vitro, with CC_{50} values ranging from 95.6 to 107.2 μg/ml (Fig. 9). Therefore, it is likely that myricetin can be further developed into a candidate for inclusion in a combination microbicide to prevent HIV-1 sexual transmission.

Increasing evidence indicates that endogenous amyloids play natural and biological roles in the human body [49]. It was recently reported that semen amyloids may play a role in reproduction by participating in sperm selection and facilitating spermatozoal clearance [50]. Positively charged SEVI fibrils exert indirect antimicrobial activity by trapping microbial pathogens [51], though it remains unclear how inhibitors of amyloid formation or action may affect the antimicrobial and sperm-clearing natural role of amyloids in semen. To date, no studies have reported the side effects of semen fibril antagonists on the normal physiological function of amyloids. Before a semen fibril-targeted agent is clinically available, its potential effects on the normal physiological function of amyloids should be tested.

Conclusions

In this study, we found that myricetin interferes with various essential steps in the formation and remodeling of SEVI fibrils. Myricetin significantly inhibited all stages of SEVI formation, including early oligomerization and fiber extension, thereby blocking SEVI-mediated enhancement of HIV-1 infection. In addition, myricetin is capable of remodeling seminal amyloid fibrils and antagonizing semen-mediated enhancement of HIV-1 infection. Notably, myricetin displays synergistic effects against HIV-1 infection in semen when combined with other ARV agents and may work synergistically to reduce the rate of HIV transmission. It is likely that myricetin can be developed into a dual-functional microbicide candidate with both anti-HIV-1 and anti-SEVI-enhancing activities for use in preventing sexual transmission of HIV-1.

Methods

Cell lines, plasmids and reagents

TZM-bl, VK2/E6E7, Ect/E6E7 cells were obtained from the National Institutes of Health AIDS Research and Reference Reagent Program. HEK-293 T cells were purchased from American Type Culture Collection (ATCC) (Manassas, VA, USA). CEMx174 5.25M7 cells were kindly provided by Dr. C. Cheng-Mayer. Plasmids encoding CXCR4-tropic HIV-1_{NL4-3} and CCR5-tropic HIV-1_{SF162} infectious clones were kindly provided by Jan Münch of Ulm University (Ulm, Baden-Württemberg, Germany). The peptides PAP248–286, SEM1_{86-107}, SEM2_{86-107} and N36 were synthesized by Scilight Biotechnology LLC (Beijing, China). Myricetin was purchased from Chengdu Must Biotechnology Co., Ltd. (Chengdu, China). ProteoStat Amyloid Plaque Detection Kits were purchased from Enzo Life Sciences (Plymouth Meeting, PA, USA). Thioflavin T (ThT), phlorizin, GST, Congo red, ammonium persulfate, DL-dithiothreitol (DTT), MTT, and XTT were purchased from Sigma (St. Louis, MO, USA). An anti-p24 monoclonal antibody (183-12H-5C), anti-HIV IgG, zidovudine (AZT), raltegravir (RAL) and maraviroc (MAR) were obtained from the National Institutes of Health AIDS Research and Reference Reagent Program. Tenofovir (TNF), nevirapine (NVP) and efavirenz (EFV) were purchased from Promega Biotech Co., Ltd. (Madison, WI, USA). Rabbit anti-PAP IgG was prepared, produced and purified by AbMax Biotechnology (Beijing, China).

Semen and seminal fluid samples

Pooled semen samples were obtained from 20 healthy donors at Nanfang Hospital (Guangzhou, China) after they had provided informed consent to participate in the study. Seminal fluid (SE-F) samples were collected by centrifugation of semen samples for 15 min at 10,000 rpm at 4 °C to remove spermatozoa and debris and were then stored at −20 °C. The research protocols for this study were approved by the Ethical Committee of Nanfang Hospital of Southern Medical University and performed in accordance with relevant guidelines and regulations.

Formation of fibrils by PAP248–286, SEM1_{86-107}, SEM2_{86-107} and formation of fibrils in semen

Solutions of the peptides PAP248–286, SEM1_{86-107} and SEM2_{86-107} (3 mg/ml) were agitated in the presence or absence of myricetin at the indicated concentrations (75, 37.5, 18.75 and 9.375 μg/ml for SEVI or 200, 100, 50 and 25 μg/ml for SEM1/2) for 48 h at 1400 rpm at 37 °C using an Eppendorf Thermomixer (Hamburg, Germany). At the indicated time points (0, 6, 12, 18, 24 and 48 h), aggregates were collected and stored at −20 °C.

SE-F samples were collected and agitated at 1400 rpm at 37 °C in the presence or absence of myricetin at the indicated concentrations (200 and 100 μg/ml) for 8 h. At the indicated time points (0, 2, 4 and 8 h), aggregates were collected and stored at −20 °C.

Remodeling of SEVI fibrils

SEVI fibrils (3 mg/ml) were agitated in the presence of myricetin (500, 250, 125 and 0 μg/ml) for 3 days at 37 °C at 1400 rpm. At various time points (1 min, 12, 24, 36, 48, 60 and 72 h), the sample mixtures were tested in ThT, Congo red and CD assays. Myricetin alone (500 μg/ml) was used as a negative control.

ThT fluorescence assay

Ten microliters of sample prepared as described above was mixed with 195 μl of ThT working solution (50 μM). The fluorescence of the mixture was measured using an RF-5301 PC spectrofluorophotometer (Shimadzu) at an excitation wavelength of 440 nm and an emission wavelength of 482 nm [23, 52].

Congo red staining assay

Five microliters of sample were mixed with 200 μl of Congo red reagent obtained from a Congo red kit for 2 min at room temperature (RT). The samples were then centrifuged at 12,000 rpm for 5 min, and the red-colored fiber precipitate was dissolved in 50 μl of dimethyl sulfoxide (DMSO). The absorbance of the solution at 490 nm was recorded using an enzyme-linked immunosorbent assay (ELISA) reader (Tecan, Research Triangle Park, NC, USA) [23, 52].

TEM

Suspensions of the collected samples (the final concentration of PAP248–286 was 500 μg/ml, and the final dilution of SE-F was 1:5) were deposited onto glow-discharged, carbon-coated grids for 2 min. The grids were then negatively stained with 2% phosphotungstic acid for an additional 2 min. The morphology of SEVI or seminal fibrils in the presence and absence of myricetin (16, 8 μg/ml for SEVI or 40, 20 μg/ml for semen) was visualized using an H-7650 TEM (Hitachi Limited, Tokyo, Japan). Myricetin (16 μg/ml) was used as a negative control.

CD

The secondary β-sheet structure of SEVI fibrils (the final concentration of PAP248–286 in the sample was 300 μg/ml) in the presence and absence of myricetin (5 μg/ml of the formation or 50, 12.5 μg/ml of the remodeling) was determined by CD spectroscopy in the "far-UV" spectral region (198–260 nm) using a Jasco 715 spectropolarimeter equipped with a thermostat-controlled cell housing and cells with a 1-mm path length (Jasco Inc., Japan). Each CD spectrum was collected three times [53].

HIV-1 infection assay

Enhancement of SEVI fibrils and SE-F samples in the presence and absence of myricetin was measured using HIV-1 infection assays as previously described [30, 54]. First, SE-F samples were collected as described above. Next, the SE-F samples were incubated with myricetin at various concentrations (200, 100, 50, 25 and 0 μg/ml) for 1 h at 37 °C; SEVI fibrils (50 μg/ml) were first incubated with graded concentrations of myricetin (50, 25, 12.5, 6.25, 3.13, 1.56, 0.78, 0.39 and 0 μg/ml) for 30 min. To eliminate the anti-HIV activity of myricetin itself as much as possible, the SEVI or SE-F and myricetin mixtures were centrifuged at 12,000 rpm for 5 min to remove free myricetin. The pellets of SE-F samples were dissolved in fresh medium at a final dilution of 1:40. Thus, the final concentrations of myricetin in 2.5% SE-F were 5, 2.5, 1.25, 0.625 and 0 μg/ml. The pellets of SE-F and SEVI fibrils in fresh medium were then incubated with CXCR4-tropic HIV-1_{NL4-3} and CCR5-tropic HIV-1_{SF162} infectious clones at RT for 10 min. Subsequently, 100 μl of the fibril–virus mixture was added to TZM-bl cells, which were cultured at 37 °C for 3 h, after which the culture supernatant was replaced with fresh medium. At 3 days post-infection, the cells were collected, washed and lysed with lysing reagent, and the luciferase activity was measured using a luciferase assay kit (Promega, Corp., Madison, WI, USA). HIV and myricetin were used as negative controls. The viability of TZM-bl cells with or without SEVI fibrils was detected in parallel with all infection assays under the exact same conditions.

The direct anti-HIV activity of myricetin on HIV-1_{SF162} infection was detected. First, 1×10^5/ml TZM-bl cells were seeded in 96-well plates at 37 °C overnight, and 100 $TCID_{50}$ HIV-1 virus was incubated with myricetin (20, 10, 5, 2.5, 1.25, 0.625, 0.3125 and 0 μg/ml) at graded concentrations at 37 °C for 30 min. The mixture was then added to cells and further incubated for 72 h. The cells were collected, washed and lysed with the lysing reagent included in a luciferase assay kit. IC_{50} values were calculated using CalcuSyn software [55], kindly provided by T. C. Chou (Sloan-Kettering Cancer Center, New York, NY).

Cytotoxicity assay in vitro

The cytotoxicity of myricetin toward TZM-bl, VK2/E6E7 and Ect/E6E7 cells was evaluated using MTT assays, and the cytotoxicity of myricetin toward CEMx174 5.25 M7 cells was evaluated using an XTT assay. Briefly, approximately 90% confluent cells were plated in 96-well plates at 1×10^5/ml, and the plates were incubated at 37 °C overnight. Different concentrations of myricetin were added, and the cells were incubated for an additional 48 h at 37 °C. For the MTT assays, the culture supernatant was discarded, and 100 μl of 0.5 mg/ml MTT solution was added to the cells. After incubating the cells for an additional 4 h, the cell-free supernatant was removed, and the formazan crystals formed were dissolved in 150 μl of DMSO. The absorbance of the resulting solution at 570 or 450 nm was measured using an ELISA reader. For the XTT assay, 50 μl of XTT solution (1 mg/ml) containing 0.02 μM phenazine methosulfate (PMS) was added. After 4 h, the absorbance at 450 nm was measured using an ELISA reader. CC_{50} was calculated using CalcuSyn software.

Fluorescence microscopy and flow cytometry assays

Lymphoid CEMx174 5.25M7 cells that had been stably transduced with a long terminal repeat (LTR)-luciferase and LTR-green fluorescent protein cassette were used as target cells [5]. SEVI (100 μg/ml) was incubated with myricetin (50 μg/ml) at 37 °C for 30 min; the samples were then centrifuged at 12,000 rpm for 5 min to remove free myricetin. The resulting pellets were dissolved in fresh medium and mixed with an equal volume of HIV-1_{SF162} for 5 min at RT. The mixture was then added to CEMx174 5.25M7 cells, which were cultured at 37 °C for 48 h. The fluorescence intensity of the resulting cell suspensions was observed by fluorescence microscopy (Nikon, Japan). The cells were washed, resuspended and analyzed by flow cytometry (BD FACSCanto TM^{II}). HIV and myricetin were used as negative controls. The viability of CEMx174 5.25M7 cells with or without SEVI fibrils was detected in parallel with all infection assays under the exact same conditions.

Seeding function of preformed SEVI

Preformed SEVI seeds (200 μg/ml) were added to 2 mg/ml PAP248–286 peptide solution in the presence or absence of myricetin (500, 200, 100, 80, 40, 20, 10 and 1 μg/ml), and the mixture was agitated at 1400 rpm at 37 °C. At the indicated time points (0, 2, 4, 6, 8, 10, and 12 h), samples were collected and subjected to ThT fluorescence assays as described above [29]. The IC_{50} values of myricetin for the formation of PAP248–286 with or without SEVI seeds at 48 h time points were calculated using CalcuSyn software.

Virus pull-down assays

SEVI fibrils (200 μg/ml) were incubated with myricetin at various concentrations (300, 60, 12, 2.4, 0.48 and 0 μg/ml) at 37 °C for 30 min. The mixture was then challenged with 100 ng/ml HIV-1 virions (HIV-1_{NL4-3} and HIV-1_{SF162}) for an additional 30 min. The samples were centrifuged at 12,000 rpm for 5 min, and the supernatants were removed. p24 antigens present in the pellets were detected by ELISA as previously described [56]. Total p24 antigens of the mixtures with myricetin at various concentrations and 100 ng/ml HIV-1 virions (HIV-1_{NL4-3} and HIV-1_{SF162}) without SEVI fibrils were detected in parallel using virus pull-down assays under the exact same conditions.

Confocal microscopy

eGFP-labeled HIV-1 virions (R5-tropic) were produced by transfecting HEK-293T cells with proviral DNA expression plasmids and eGFP-Vpr plasmids by polyethylenimine (PEI) transfection. Semen samples were stained with a Proteostat dye contained in ProteoStat Amyloid Plaque Detection Kit at RT [10]. After staining, the fibril samples were incubated 1:1 with eGFP-labeled HIV-1 virions (100 ng/ml p24) with or without myricetin (200 μg/ml) at 37 °C for 3 h. The samples were then centrifuged at 5000 rpm for 3 min and imaged using a laser scanning Nikon A1 confocal microscope (Nikon, Japan). Proteostat and eGFP were excited using a 561 nm and a 488 nm laser line, respectively, and the emissions were collected using appropriate beam splitters.

Zeta potential

SEVI (200 μg/ml) was incubated in the presence or absence of myricetin at various concentrations (300, 150 and 30 μg/ml) for 30 min at RT. The samples were then centrifuged at 12,000 rpm for 5 min to remove free myricetin. The resulting pellets were resuspended in 1 mM KCl buffer, and the surface zeta potential of the resuspended samples was measured using a Zeta Nanosizer (Malvern Instruments, UK) [29].

WB assay

PAP248–286 monomers (100 μg/ml) were incubated with serially diluted myricetin (12.5, 6.25, 3.13 and 1.56 μg/ml) at 37 °C for 30 min, and the samples were centrifuged at 5000 rpm for 15 min. The PAP248–286 peptides remaining in the supernatant were analyzed by electrophoresis through 10% polyacrylamide continuous native gels as previously described [52]. The gels were subjected to WB analysis to specifically identify the remaining PAP248–286 peptides using rabbit anti-PAP IgG.

Photo-induced cross-linking of unmodified proteins (PICUP)

Samples were chemically cross-linked using PICUP as previously reported [28]. First, PAP248–286 monomer (500 μg/ml) was incubated with various concentrations of myricetin (20, 10, 5 and 2.5 μg/ml) at 37 °C for 30 min. Next, 1 μl of 40 mM tris(2,2-bipyridyl) dichlororuthenium(II)($Ru(bpy)_3^{2+}$) and 1 μl of 800 mM ammonium persulfate were added to 18 μl of the mixed sample. The mixture was then exposed to visible light for 5 s, and the cross-linking reaction was terminated by the addition of 2 μl of 5 M DTT. Non-cross-linked PAP248–286 monomer served as a negative control. The samples were separated by 10–20% gradient tricine sodium dodecyl sulfate-polyacrylamide gel electrophoresis (SDS-PAGE), and Coomassie blue staining was used to determine the frequency distribution of monomers and oligomers of PAP248–286. An irrelevant protein (GST) and an irrelevant flavonoid (phlorizin) were used as negative controls.

Computational docking analysis

The binding of PAP248–286 (RCSB Protein Data Bank (PDB) No. 2L3H) to myricetin was analyzed using the extensible molecular visualization UCSF Chimera program of AutoDock software [42, 57]. High-quality atomic charges for computer simulations of organic molecules in polar media can be generated by a novel charge model called AM1-BCC [58]. Hydrogen atoms were added using the Dock Prep module. The protein–ligand interaction online analysis tool Protein–Ligand Interaction Profiler [59] was applied, and images were generated by PyMOL software [31].

SPR analysis

The affinity of myricetin binding to PAP248–286 was measured using the BIAcore T100 system (GE Healthcare, Sweden) as previously described [60]. After PAP248–286 (50 μg/ml) was attached to the surface of a CM5 sensor chip as a cross-linker, myricetin (6.36, 3.18, 1.59, 0.80, 0.40 and 0.20 μg/ml) was injected at a flow rate of 20 μl/min, with a contact time of 2 min and a dissociation time of 2.5 min. The running buffer was water. The chip platform was regenerated with 10 mM HEPES, 150 mM NaCl, and 0.01% vol/vol Tween 20 (pH 7.4) and washed with the running buffer. A binding affinity (K_D) value was calculated using BIAcore software. An irrelevant peptide (N36) and an irrelevant flavonoid (phlorizin) were used as negative controls.

Combination treatment in SE-F

The inhibitory activity of myricetin and other ARV drugs and combinations on infection by HIV-1_{SF162} infectious clones was determined as previously described [30, 31, 61]. Briefly, HIV-1_{SF162} at 100 $TCID_{50}$ (50% tissue culture infective dose) was incubated in the presence or absence of various concentrations of myricetin, ARV drugs or combinations thereof in SE-F at 37 °C for 30 min. SE-F was diluted to 1:100 to prevent cytotoxicity, which slightly enhanced HIV infection. The mixture of myricetin, ARV drugs or combinations in 1% SE-F was added to TZM-bl cells, and the cells were incubated for 3 h at 37 °C; the medium was then replaced with fresh medium. Compounds were examined at fixed molar ratios in combination according to their individual IC_{50} values. At 72 h post-infection, the luciferase activity of the cell cultures was measured as described above. The effective concentrations resulting in 50 and 90% inhibition (EC_{50} and EC_{90}) were calculated using CalcuSyn software. The data were analyzed for cooperative effects using the Chou-Talalay method. Combination index (CI) values were calculated using the CalcuSyn program. The CI value reflects the nature of the interaction between compounds. CI values of <1, $=1$ and >1 indicate synergy, additivity and antagonism, respectively. In detail, CI < 0.10 indicates very strong synergism, 0.10–0.30 indicates strong synergism, 0.30–0.70 indicates synergism, 0.70–0.85 indicates moderate synergism, and 0.85–0.90 indicates slight synergism. Dose reductions were calculated as the ratio of the EC_{50} values of the compounds when used alone and when used in combination.

Statistical analysis

Statistical analysis of the experimental data was performed using a one-way analysis of variance (ANOVA) test in GraphPad Prism 5.0 (San Diego, CA, USA). A p value of <0.05 was regarded as statistically significant; the probability level is indicated by single or multiple asterisks (*) (*$p<0.05$; **$p<0.01$; ***$p<0.001$). All values represent the mean ± SD (standard deviation) of at least three measurements.

Additional file

Additional file 1: Figure S1. Myricetin inhibits other seminal amyloid fibril formation, as shown by ThT assays. (**a**) SEM186-107; (**b**) SEM286-107. Peptide (3 mg/ml) was incubated with myricetin (200, 100, 50 and 10 µg/ml) and agitated at 1,400 rpm at 37 °C. Then samples were collected and monitored by ThT. Average values (± SD) were calculated from triplicate measurements, and the data represent one representative trial of three independent experiment. **Figure S2.** Amyloid fibril samples display loss of enhancement of HIV-1 infection in the presence of myricetin. The raw luciferase activities of mixed SEVI fibril samples prepared in the presence or absence of myricetin with HIV-1_{SF162} (**a**) and HIV-1_{NL4-3} (**b**) infectious clones. The values shown here represent the mean ± SD (n = 3). One-way ANOVA with Dunnett's post hoc multiple comparisons test was used to statistically analyze the differences between samples containing PAP248-286 alone and samples containing PAP248-286 and myricetin (*$p < 0.05$; **$p < 0.01$, ***$p < 0.001$). **Figure S3.** SEVI (50 µg/ml) was incubated with myricetin at various concentrations (50, 25, 12.5, 6.25, 3.13, 1.56, 0.78 and 0.39 µg/ml). The mixtures were washed one to five times with PBS buffer and centrifuged to remove soluble myricetin. The pellets were resuspended in the original volume of medium and mixed with CCR5-tropic HIV-1_{SF162} (**a**) or CXCR4-tropic HIV-1_{NL4-3} (**b**). The luciferase activities of the cultures were measured at 72 h post-infection. Average values (± SD) were calculated from triplicate measurements; the data shown here represent one representative trial of three independent experiments. One-way ANOVA with Dunnett's post hoc multiple comparisons test was used to statistically analyze the differences between samples containing SEVI alone and samples containing SEVI and myricetin (*$p < 0.05$; **$p < 0.01$, ***$p < 0.001$). **Figure S4.** The total p24 antigens of the mixtures of myricetin at various concentrations and 100 ng/ml HIV-1 virions without SEVI fibrils were detected in parallel with virus pull-down assays under the exact same conditions. (**a**) HIV-1_{SF162} and (**b**) HIV-1_{NL4-3}. Average values (± SD) were calculated from triplicate measurements, and the data represent one representative trial of three independent experiment.

Authors' contributions

LL and SL conceived the idea and designed the research. RR, SY, BL, LM, JW, XZ and FL performed the research. RR, LM, SL and LL analyzed the data and wrote the manuscript. All authors read and approved the final manuscript.

Author details

[1] Guangdong Provincial Key Laboratory of New Drug Screening, Guangzhou Key Laboratory of Drug Research for Emerging Virus Prevention and Treatment, School of Pharmaceutical Sciences, Southern Medical University, 1838 Guangzhou Avenue North, Guangzhou 510515, Guangdong, China. [2] Department of Pharmacy, The Seventh Affiliated Hospital of Sun Yat-sen University, Shenzhen 518107, China.

Acknowledgements

We thank Dr. Shibo Jiang of Fudan University, China for providing technical help. We thank Chunyan Wang of Nanfang Hospital of Southern Medical University, China, for the kind gift of human semen samples. This study was supported by the National Natural Science Foundation of China (81673481 to Lin Li).

Competing interests

The authors declare that they have no competing interests.

Consent for publication

Not applicable.

References

1. Lu L, Yu F, Cai L, Debnath AK, Jiang S. Development of small-molecule HIV entry inhibitors specifically targeting gp120 or gp41. Curr Top Med Chem. 2016;16:1074–90.
2. UNAIDS. Global AIDS Update 2017; http://aidsinfo.unaids.org/. UNAIDS. Core Epidemiology Slides; June 2017.
3. Sabatte J, Remes Lenicov F, Cabrini M, Rodriguez Rodrigues C, Ostrowski M, Ceballos A, Amigorena S, Geffner J. The role of semen in sexual transmission of HIV: beyond a carrier for virus particles. Microbes Infect. 2011;13:977–82.
4. Zirafi O, Kim KA, Roan NR, Kluge SF, Muller JA, Jiang S, Mayer B, Greene WC, Kirchhoff F, Munch J. Semen enhances HIV infectivity and impairs the antiviral efficacy of microbicides. Sci Transl Med. 2014;6:262ra157.

5. Munch J, Rucker E, Standker L, Adermann K, Goffinet C, Schindler M, Wildum S, Chinnadurai R, Rajan D, Specht A, et al. Semen-derived amyloid fibrils drastically enhance HIV infection. Cell. 2007;131:1059–71.
6. Roan NR, Münch J, Arhel N, Mothes W, Neidleman J, Kobayashi A, Smith-McCune K, Kirchhoff F, Greene WC. The cationic properties of SEVI underlie its ability to enhance human immunodeficiency virus infection. J Virol. 2009;83:73–80.
7. Arnold F, Schnell J, Zirafi O, Sturzel C, Meier C, Weil T, Standker L, Forssmann WG, Roan NR, Greene WC, et al. Naturally occurring fragments from two distinct regions of the prostatic acid phosphatase form amyloidogenic enhancers of HIV infection. J Virol. 2012;86:1244–9.
8. Roan NR, Muller JA, Liu H, Chu S, Arnold F, Sturzel CM, Walther P, Dong M, Witkowska HE, Kirchhoff F, et al. Peptides released by physiological cleavage of semen coagulum proteins form amyloids that enhance HIV infection. Cell Host Microbe. 2011;10:541–50.
9. Castellano LM, Shorter J. The surprising role of amyloid fibrils in HIV infection. Biology (Basel). 2012;1:58–80.
10. Usmani SM, Zirafi O, Muller JA, Sandi-Monroy NL, Yadav JK, Meier C, Weil T, Roan NR, Greene WC, Walther P, et al. Direct visualization of HIV-enhancing endogenous amyloid fibrils in human semen. Nat Commun. 2014;5:3508.
11. Roan NR, Liu H, Usmani SM, Neidleman J, Muller JA, Avila-Herrera A, Gawanbacht A, Zirafi O, Chu S, Dong M, et al. Liquefaction of semen generates and later degrades a conserved semenogelin peptide that enhances HIV infection. J Virol. 2014;88:7221–34.
12. Hauber I, Hohenberg H, Holstermann B, Hunstein W, Hauber J. The main green tea polyphenol epigallocatechin-3-gallate counteracts semen-mediated enhancement of HIV infection. Proc Natl Acad Sci USA. 2009;106:9033–8.
13. Castellano LM, Hammond RM, Holmes VM, Weissman D, Shorter J. Epigallocatechin-3-gallate rapidly remodels PAP85-1, SEM1 (45–107), and SEM2 (49–107) seminal amyloid fibrils. Biol Open. 2015;4:1206–12.
14. Devi KP, Rajavel T, Habtemariam S, Nabavi SF, Nabavi SM. Molecular mechanisms underlying anticancer effects of myricetin. Life Sci. 2015;142:19–25.
15. Semwal DK, Semwal RB, Combrinck S, Viljoen A. Myricetin: a dietary molecule with diverse biological activities. Nutrients. 2016;8:90.
16. Shih YW, Wu PF, Lee YC, Shi MD, Chiang TA. Myricetin suppresses invasion and migration of human lung adenocarcinoma A549 cells: possible mediation by blocking the ERK signaling pathway. J Agric Food Chem. 2009;57:3490–9.
17. Pasetto S, Pardi V, Murata RM. Anti-HIV-1 activity of flavonoid myricetin on HIV-1 infection in a dual-chamber in vitro model. PLoS ONE. 2014;9:e115323.
18. Chaniad P, Wattanapiromsakul C, Pianwanit S, Tewtrakul S. Anti-HIV-1 integrase compounds from *Dioscorea bulbifera* and molecular docking study. Pharm Biol. 2016;54:1077–85.
19. Hirohata M, Hasegawa K, Tsutsumi-Yasuhara S, Ohhashi Y, Ookoshi T, Ono K, Yamada M, Naiki H. The anti-amyloidogenic effect is exerted against Alzheimer's beta-amyloid fibrils in vitro by preferential and reversible binding of flavonoids to the amyloid fibril structure. Biochemistry. 2007;46:2488–99.
20. Zelus C, Fox A, Calciano A, Faridian BS, Nogaj LA, Moffet DA. Myricetin inhibits islet amyloid polypeptide (IAPP) aggregation and rescues living mammalian cells from IAPP toxicity. Open Biochem J. 2012;6:66–70.
21. He J, Wang Y, Chang AK, Xu L, Wang N, Chong X, Li H, Zhang B, Jones GW, Song Y. Myricetin prevents fibrillogenesis of hen egg white lysozyme. J Agric Food Chem. 2014;62:9442–9.
22. Duan JM, Qiu JY, Tan SY, Liu SW, Li L. Semen-derived enhancer of viral infection—a key factor in sexual transmission of HIV. Bing Du Xue Bao. 2012;28:84–8.
23. Chen J, Ren R, Yu F, Wang C, Zhang X, Li W, Tan S, Jiang S, Liu S, Li L. A degraded fragment of HIV-1 Gp120 in rat hepatocytes forms fibrils and enhances HIV-1 infection. Biophys J. 2017;113:1425–39.
24. Frid P, Anisimov SV, Popovic N. Congo red and protein aggregation in neurodegenerative diseases. Brain Res Rev. 2007;53:135–60.
25. Olsen JS, DiMaio JTM, Doran TM, Brown C, Nilsson BL, Dewhurst S. Seminal plasma accelerates semen-derived enhancer of viral infection (SEVI) fibril formation by the prostatic acid phosphatase (PAP248–286) peptide. J Biol Chem. 2012;287(15):11842–9.
26. Le Y, Jiang S, Hu J, Gong W, Su S, Dunlop NM, Shen W, Li B, Ming Wang J. N36, a synthetic N-terminal heptad repeat domain of the HIV-1 envelope protein gp41, is an activator of human phagocytes. Clin Immunol. 2000;96(3):236–42.
27. Elias AK, Scanlon D, Musgrave IF, Carver JA. SEVI, the semen enhancer of HIV infection along with fragments from its central region, form amyloid fibrils that are toxic to neuronal cells. Biochim Biophys Acta. 2014;1844:1591–8.
28. Ono K, Li L, Takamura Y, Yoshiike Y, Zhu L, Han F, Mao X, Ikeda T, Takasaki J, Nishijo H, et al. Phenolic compounds prevent amyloid beta-protein oligomerization and synaptic dysfunction by site-specific binding. J Biol Chem. 2012;287:14631–43.
29. Lump E, Castellano LM, Meier C, Seeliger J, Erwin N, Sperlich B, Sturzel CM, Usmani S, Hammond RM, von Einem J, et al. A molecular tweezer antagonizes seminal amyloids and HIV infection. Elife. 2015;18:4.
30. Xun T, Li W, Chen J, Yu F, Xu W, Wang Q, Yu R, Li X, Zhou X, Lu L, et al. ADS-J1 inhibits semen-derived amyloid fibril formation and blocks fibril-mediated enhancement of HIV-1 infection. Antimicrob Agents Chemother. 2015;59:5123–34.
31. Osipovitch M, Lambrecht M, Baker C, Madha S, Mills JL, Craig PA. Automated protein motif generation in the structure-based protein function prediction tool. ProMOL. 2015;16:101–11.
32. Li L, Tan S, Lu H, Lu L, Yang J, Jin H, Liu S, Jiang S. Combinations of 3-hydroxyphthalic anhydride-modified ovalbumin with antiretroviral drug-based microbicide candidates display synergistic and complementary effects against HIV-1 infection. J Acquir Immune Defic Syndr. 2011;56:384–92.
33. Solomon MM, Lama JR, Glidden DV, Mulligan K, McMahan V, Liu AY, Guanira JV, Veloso VG, Mayer KH, Chariyalertsak S, et al. Changes in renal function associated with oral emtricitabine/tenofovir disoproxil fumarate use for HIV pre-exposure prophylaxis. Aids. 2014;28:851–9.
34. Karim SS, Kashuba AD, Werner L, Karim QA. Drug concentrations after topical and oral antiretroviral pre-exposure prophylaxis: implications for HIV prevention in women. Lancet. 2011;378:279–81.
35. LoRicco JG, Xu CS, Neidleman J, Bergkvist M, Greene WC, Roan NR, Makhatadze GI. Gallic acid is an antagonist of semen amyloid fibrils that enhance HIV-1 infection. J Biol Chem. 2016;291:14045–55.
36. Landis-Piwowar K, Chen D, Foldes R, Chan TH, Dou QP. Novel epigallocatechin gallate analogs as potential anticancer agents: a patent review (2009–present). Expert Opin Ther Pat. 2013;23:189–202.
37. Chen Z, Zhu QY, Tsang D, Huang Y. Degradation of green tea catechins in tea drinks. J Agric Food Chem. 2001;49:477–82.
38. Sinha S, Du Z, Maiti P, Klärner FG, Schrader T, Wang C, Bitan G. Comparison of three amyloid assembly inhibitors: the sugar scyllo-inositol, the polyphenol epigallocatechin gallate, and the molecular tweezer CLR01. ACS Chem Neurosci. 2012;3:451–8.
39. Röcker AE, Müller JA, Dietzel E, Harms M, Krüger F, Heid C, Sowislok A, Riber CF, Kupke A, Lippold S, von Einem J, et al. The molecular tweezer CLR01 inhibits Ebola and Zika virus infection. Antiviral Res. 2018;152:26–35.
40. Kim KA, Yolamanova M, Zirafi O, Roan NR, Staendker L, Forssmann WG, Burgener A, Dejucq-Rainsford N, Hahn BH, Shaw GM, et al. Semen-mediated enhancement of HIV infection is donor-dependent and correlates with the levels of SEVI. Retrovirology. 2010;7:55.
41. Sievers SA, Karanicolas J, Chang HW, Zhao A, Jiang L, Zirafi O, Stevens JT, Munch J, Baker D, Eisenberg D. Structure-based design of non-natural amino-acid inhibitors of amyloid fibril formation. Nature. 2011;475:96–100.
42. Nanga RP, Brender JR, Vivekanandan S, Popovych N, Ramamoorthy A. NMR structure in a membrane environment reveals putative amyloidogenic regions of the SEVI precursor peptide PAP(248–286). J Am Chem Soc. 2009;131:17972–9.
43. Straub JE, Thirumalai D. Toward a molecular theory of early and late events in monomer to amyloid fibril formation. Annu Rev Phys Chem. 2011;62:437–63.
44. Sadasivam K, Kumaresan R. Antioxidant behavior of mearnsetin and myricetin flavonoid compounds–a DFT study. Spectrochim Acta A Mol Biomol Spectrosc. 2011;79:282–93.
45. Roberts L, Liebenberg L, Barnabas S, Passmore JA. Vaginal microbicides to prevent human immunodeficiency virus infection in women: perspec-

tives on the female genital tract, sexual maturity and mucosal inflammation. Best Pract Res Clin Obstet Gynaecol. 2012;26(4):441–9.
46. Ross JA, Kasum CM. Dietary flavonoids: bioavailability, metabolic effects, and safety. Annu Rev Nutr. 2002;22:19–34.
47. Rahimifard M, Maqbool F, Moeini-Nodeh S, Niaz K, Abdollahi M, Braidy N, Nabavi SM, Nabavi SF. Targeting the TLR4 signaling pathway by polyphenols: a novel therapeutic strategy for neuroinflammation. Ageing Res Rev. 2017;36:11–9.
48. Vuorinen H, Maatta K, Torronen R. Content of the flavonols myricetin, quercetin, and kaempferol in finnish berry wines. J Agric Food Chem. 2000;48:2675–80.
49. Myers C, Muthusubramanian A, Sutton RB, Wylie BJ, Cornwall G. Functional amyloids in reproduction. Biomolecules. 2017. https://doi.org/10.3390/biom7030046.
50. Roan NR, Sandi-Monroy N, Kohgadai N, Usmani SM, Hamil KG, Neidleman J, Montano M, Ständker L, Röcker A, Cavrois M, et al. Semen amyloids participate in spermatozoa selection and clearance. Elife. 2017. https://doi.org/10.7554/eLife.24888.
51. Easterhoff D, Ontiveros F, Brooks LR, Kim Y, Ross B, Silva JN, Olsen JS, Feng C, Hardy DJ, Dunman P. Semen-derived enhancer of viral infection (SEVI) binds bacteria, enhances bacterial phagocytosis by macrophages, and can protect against vaginal infection by a sexually transmitted bacterial pathogen. Antimicrob Agents Chemother. 2013;57:2443–50.
52. Tan S, Lu L, Li L, Liu J, Oksov Y, Lu H, Jiang S, Liu S. Polyanionic candidate microbicides accelerate the formation of semen-derived amyloid fibrils to enhance HIV-1 infection. PLoS ONE. 2013;8:e59777.
53. Ye Z, French KC, Popova LA, Lednev IK, Lopez MM, Makhatadze GI. Mechanism of fibril formation by a 39-residue peptide (PAPf39) from human prostatic acidic phosphatase. Biochemistry. 2009;48:11582–91.
54. Chen J, Ren R, Tan S, Zhang W, Zhang X, Yu F, Xun T, Jiang S, Liu S, Li L. A peptide derived from the HIV-1 gp120 coreceptor-binding region promotes formation of PAP248–286 amyloid fibrils to enhance HIV-1 infection. PLoS ONE. 2015;10:e0144522.
55. Chang TT, Chou TC. Rational approach to the clinical protocol design for drug combinations: a review. Acta Paediatr Taiwan. 2000;41:294–302.
56. Tan S, Li L, Lu L, Pan C, Lu H, Oksov Y, Tang X, Jiang S, Liu S. Peptides derived from HIV-1 gp120 co-receptor binding domain form amyloid fibrils and enhance HIV-1 infection. FEBS Lett. 2014;588:1515–22.
57. Pettersen EF, Goddard TD, Huang CC, Couch GS, Greenblatt DM, Meng EC, Ferrin TE. UCSF Chimera—a visualization system for exploratory research and analysis. J Comput Chem. 2004;25:1605–12.
58. Jakalian A, Jack DB, Bayly CI. Fast, efficient generation of high-quality atomic charges. AM1-BCC model: II. Parameterization and validation. J Comput Chem. 2002;23:1623–41.
59. Salentin S, Schreiber S, Haupt VJ, Adasme MF, Schroeder M. PLIP: fully automated protein–ligand interaction profiler. Nucleic Acids Res. 2015;43:W443–7.
60. Li R, Liu T, Liu M, Chen F, Liu S. Anti-influenza A virus activity of dendrobine and its mechanism of action. J Agric Food Chem. 2017;65:3665–74.
61. Jiang S, Lu H, Liu S, Zhao Q, He Y, Debnath AK. N-substituted pyrrole derivatives as novel human immunodeficiency virus type 1 entry inhibitors that interfere with the gp41 six-helix bundle formation and block virus fusion. Antimicrob Agents Chemother. 2004;48:4349–59.

6

Moloney leukemia virus 10 (MOV10) inhibits the degradation of APOBEC3G through interference with the Vif-mediated ubiquitin–proteasome pathway

Cancan Chen[1,2,3], Xiaocao Ma[1,2], Qifei Hu[1,2], Xinghua Li[4], Feng Huang[1,2], Junsong Zhang[1,2], Ting Pan[1,2], Jinyu Xia[4], Chao Liu[1,2*] and Hui Zhang[1,2]

Abstract

Background: MOV10 protein has ATP-dependent 5′–3′ RNA helicase activity and belongs to the UPF1p superfamily. It can inhibit human immunodeficiency virus type 1 (HIV-1) replication at multiple stages and interact with apolipoprotein-B-mRNA-editing enzyme catalytic polypeptide-like 3G (APOBEC3G or A3G), a member of the cytidine deaminase family that exerts potent inhibitory effects against HIV-1 infection. However, HIV-1-encoded virion infectivity factor (Vif) protein specifically mediates the degradation of A3G via the ubiquitin–proteasome system (UPS).

Results: We demonstrate that MOV10 counteracts Vif-mediated degradation of A3G by inhibiting the assembly of the Vif-CBF-β-Cullin 5-ElonginB-ElonginC complex. Through interference with UPS, MOV10 enhances the level of A3G in HIV-1-infected cells and virions, and synergistically inhibits the replication and infectivity of HIV-1. In addition, the DEAG-box of MOV10 is required for inhibition of Vif-mediated A3G degradation as the DEAG-box mutant significantly loses this ability.

Conclusions: Our results demonstrate a novel mechanism involved in the anti-HIV-1 function of MOV10. Given that both MOV10 and A3G belong to the interferon antiviral system, their synergistic inhibition of HIV-1 suggests that these proteins may play complicated roles in antiviral functions.

Keywords: MOV10, A3G, Vif, HIV-1, Ubiquitin–proteasome system (UPS)

Background

Cellular apolipoprotein-B-mRNA-editing enzyme catalytic polypeptide-like 3G (APOBEC3G or A3G) is a potent antiviral host factor that can be packaged into HIV-1 virions and induces a C–U conversion in the newly synthesized minus-stranded viral DNA, thereby triggering the breakage of viral DNA or generating G-to-A hypermutations that result in a premature stop codon or mutated viral protein [1–5]. HIV-1 virion infectivity factor (Vif) can effectively counteract the antiviral activity of A3G by inducing its degradation through the ubiquitin–proteasome system (UPS) [6–10]. Vif interacts with A3G through its N-terminal domain and has a SOCS-box motif within its C-terminal domain, which includes a BC-box and Cullin-box and interacts with Cullin 5, ElonginB, and ElonginC to form an E3 ubiquitin ligase complex and subsequently mediate the ubiquitination of A3G. CBF-β can bind with Vif directly and facilitate the degradation of A3G. Moreover, CBF-β can increase the stability of HIV-1 Vif and promote assembly of Vif-Cullin 5-E3-ubiquitin-ligase complex; however, ElonginB and ElonginC facilitate the binding of CBF-β with Vif [11–16].

MOV10 is originally identified in the MOV-10 mouse strain, which carries the Moloney murine leukemia

*Correspondence: liuchao9@mail.sysu.edu.cn
[1] Institute of Human Virology, Zhongshan School of Medicine, Sun Yat-sen University, Guangzhou 510080, China
Full list of author information is available at the end of the article

virus. It is a member of the UPF1p family and has ATP-dependent 5′–3′ RNA helicase activity [17, 18]. MOV10 has complicated functions and features. For example, MOV10 is found to interact with Argonaute proteins and plays a role in microRNA (miRNA)-mediated regulation [19, 20]. MOV10 is also involved in polycomb-mediated repression of the tumor-suppressor INK4a [21]. Moreover, it has been reported that MOV10 is a type I interferon stimulated gene and several reports have indicated that this protein has broad antiretroviral activity against various viruses, such as HIV-1, murine leukemia virus (MLV), and equine infectious anemia virus (EIAV) [22–25]. Its inhibitory effect on LINE-1 retrotransposition has also been investigated [26, 27]. MOV10 can also be packaged into HIV-1 particles and affects HIV-1 replication at multiple stages [22, 28–30]. MOV10 expresses in varieties of human cells. And according to the data from GEO profile, we found that the expression profile of MOV10 is in moderate or high level in CD4 + T cells and monocytes (https://www.ncbi.nlm.nih.gov/geoprofiles, GEO Profiles ID: 89710126, 106167926, 52933168, 51070326).

Recently, several studies have reported that MOV10 interacts with A3G and both of these proteins are located in P-bodies and can be induced by interferon-α [23, 29, 31]. Based on the similar features of MOV10 and A3G, a research group has studied the possible relationship between these two restriction factors in HIV-1 infection [29]. They co-expressed MOV10 with A3G, but failed to find any functional synergistic effects on viral replication. Conversely, after knocking down endogenous MOV10 by siRNA in the presence of A3G, they did not find any significant impact on HIV-1 infectivity. Nevertheless, they detected the possible synergy of these two inhibitors in the absence of HIV-1 Vif protein. Given that MOV10 can bind with A3G, we hypothesize that MOV10 may affect the process of Vif-mediated degradation of A3G.

Thus, in this study, we aim to elucidate the correlations between MOV10 and A3G in the presence of Vif, which could occur during natural infection by HIV-1. Our findings provide important insights into the role of MOV10 in Vif-mediated A3G degradation and the mechanism through which MOV10 mediates the functional assembly of the Vif-CBF-β-Cullin 5-ElonginB-ElonginC complex to affect the Vif-medicated ubiquitin–proteasome pathway.

Results

MOV10 counteracts Vif-mediated degradation of A3G by interfering with the ubiquitin–proteasome pathway

In order to examine the relationships among MOV10, A3G, and Vif, we co-transfected MOV10-FLAG-, A3G-HA-, and Vif-HA-expressing plasmids into 293T cells, and then evaluated the expression levels of A3G and Vif. Interestingly, significant increases in A3G and Vif protein expression were observed in cells overexpressing MOV10 (Fig. 1a). To confirm this phenomenon, we analyzed changes in expression of A3G and Vif in the presence of different levels of MOV10. We found that the enhancement of A3G and Vif expression was correlated with the level of MOV10 (Fig. 1b). We also observed the same phenotype by depleting endogenous MOV10 with *MOV10*-specific siRNA (Fig. 1c) [32]. To exclude the possibility of off-target effects of siRNA, a restoration experiment was conducted. Co-transfection of rMOV10-FLAG-expressing plasmid, a MOV10 construct that is resistant to siRNA-targeting, with the *MOV10*-specific siRNA restored the expression of A3G (Fig. 1d), indicating that *MOV10*-specific siRNA does not have off-target effects.

To further validate this phenotype, a similar experiment was performed in H9 cells infected with wild-type HIV-1. MOV10-knockdown H9 cells were constructed by the infection of MOV10-specific shRNA-expressing lentivirus. The cells were then infected with wild-type HIV-1 viruses. The culture supernatants were collected at different days after infection, and HIV-1 p24 was detected by ELISA kit. After culture for 12 days, HIV-1 p24-positive cells were sorted by flow cytometry (Fig. 2). In these HIV-1 p24-positive cells, the expression of endogenous A3G was significantly down-regulated by endogenous MOV10 depletion. HIV-1 replication and the expression of Gag protein were also enhanced by the depletion of MOV10 (Fig. 2d). The results demonstrate that MOV10 and A3G synergistically inhibit the replication of HIV-1.

(See figure on next page.)

Fig. 1 MOV10 counteracts Vif-mediated degradation of A3G. **a, b** MOV10 overexpression inhibits Vif-mediated A3G degradation. Human 293T cells were transfected with pcDNA3.1-A3G-HA (0.8 μg), pcDNA3.1-Vif-HA (0.5 μg), and pcDNA3.1-MOV10-FLAG (1.5 μg) (**a**) or different amounts of pcDNA3.1-MOV10-FLAG (from 0.5 to 2 μg) (**b**). Then, cells were collected and lysed at 48 h, and analyzed by western blotting with anti-HA, anti-FLAG, and anti-GAPDH antibodies. **c, d** The effect of MOV10 depletion on Vif-mediated degradation of A3G. Cells were transfected with pcDNA3.1-A3G-HA (0.8 μg), pcDNA3.1-Vif-HA (0.5 μg), *MOV10*-specific siRNA (50 nM) (**c**) and/or siRNA-resistant MOV10 construct (rMOV10 FLAG) (0.4 μg) (**d**). After 48 h, cells were collected and analyzed by western blotting assay with anti-HA, anti-FLAG, anti-MOV10 and anti-GAPDH antibodies. Empty vector pcDNA3.1 was used in each transfection to normalize DNA amounts. Values in **a–d** represent percentages of A3G or MOV10 normalized against GAPDH and compared with control. The *bar graphs* represent the average expression of A3G with different treatment and relative to the A3G-only reaction control (set to 100%). All the data represent mean ± SD from three independent experiments. Statistical significance was determined using *t* test: $^{*}p \le 0.05$; $^{**}p \le 0.01$; $^{***}p \le 0.001$

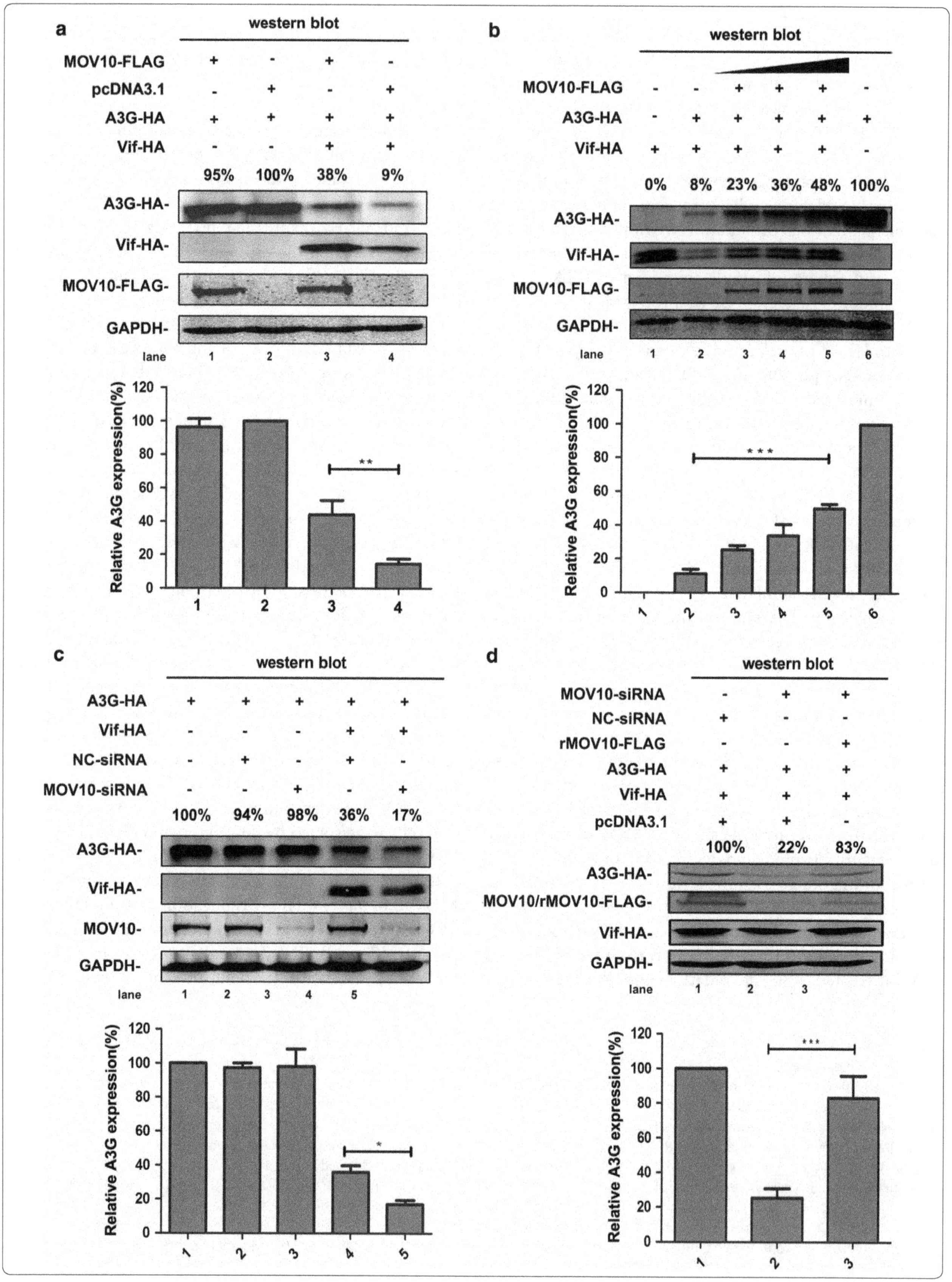
a
western blot
MOV10-FLAG + - + -
pcDNA3.1 - + - +
A3G-HA + + + +
Vif-HA - - + +
95% 100% 38% 9%
A3G-HA-
Vif-HA-
MOV10-FLAG-
GAPDH-
lane 1 2 3 4
Relative A3G expression(%)
**
b
western blot
MOV10-FLAG - - + + + -
A3G-HA - + + + + +
Vif-HA + + + + + -
0% 8% 23% 36% 48% 100%
A3G-HA-
Vif-HA-
MOV10-FLAG-
GAPDH-
lane 1 2 3 4 5 6
Relative A3G expression(%)
***
c
western blot
A3G-HA + + + + +
Vif-HA - - - + +
NC-siRNA - + - + -
MOV10-siRNA - - + - +
100% 94% 98% 36% 17%
A3G-HA-
Vif-HA-
MOV10-
GAPDH-
lane 1 2 3 4 5
Relative A3G expression(%)
*
d
western blot
MOV10-siRNA - + +
NC-siRNA + - -
rMOV10-FLAG - - +
A3G-HA + + +
Vif-HA + + +
pcDNA3.1 + + -
100% 22% 83%
A3G-HA-
MOV10/rMOV10-FLAG-
Vif-HA-
GAPDH-
lane 1 2 3
Relative A3G expression(%)
***

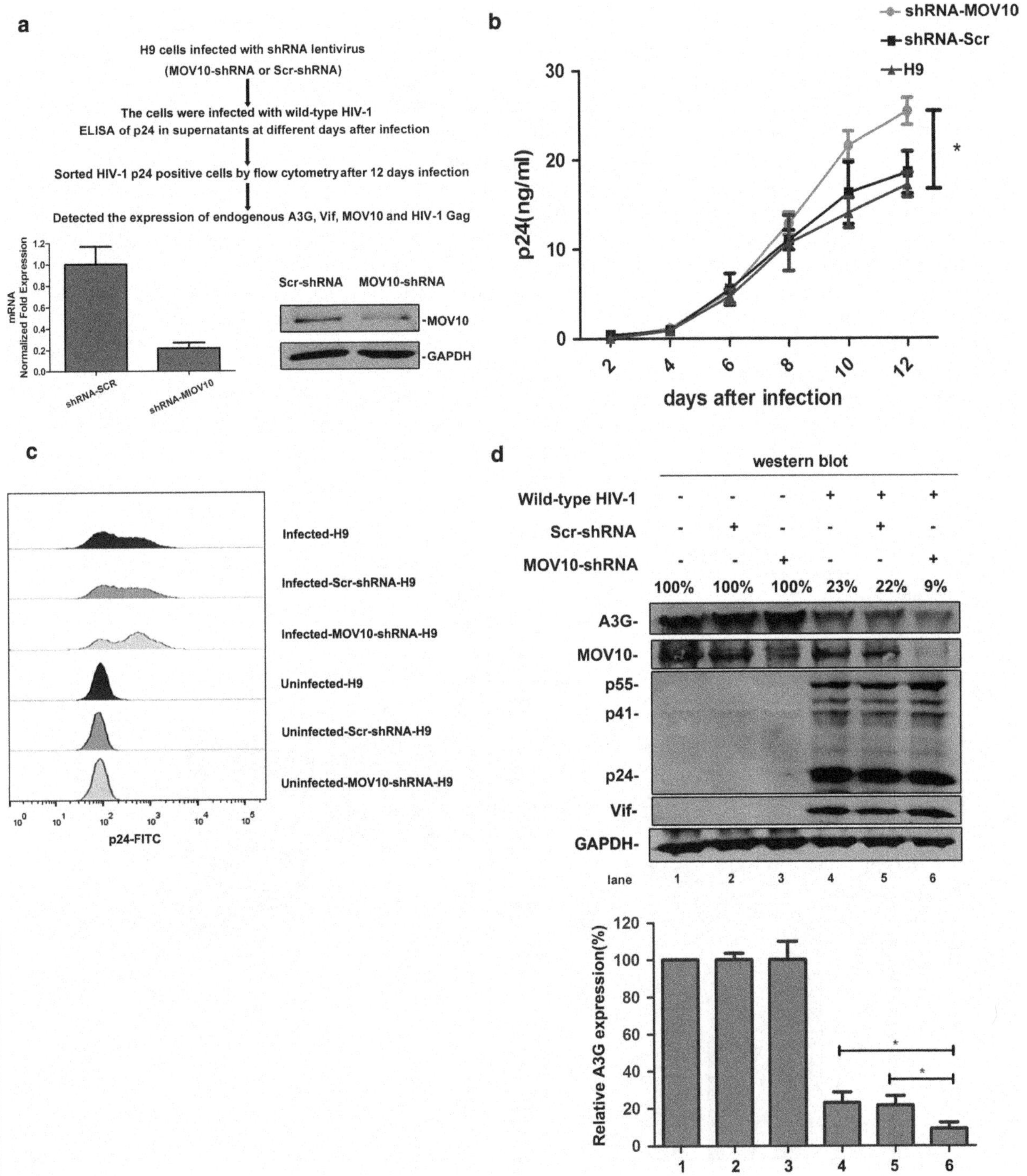

Fig. 2 MOV10 protects A3G from Vif-mediated degradation in wild-type HIV-1. **a** H9 cells were infected with pLKO.1-MOV10-shRNA or pLKO.1-Scr-shRNA lentivirus for 8 h and then selected with puromycin for 2 weeks. MOV10-knockdown H9 cells and control cells were infected with wild-type HIV-1 for 3 h and cultured with fresh medium for 12 days. The culture supernatants were collected at the indicated time points. Then HIV-1 p24 was detected using HIV-1 p24 ELISA kit at different time points (**b**). And at 12th day, these cells were analyzed by flow cytometer (**c**). HIV-1 p24 positive H9 cells were sorted and detected by western blotting with anti-MOV10, anti-A3G, anti-Vif, anti-HIV-1 p24, and anti-GAPDH antibodies (**d**). Values in **d** represent percentages of A3G or MOV10 normalized against GAPDH and compared with control. The *bar graphs in **d*** represent the average expression of A3G with different treatment and relative to the A3G-only reaction control (set to 100%). Data in **a**, **b**, and **d** represent mean ± SD from three independent experiments. *, statistically significant, $p \leq 0.05$ (*t* test). All the results are representative of at least three independent experiments

Considering the relationships between Vif/A3G and the ubiquitin–proteasome pathway [6–8, 10, 33], we evaluated the effects of MOV10 on the expression levels of A3G and Vif in the presence of the proteasome inhibitor MG132. After treatment with MG132 for 16 h, the expression levels of A3G and Vif were not affected by MOV10 overexpression (Fig. 3a). Further study showed that MOV10 could decrease the ubiquitination of A3G directly (Fig. 3b). Taken together, these data indicate that MOV10 can protect A3G from Vif-mediated degradation by interfering with the ubiquitin–proteasome pathway.

MOV10 affects the assembly of the Vif-CBF-β-Cullin 5-ElonginB-ElonginC complex

Previous studies have reported that A3G can bind with Vif (Fig. 4a) [10]. And, A3G protein contains two domains: the N-terminal domain is responsible for encapsidation and the C-terminal domain is responsible for deamination activity [34, 35]. Only the N-terminal domain of A3G can bind with Vif and the binding initiates the degradation process of A3G [36]. Because A3G can also interact with MOV10 (Fig. 4b) [20], the interaction of Vif with A3G may be affected by MOV10. To this end, we investigated the effects of MOV10 on the interaction between Vif and A3G in the presence of MG132. Human 293T cells were transfected with MOV10-FLAG-, A3G-HA-, and Vif-FLAG-expressing plasmids and then treated with MG132 for 16 h. However, we did not detect any changes in the levels of Vif-FLAG in the A3G-HA-immunoprecipitated samples with or without MOV10 (Fig. 4c). These results suggest that MOV10 does not affect the binding of Vif with A3G and therefore it may interfere with other steps in the A3G degradation process.

The interaction of Vif with CBF-β, Cullin 5, ElonginB, and ElonginC can facilitate the formation of a ubiquitin ligase complex, which is required for Vif to induce the degradation of A3G [11–13, 37, 38]. Therefore, we next examined

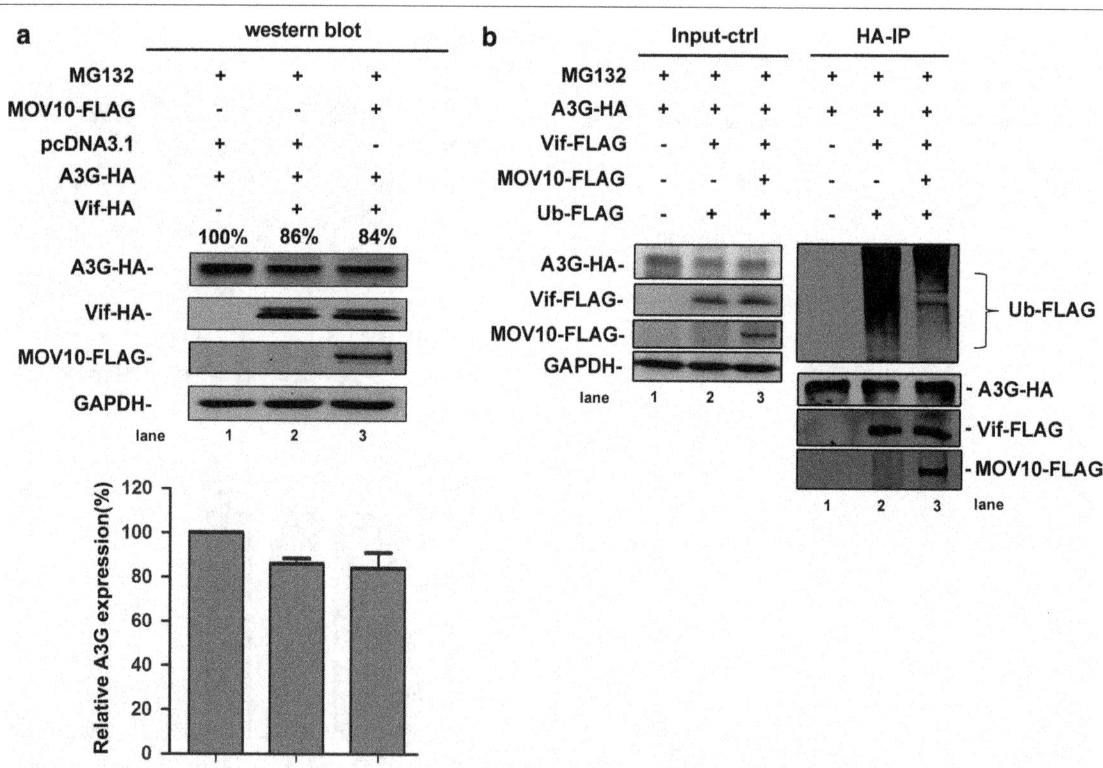

Fig. 3 MOV10 prevents A3G from Vif-induced degradation by decreasing the ubiquitination of A3G. **a** Human 293T cells were transfected with pcDNA3.1-A3G-HA (0.8 μg), pcDNA3.1-Vif-HA (0.5 μg), and pcDNA3.1-MOV10-FLAG (1.5 μg) and then treated with MG132 (4 μM) for 16 h. Lysed cells were collected at 48 h and detected by western blotting with anti-HA, anti-FLAG, and anti-GAPDH antibodies. **b** 293T cells were transfected with pcDNA3.1-A3G-HA (2 μg), pcDNA3.1-Vif-FLAG (1.25 μg), pcDNA3.1-MOV10-FLAG (2.5 μg), and pcDNA3.1-Ub-FLAG (3 μg). Cells were treated with MG132 (4 μM) for 16 h and analyzed by co-immunoprecipitation with anti-HA agarose beads. And then, samples were detected by western-blotting using anti-HA, anti-FLAG, and anti-GAPDH. Values in **a** represent percentages of A3G normalized against GAPDH and compared with control. The *bar graphs in* **a** represent the average expression of A3G with different treatment and relative to the A3G-only reaction control (set to 100%). Data in **a** represent mean ± SD from three independent experiments. Empty vector pcDNA3.1 was used to equalize DNA amounts in each transfection. Data in **a** and **b** are representative of at least three independent experiments

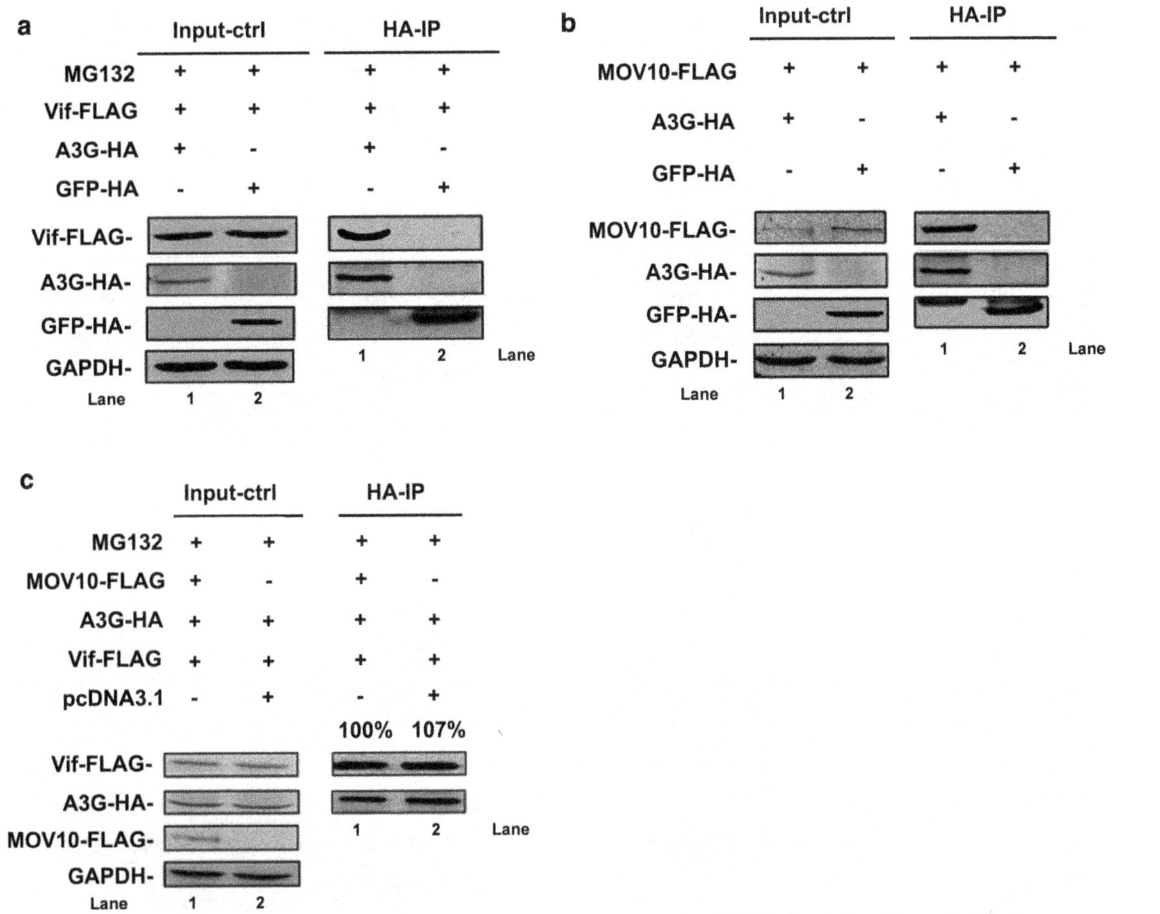

Fig. 4 MOV10 has no influence on the binding of A3G with Vif. **a**, **b** A3G interacts with Vif or MOV10 effectively. **a** Human 293T cells were transfected with 2 μg of pcDNA3.1-A3G-HA (pcDNA3.1-GFP-HA as a control) and 1 μg of pcDNA3.1-Vif-FLAG and then treated with MG132 for 16 h. **b** Human 293T cells were transfected with 1 μg of pcDNA3.1-Vif-FLAG and 2 μg of pcDNA3.1-A3G-HA or pcDNA3.1-GFP-HA. **a**, **b** lysates from these transfected cell samples were subjected to co-immunoprecipitation analysis using anti-HA agarose beads and then detected by western blotting. **c** The effect of MOV10 on the interaction between A3G and Vif. 293T cells were transfected with 2 μg of pcDNA3.1-A3G-HA together with 1 μg of pcDNA3.1-Vif-FLAG, and 2 μg of pcDNA3.1-MOV10-FLAG and then treated with MG132 (4 μM) for 16 h. Samples were immunoprecipitated with anti-HA agarose beads and analyzed by western blotting. Empty vector pcDNA3.1 was used to equalize DNA amounts in each transfection. Values in **c** represent portions of Vif-FLAG normalized against A3G-HA relative to control values. Data in **a**, **b**, and **c** are representative of at least three independent experiments

the effects of MOV10 on the interaction between Vif and different components in the complex. We transfected Vif-HA- and MOV10-FLAG-expressing plasmids with pcDNA3.1-ElonginB-FLAG, pcDNA3.1-ElonginC-FLAG, pcDNA3.1-Cullin 5-FLAG, or pcDNA3.1-CBF-β-FLAG into 293T cells. After immunoprecipitation, we found that the interactions of Vif with ElonginB, ElonginC, Cullin 5, and CBF-β significantly decreased when MOV10 was overexpressed (Fig. 5a–d), indicating that MOV10 affects Vif-ElonginB, Vif-ElonginC, Vif-Cullin 5, and Vif-CBF-β interactions during the assembly of the Vif-CBF-β-Cullin 5-ElonginB-ElonginC complex.

According to the above results, we suspected that MOV10 could interact with ElonginB, ElonginC, Cullin 5, or CBF-β. To test this hypothesis, we co-transfected 293T cells with pcDNA3.1-MOV10-HA plus pcDNA3.1-ElonginB-FLAG, pcDNA3.1-ElonginC-FLAG, pcDNA3.1-Cullin 5-FLAG or pcDNA3.1-CBF-β-FLAG. Previous study has demonstrated that ElonginC, ElonginB, and Cullin 5 can interact with each other [39]. To eliminate the influence of these endogenous proteins, siRNAs specific to *ElonginB*, *ElonginC*, and *Cullin 5* mRNA were also co-transfected into cells at the same time (Fig. 6a). After immunoprecipitation and western blotting, significant binding was found between MOV10 and ElonginC or Cullin 5 (Fig. 6c, d). To further confirm the binding, we detected the interaction between MOV10-HA and endogenous ElonginC or Cullin 5. As shown in the Fig. 6f, g, the same phenomenon was observed. After treatment with an RNase mixture,

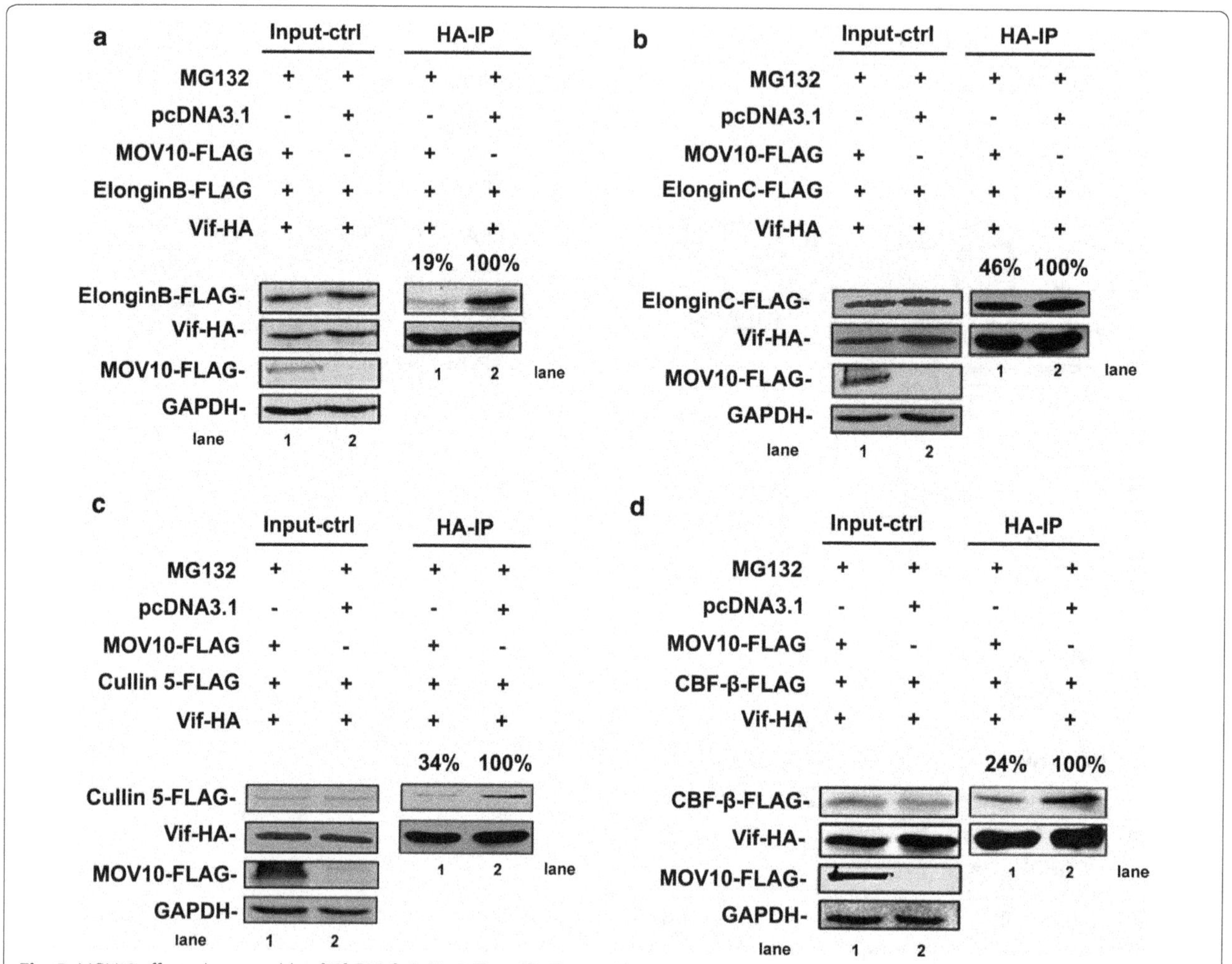

Fig. 5 MOV10 affects the assembly of Vif-CBF-β-Cullin 5-ElonginB-ElonginC Complex. **a–d** The effect of MOV10 on the interaction between Vif and ElonginB (**a**), ElonginC (**b**), Cullin 5 (**c**), or CBF-β (**d**). 293T cells were transfected with pcDNA3.1-MOV10-FLAG (2 μg), pcDNA3.1-Vif-HA (1 μg), and 4 μg of ElonginB-FLAG (**a**) or ElonginC-FLAG (**b**) or pcDNA3.1-Cullin 5-FLAG (**c**) or pcDNA3.1-CBF-β-FLAG (**d**). After treated with MG132 (4 μM) for 16 h, cell lysates were immunoprecipitated with anti-HA agarose beads and analyzed by immunoblotting using anti-FLAG, anti-HA, and anti-GAPDH antibodies. In each transfection, empty vector pcDNA3.1 was used to normalize DNA amounts. Values in **a–d** represent percentages of ElonginB-FLAG/ElonginC-FLAG/Cullin 5-FLAG/CBF-β-FLAG normalized against Vif-HA relative to control values. All the data is representative of at least three independent experiments

we found that the binding of MOV10 with Cullin 5 was partially dependent on RNA, whereas the interaction between MOV10 and ElonginC was not (Fig. 6h, i). However, the interaction between MOV10 and ElonginB or CBF-β was not detected (Fig. 6b, e).

The helicase activity center of MOV10 is required for its inhibitory effects on Vif-mediated A3G degradation

MOV10 contains a DEAG-box (D-E-A-G = Asp-Glu-Ala-Gly) motif and the DEAG-box mutant impairs the helicase activity of MOV10 [18, 28]. To examine whether the DEAG-box motif was required for the effects of MOV10 on Vif-mediated degradation of A3G, we used a MOV10-DEAG mutant (a point mutation in the DEAG-box motif, from DEAG to DQAG) to repeat the experiment shown as Fig. 1a [18, 22, 23, 32, 40]. Compared with wild-type MOV10, the MOV10-DEAG mutant almost lost the ability to prevent the degradation of A3G mediated by Vif (Fig. 7a), suggesting that the DEAG-box motif is involved in regulating this inhibitory effects of MOV10. To confirm this conclusion, we further detected the binding of the MOV10-DEAG mutant with ElonginC or Cullin 5. Compared with wild-type MOV10, the bindings of MOV10-DEAG mutant with ElonginC or Cullin 5 decreased significantly (Fig. 7b, c), suggesting that the DEAG-box motif of MOV10 plays an important role in the interaction between MOV10 with ElonginC or Cullin 5.

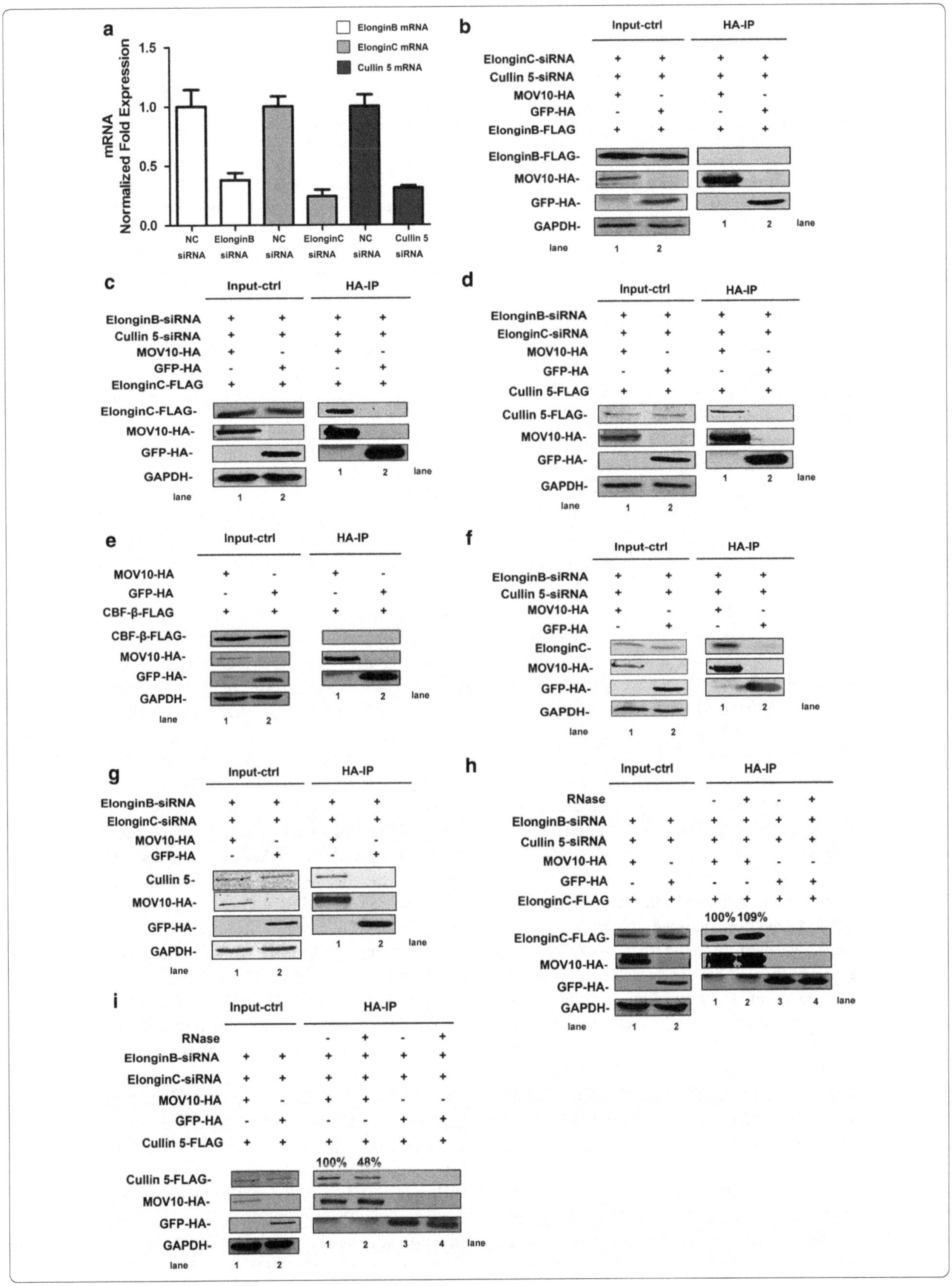
a
ElonginB mRNA
ElonginC mRNA
Cullin 5 mRNA
mRNA Normalized Fold Expression
1.5
1.0
0.5
0.0
NC siRNA
ElonginB siRNA
NC siRNA
ElonginC siRNA
NC siRNA
Cullin 5 siRNA
b
Input-ctrl
HA-IP
ElonginC-siRNA
Cullin 5-siRNA
MOV10-HA
GFP-HA
ElonginB-FLAG
ElonginB-FLAG-
MOV10-HA-
GFP-HA-
GAPDH-
lane
c
Input-ctrl
HA-IP
ElonginB-siRNA
Cullin 5-siRNA
MOV10-HA
GFP-HA
ElonginC-FLAG
ElonginC-FLAG-
MOV10-HA-
GFP-HA-
GAPDH-
lane
d
Input-ctrl
HA-IP
ElonginB-siRNA
ElonginC-siRNA
MOV10-HA
GFP-HA
Cullin 5-FLAG
Cullin 5-FLAG-
MOV10-HA-
GFP-HA-
GAPDH-
lane
e
Input-ctrl
HA-IP
MOV10-HA
GFP-HA
CBF-β-FLAG
CBF-β-FLAG-
MOV10-HA-
GFP-HA-
GAPDH-
lane
f
Input-ctrl
HA-IP
ElonginB-siRNA
Cullin 5-siRNA
MOV10-HA
GFP-HA
ElonginC-
MOV10-HA-
GFP-HA-
GAPDH-
lane
g
Input-ctrl
HA-IP
ElonginB-siRNA
ElonginC-siRNA
MOV10-HA
GFP-HA
Cullin 5-
MOV10-HA-
GFP-HA-
GAPDH-
lane
h
Input-ctrl
HA-IP
RNase
ElonginB-siRNA
Cullin 5-siRNA
MOV10-HA
GFP-HA
ElonginC-FLAG
100% 109%
ElonginC-FLAG-
MOV10-HA-
GFP-HA-
GAPDH-
lane
i
Input-ctrl
HA-IP
RNase
ElonginB-siRNA
ElonginC-siRNA
MOV10-HA
GFP-HA
Cullin 5-FLAG
100% 48%
Cullin 5-FLAG-
MOV10-HA-
GFP-HA-
GAPDH-
lane

(See figure on previous page.)
Fig. 6 MOV10 binds with ElonginC or Cullin 5. **a** The knockdown efficiency of siElonginB, siElonginC and siCullin 5. 293T cells were transfected with siElonginB, siElonginC or siCullin 5, after 48 h, the cells were collected and detected with qRT-PCR. Data in A represents mean ± SD (*error bars*). **b–i** Co-immunoprecipitated analysis of the interaction between MOV10 and ElonginB (**b**), ElonginC (**c**, **f**, **h**), Cullin 5 (**d**, **g**, **i**), or CBF-β (**e**). pcDNA3.1-ElonginB-FLAG plus siElonginC and siCullin 5 (**b**), pcDNA3.1-ElonginC-FLAG plus siElonginB and siCullin 5 (**c**, **h**), pcDNA3.1-Cullin 5-FLAG plus siElonginB and siEloingC (**d**, **i**) or pcDNA3.1-CBF-β-FLAG (**e**) was transfected into 293T cells with pcDNA3.1-MOV10-HA or pcDNA3.1-GFP-HA. 293T cells were transfected with pcDNA3.1-MOV10-HA (pcDNA3.1-GFP-HA as a control) plus siElonginB and siCullin 5 (**f**) or siElonginB and siEloingC (**g**). After 48 h, the cells were collected and immunoprecipitated with anti-HA agarose beads (**b–i**). The samples in **h** and **i** were treated with RNase mixture. And then, immunoprecipitated samples were analyzed by immunoblotting with anti-FLAG, anti-HA, anti-GAPDH, anti-MOV10, anti-ElonginC, and anti-Cullin 5 antibodies. Empty vector pcDNA3.1 was used to equalize DNA amounts in each transfection. Values in **h** and **i** represent portions of ElonginC-FLAG/Cullin 5-FLAG normalized against MOV10-HA and compared with control. Data in **a–i** is representative of at least three independent experiments

MOV10 counteracts Vif-mediated A3G degradation in the context of HIV-1 replication

All of the above experiments were performed in the context of lack of other HIV-1 proteins. To verify whether the effect of MOV10 on Vif-mediated A3G degradation could be observed in the context of HIV-1 replication, we used two types of HIV-1 pNL4-3ΔEnv-GFP clones. Human 293T cells were transfected with pNL4-3ΔEnv-GFP-ΔVif, pcDNA3.1-Vif-HA, pcDNA3.1-A3G-HA, and different amounts of MOV10-FLAG-expressing plasmid. Consistent with Fig. 1b, the expression levels of A3G and Vif were correlated with the expression levels of MOV10 in the presence of other HIV-1 proteins (Fig. 8a). Moreover, the same results were observed when we co-transfected 293T cells with pNL4-3ΔEnv-GFP, pcDNA3.1-A3G-HA, and different amounts of pcDNA3.1-MOV10-FLAG (Fig. 8b). To further confirm this, we examined the effect of MOV10 depletion on Vif-mediated A3G degradation in the context of HIV-1 replication. The same phenotypes as shown in Fig. 1c were recapitulated by *MOV10*-specific siRNAs (Fig. 8c, d). These data indicate that MOV10 can inhibit Vif-mediated degradation of A3G in the context of HIV-1 replication.

MOV10 increases the quantity of A3G in HIV-1 virions by protecting A3G from Vif-mediated degradation

As noted above, we demonstrated that MOV10 could increase the quantity of A3G by interfering with the proteasome pathway in virus-producing cells. Given that A3G can be packaged into HIV-1 virions and exert anti-HIV-1 activity [2, 35, 41–44]. We next evaluated the effects of MOV10 on the quantity of A3G in HIV-1 virions. We co-transfected 293T cells with pNL4-3ΔEnv-GFP, pcDNA3.1-A3G-HA, and pcDNA3.1-MOV10-FLAG, subsequently collected the supernatants and cells of each sample. Interestingly, although MOV10 enhanced A3G levels in cell lysates, the quantity of A3G was also increased in the supernatant viral particles (Fig. 9a). To further confirm this, the A3G levels were analyzed in virus-producing cells and viral particles following depletion of endogenous MOV10 with *MOV10*-specific siRNAs. MOV10 knockdown could reduce the quantity of A3G in both virus producing cells and viral particles (Fig. 9b). Previous studies showed that MOV10 can be packaged into virions and affects HIV-1 replication at multiple stages. Here, we overexpressed MOV10 in a dose-dependent manner, and we found that there was a synergy between the packaging levels of MOV10 and A3G (Fig. 9c). Taken together, these results demonstrate that MOV10 increases the quantity of A3G in HIV-1 virions by interfering with the Vif-mediated ubiquitin–proteasome pathway.

Furthermore, to explore the potential synergistic effect of MOV10 and A3G on the infectivity of the newly-produced virions, NL4-3-ΔEnv-GFP and NL4-3-ΔEnv-GFP-ΔVif particles were produced with increasing amounts of MOV10 in either the presence or absence of A3G. After normalization for HIV-1 p24, TZM-bl cells were infected with these viral particles and then the infectivity of viruses was determined (Fig. 10a, b). For NL4-3-ΔEnv-GFP-ΔVif particles, compared with the group of single MOV10 or A3G treatment, the inhibitory effect of MOV10 or MOV10-DEAG mutant plus A3G group was equal to the effect of single A3G treatment group. And, the depletion of endogenous MOV10 with siRNA in the presence of A3G also has no impact on HIV-1 infectivity. It is consistent with previous study that co-expression of MOV10 did not enhance the inhibitory effect of A3G on the infectivity of ΔVif HIV-1 (Fig. 10a) [29]. However, for NL4-3-ΔEnv-GFP particles, the fold reductions in infectivity at various amounts of MOV10 were different in the presence or absence of A3G and endogenous MOV10 was helpful for A3G to decrease the infectivity of HIV-1. But, MOV10-DEAG mutant lost its ability to help A3G decrease the infectivity of HIV-1 (Fig. 10b). This data further confirms that DEAG-box motif of MOV10 plays an important role in protecting A3G from Vif-mediated degradation. These results indicate that the inhibitory effect of A3G on the infectivity of HIV-1 can be synergistically enhanced by MOV10.

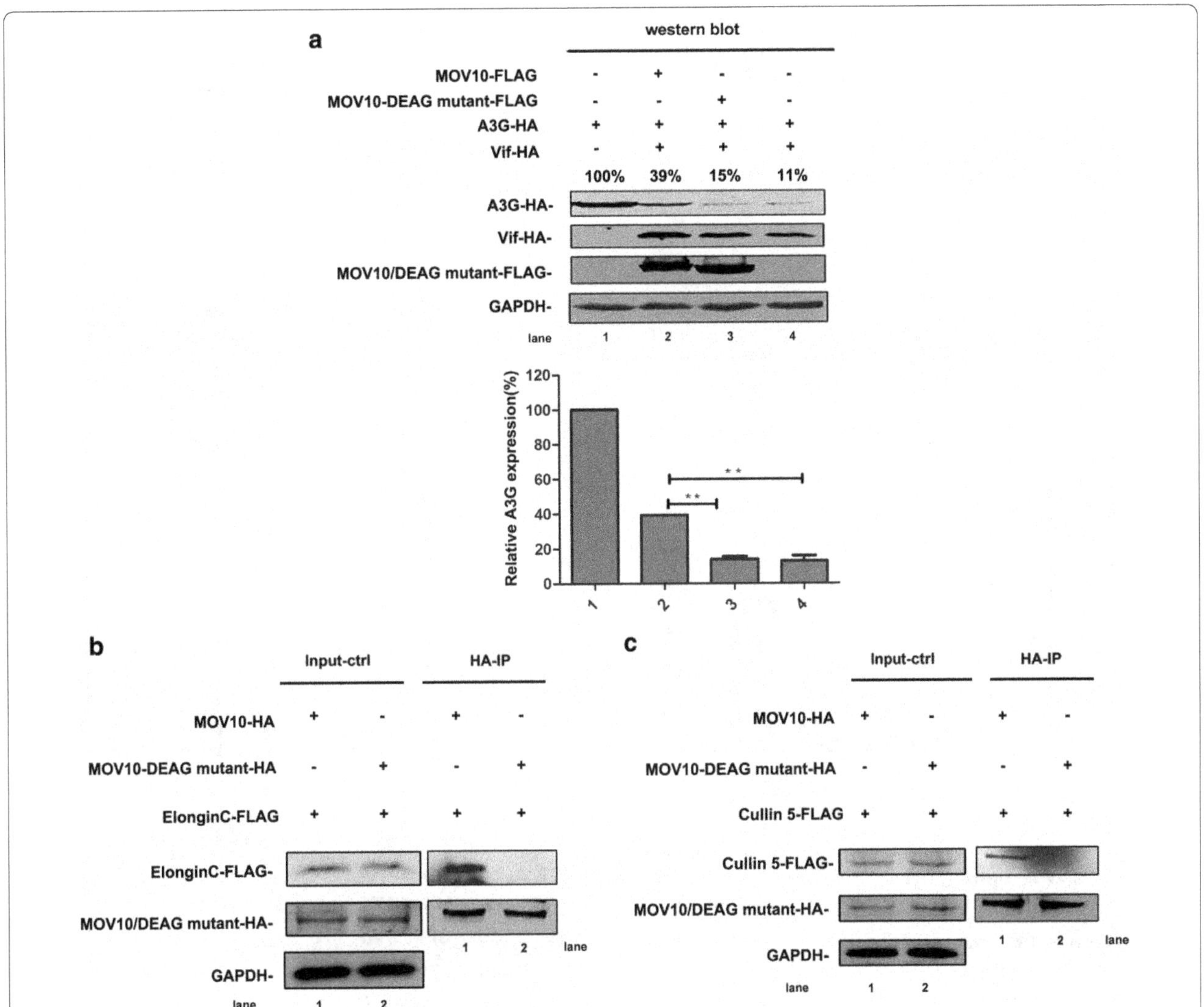

Fig. 7 The DEAG-box motif of MOV10 is required for the binding of MOV10 with ElonginC or Cullin 5. **a** The effect of MOV10-DEAG mutant on Vif-mediated A3G degradation. 293T cells were transfected with 0.4 μg of pcDNA3.1-Vif-HA, 0.8 μg of pcDNA3.1-A3G-HA, and 1.5 μg of pcDNA3.1-MOV10-FLAG or pcDNA3.1-MOV10-DEAG-mutant-FLAG as indicated. After 48 h, cell lysates were detected by western blotting assay with anti-HA, anti-FLAG, and anti-GAPDH antibodies. Values represent portions of A3G-HA normalized against GAPDH and compared with control. **b**, **c** Co-immunoprecipitated analysis of the interaction between MOV10-DEAG mutant and ElonginC or Cullin 5. Human 293T cells were transfected with 2 μg of pcDNA3.1-MOV10-HA or pcDNA3.1-MOV10-DEAG-HA and 6 μg of pcDNA3.1-ElonginC-FLAG or pcDNA3.1-Cullin 5-FLAG. After 24 h, MG132 were added in the transfected cells for 16 h. Then, the cells were collected for co-immunoprecipitation analysis with anti-HA agarose beads and detected by western blotting with anti-HA, anti-FLAG, and anti-GAPDH antibodies. In each transfection, empty vector pcDNA3.1 was used to equalize DNA amounts. Values in **a** represent percentages of A3G-HA normalized against GAPDH relative to control. The *bar graphs* in **a** represent the average expression of A3G with different treatments and relative to the A3G-only reaction control (set to 100%). Data in **a** represent mean ± SD from three independent experiments. Statistical significance was determined using *t* test: $^{**}p \leq 0.01$. All the data in **a**, **b**, and **c** is representative of at least three independent experiments

Discussion

In this report, we studied the relationship between MOV10 and A3G in the presence of HIV-1 Vif. Interestingly, we observed that MOV10 increased the levels of A3G and Vif in a concentration-dependent manner. The phenomenon is consistent with the previous study that the expression levels of A3G and Vif can be increased simultaneously in the presence of the proteasome inhibitor MG132 [45, 46]. Vif is an E3 ubiquitin ligase substrate receptor that interacts with host factors ElonginB, ElonginC, Cullin 5, and CBF-β to form an E3 ubiquitin ligase complex, which results in the polyubiquitylation of both Vif and A3G. And then, the Vif-A3G complex can be degraded together via the ubiquitin–proteasome system

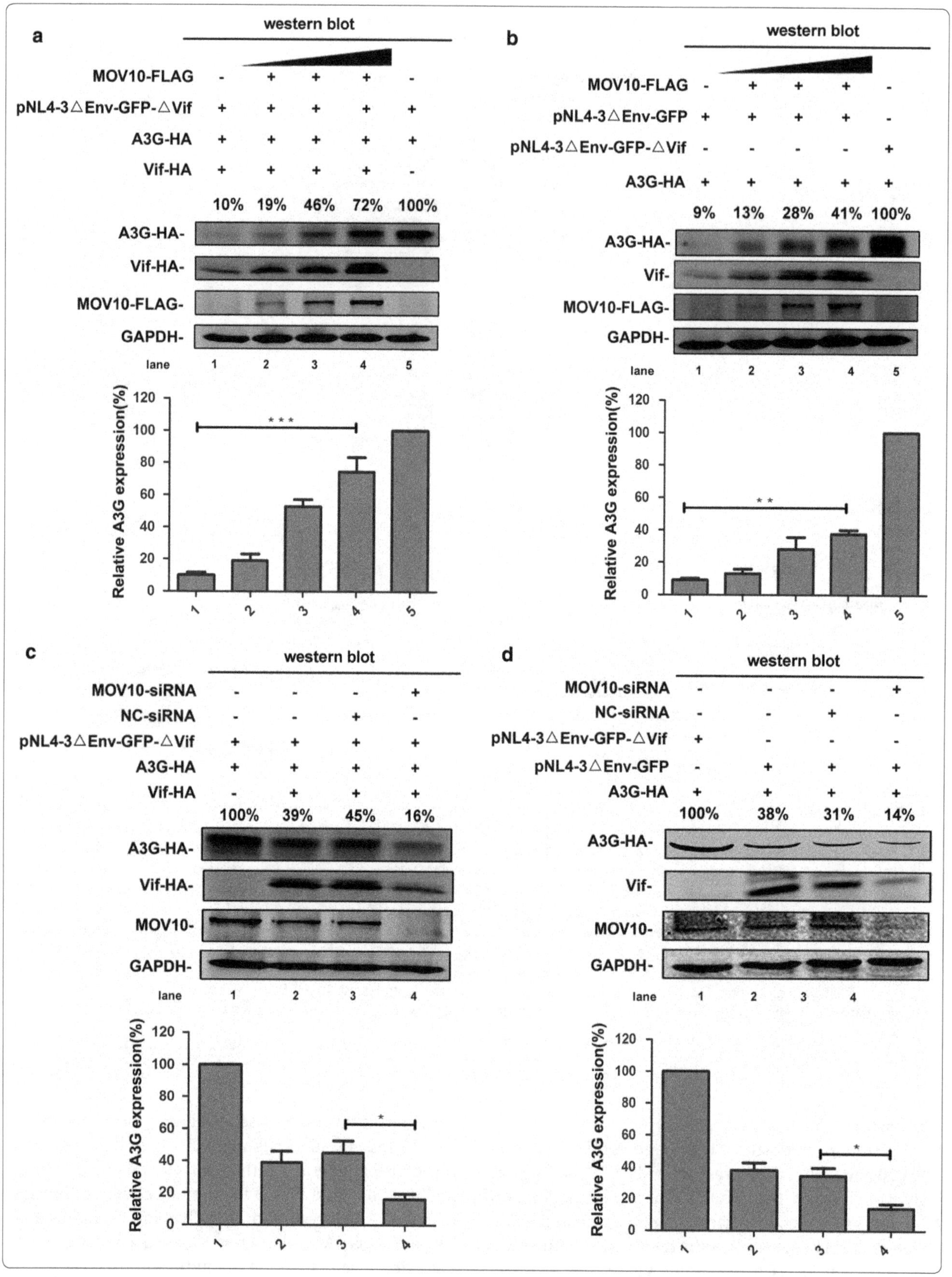
a
western blot
MOV10-FLAG - + + + -
pNL4-3△Env-GFP-△Vif + + + + +
A3G-HA + + + + +
Vif-HA + + + + -
10% 19% 46% 72% 100%
A3G-HA-
Vif-HA-
MOV10-FLAG-
GAPDH-
lane 1 2 3 4 5
Relative A3G expression(%)
***
b
western blot
MOV10-FLAG - + + + -
pNL4-3△Env-GFP + + + + -
pNL4-3△Env-GFP-△Vif - - - - +
A3G-HA + + + + +
9% 13% 28% 41% 100%
A3G-HA-
Vif-
MOV10-FLAG-
GAPDH-
lane 1 2 3 4 5
Relative A3G expression(%)
**
c
western blot
MOV10-siRNA - - - +
NC-siRNA - - + -
pNL4-3△Env-GFP-△Vif + + + +
A3G-HA + + + +
Vif-HA - + + +
100% 39% 45% 16%
A3G-HA-
Vif-HA-
MOV10-
GAPDH-
lane 1 2 3 4
Relative A3G expression(%)
*
d
western blot
MOV10-siRNA - - - +
NC-siRNA - - + -
pNL4-3△Env-GFP-△Vif + - - -
pNL4-3△Env-GFP - + + +
A3G-HA + + + +
100% 38% 31% 14%
A3G-HA-
Vif-
MOV10-
GAPDH-
lane 1 2 3 4
Relative A3G expression(%)
*

(See figure on previous page.)
Fig. 8 MOV10 reduces A3G proteasomal degradation significantly in the context of HIV-1 replication. **a**, **b** Overexpression of MOV10 inhibits Vif-induced A3G degradation in the context of HIV-1 replication. Human 293T cells were transfected with 0.8 μg of pcDNA3.1-A3G-HA, different amounts of pcDNA3.1-MOV10-FLAG (from 0.5 μg to 2 μg), 0.5 μg of pcDNA3.1-Vif-HA, 1 μg of pNL4-3ΔEnv-GFP-ΔVif (**a**) and/or 1 μg of pNL4-3ΔEnv-GFP (**b**) as indicated. Then, cells were collected at 48 h for western blotting assay with anti-FLAG, anti-HA, anti-Vif, and anti-GAPDH antibodies. **c**, **d** The effect of MOV10 depletion on the proteasomal degradation of A3G in the context of other HIV-1 proteins. 293T cells were transfected with pcDNA3.1-A3G-HA (0.8 μg), pcDNA3.1-Vif-HA (0.5 μg), *MOV10*-specific siRNA (or negative control-siRNA), 1 μg of pNL4-3ΔEnv-GFP-ΔVif (**c**) and/or 1 μg of pNL4-3ΔEnv-GFP (**d**). Cell lysates were detected by immunoblotting with anti-HA, anti-FLAG, anti-MOV10, anti-Vif, and anti-GAPDH antibodies. The *bar graphs* represent the average expression of A3G with different treatment and relative to the A3G-only reaction control (set to 100%). Data in **a–d** represent mean ± SD from three independent experiments. Statistical significance was determined using *t* test: $^*p \leq 0.05$; $^{**}p \leq 0.01$; $^{***}p \leq 0.001$. Empty vector pcDNA3.1 was used to equalize DNA amounts in each transfection. Values in **a–d** represent portions of A3G-HA normalized against GAPDH and compared with control. Each data is representative of at least three independent experiments

[6, 12, 28, 37, 47, 48]. HIV-1 Vif has at least 4 conserved motifs, which are required for interactions with host proteins. The HCCH motif can bind to Cullin 5, the BC-box motif (144-SLQYLA-149) binds to ElonginC, 101-DVMK-104 binds to ElonginB, and 88-EW-89 is crucial for binding with CBF-β [38, 49–51]. Through a series of co-immunoprecipitation analyses, we found that MOV10 can disrupt the interaction of Vif with ElonginB, ElonginC, Cullin 5, or CBF-β and then decrease the ubiquitination of A3G. Finally, the degradations of A3G and Vif are blocked and their expression levels in cells are increased subsequently. These results suggest that MOV10 functions to mediate assembly of the Vif-CBF-β-ElonginB-ElonginC-Cullin 5 complex.

Previous studies have shown that Cullin 5 functions as a scaffold protein for the *E3 ubiquitin ligase* [52, 53] ubiquitin–proteasome . ElonginB, ElonginC, and CBF-β are adaptor proteins that function to maintain this complex. Moreover, Vif acts as a substrate acceptor to modulate the degradation of A3G [52, 54]. Therefore, reduced binding of Vif with Cullin 5 could affect the complex assembly efficiency. Moreover, researchers have verified the interactions between the different components of the complex. The binding of Cullin 5 to Vif enhances the stability of the Vif-CBF-β interaction [55]. Conversely, CBF-β is also crucial for the binding of Vif with Cullin 5, ElonginB, and ElonginC [37, 56, 57]. ElonginB and ElonginC play important roles in the interaction between Vif and CBF-β [16]. To clarify the mechanisms through which MOV10 disrupts the assembly of the Vif-CBF-β-ElonginB-ElonginC-Cullin 5 complex, we examined whether there were direct interactions between MOV10 and different components of the CBF-β-Cullin 5-ElonginB-ElonginC complex. The results demonstrate that MOV10 can bind with ElonginC or Cullin 5 and that binding between MOV10 and Cullin 5 is partially dependent on RNA. Our own study and previous studies have shown that MOV10 usually interacts with numerous RNA-associated proteins, such as AGO1/2, A3G, and HIV-1 Rev [20, 32]. Thus, it is not surprising that MOV10 interacts with Cullin 5 in an RNA-dependent manner. Accordingly, significant decreases in the binding of Vif with ElonginB, ElonginC, Cullin 5, and CBF-β were observed when MOV10 was overexpressed. For the inhibitory effects of MOV10 on the binding of Vif with ElonginB or CBF-β, the interactions of MOV10 with ElonginC, Cullin 5, and Vif may induce structural changes in the Vif-CBF-β-ElonginB-ElonginC-Cullin 5 complex, subsequently disrupting the interactions between Vif and ElonginB and between Vif and CBF-β. Several studies have shown that DEAG-box motif of MOV10 is crucial for its helicase activity [18, 28]. In our report, we also explored the

(See figure on next page.)
Fig. 9 MOV10 increases the quantity of A3G in HIV-1 virions. **a** Overexpression of MOV10 increases the quantity of A3G in virions. Human 293T cells were transfected with 0.8 μg of pcDNA3.1-A3G-HA, 1.5 μg of pcDNA3.1-MOV10-FLAG, 1 μg of pNL4-3ΔEnv-GFP or pNL4-3ΔEnv-GFP-ΔVif as indicated. **b** The effect of endogenous MOV10 knockdown on Vif-induced A3G degradation in HIV-1 virions. Cells were transfected with 0.8 μg of pcDNA3.1-A3G-HA, 50 nM of *MOV10*-specific siRNA (or negative control siRNA), and 1 μg of pNL4-3ΔEnv-GFP or pNL4-3ΔEnv-GFP-ΔVif. **c** MOV10 can be packaged into HIV-1 virions and its packaging level increases with the survival level of A3G. 293T cells were transfected with 0.8 μg of pcDNA3.1-A3G-HA, different amounts of pcDNA3.1-MOV10-FLAG (from 0.5 to 1.5 μg), 1 μg of pNL4-3ΔEnv-GFP or pNL4-3ΔEnv-GFP-ΔVif. **a**, **b** and **c** After 48 h, cell pellets and supernatants were collected respectively. Cell pellets were lysed and subjected to immunoblotting with anti-HA, anti-FLAG, anti-MOV10, anti-Vif, and anti-GAPDH antibodies. VLPs were collected from filtered supernatants by ultracentrifugation. The pelleted VLPs were lysed and detected by western blotting with anti-HA, anti-FLAG, and anti-p24 antibodies. The *bar graphs* represent the average expression of A3G with different treatments and relative to the A3G-only reaction control (set to 100%). Data in **a–c** represent mean ± SD from three independent experiments. Statistical significance was determined using *t* test: $^*p \leq 0.05$; $^{**}p \leq 0.01$; $^{***}p \leq 0.001$. In each transfection, empty vector pcDNA3.1 was used to equalize DNA amounts. Values in **a–c** represent percentages of A3G-HA normalized against GAPDH or p24 and compared with control. Results are representative of at least three independent experiments

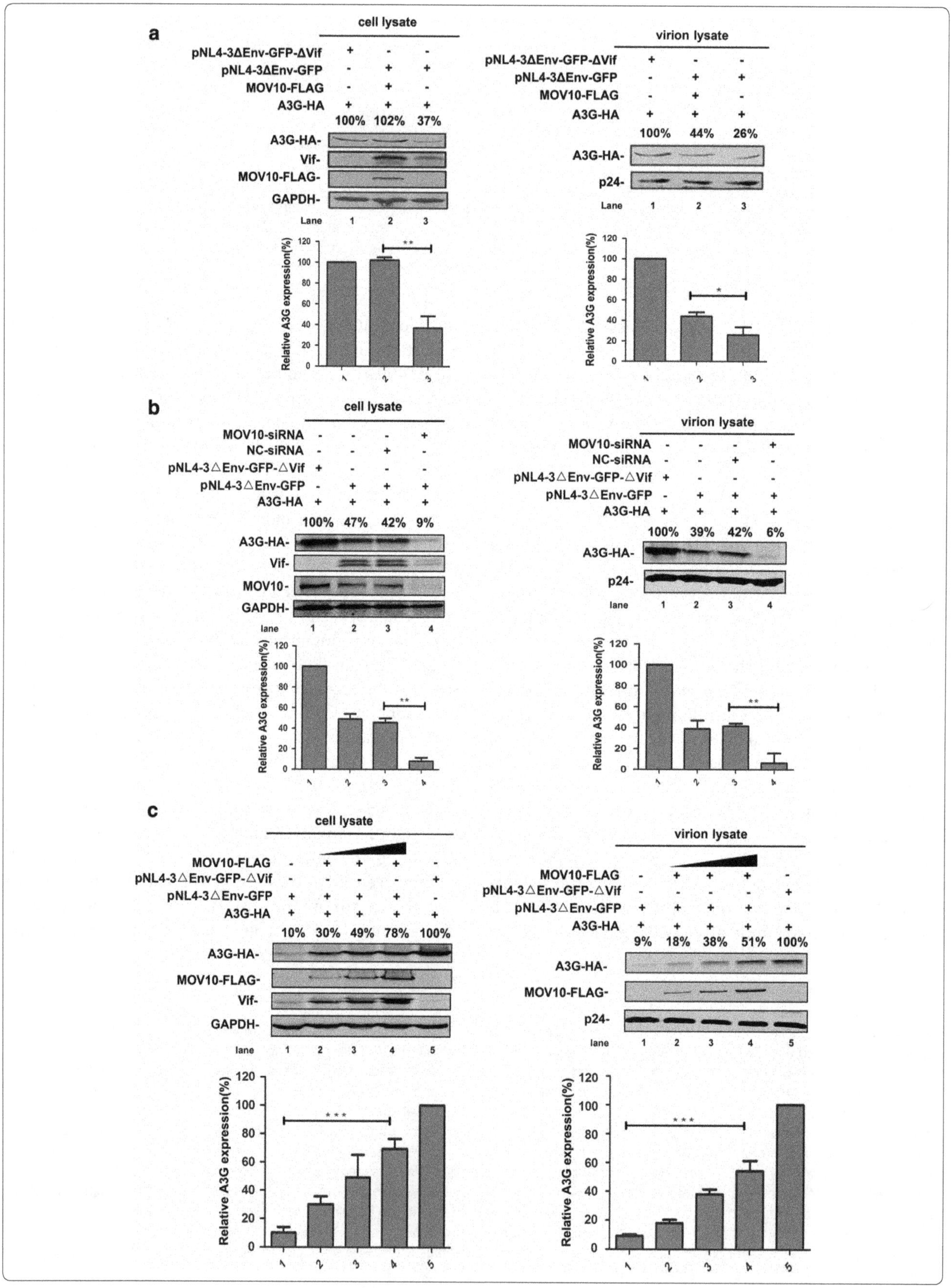
a
cell lysate
pNL4-3ΔEnv-GFP-ΔVif + - -
pNL4-3ΔEnv-GFP - + +
MOV10-FLAG - + -
A3G-HA + + +
100% 102% 37%
A3G-HA-
Vif-
MOV10-FLAG-
GAPDH-
Lane 1 2 3
Relative A3G expression(%)
120 100 80 60 40 20 0
**
1 2 3
virion lysate
pNL4-3ΔEnv-GFP-ΔVif + - -
pNL4-3ΔEnv-GFP - + +
MOV10-FLAG - + -
A3G-HA + + +
100% 44% 26%
A3G-HA-
p24-
Lane 1 2 3
Relative A3G expression(%)
*
b
cell lysate
MOV10-siRNA - - - +
NC-siRNA - - + -
pNL4-3△Env-GFP-△Vif + - - -
pNL4-3△Env-GFP - + + +
A3G-HA + + + +
100% 47% 42% 9%
A3G-HA-
Vif-
MOV10-
GAPDH-
lane 1 2 3 4
Relative A3G expression(%)
**
virion lysate
MOV10-siRNA - - - +
NC-siRNA - - + -
pNL4-3△Env-GFP-△Vif + - - -
pNL4-3△Env-GFP - + + +
A3G-HA + + + +
100% 39% 42% 6%
A3G-HA-
p24-
lane 1 2 3 4
Relative A3G expression(%)
**
c
cell lysate
MOV10-FLAG - + + + -
pNL4-3△Env-GFP-△Vif - - - - +
pNL4-3△Env-GFP + + + + -
A3G-HA + + + + +
10% 30% 49% 78% 100%
A3G-HA-
MOV10-FLAG-
Vif-
GAPDH-
lane 1 2 3 4 5
Relative A3G expression(%)
***
virion lysate
MOV10-FLAG - + + + -
pNL4-3△Env-GFP-△Vif - - - - +
pNL4-3△Env-GFP + + + + -
A3G-HA + + + + +
9% 18% 38% 51% 100%
A3G-HA-
MOV10-FLAG-
p24-
lane 1 2 3 4 5
Relative A3G expression(%)
***

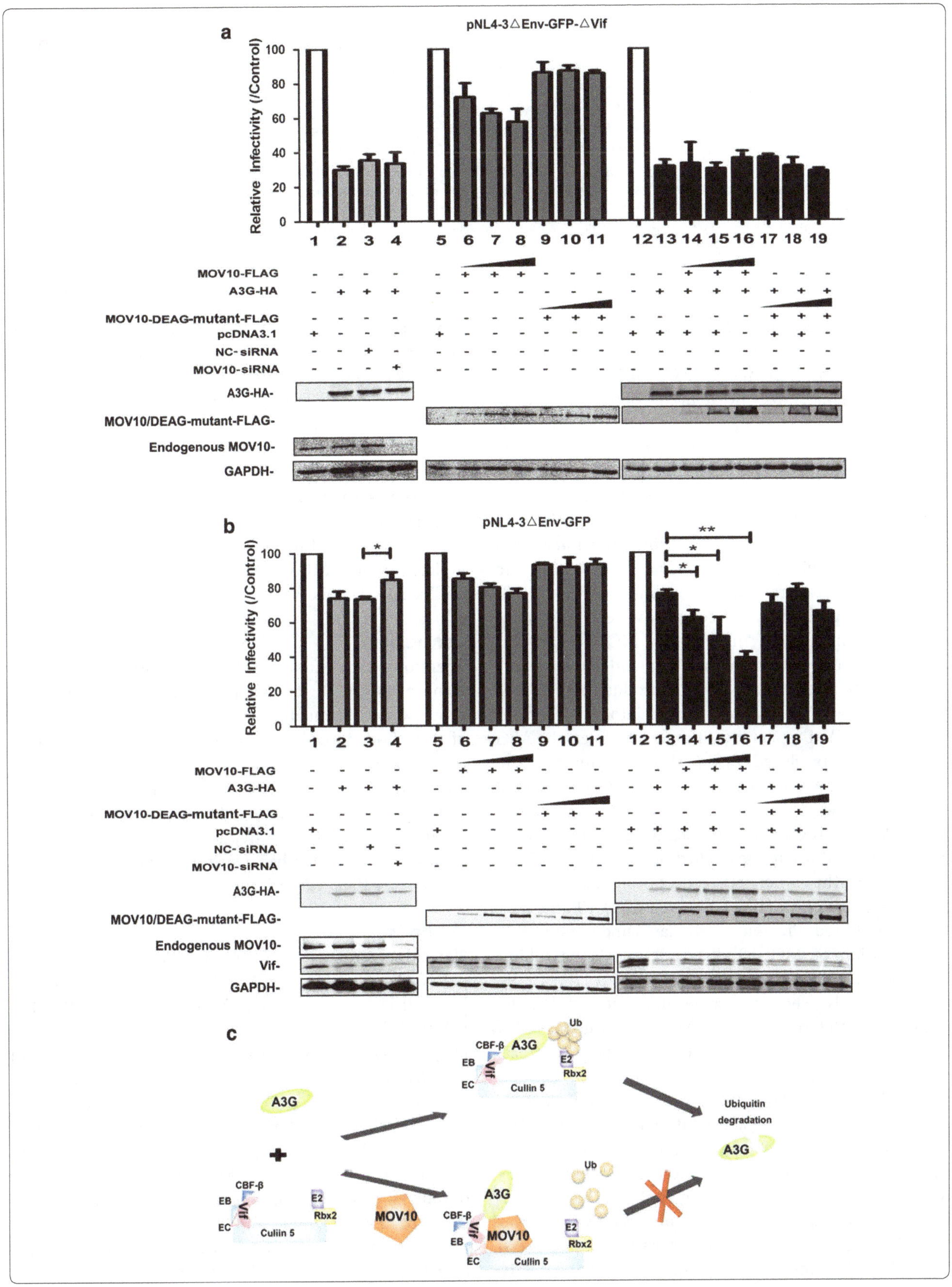
a
pNL4-3△Env-GFP-△Vif
Relative Infectivity (/Control)
MOV10-FLAG
A3G-HA
MOV10-DEAG-mutant-FLAG
pcDNA3.1
NC-siRNA
MOV10-siRNA
A3G-HA-
MOV10/DEAG-mutant-FLAG-
Endogenous MOV10-
GAPDH-
b
pNL4-3△Env-GFP
Vif-
c
A3G
MOV10
CBF-β
EB
EC
Vif
Cullin 5
E2
Rbx2
Ub
Ubiquitin degradation

(See figure on previous page.)

Fig. 10 MOV10 synergistically enhances the inhibitory effect of A3G on the infectivity of HIV-1. **a**, **b** *MOV10*-specific siRNA was transfected in 293T cells with pcDNA3.1-A3G-HA (0.8 μg), pCMV-VSV-G (2.5 μg), and pNL4-3ΔEnv-GFP-ΔVif (7.5 μg) or pNL4-3ΔEnv-GFP (7.5 μg). 293T cells were co-transfected with pCMV-VSV-G (2.5 μg), pNL4-3ΔEnv-GFP-ΔVif (7.5 μg) or pNL4-3ΔEnv-GFP (7.5 μg), and increasing amounts of pcDNA3.1-MOV10-FLAG (0.5–1.5 μg) or pcDNA3.1-MOV10-DEAG-mutant-FLAG (0.5–1.5 μg) in the presence or absence of pcDNA3.1-A3G-HA (0.8 μg). Culture supernatants containing 5 ng of p24 were used to infect TZM-bl cells and luciferase activity was determined at 72 h post infection. For viruses with different amounts of MOV10 but no A3G, and viruses with a fixed amount of A3G combined with different amounts of MOV10 or not, the data are plotted as relative infectivity, with the control virus (pcDNA3.1) set to 100%. Error bars represent standard errors from three independent experiments. Statistical significance was determined using *t* test: *$p \leq 0.05$; **$p \leq 0.01$. **c** A cartoon to show the interaction between MOV10, A3G, and Vif. In the absence of MOV10, the Vif-CBF-β-Cullin 5-ElonginB-ElonginC complex is stable and triggers the proteasomal degradation of A3G. In the presence of MOV10, the assembly of Vif-CBF-β-Cullin 5-ElonginB-ElonginC complex can be disturbed by MOV10, which leads to A3G escaping from Vif-induced proteasomal degradation

correlation between the helicase activity and anti-HIV-1 function of MOV10. Because the binding of MOV10-DEAG mutant with ElonginC or Cullin 5 decreased significantly, it almost lost the ability to protect A3G from Vif-mediated degradation, indicating that the helicase activity center of MOV10 is required for its inhibitory effect on Vif-mediated A3G degradation.

According to these data, we propose a model that, during the process of HIV-1 infection, MOV10 can interact with ElonginC and Cullin 5 to disturb the interaction of Vif with ElonginB, ElonginC, Culiin 5 or CBF-β and subsequently interfere with the assembly of Vif-CBF-β-Cullin 5-ElonginB-ElonginC complex which induces the ubiquitination of A3G. In this way, MOV10 prevents A3G from proteasomal degradation and subsequently enhances the level of A3G in virus-producing cells. It is well known that A3G can be packaged into HIV-1 virions and inhibit HIV-1 replication at multiple stages, the A3G level in newly-produced virions should be increased in the presence of MOV10 (Fig. 10c) [44]. Indeed, we found that the A3G level was significantly enhanced in virions by MOV10 overexpression and significantly reduced by MOV10 knockdown in the context of HIV-1 replication.

Moreover, the synergistic effects on the infectivity and replication of HIV-1 between MOV10 and A3G have been tested. Previous study has demonstrated that co-expression of MOV10 does not affect the inhibitory effect of A3G on the infectivity of ΔVif HIV-1 [29]. Our results also show the same phenomenon that the inhibitory effect of A3G plus MOV10 group on ΔVif HIV-1 is consistent with the effect of single A3G treatment group. As the anti-HIV-1 activity of A3G is more potent than that of MOV10, it will overspread the anti-HIV-1 effect of MOV10 when Vif deficiency. Nevertheless, consistent with our hypothesis, the infectivity of Vif-positive viral particles is synergistically inhibited by MOV10 and A3G. In 2010, Wang et al. [22] have shown that the replication of HIV-1 was enhanced by the depletion of endogenous MOV10 in permissive human T cell line (CEM-SS). Previous study also showed that MOV10 can be packaged into HIV-1 virions and inhibit viral replication at a postentry step [28]. However, little effect of MOV10 on HIV-1 replication in Hut78 T cells was reported by another group [58]. Considering that high concentration of the virus and short term infection will cover up the true effect of antiviral factors, we used low dose virus (5 ng of HIV-1 p24) to perform the experiment and extended the observation time. We found that the replication of wild-type HIV-1 was enhanced in MOV10-shRNA transduced non-permissive human T cells (H9), indicating that MOV10 and A3G can synergistically inhibit HIV-1 replication.

Conclusions

Therefore, our results reveal a novel anti-HIV-1 mechanism of MOV10: it prevents A3G from Vif-induced proteasomal degradation and then increases the levels of A3G both in cells and in newly-synthesized virions. In addition, because both MOV10 and A3G have anti-HIV-1 activity and belong to the interferon antiviral system, our findings suggested that these proteins are of synergistic anti-HIV-1 activities. These results will help us to get more comprehensive and profound understanding of MOV10 and A3G.

Methods

Plasmid construction and siRNAs synthesis

pcDNA3.1-A3G-HA, pcDNA3.1-GFP-HA, pcDNA3.1-MOV10-FLAG, pcDNA3.1-MOV10-HA, pcDNA3.1-rMOV10-FLAG, pcDNA3.1-MOV10-DEAG-mutant-HA, pcDNA3.1-MOV10-DEAG-mutant-FLAG and pcDNA3.1-Ub-FLAG were constructed as described previously [20, 32, 59]. pcDNA3.1-Vif-HA, pcDNA3.1-Vif-FLAG, pcDNA3.1-ElonginB-FLAG, pcDNA3.1-Cullin 5-FLAG, pcDNA3.1-CBF-β-FLAG, and were constructed by our lab [60]. HA or FLAG epitope tagged codon-optimized HIV-1 *vif* was constructed by chemically-synthesis of DNA fragment and subcloned into pcDNA3.1. Codon optimization was performed using the sequence of HIV-1_{NL4-3} [61]. FLAG epitope tag sequence at 3′

terminus of ElonginB, Cullin 5, or CBF-β was amplified through reverse transcription-polymerase chain reaction (RT-PCR) with the mRNA of 293T cells as the template. ElonginC with FLAG tag sequence at its 3′ terminus was amplified via PCR from the pElonginC-HA plasmid, which was generously provided by Dr. Xianghui Yu in Jilin University [62]. Then, the tagged ElonginB, ElonginC, Cullin 5, or CBF-β was inserted into pcDNA3.1 vector. HIV-1 proviral construct pNL4-3-ΔEnv-GFP has been described in our previous reports [63, 64]. pNL4-3-ΔEnv-GFP-ΔVif, a *vif* defective construct, was generated from pNL4-3-ΔEnv-GFP. The *vif* gene in pNL4-3-ΔEnv-GFP-ΔVif was disrupted by PCR-mediated site-directed mutagenesis and introduced nonsense mutations at codon positions 26, 27 (AAA, CAC → TAA, TAG) and/or 33, 34 (ACT, AAA → TAA, TAG) [65]. The vector pLKO.1-TRC, which contains a U6 promoter and *puromycin* selection gene and was obtained from Addgene (plasmid # 10878), was used for expression of MOV10-shRNA or scrambled control (Scr)-shRNA. Forward oligo of MOV10-shRNA (TRCN0000425452): 5′-CCGGGGCCAGTGTTTCGAGAGTTTCCTCGAGG AAACTCTCGAAACACTGGCCTTTTTG-3′; reverse oligo of MOV10-shRNA: 5′-AATTCAAAAAGGCCAG TGTTTCGAGAGTTTCCTCGAGGAAACTCTCGAA ACACTGGCC-3′. Scr-shRNA forward oligo: 5′- CCGGA ACGTACGCGGAATACTTCGACTCGAGTCGAAGTA TTCCGCGTACGTTTTTTTG-3′; Scr-shRNA reverse oligo: 5′- AATTCAAAAAAACGTACGCGGAATACTT CGACTCGAGTCGAAGTATTCCGCGTACGTT-3′ [40]. All oligos were synthesized from Ribobio (Guangzhou, China).

The siGENOME SMART pool against MOV10 and siRNA for negative control were designed by Dharmacon and the target sequences for *MOV10*-specific siRNAs were chosen as described previously [20, 32]. *ElonginB*-specific siRNA, *ElonginC*-specific siRNA, and *Cullin 5*-specific siRNA were designed and synthesized by Ribobio (Guangzhou, China).

Cell culture and transfection

Human 293T cells were obtained from American Type Culture Collection (ATCC) and grown at 37 °C with 5% CO2 in Dulbecco's modified Eagle's medium (DMEM) (Invitrogen) supplemented with 10% fetal bovine serum (FBS) (Invitrogen) and 1% penicillin–streptomycin (Invitrogen). The cells were transfected with the indicated plasmids or siRNAs by lipofectamine 2000 (Invitrogen). The procedures described by the manufacturer were followed.

Co-immunoprecipitation and western blotting

Co-immunoprecipitation and western blotting assays were performed as previously described [20, 32]. In brief, human 293T cells were lysed with the lysis buffer (150 mM NaCl, 50 mM Tris–HCl [pH 7.5], 1 mM EDTA, 1% Triton X-100, 0.5% NP-40, plus PMSF and protease inhibitor cocktail [Sigma]) for 30 min at 4 °C. The cell lysates were clarified by centrifugation at 18,000*g* for 30 min at 4 °C, then mixed with anti-HA agarose beads (Sigma) and incubated at 4 °C for 4 h, followed by washing four times with cold lysis buffer and eluting in gel loading buffer. As indicated, the beads were treated with RNase mixture (DNase-free, Roche) (20 μg/ml) and incubated at 37 °C for 30 min. The immunoprecipitated samples were analyzed by SDS-PAGE and detected by western blotting. Anti-HA antibody (mouse monoclonal, Covance), anti-FLAG antibody (rabbit polyclonal, MBL), anti-GAPDH antibody (rabbit polyclonal, MBL), anti-MOV10 antibody (rabbit polyclonal, Abcam), anti-ElonginC antibody (rabbit polyclonal, Abcam), anti-Cullin 5 antibody (rabbit polyclonal, Abcam), anti-A3G antibody (rabbit polyclonal, Abcam), anti-Vif antibody (mouse monoclonal, Abcam), and anti-HIV-1 p24 antibody (rabbitpolyclonal antibodies made by our lab) were used as primary antibodies [64]. Quantity One program (Bio-rad) was used to quantify the western blotting results.

HIV-1 virus-like particle (VLP) purification

Human 293T cells were transfected with pNL4-3-ΔEnv-GFP or pNL4-3-ΔEnv-GFP-ΔVif and other indicated plasmids. After 48 h of transfection, cell supernatants were collected, centrifuged at 4 °C for 10 min at 8000 rpm (≈ 7000 g) and filtered through a 0.45 μm filter to remove cellular debris. And the cell-free supernatants were concentrated by ultracentrifugation through a 20% sucrose cushions at 4 °C for 2 h at 45,000 rpm (≈ 40,000 g) (HITACHI Preparative Ultracentrifuge, CP80WX). Then, the pellets were re-suspended in RIPA buffer containing protease inhibitor cocktail and subjected to immunoblotting.

Construction of MOV10-knockdown H9 cells

The pLKO.1-MOV10-shRNA or pLKO.1-Scr-shRNA was co-transfected with psPAX2 and pCMV-VSV-G into 293T cells. After 48 h, the supernatants were harvested and filtered with 0.45 μm filters (Millipore). Then, H9 cells were infected with MOV10-specific-shRNA-expressing or Scr-shRNA-expressing lentivirus respectively for 8 h and cultured with fresh medium. After 48 h, virus-infected H9 cells were selected by puromycin (1 μg/ml) for 2 weeks and subjected to the following experiments.

Wild-type HIV-1 infection

MOV10-knockdown H9 cells and negative control cells were infected with HIV-1_{NL4-3} (p24 titer of 5 ng ml^{-1}) for 3 h and then cultured with fresh medium and detected p24 in culture supernatant at different days. After 12 days, cells were collected and treated with the transcription factor buffer set including fixation/permeabilization and fixation/wash buffers (BD Biosciences) according to the manufacturer supernatant at different days. After 12 days, cellFITC-conjugated anti-HIV-1 p24 antibody (Santa Cruz Biotechnology) for intracellular HIV-1 Gag (p24) expression. Then, p24 positive cells were sorted with BD Aria sorter for further analysis. Data was analyzed with FlowJo software (Tree Star, Ashland, OR).

Virus infectivity assay

Human 293T cells were co-transfected with pCMV-VSV-G, pNL4-3ΔEnv-GFP-ΔVif (or pNL4-3ΔEnv-GFP), and increasing amounts of pcDNA3.1-MOV10-FLAG or pcDNA31-MOV10-mutant-FLAG in the presence or absence of A3G-HA expressing plasmid. The virus-containing supernatant were collected at 48 h after transfection and filtered by a 0.45 μm filter. After normalization for HIV-1 p24 by enzyme-linked immunosorbent assay (ELISA, Clonetech), TZM-bl cells (2.5×10^5 cells per well in 24-well plates) were infected with virus which containing 5 ng of p24 capsid. And then, luciferase enzyme activity determinations at 72 h post infection were carried out.

Abbreviations

MOV10: Moloney leukemia virus 10; A3G: Apolipoprotein-B-mRNA-editing enzyme catalytic polypeptide-like 3G; USP: Ubiquitin–proteasome system.

Authors' contributions

CL and HZ designed this project. CC, XM, QH, and CL performed most of the experiments. CC, CL, and HZ analyzed the results and wrote the paper. XL, FH, JZ, and JX participated in some experiments. TP constructed ElonginB- and Cullin 5-expressing plasmids. All authors read and approved the final manuscript.

Author details

[1] Institute of Human Virology, Zhongshan School of Medicine, Sun Yat-sen University, Guangzhou 510080, China. [2] Key Laboratory of Tropical Disease Control of Ministry of Education, Zhongshan School of Medicine, Sun Yat-sen University, Guangzhou 510080, China. [3] Department of Pathology, The First Affiliated Hospital, Sun Yat-sen University, Guangzhou 510080, China. [4] Department of Infectious Diseases, The Fifth Affiliated Hospital, Sun Yat-sen University, Zhuhai 519000, China.

Acknowledgements

We obtained pNL4-3ΔEnv-GFP from the National Institutes of Health AIDS Research and Reference Reagent Program. We thank Dr. Xianghui Yu (Jinli University, China) for the ElonginC-HA-expressing plasmid, pElonginC-HA.

Competing interests

The authors declare that they have no competing interests.

Funding

This work was supported by the National Natural Science Foundation of China (No. 81471935), the National Natural Science Foundation of China (NSFC-NIH project) (No. 81561128007), the Important Key Program of Natural Science Foundation of China (No. 81590765), the Introduction of Innovative R&D Team Program of Guangdong Province (No. 2009010058), and the Joint-innovation Program in Healthcare for Special Scientific Research Projects of Guangzhou, China (No. 201508020256).

References

1. Zhang H, Yang B, Pomerantz RJ, Zhang C, Arunachalam SC, Gao L. The cytidine deaminase CEM15 induces hypermutation in newly synthesized HIV-1 DNA. Nature. 2003;424:94–8.
2. Mariani R, Chen D, Schrofelbauer B, Navarro F, Konig R, Bollman B, Munk C, Nymark-McMahon H, Landau NR. Species-specific exclusion of APOBEC3G from HIV-1 virions by Vif. Cell. 2003;114:21–31.
3. Lecossier D, Bouchonnet F, Clavel F, Hance AJ. Hypermutation of HIV-1 DNA in the absence of the Vif protein. Science. 2003;300:1112.
4. Mangeat B, Turelli P, Caron G, Friedli M, Perrin L, Trono D. Broad antiretroviral defence by human APOBEC3G through lethal editing of nascent reverse transcripts. Nature. 2003;424:99–103.
5. Bourara K, Liegler TJ, Grant RM. Target cell APOBEC3C can induce limited G-to-A mutation in HIV-1. PLoS Pathog. 2007;3:1477–85.
6. Mehle A, Strack B, Ancuta P, Zhang C, McPike M, Gabuzda D. Vif overcomes the innate antiviral activity of APOBEC3G by promoting its degradation in the ubiquitin-proteasome pathway. J Biol Chem. 2004;279:7792–8.
7. Sheehy AM, Gaddis NC, Malim MH. The antiretroviral enzyme APOBEC3G is degraded by the proteasome in response to HIV-1 Vif. Nat Med. 2003;9:1404–7.
8. Conticello SG, Harris RS, Neuberger MS. The Vif protein of HIV triggers degradation of the human antiretroviral DNA deaminase APOBEC3G. Curr Biol. 2003;13:2009–13.
9. Stopak K, de Noronha C, Yonemoto W, Greene WC. HIV-1 Vif blocks the antiviral activity of APOBEC3G by impairing both its translation and intracellular stability. Mol Cell. 2003;12:591–601.
10. Marin M, Rose KM, Kozak SL, Kabat D. HIV-1 Vif protein binds the editing enzyme APOBEC3G and induces its degradation. Nat Med. 2003;9:1398–403.
11. Yu X, Yu Y, Liu B, Luo K, Kong W, Mao P, Yu XF. Induction of APOBEC3G ubiquitination and degradation by an HIV-1 Vif-Cul5-SCF complex. Science. 2003;302:1056–60.
12. Kobayashi M, Takaori-Kondo A, Miyauchi Y, Iwai K, Uchiyama T. Ubiquitination of APOBEC3G by an HIV-1 Vif-Cullin5-Elongin B-Elongin C complex is essential for Vif function. J Biol Chem. 2005;280:18573–8.
13. Jager S, Kim DY, Hultquist JF, Shindo K, LaRue RS, Kwon E, Li M, Anderson BD, Yen L, Stanley D, et al. Vif hijacks CBF-beta to degrade APOBEC3G and promote HIV-1 infection. Nature. 2012;481:371–5.
14. Yu Y, Xiao Z, Ehrlich ES, Yu X, Yu XF. Selective assembly of HIV-1 Vif-Cul5-ElonginB-ElonginC E3 ubiquitin ligase complex through a novel SOCS box and upstream cysteines. Genes Dev. 2004;18:2867–72.
15. Zhang W, Du J, Evans SL, Yu Y, Yu XF. T-cell differentiation factor CBF-beta regulates HIV-1 Vif-mediated evasion of host restriction. Nature. 2012;481:376–9.
16. Wang X, Wang X, Zhang H, Lv M, Zuo T, Wu H, Wang J, Liu D, Wang C, Zhang J, et al. Interactions between HIV-1 Vif and human ElonginB-ElonginC are important for CBF-beta binding to Vif. Retrovirology. 2013;10:94.
17. Koonin EV. A new group of putative RNA helicases. Trends Biochem Sci. 1992;17:495–7.
18. Gregersen LH, Schueler M, Munschauer M, Mastrobuoni G, Chen W, Kempa S, Dieterich C, Landthaler M. MOV10 Is a 5′ to 3′ RNA helicase contributing to UPF1 mRNA target degradation by translocation along 3′ UTRs. Mol Cell. 2014;54:573–85.
19. Meister G, Landthaler M, Peters L, Chen PY, Urlaub H, Luhrmann R, Tuschl T. Identification of novel argonaute-associated proteins. Curr Biol. 2005;15:2149–55.

20. Liu C, Zhang X, Huang F, Yang B, Li J, Liu B, Luo H, Zhang P, Zhang H. APOBEC3G inhibits microRNA-mediated repression of translation by interfering with the interaction between Argonaute-2 and MOV10. J Biol Chem. 2012;287:29373–83.
21. El Messaoudi-Aubert S, Nicholls J, Maertens GN, Brookes S, Bernstein E, Peters G. Role for the MOV10 RNA helicase in polycomb-mediated repression of the INK4a tumor suppressor. Nat Struct Mol Biol. 2010;17:862–8.
22. Wang X, Han Y, Dang Y, Fu W, Zhou T, Ptak RG, Zheng YH. Moloney leukemia virus 10 (MOV10) protein inhibits retrovirus replication. J Biol Chem. 2010;285:14346–55.
23. Furtak V, Mulky A, Rawlings SA, Kozhaya L, Lee K, Kewalramani VN, Unutmaz D. Perturbation of the P-body component Mov10 inhibits HIV-1 infectivity. PLoS One. 2010;5:e9081.
24. Schoggins JW, Wilson SJ, Panis M, Murphy MY, Jones CT, Bieniasz P, Rice CM. A diverse array of gene products are effectors of the type I interferon antiviral response. Nature. 2011;472:481–5.
25. Schoggins JW, Rice CM. Interferon-stimulated genes and their antiviral effector functions. Curr Opin Virol. 2011;1:519–25.
26. Goodier JL, Cheung LE, Kazazian HH Jr. MOV10 RNA helicase is a potent inhibitor of retrotransposition in cells. PLoS Genet. 2012;8:e1002941.
27. Li X, Zhang J, Jia R, Cheng V, Xu X, Qiao W, Guo F, Liang C, Cen S. The MOV10 helicase inhibits LINE-1 mobility. J Biol Chem. 2013;288:21148–60.
28. Abudu A, Wang X, Dang Y, Zhou T, Xiang SH, Zheng YH. Identification of molecular determinants from Moloney leukemia virus 10 homolog (MOV10) protein for virion packaging and anti-HIV-1 activity. J Biol Chem. 2012;287:1220–8.
29. Burdick R, Smith JL, Chaipan C, Friew Y, Chen J, Venkatachari NJ, Delviks-Frankenberry KA, Hu WS, Pathak VK. P body-associated protein Mov10 inhibits HIV-1 replication at multiple stages. J Virol. 2010;84:10241–53.
30. Zheng YH, Jeang KT, Tokunaga K. Host restriction factors in retroviral infection: promises in virus-host interaction. Retrovirology. 2012;9:112.
31. Izumi T, Burdick R, Shigemi M, Plisov S, Hu WS, Pathak VK. Mov10 and APOBEC3G localization to processing bodies is not required for virion incorporation and antiviral activity. J Virol. 2013;87:11047–62.
32. Huang F, Zhang J, Zhang Y, Geng G, Liang J, Li Y, Chen J, Liu C, Zhang H. RNA helicase MOV10 functions as a co-factor of HIV-1 Rev to facilitate Rev/RRE-dependent nuclear export of viral mRNAs. Virology. 2015;486:15–26.
33. Donahue JP, Vetter ML, Mukhtar NA, D'Aquila RT. The HIV-1 Vif PPLP motif is necessary for human APOBEC3G binding and degradation. Virology. 2008;377:49–53.
34. Lu X, Zhang T, Xu Z, Liu S, Zhao B, Lan W, Wang C, Ding J, Cao C. Crystal structure of DNA cytidine deaminase ABOBEC3G catalytic deamination domain suggests a binding mode of full-length enzyme to single-stranded DNA. J Biol Chem. 2015;290:4010–21.
35. Feng Y, Baig TT, Love RP, Chelico L. Suppression of APOBEC3-mediated restriction of HIV-1 by Vif. Front Microbiol. 2014;5:450.
36. Kouno T, Luengas EM, Shigematsu M, Shandilya SM, Zhang J, Chen L, Hara M, Schiffer CA, Harris RS, Matsuo H. Structure of the Vif-binding domain of the antiviral enzyme APOBEC3G. Nat Struct Mol Biol. 2015;22:485–91.
37. Wang H, Liu B, Liu X, Li Z, Yu XF, Zhang W. Identification of HIV-1 Vif regions required for CBF-beta interaction and APOBEC3 suppression. PLoS One. 2014;9:e95738.
38. Wang H, Lv G, Zhou X, Li Z, Liu X, Yu XF, Zhang W. Requirement of HIV-1 Vif C-terminus for Vif-CBF-beta interaction and assembly of CUL5-containing E3 ligase. BMC Microbiol. 2014;14:290.
39. Kamura T, Burian D, Yan Q, Schmidt SL, Lane WS, Querido E, Branton PE, Shilatifard A, Conaway RC, Conaway JW. Muf1, a novel Elongin BC-interacting leucine-rich repeat protein that can assemble with Cul5 and Rbx1 to reconstitute a ubiquitin ligase. J Biol Chem. 2001;276:29748–53.
40. He Z, Zhang W, Chen G, Xu R, Yu XF. Characterization of conserved motifs in HIV-1 Vif required for APOBEC3G and APOBEC3F interaction. J Mol Biol. 2008;381:1000–11.
41. Schafer A, Bogerd HP, Cullen BR. Specific packaging of APOBEC3G into HIV-1 virions is mediated by the nucleocapsid domain of the gag polyprotein precursor. Virology. 2004;328:163–8.
42. Svarovskaia ES, Xu H, Mbisa JL, Barr R, Gorelick RJ, Ono A, Freed EO, Hu WS, Pathak VK. Human apolipoprotein B mRNA-editing enzyme-catalytic polypeptide-like 3G (APOBEC3G) is incorporated into HIV-1 virions through interactions with viral and nonviral RNAs. J Biol Chem. 2004;279:35822–8.
43. Zennou V, Perez-Caballero D, Gottlinger H, Bieniasz PD. APOBEC3G incorporation into human immunodeficiency virus type 1 particles. J Virol. 2004;78:12058–61.
44. Martin KL, Johnson M, D'Aquila RT. APOBEC3G complexes decrease human immunodeficiency virus type 1 production. J Virol. 2011;85:9314–26.
45. Shao Q, Wang Y, Hildreth JE, Liu B. Polyubiquitination of APOBEC3G is essential for its degradation by HIV-1 Vif. J Virol. 2010;84:4840–4.
46. Baig TT, Feng Y, Chelico L. Determinants of efficient degradation of APOBEC3 restriction factors by HIV-1 Vif. J Virol. 2014;88:14380–95.
47. Doerks T, Copley RR, Schultz J, Ponting CP, Bork P. Systematic identification of novel protein domain families associated with nuclear functions. Genome Res. 2002;12:47–56.
48. Dang Y, Siew LM, Zheng YH. APOBEC3G is degraded by the proteasomal pathway in a Vif-dependent manner without being polyubiquitylated. J Biol Chem. 2008;283:13124–31.
49. Bergeron JR, Huthoff H, Veselkov DA, Beavil RL, Simpson PJ, Matthews SJ, Malim MH, Sanderson MR. The SOCS-box of HIV-1 Vif interacts with ElonginBC by induced-folding to recruit its Cul5-containing ubiquitin ligase complex. PLoS Pathog. 2010;6:e1000925.
50. Wang J, Zhang W, Lv M, Zuo T, Kong W, Yu X. Identification of a Cullin5-ElonginB-ElonginC E3 complex in degradation of feline immunodeficiency virus Vif-mediated feline APOBEC3 proteins. J Virol. 2011;85:12482–91.
51. Guo Y, Dong L, Qiu X, Wang Y, Zhang B, Liu H, Yu Y, Zang Y, Yang M, Huang Z. Structural basis for hijacking CBF-beta and CUL5 E3 ligase complex by HIV-1 Vif. Nature. 2014;505:229–33.
52. Iwatani Y, Chan DS, Liu L, Yoshii H, Shibata J, Yamamoto N, Levin JG, Gronenborn AM, Sugiura W. HIV-1 Vif-mediated ubiquitination/degradation of APOBEC3G involves four critical lysine residues in its C-terminal domain. Proc Natl Acad Sci USA. 2009;106:19539–44.
53. Mehle A, Thomas ER, Rajendran KS, Gabuzda D. A zinc-binding region in Vif binds Cul5 and determines cullin selection. J Biol Chem. 2006;281:17259–65.
54. Skowyra D, Craig KL, Tyers M, Elledge SJ, Harper JW. F-box proteins are receptors that recruit phosphorylated substrates to the SCF ubiquitin-ligase complex. Cell. 1997;91:209–19.
55. Fribourgh JL, Nguyen HC, Wolfe LS, Dewitt DC, Zhang W, Yu XF, Rhoades E, Xiong Y. Core binding factor beta plays a critical role by facilitating the assembly of the Vif-cullin 5 E3 ubiquitin ligase. J Virol. 2014;88:3309–19.
56. Zhou X, Evans SL, Han X, Liu Y, Yu XF. Characterization of the interaction of full-length HIV-1 Vif protein with its key regulator CBFbeta and CRL5 E3 ubiquitin ligase components. PLoS One. 2012;7:e33495.
57. Kim DY, Kwon E, Hartley PD, Crosby DC, Mann S, Krogan NJ, Gross JD. CBF-beta stabilizes HIV Vif to counteract APOBEC3 at the expense of RUNX1 target gene expression. Mol Cell. 2013;49:632–44.
58. Arjan-Odedra S, Swanson CM, Sherer NM, Wolinsky SM, Malim MH. Endogenous MOV10 inhibits the retrotransposition of endogenous retroelements but not the replication of exogenous retroviruses. Retrovirology. 2012;9:53.
59. Li J, Chen C, Ma X, Geng G, Liu B, Zhang Y, Zhang S, Zhong F, Liu C, Yin Y, et al. Long noncoding RNA NRON contributes to HIV-1 latency by specifically inducing tat protein degradation. Nat Commun. 2016;7:11730.
60. Zhang S, Zhong L, Chen B, Pan T, Zhang X, Liang L, Li Q, Zhang Z, Chen H, Zhou J, et al. Identification of an HIV-1 replication inhibitor which rescues host restriction factor APOBEC3G in Vif-APOBEC3G complex. Antiviral Res. 2015;122:20–7.
61. Nguyen KL, Llano M, Akari H, Miyagi E, Poeschla EM, Strebel K, Bour S. Codon optimization of the HIV-1 vpu and vif genes stabilizes their mRNA and allows for highly efficient Rev-independent expression. Virology. 2004;319:163–75.
62. Zuo T, Liu D, Lv W, Wang X, Wang J, Lv M, Huang W, Wu J, Zhang H, Jin H, et al. Small-molecule inhibition of human immunodeficiency virus type 1 replication by targeting the interaction between Vif and ElonginC. J Virol. 2012;86:5497–507.
63. Huang J, Wang F, Argyris E, Chen K, Liang Z, Tian H, Huang W, Squires K, Verlinghieri G, Zhang H. Cellular microRNAs contribute to HIV-1 latency in resting primary CD4 + T lymphocytes. Nat Med. 2007;13:1241–7.

7

Natural APOBEC3C variants can elicit differential HIV-1 restriction activity

Brett D. Anderson[1†], Terumasa Ikeda[1,2†], Seyed Arad Moghadasi[1†], Amber St. Martin[1], William L. Brown[1] and Reuben S. Harris[1,2*]

Abstract

Background: The APOBEC3 (A3) family of DNA cytosine deaminases provides an innate barrier to infection by retroviruses including HIV-1. A total of five enzymes, A3C, A3D, A3F, A3G and A3H, are degraded by the viral accessory protein Vif and expressed at high levels in CD4+ T cells, the primary reservoir for HIV-1 replication in vivo. Apart from A3C, all of these enzymes mediate restriction of Vif-deficient HIV-1. However, a rare variant of human A3C (Ile188) was shown recently to restrict Vif-deficient HIV-1 in a 293T-based single cycle infection system. The potential activity of this naturally occurring A3C variant has yet to be characterized in a T cell-based spreading infection system. Here we employ a combination of Cas9/gRNA disruption and transient and stable protein expression to assess the roles of major Ser188 and minor Ile188 A3C variants in HIV-1 restriction in T cell lines.

Results: Cas9-mediated mutation of endogenous A3C in the non-permissive CEM2n T cell line did not alter HIV-1 replication kinetics, and complementation with A3C-Ser188 or A3C-Ile188 was similarly aphenotypic. Stable expression of A3C-Ser188 in the permissive T cell line SupT11 also had little effect. However, stable expression of A3C-Ile188 in SupT11 cells inhibited Vif-deficient virus replication and inflicted G-to-A mutations.

Conclusions: A3C-Ile188 is capable of inhibiting Vif-deficient HIV-1 replication in T cells. Although A3C is eclipsed by the dominant anti-viral activities of other A3s in non-permissive T cell lines and primary T lymphocytes, this enzyme may still be able to contribute to HIV-1 diversification in vivo. Our results highlight the functional redundancy in the human A3 family with regards to HIV-1 restriction and the need to consider naturally occurring variants.

Keywords: APOBEC3C, DNA cytosine deaminase, HIV-1, Human genetic variation, Innate immunity, Retrovirus restriction factor

Background

Retroviruses, including the AIDS virus human immunodeficiency virus type-1 (HIV-1), must evade destruction by an extensive array of antiviral host proteins known as restriction factors [1–3]. The seven human APOBEC3 (A3) enzymes constitute an important arm of this innate network of restriction factors. These enzymes catalyze the deamination of cytosine to uracil in single-stranded (ss) DNA substrates, and APOBEC3D (A3D), APOBEC3F (A3F), APOBEC3G (A3G) and APOBEC3H (A3H) are known to contribute to HIV-1 restriction (reviewed by [4, 5]). These four enzymes package into budding virions and, following virus entry into a new target cell, catalyze the formation of uracil lesions in reverse-transcription intermediates. These uracils template the incorporation of adenines during synthesis of the viral genomic strand, resulting in G-to-A mutations.

HIV-1 and related lentiviruses avoid lethal levels of A3 induced mutation through the virion infectivity factor (Vif), which is a small viral protein required for virus infectivity in most cell types. Vif functions as a molecular adapter to recruit A3 enzymes to a host ubiquitin ligase comprised of CBF-β, CUL5, RBX2, ELOB, and ELOC for polyubiquitination and subsequent degradation by

*Correspondence: rsh@umn.edu
†Brett D. Anderson, Terumasa Ikeda and Seyed Arad Moghadasi have contributed equally to this work
[2] Howard Hughes Medical Institute, University of Minnesota, 2231 6th St. S.E., Minneapolis, MN 55455, USA
Full list of author information is available at the end of the article

the proteasome [6–8]. HIV-1 Vif effectively triggers the degradation of human A3D, A3F, A3G and A3H, which would otherwise restrict virus replication [9]. Through CBF-β recruitment, Vif also downregulates the transcriptional activity of these *A3* genes and thereby contributes to a permissive environment for virus replication [10].

Curiously, although human APOBEC3C (A3C) elicits little Vif-deficient HIV-1 restriction activity, Vif efficiently targets it for degradation [9, 11–13]. Human A3C also is highly expressed in the primary cellular reservoir for HIV-1 replication, CD4+ T cells, and is upregulated upon HIV-1 infection similar to the other restrictive A3 proteins [9]. Moreover, human A3C also potently restricts a strain of simian immunodeficiency virus (SIV) isolated from African Green Monkeys (SIVagm) in the absence of Vif in 293T-based single cycle infection experiments [11, 14]. These observations combine to indicate that human A3C is a *bona fide* retrovirus restriction enzyme and, further, that HIV-1 (or a SIV precursor from chimpanzees) may have recently evolved a Vif-independent mechanism for evading restriction by this enzyme (since little restriction is observed in the absence of Vif).

The *A3* gene family exhibits significant variation within the human population [15]. First, a deletion spanning the entire *A3B* coding sequence occurs at 37% frequency worldwide with clear geographic biases [16, 17]. Second, *A3H* has at least 7 distinct haplotypes that exhibit drastically different expression levels and HIV-1 restriction phenotypes [17–21]. The most common A3H haplotype (hap I) is a poorly expressed protein with little to no Vif-deficient HIV-1 restriction activity, and it is found at a 48% frequency worldwide and implicated in cancer mutagenesis [17–22]. In contrast, the next most common A3H haplotype (hap II) is well expressed and shows strong Vif-deficient HIV-1 restriction activity [17–22]. Third, two recent reports have characterized a rare A3C-Ile188 variant (~2% global allele frequency) with enhanced restriction activity against Vif-deficient HIV-1 in a 293T based single cycle infection model in comparison to the predominant A3C-Ser188 enzyme [11, 23]. Biochemical studies have indicated that the enhanced restriction activity of A3C-Ile188 may be due to increased enzyme processivity for cytosine deamination and/or to an increased propensity to homodimerize [11, 23].

The impact of the rare A3C-Ile188 enzyme on HIV-1 restriction in T lymphocytes has yet to be studied. Here, using a combination of Cas9/gRNA and stable expression approaches in T cell lines, we show that neither genetic disruption of *A3C* in the non-permissive CEM2n T cell line, nor complementation with natural A3C variants is sufficient to alter HIV-1 replication kinetics in the presence or absence of Vif. However, stable expression of A3C-Ile188 can slow Vif-deficient virus replication and inflict G-to-A mutations in the permissive SupT11 T cell line in the absence of other A3 proteins. Taken together, these findings confirm prior reports identifying the rare Ile188 allele as a more restrictive form of A3C. Our studies underscore the possibility that this naturally occurring A3C variation may influence HIV-1 genetic variation and pathogenesis in vivo.

Results

Immunoblot comparisons of epitope-tagged and untagged A3C variants

The first study comparing A3C-Ser188 and A3C-Ile188 used a C-terminal HA tag and reported lower expression levels of the C-terminally tagged Ile188 variant [15]. A more recent report using C-terminally HA-tagged A3C variants in a different vector showed no significant expression difference [11]. Our results using different C-terminally triple-HA-tagged variants mirror the original report with the Ile188 variant expressing at lower levels (Fig. 1, top blot, lanes 2/3 vs 4/5). Another recent study used N-terminally HA-tagged variants and found no expression difference [23]. Our results with different N-terminally triple-HA-tagged A3C variants confirm this observation (Fig. 1, top blot, lanes 6/7 vs 8/9). We postulate that the observed instability of the Ile188 variant may be due to an interaction with the C-terminal HA-epitope tag in some, but not all, vectors (i.e., a potential epitope tag-associated artifact).

In an attempt to resolve this issue, we tested all available commercial anti-A3C antibodies in immunoblot experiments and found only one reagent that

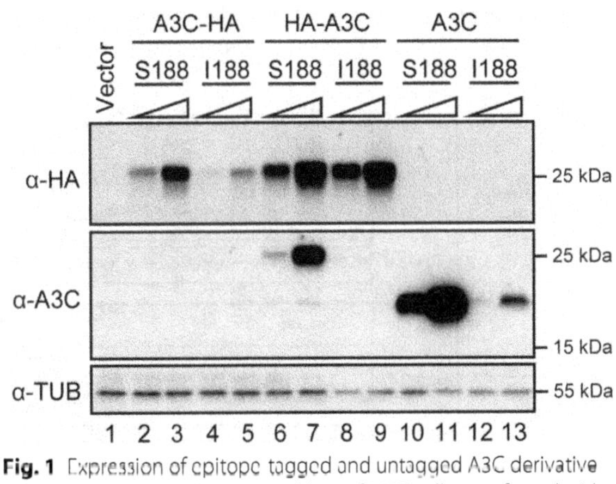

Fig. 1 Expression of epitope tagged and untagged A3C derivative constructs in 293T cells. Immunoblots of 293T cells transfected with 100 ng or 400 ng of C-terminally tagged (A3C-HA), N-terminally tagged (HA-A3C), or untagged A3C variants using either an anti-HA or anti-A3C antibody for detection. Tubulin was used as a loading control

reproducibly detected both transfected and endogenous A3C (Proteintech #10591-1-AP; see validation in Cas9/gRNA disruption studies below and "Methods" for a list of reagents that failed). This polyclonal antibody was raised against full-length A3C, and it enabled confirmation of a subset of the anti-epitope blots above and yielded an additional new consideration. First, it showed that the N-terminally HA-tagged A3C-Ser188 construct expresses much better than the reciprocal C-terminally HA-tagged construct (Fig. 1, middle blot, lanes 6/7 vs 2/3). Second, it strongly detected HA-A3C-Ser188 but only weakly the Ile188 variant (Fig. 1, middle blot, lanes 6/7 vs 8/9). This expression difference was particularly clear for untagged A3C proteins (Fig. 1, middle blot, lanes 10/11 vs 12/13). Because these proteins only differ by one amino acid, these data strongly indicate that the dominant antibody in the polyclonal mixture recognizes an epitope that requires Ser188 and is somehow compromised by Ile188. Consistent with this interpretation, a different single amino acid substitution, Ser188 to Leu, also renders A3C undetectable in immunoblot experiments with this antibody (data not shown). These epitope tag and immunoblot limitations impose significant constraints on studies with A3C.

Single cycle infectivity results using 293T cells

Because untagged, and to a lesser degree N-terminally HA-tagged, A3C-Ser188 displays visibly higher expression levels than C-terminally HA-tagged derivatives, untagged and N-terminally HA-tagged constructs were used for functional studies from here onward. In agreement with prior reports [11, 23], untagged A3C-Ser188 caused a modest twofold reduction in the infectivity of Vif-deficient HIV-1 in the 293T-based single cycle infection system (Fig. 2a). A comparison of this A3C-Ser188 construct and a Glu68-to-Gln derivative indicated that effect is largely independent of the conserved catalytic glutamate. In contrast, A3C-Ile188 caused a larger fourfold infectivity decline that partly required deaminase activity (Fig. 2a). Parallel reactions included A3G as a strong positive control and an empty expression vector as a negative control (Fig. 2a). In comparison, similar restriction activities were observed using N-terminally tagged A3C Ser188 and Ile188 expression constructs with a trend toward greater restriction by the Ile188 protein at higher expression levels (Fig. 2b). The untagged and N-terminally tagged constructs were not analyzed in the same experiments in parallel so head-to-head comparisons of restriction activity are not possible. The most important point from these single cycle infectivity

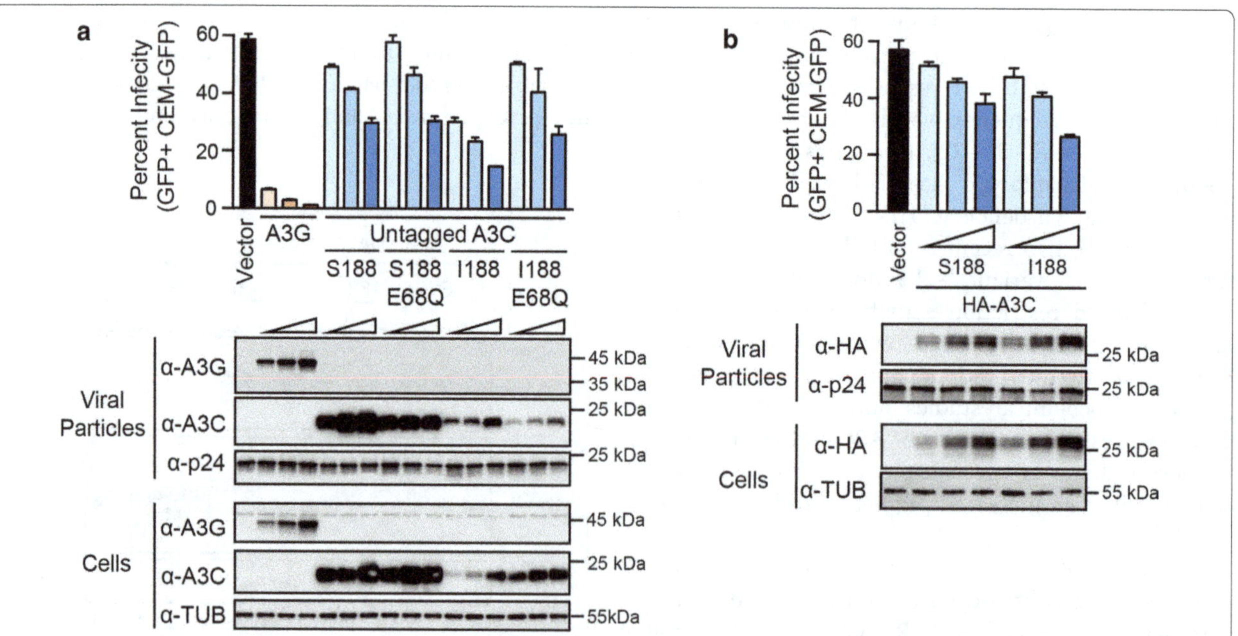

Fig. 2 A3C-Ile188 exhibits enhanced HIV-1 restriction activity in 293T cells. **a** Single cycle infectivity data for Vif-deficient HIV-1 viruses produced in the presence of untagged A3C-S188, A3C-I188, or catalytic mutant derivatives (E68Q). Immunoblots are shown below for viral particles (anti-A3G, anti-A3C, and anti-p24) and producer cells (anti-A3G, anti-A3C, and anti-tubulin). **b** Single cycle infectivity data for Vif-deficient HIV-1 viruses produced in the presence of N-terminally HA-tagged A3C-S188 or A3C-I188. Immunoblots are shown below for viral particles (anti-HA and anti-p24) and producer cells (anti-HA and anti-tubulin). All single cycle experiments were repeated at least 3 times, with representative infectivity data (mean ± SD) and immunoblots shown for one experiment

experiments is that the Ile188 enzyme is approximately twofold more restrictive than the Ser188 enzyme.

A3C disruption and variant complementation has little effect on HIV-1 replication kinetics in non-permissive CEM2n cells

Little is known about the role of endogenous A3C in HIV-1 restriction in T cells. Endogenous A3C is detected readily in non-permissive H9 and CEM2n T cell lines, but not in the permissive SupT11 T cell line using the commercial anti-A3C antibody described above (Fig. 3a). Cas9/gRNA-mediated genome editing was used to disrupt endogenous *A3C* in CEM2n [24]. The parental CEM2n line was transduced with a vector expressing Cas9 and an *A3C*-specific guide RNA, and limiting dilution was used to isolate clones harboring no detectable A3C protein by immunoblot (e.g., clone 1 in Fig. 3b). Genomic DNA sequencing of the gRNA-targeted exon 3 region of *A3C* revealed multiple mutant alleles and, importantly, no wild-type sequences (Fig. 3c). The presence of 3 distinct alleles indicates that the original near-diploid cell line CEM2n [24] may have increased ploidy during culture or duplicated the *A3C* region of the genome. Regardless, A3C-disrupted clones showed no overt phenotypes such as morphology or growth rates in comparison to non-targeted sister clones (data not shown).

Next, a complementation experiment was set-up by transducing the *A3C* mutant clone described above with viruses expressing A3C-Ser188, A3C-Ile188, or an empty vector as a negative control. Each construct also expressed eGFP to control for transduction efficiency (and to help mitigate the fact that the untagged A3C-Ile188 variant cannot be detected with available antibodies, as shown in Fig. 1). In all instances, near complete transduction efficiency was confirmed by GFP+ flow cytometry (Fig. 3d). Moreover, immunoblotting showed that complemented A3C-Ser188 expression levels are similar to endogenous levels normally found in the parental CEM2n line (Fig. 3e). As explained above, protein-level expression of A3C-Ile188 could not be measured accurately due to Ser188 likely being part of the main epitope recognized by the antibody. Nevertheless, these two variants are expected to express at similar levels because the vectors are otherwise isogenic, the Ser188 and Ile188 variants express similarly with N-terminal tags, and the eGFP levels of the complemented pools are almost indistinguishable (and results below show a phenotype with the same A3C-Ile188 construct).

To address the role of endogenous A3C (Ser188/Ser188) and to compare complemented A3C-Ser188 and A3C-Ile188 variants in parallel in HIV-1 spreading infections, transduced cell pools were infected with HIV-1_{IIIB} or a Vif-deficient derivative and virus replication was monitored over time. Neither the Cas9/gRNA-mediated disruption of endogenous A3C nor complementation with either natural A3C variant altered the kinetics of Vif-proficient virus replication with all CEM2n ΔA3C derivatives showing peak infectivity at 7 days post-infection (Fig. 3f). In contrast, CEM2n ΔA3C derivatives remained fully non-permissive to Vif-deficient virus replication presumably due to expression of other endogenous restrictive A3 enzymes including A3D, A3F, A3G and A3H [9, 24] (Fig. 3g). Vif-deficient virus replication in the permissive SupT11 cell line confirmed infectivity of the viral stock (Fig. 3f, g). Taken together, these results demonstrate that A3C disruption is not sufficient to render CEM2n cells detectably permissive to Vif-deficient HIV-1 replication, and that complementation with A3C-Ser188 or A3C-Ile188 is unable to alter the kinetics of virus replication in the presence of Vif, potentially because both variants are targeted for degradation by Vif at similar efficiencies [11].

Stable expression of A3C-Ser188 and A3C-Ile188 in SupT11 cells restricts Vif-deficient HIV-1 replication and inflicts G-to-A mutations

The phenotypically redundant antiviral activities of A3D, A3F, A3G, and A3H make it challenging to study A3C activity in non-permissive T cell lines such as CEM2n. Therefore, permissive, non-A3 expressing T cell lines such as SupT11 provide robust systems for comparing the antiviral activities of single DNA deaminases and mutant derivatives [9, 20, 25–27]. Prior

(See figure on next page.)
Fig. 3 HIV-1 replication phenotypes following *A3C* disruption and variant complementation in non-permissive CEM2n cells. **a** Immunoblots of endogenous A3C in SupT11, H9, and CEM2n cells. Tubulin was used as a loading control. **b** Immunoblots of endogenous A3C in CEM2n clones following targeted disruption of *A3C* exon 3 by Cas9/gRNA complexes. Tubulin was used as a loading control. **c** *A3C* exon 3 sequences from parental CEM2n and an A3C-disrupted clone (CEM2n Δ*A3C*). **d** Flow cytometry plots for CEM2n Δ*A3C* cell pools 72 h post-transduction with GFP-reporter complementation vectors. The percentage of GFP+ cells is indicated for each population. **e** Immunoblots of A3C in the parental CEM2n line, CEM2n Δ*A3C*, and complemented CEM2n Δ*A3C* derivatives. Tubulin was used as a loading control. **f, g** Spreading infection kinetics of Vif-proficient and Vif-deficient HIV-1 (initial MOI = 0.02) in CEM2n, CEM2n Δ*A3C*, and the indicated complemented conditions. SupT11 cells are included as a control permissive cell type. Virus infectivity was determined by infection of CEM-GFP with culture supernatants followed 48 h later by quantification with flow cytometry. Each spreading infection experiment was performed 3 times and representative data are shown

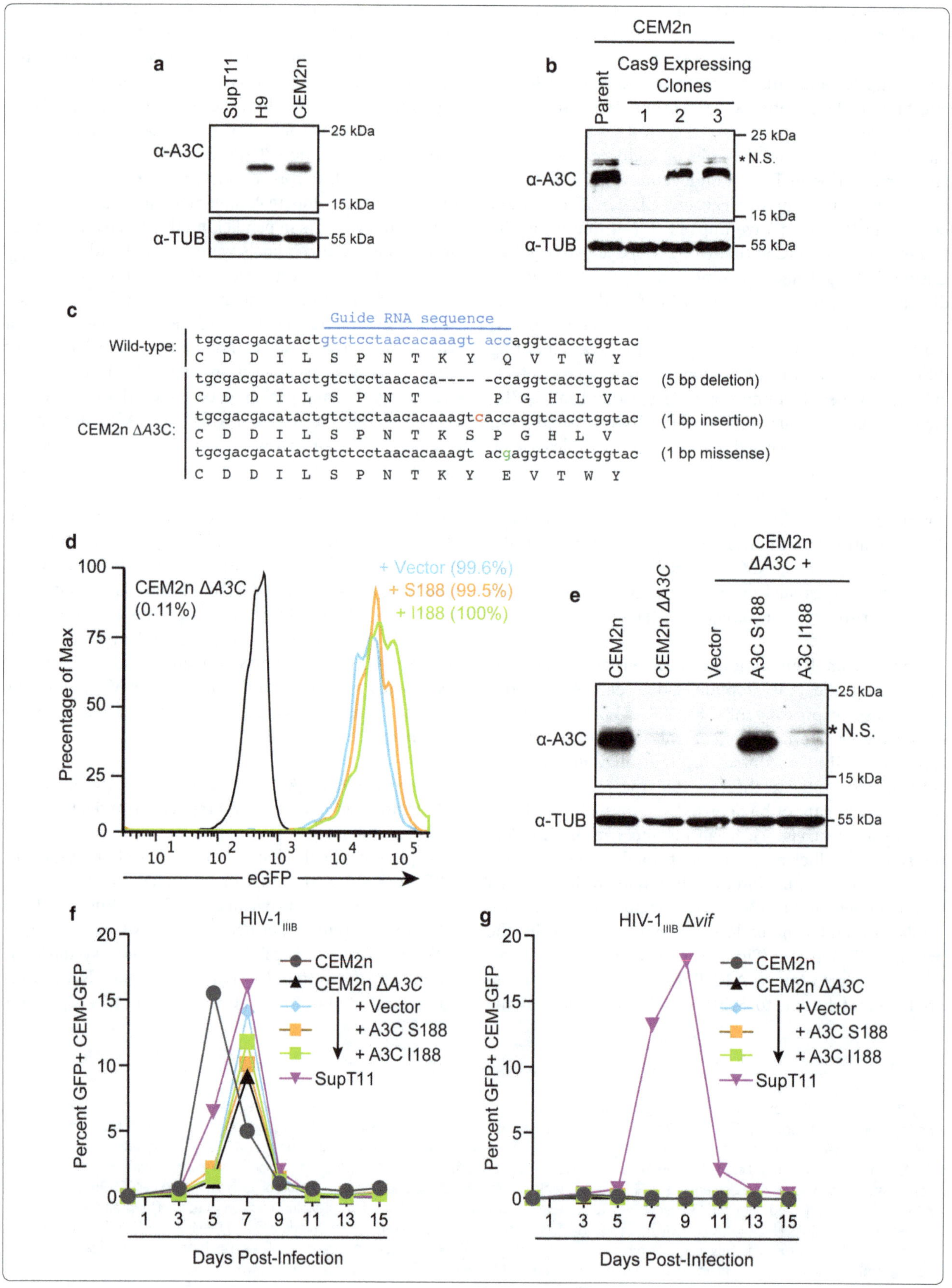
a
SupT11
H9
CEM2n
α-A3C
α-TUB
25 kDa
15 kDa
55 kDa
b
CEM2n
Parent
Cas9 Expressing Clones
1 2 3
α-A3C
α-TUB
25 kDa
* N.S.
15 kDa
55 kDa
c
Guide RNA sequence
Wild-type:
tgcgacgacatactgtctcctaacacaaagt accaggtcacctggtac
C D D I L S P N T K Y Q V T W Y
CEM2n ΔA3C:
tgcgacgacatactgtctcctaacaca---- -ccaggtcacctggtac
C D D I L S P N T P G H L V
(5 bp deletion)
tgcgacgacatactgtctcctaacacaaagtcaccaggtcacctggtac
C D D I L S P N T K S P G H L V
(1 bp insertion)
tgcgacgacatactgtctcctaacacaaagt acgaggtcacctggtac
C D D I L S P N T K Y E V T W Y
(1 bp missense)
d
CEM2n ΔA3C (0.11%)
+ Vector (99.6%)
+ S188 (99.5%)
+ I188 (100%)
Precentage of Max
eGFP
e
CEM2n
CEM2n ΔA3C
CEM2n ΔA3C +
Vector
A3C S188
A3C I188
α-A3C
α-TUB
25 kDa
* N.S.
15 kDa
55 kDa
f
HIV-1 IIIB
CEM2n
CEM2n ΔA3C
+ Vector
+ A3C S188
+ A3C I188
SupT11
Percent GFP+ CEM-GFP
Days Post-Infection
g
HIV-1 IIIB Δvif
CEM2n
CEM2n ΔA3C
+Vector
+ A3C S188
+ A3C I188
SupT11
Percent GFP+ CEM-GFP
Days Post-Infection

to knowledge of A3C-Ile188, we reported that A3C-Ser188 is unable to restrict Vif-deficient HIV-1 when expressed stably in this system [9]. These studies were additionally limited by the utilization of a C-terminal triple-HA epitope tag, which based on results above is detrimental to A3C expression (Fig. 1). Therefore, the intrinsic restriction activities of untagged A3C-Ser188 and A3C-Ile188 were compared following stable transduction and expression in SupT11 cells. As above, efficient transduction was confirmed by GFP+ flow cytometry for all constructs (Fig. 4a) and by immunoblotting for A3C-Ser188 (Fig. 4b).

The resulting SupT11 pools were infected with Vif-deficient HIV-1 (MOI = 0.01) and virus replication kinetics were monitored over time. Both A3C-Ser188 and A3C-Ile188 suppressed virus replication, but the latter variant imposed a reproducibly but not significantly greater delay (Fig. 4c). Moreover, in addition to imposing replication delays, the spreading infection peaks were lower than those in the vector control condition (28 ± 14, 12 ± 10, and 9 ± 6% GFP-positive reporter cells for vector, A3C-Ser188, and A3C-Ile188 conditions, respectively; n = 4 biologically independent experiments).

To ask whether A3C inflicts G-to-A mutations in HIV-1 genomic sequences, high-fidelity PCR was used to recover proviral DNA from the HIV-1 infected SupT11 derivatives after 9 and 13 days of spreading infection. A minimum of 10 independent 564 bp *pol* region DNA amplicons were cloned and sequenced for each condition. These analyses revealed that both A3C variants caused significant increases in overall G-to-A mutation loads (twofold for Ser188 and fourfold for Ile188; Fig. 4d, e). In both instances, the underlying mutation spectra showed evidence for both 5′-GA-to-AA and 5′-GG-to-AG mutations due to cDNA strand 5′-TC-to-TU and 5′-CC-to-CU deamination events, respectively (Fig. 4e). These proviral DNA sequencing analyses showed that both A3C-Ser188 and A3C-Ile188 have the capacity to inflict G-to-A mutations in viral DNA during a spreading infection in a T cell model system. However, in direct comparison to A3G, much lower frequencies of G-to-A mutation are observed suggesting some of the antiviral activity of A3C may also be due to a deamination-independent mechanism (supported by data in Fig. 2a).

Discussion

The A3 DNA cytosine deaminase family imposes a significant barrier to productive HIV-1 infection, especially for viruses lacking Vif. Prior studies converged on A3D, A3F, A3G and A3H as the HIV-1 restrictive A3 repertoire in CD4+ T cells (e.g., [9, 20, 28, 29]; reviewed by [2, 4, 30]). Our current studies indicate that A3C-Ile188 has the potential to contribute to HIV-1 restriction in human T cells. Our studies confirm prior 293T-based single cycle results [11, 23] and, importantly, extend this conclusion to a relevant model T cell line SupT11. Importantly, our studies show that both the common (Ser188) and the rare (Ile188) A3C variants are capable of inflicting G-to-A mutations during spreading infection in T lymphocytes with the rare variant inflicting a twofold higher mutation frequency. This increase in G-to-A mutations may be explained by an increased capacity to dimerize and to function in a processive manner [23]. This result raises the possibility that both variants may contribute to HIV-1 diversity. However, additional work will be needed to extend these findings to primary T lymphocytes, for instance, through an analysis of endogenous A3C variants following Cas9/gRNA-mediated deletion of A3D/F/G/H *ex vivo*. Such studies would also be facilitated by an anti-A3C monoclonal antibody (yet to be developed to our knowledge) that binds equally well to both naturally occurring A3C-Ser188 and A3C-Ile188 variants.

The seven gene human *A3* locus is variable in the human population, consistent with an overall function of the encoded enzymes in innate immunity and previously documented evidence for positive selection [11, 31–34]. Known variations include complete inactivation of *A3* family members (29.5 kb *A3B* deletion [16] and A3H-ΔAsn15 [18]), hypomorphic alleles (A3H-G105 [18]), and hypermorphic alleles such as the A3C-Ile188 variant (prior work [11, 23] and this study). Several other coding and non-coding variants have been documented but, thus far, their functional relevance is less clear.

A3C is conserved in primates and phylogenetic comparisons show evidence for positive selection, including

(See figure on next page.)

Fig. 4 Stable Expression of both A3C-Ile188 and A3C-Ser188 in SupT11 cells provides a partial block to Vif-deficient HIV-1 replication. **a** Flow cytometry plots for SupT11 cells 72 h after transduction with the indicated GFP-marked complementation vectors. The percentage of GFP+ cells is indicated for each condition. **b** Immunoblots of A3C in SupT11 cells transduced with constructs expressing A3C-Ser188, A3C-Ile188, or an empty control vector. Tubulin was used as a loading control. **c** Representative spreading infection data for Vif-deficient HIV-1 (MOI = 0.01) in SupT11 cells expressing A3C-Ser188, A3C-Ile188, or an empty control vector. Virus infectivity was determined by infection of CEM-GFP with culture supernatants followed 48 h later by quantification with flow cytometry. These data are from one of four biologically independent experiments. **d** Average G-to-A mutation loads for each condition. Error bars show ± SD of 4 independent experiments. Statistical comparisons were done using Student's *t* test. *p* values above each panel are in comparison to vector control or A3C-Ser188. **e** Mutation data for each condition

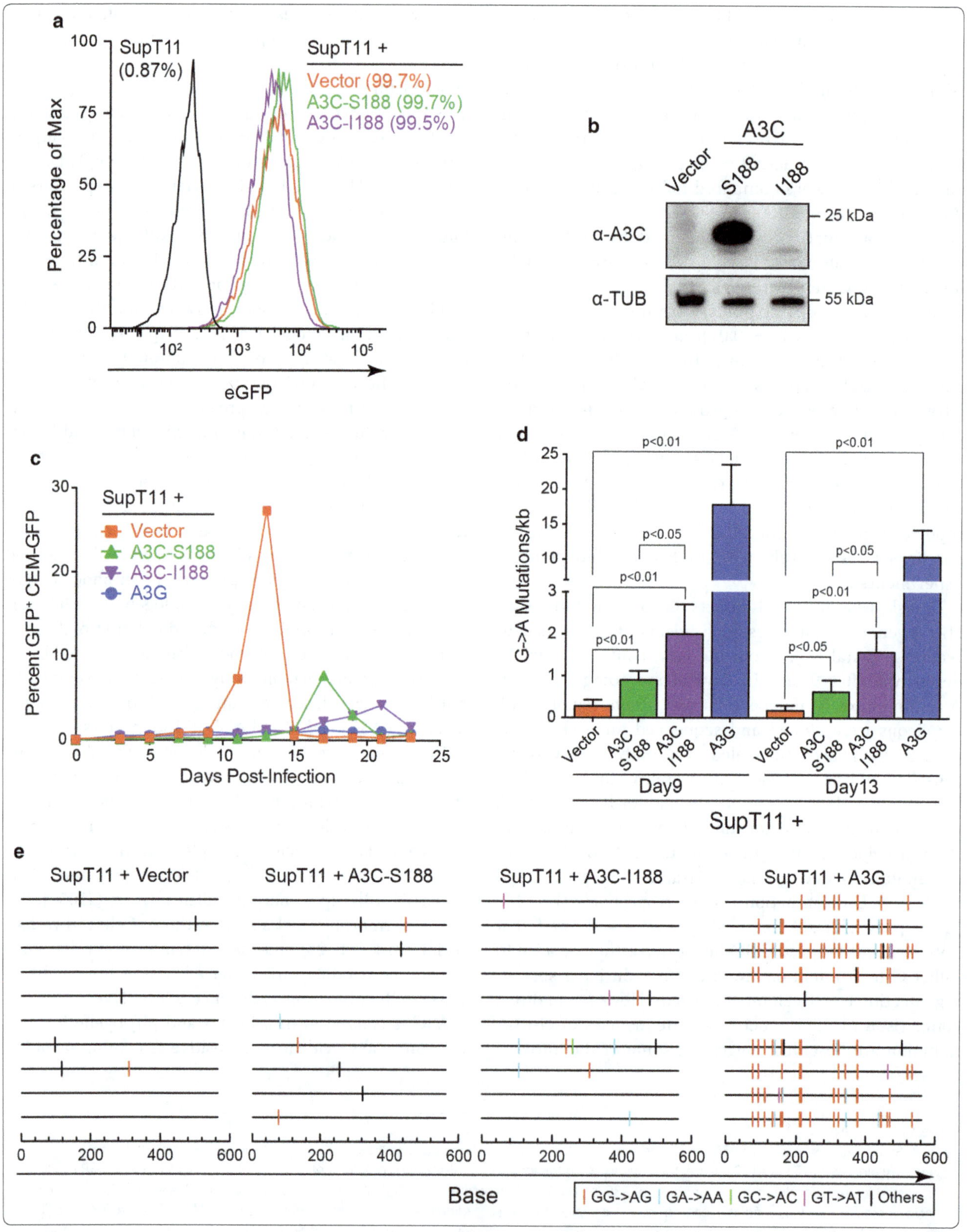
a
Percentage of Max
SupT11
(0.87%)
SupT11 +
Vector (99.7%)
A3C-S188 (99.7%)
A3C-I188 (99.5%)
eGFP
b
A3C
Vector
S188
I188
α-A3C
α-TUB
25 kDa
55 kDa
c
Percent GFP⁺ CEM-GFP
SupT11 +
Vector
A3C-S188
A3C-I188
A3G
Days Post-Infection
d
G->A Mutations/kb
p<0.01
p<0.05
Vector
A3C S188
A3C I188
A3G
Day9
Day13
SupT11 +
e
SupT11 + Vector
SupT11 + A3C-S188
SupT11 + A3C-I188
SupT11 + A3G
Base
GG->AG
GA->AA
GC->AC
GT->AT
Others

multiple amino acid substitutions within the predicted Vif-binding interface [11, 13, 35]. This suggests an ongoing conflict between ancestral lentiviruses with Vif and ancestral A3C enzymes. Consistent with this idea, cross species comparisons have shown restriction of SIVagm by human A3C [11, 14]. Taken together with the fact that HIV-1 Vif still degrades A3C and HIV-1 infection induces *A3C* expression [9, 14, 35–37], it is possible that A3C-Ser188 and A3C-Ile188 still impose some (albeit low) selective pressure on present day HIV-1 (or SIV in recent evolutionary history) [9]. Alternatively, A3D and/or A3F are still so similar to A3C that molecular mimicry within the Vif binding surface may be sufficient to maintain Vif function against A3C.

A3C amino acid 188 is located in the C-terminal helix, which is physically separated from the Vif binding interface [13]. However, phylogenetic comparisons for this region are interesting. Within the human A3 repertoire, most enzymes have Ile at the analogous structural position suggesting a need for greater restriction activity [11]. Sequence comparisons across divergent primate species indicate that Ile188 is the ancestral primate residue, occurring in divergent primate species including the orangutan, gibbon, siamang, and all sequenced Old World monkeys [11]. In contrast, Ser188 is found in humans and other hominids including chimpanzees, gorillas, and bonobos. Curiously, chimpanzee and gorilla A3C proteins have a HIV-1 restriction activity that is comparable to human A3C-Ile188, despite harboring Ser at the analogous amino acid position, suggesting that additional amino acid differences between hominid species may play a role in defining intrinsic restriction activity [23]. Biochemical studies of these enzymes have revealed a correlation between self-dimerization of A3C in solution and HIV-1 restriction activity (observed for human A3C-Ile188, chimp A3C, and gorilla A3C, but not human A3C-Ser188) [23]. The biochemical characterization of additional primate A3C proteins will be needed to test the broad relevance of dimerization in A3C-dependent restriction of primate lentiviruses, though these findings strongly suggest that Ile188 is not the only determinant for enhanced restriction activity.

The overall contribution of A3C-Ile188 to HIV-1 replication in vivo remains unclear at this time, though relative comparisons in single cycle infection experiments suggest that its antiviral activity will pale in comparison to the potent restriction activity of A3G (e.g., Fig. 2a). Nevertheless, studies here and recent reports [11, 23] on A3C combine to reveal an additional example of genetic variation within the human A3 repertoire that may influence HIV-1 evolvability and pathogenesis in the primary cellular reservoir, CD4+ T cells. Future studies leveraging large patient cohorts of known *A3C* genotypes, with due attention to other variations within the locus, may be able to address this possibility.

Conclusions

We show that the rare human A3C-Ile188 variant exhibits enhanced restriction activity against Vif-deficient HIV-1 in a 293T-based single cycle infection system, providing independent verification of prior work [11, 23]. Using Cas9-mediated gene disruption in non-permissive CEM2n T cells, we found that endogenous A3C is not required for Vif-deficient HIV-1 restriction, most likely due to the dominant restriction activities of A3D, A3F, A3G and A3H. Complementation of A3C-disrupted CEM2n cells also had no effect on virus replication kinetics. However, stable expression of A3C-Ile188 can convert the normally permissive SupT11 T cells to a less permissive phenotype with respect to Vif-deficient virus replication. Vif-deficient viruses also accumulate G-to-A mutations from A3C-Ile188, as well as from A3C-Ser188, under these spreading infection conditions. Taken together, these findings support a role for A3C in HIV-1 restriction and G-to-A mutation in T cells, and highlight an additional level of variation within the human *A3* gene locus that may effect HIV-1 adaptation and pathogenesis in vivo.

Methods

Cell lines

SupT11 [9], SupT11-A3G [38], CEM2n [24] and derivative cell lines were maintained in RPMI (Hyclone) supplemented with 10% FBS (Thermo Fisher) and 1% penicillin/streptomycin. 293T cells were maintained in DMEM (Hyclone) supplemented with 10% FBS (Gibco) and 1% penicillin/streptomycin. 293T cells were transfected with TransIT LTI (Mirus) according to the manufacture's protocol. All experiments were harvested 48 h post-transfection.

A3C expression plasmids

The C-terminally triple-HA tagged A3C construct in the pcDNA3.1(+) backbone has been described [9]. The N-terminally triple-HA tagged A3C constructs were ligated into the pcDNA3.1(+) backbone with a triple-HA epitope sequence between the *Nhe*I and *Hind*III cloning sites. Untagged A3C constructs were assembled similarly into an epitope tag-free pcDNA3.1(+) backbone. Untagged A3C variants were subcloned into the CSU6-IDR2 IRES-eGFP lentiviral expression vector (previously described [10] but modified here to express eGFP instead of dsRed) for stable transduction experiments in CEM2n and SupT11 cells.

Immunoblots

Cell pellets were lysed directly in 2.5x Laemmli sample buffer, separated by 12.5% SDS-PAGE, and transferred to PVDF-FL membranes (Millipore). Membranes were blocked in 4% milk in PBS and incubated with primary antibodies diluted in 4% milk in PBS supplemented with 0.1% Tween20. Secondary antibodies were diluted in 4% milk in PBS supplemented with 0.1% Tween20 and 0.01% SDS. Membranes were imaged on a LI-COR Odyssey instrument. Primary antibodies used in these studies were rabbit anti-A3C (Proteintech 10591-1-AP), rabbit anti-A3G (NIH AIDS Reagent Program 10201 courtesy of J. Lingappa), rabbit anti-HA (Cell signaling C29F4), mouse anti-Tubulin (Sigma T5168), mouse anti-HIV-1 p24/CA (NIH AIDS Reagent Program 3537 courtesy of B. Chesebro and K. Wehrly), and mouse anti-HIV-1 Vif (NIH AIDS Reagent Program 6459 courtesy of M. Malim). Secondary antibodies employed were IRdye 800CW goat anti-rabbit (LI-COR 827-08365) and Alexa Fluor 680 goat anti-mouse (Molecular Probes A-21057), except when detecting the anti-A3C antibody, which was probed with an anti-Rb HRP-conjugated secondary antibody (Bio Rad 1706515) and visualized using Super Signal West Femto Maximum Sensitivity Substrate (ThermoFisher Scientific). The following A3C antibodies failed to detect overexpressed N-terminally tagged A3C variants by immunoblotting: rabbit anti-A3C (Abcam ab181356), rabbit anti-A3C (Abcam ab209560), rabbit anti-A3C (LSBio LS-B14538), and mouse anti-A3C (Sigma SAB1403204).

HIV-1 single-cycle infectivity experiments

293T-based HIV-1 single cycle infectivity assays were performed by co-transfecting 293T cells with an HIV-1_{IIIB} A200C *vif*-deficient molecular clone derivative (pIIIB Δ*vif*) along with the indicated APOBEC3 expression construct or an empty vector control [39, 40]. 48 h later, cell free supernatants containing progeny viruses were used to infect CEM-GFP reporter cells. Reporter cell infectivity was assayed 48 h post infection by flow cytometry using a Becton Dickinson FacsCanto II instrument.

***A3C* disruption by Cas9/gRNA**

A gRNA sequence homologous to 5′-GGG GCT CCG CAG CCT GAG TC-3′ in *A3C* exon 3 was generated using the CRISPR Design Tool (Massachusetts Institute of Technology) and cloned into the lentiCRISPRv1 vector (Addgene #49535) according to the accompanying Zhang lab protocol [41]. This targeting construct was generated by annealing oligos 5′-ACA CCG TCT CCT AAC ACA AAG TAC CG-3′ and 5′-AAA ACG GTA CTT TGT GTT AGG AGA CG-3′ and ligating the resulting dsDNA fragment into *Bsm*BI-digested lentiCRISPRv1. 293T cells were transfected with the lentiCRISPRv1 targeting construct along with pΔ-NRF (HIV-1 *gag, pol, rev, tat* genes) and pMDG (VSV-G) expression vectors and, 48 h later, virus-containing supernatants were filtered (0.45 μm) and concentrated by centrifugation (22,000×*g*, 2 h, 10 °C). Viral pellets were resuspended in complete RPMI and incubated with cells for 48 h before being placed under drug selection (1 μg/mL puromycin). Clones were isolated by limiting dilution of the drug resistant cell pool, and A3C expression was analyzed by immunoblotting lysates from individual clones.

A3C complementation

CSU6-IDR2 IRES-eGFP vectors encoding untagged and N-terminally HA-tagged A3C variants were co-transfected into 293T cells with pΔ-NRF (HIV-1 *gag-pol-rev-tat*) and pMDG (VSV-G). 48 h post transfection, culture supernatants were filtered (0.45 μm) and concentrated by centrifugation (22,000×*g*, 2 h, 10 °C). Viral pellets were resuspended in complete RPMI and used to directly infect SupT11 and CEM2n target cells. Transduced cell populations were assayed for GFP expression by flow cytometry and A3C expression was measured by immunoblotting 72 h post-transduction.

HIV-1 spreading infections

HIV-1 spreading infections were performed as described [9]. Briefly, replication-competent HIV-1 stocks were produced by transfecting 293T cells with HIV-1_{IIIB} A200C wild-type (pIIIB) or *vif*-deficient (pIIIB Δ*vif*) molecular clone derivatives [39, 40]. 48 h post-transfection, culture supernatants were filtered (0.45 μm), and viral titers were determined by serial titration on CEM-GFP reporter cells. CEM2n and SupT11 derivative cell lines were infected accordingly, and virus replication was monitored every other day (beginning on day 3 post-infection) by infecting CEM-GFP reporter cells with culture supernatants. Infected CEM-GFP cells were fixed at 48 h post-infection and assayed for GFP expression by flow cytometry.

Proviral DNA G-to-A mutation analyses

SupT11 cells at day 9 and 13 of spreading infections were harvested and genomic DNA (including proviral DNA) was prepared (Puregene) for sequencing analysis as described [42]. The viral *pol* region was amplified by nested PCR with RSH4196/4197 (5′-TCC ART ATT TRC CAT AAA RAA AAA-3′ and 5′-TTY AGA TTT TTA AAT GGY TYT TGA-3′) as outer primers (876 bp) and RSH4205/4206 (5′-AAT ATT CCA RTR TAR CAT RAC AAA AAT-3′ and 5′-AAT GGY TYT TGA TAA ATT TGA TAT GT-3′) as inner primers (564 bp) and then the amplified products were subjected to pJET cloning and

Sanger DNA sequencing (GENEWIZ). Sequences were aligned using Sequencher (Gene Codes Corporation) and analyzed for hypermutation by the Los Alamos Hypermut algorithm [43].

Authors' contributions

BDA, TI, RSH: conceived and designed the experiments; BDA, TI, SAM: Performed the experiments; BDA, TI, SAM, RSH: Analyzed the data; BDA, TI, AS, WLB: Contributed key reagents; RSH: Wrote the paper. All authors read and approved the final manuscript.

Author details

[1] Department of Biochemistry, Molecular Biology and Biophysics, Masonic Cancer Center, Center for Genome Engineering, Institute for Molecular Virology, University of Minnesota, Minneapolis, MN, USA. [2] Howard Hughes Medical Institute, University of Minnesota, 2231 6th St. S.E., Minneapolis, MN 55455, USA.

Acknowledgements

We thank Christopher Richards, Nadine Shaban, Gabriel Starrett, and Diako Ebrahimi for helpful comments throughout this study. R.S.H. is the Margaret Harvey Schering Land Grant Chair for Cancer Research, a Distinguished McKnight University Professor, and an Investigator of the Howard Hughes Medical Institute.

Competing interests

RSH is a co-founder, shareholder, and consultant of ApoGen Biotechnologies Inc. The other authors declare no competing interests.

Consent for publication

Not applicable.

Funding

This work was supported by NIAID R37 AI064046 and NCI R21 CA206309. BDA received salary support from NIAID F31 AI116305, and ASM from NSF GRFP 00039202.

References

1. Altfeld M, Gale M Jr. Innate immunity against HIV-1 infection. Nat Immunol. 2015;16:554–62.
2. Simon V, Bloch N, Landau NR. Intrinsic host restrictions to HIV-1 and mechanisms of viral escape. Nat Immunol. 2015;16:546–53.
3. Harris RS, Hultquist JF, Evans DT. The restriction factors of human immunodeficiency virus. J Biol Chem. 2012;287:40875–83.
4. Harris RS, Dudley JP. APOBECs and virus restriction. Virology. 2015;479-480C:131–45.
5. Stavrou S, Ross SR. APOBEC3 proteins in viral immunity. J Immunol. 2015;195:4565–70.
6. Yu Y, Xiao Z, Ehrlich ES, Yu X, Yu XF. Selective assembly of HIV-1 Vif-Cul5-ElonginB-ElonginC E3 ubiquitin ligase complex through a novel SOCS box and upstream cysteines. Genes Dev. 2004;18:2867–72.
7. Jäger S, Kim DY, Hultquist JF, Shindo K, LaRue RS, Kwon E, Li M, Anderson BD, Yen L, Stanley D, et al. Vif hijacks CBF-beta to degrade APOBEC3G and promote HIV-1 infection. Nature. 2011;481:371–5.
8. Zhang W, Du J, Evans SL, Yu Y, Yu XF. T-cell differentiation factor CBF-beta regulates HIV-1 Vif-mediated evasion of host restriction. Nature. 2011;481:376–9.
9. Hultquist JF, Lengyel JA, Refsland EW, LaRue RS, Lackey L, Brown WL, Harris RS. Human and rhesus APOBEC3D, APOBEC3F, APOBEC3G, and APOBEC3H demonstrate a conserved capacity to restrict Vif-deficient HIV-1. J Virol. 2011;85:11220–34.
10. Anderson BD, Harris RS. Transcriptional regulation of APOBEC3 antiviral immunity through the CBF-beta/RUNX axis. Sci Adv. 2015;1:e1500296.
11. Wittkopp CJ, Adolph MB, Wu LI, Chelico L, Emerman M. A single nucleotide polymorphism in human APOBEC3C enhances restriction of lentiviruses. PLoS Pathog. 2016;12:e1005865.
12. Zhang Z, Gu Q, Jaguva Vasudevan AA, Jeyaraj M, Schmidt S, Zielonka J, Perkovic M, Heckel JO, Cichutek K, Haussinger D, et al. Vif proteins from diverse human immunodeficiency virus/simian immunodeficiency virus lineages have distinct binding sites in A3C. J Virol. 2016;90:10193–208.
13. Kitamura S, Ode H, Nakashima M, Imahashi M, Naganawa Y, Kurosawa T, Yokomaku Y, Yamane T, Watanabe N, Suzuki A, et al. The APOBEC3C crystal structure and the interface for HIV-1 Vif binding. Nat Struct Mol Biol. 2012;19:1005–10.
14. Yu Q, Chen D, Konig R, Mariani R, Unutmaz D, Landau NR. APOBEC3B and APOBEC3C are potent inhibitors of simian immunodeficiency virus replication. J Biol Chem. 2004;279:53379–86.
15. Duggal NK, Fu W, Akey JM, Emerman M. Identification and antiviral activity of common polymorphisms in the APOBEC3 locus in human populations. Virology. 2013;443:329–37.
16. Kidd JM, Newman TL, Tuzun E, Kaul R, Eichler EE. Population stratification of a common APOBEC gene deletion polymorphism. PLoS Genet. 2007;3:e63.
17. Starrett GJ, Luengas EM, McCann JL, Ebrahimi D, Temiz NA, Love RP, Feng Y, Adolph MB, Chelico L, Law EK, et al. The DNA cytosine deaminase APOBEC3H haplotype I likely contributes to breast and lung cancer mutagenesis. Nat Commun. 2016;7:12918.
18. OhAinle M, Kerns JA, Li MM, Malik HS, Emerman M. Antiretroelement activity of APOBEC3H was lost twice in recent human evolution. Cell Host Microbe. 2008;4:249–59.
19. Wang X, Abudu A, Son S, Dang Y, Venta PJ, Zheng YH. Analysis of human APOBEC3H haplotypes and anti-human immunodeficiency virus type 1 activity. J Virol. 2011;85:3142–52.
20. Refsland EW, Hultquist JF, Luengas EM, Ikeda T, Shaban NM, Law EK, Brown WL, Reilly C, Emerman M, Harris RS. Natural polymorphisms in human APOBEC3H and HIV-1 Vif combine in primary T lymphocytes to affect viral G-to-A mutation levels and infectivity. PLoS Genet. 2014;10:e1004761.
21. Ebrahimi D, Richards CM, Carpenter MA, Wang J, Ikeda T, Becker JT, Cheng AZ, McCann JL, Shaban NM, Salamango DJ, et al. Genetic and mechanistic basis for APOBEC3H alternative splicing, retrovirus restriction, and counteraction by HIV-1 protease. Nat Commun. 2018;9:4137.
22. Dang Y, Siew LM, Wang X, Han Y, Lampen R, Zheng YH. Human cytidine deaminase APOBEC3H restricts HIV-1 replication. J Biol Chem. 2008;283:11606–14.
23. Adolph MB, Ara A, Feng Y, Wittkopp CJ, Emerman M, Fraser JS, Chelico L. Cytidine deaminase efficiency of the lentiviral viral restriction factor APOBEC3C correlates with dimerization. Nucleic Acids Res. 2017;45:3378–94.
24. Refsland EW, Hultquist JF, Harris RS. Endogenous origins of HIV-1 G-to-A hypermutation and restriction in the nonpermissive T cell line CEM2n. PLoS Pathog. 2012;8:e1002800.
25. Ooms M, Brayton B, Letko M, Maio SM, Pilcher CD, Hecht FM, Barbour JD, Simon V. HIV-1 Vif adaptation to human APOBEC3H haplotypes. Cell Host Microbe. 2013;14:411–21.
26. Letko M, Booiman T, Kootstra N, Simon V, Ooms M. Identification of the HIV-1 Vif and human APOBEC3G protein interface. Cell Rep. 2015;13:1789–99.
27. Ooms M, Letko M, Simon V. The structural interface between HIV-1 Vif and human APOBEC3H. J Virol. 2017;91:e02289.
28. Nakano Y, Misawa N, Juarez-Fernandez G, Moriwaki M, Nakaoka S, Funo T, Yamada E, Soper A, Yoshikawa R, Ebrahimi D, et al. HIV-1 competition experiments in humanized mice show that APOBEC3H imposes selective pressure and promotes virus adaptation. PLoS Pathog. 2017;13:e1006348.
29. Krisko JF, Martinez-Torres F, Foster JL, Garcia JV. HIV restriction by APOBEC3 in humanized mice. PLoS Pathog. 2013;9:e1003242.
30. Malim MH, Bieniasz PD. HIV restriction factors and mechanisms of evasion. Cold Spring Harb Perspect Med. 2012;2:a006940.
31. McLaughlin RN Jr, Gable JT, Wittkopp CJ, Emerman M, Malik HS. Conservation and innovation of APOBEC3A restriction functions during primate evolution. Mol Biol Evol. 2016;33:1889–901.
32. Duggal NK, Malik HS, Emerman M. The breadth of antiviral activity of APOBEC3DE in chimpanzees has been driven by positive selection. J Virol. 2011;85:11361–71.
33. Sawyer SL, Emerman M, Malik HS. Ancient adaptive evolution of the primate antiviral DNA-editing enzyme APOBEC3G. PLoS Biol. 2004;2:E275.

34. OhAinle M, Kerns JA, Malik HS, Emerman M. Adaptive evolution and antiviral activity of the conserved mammalian cytidine deaminase APOBEC3H. J Virol. 2006;80:3853–62.
35. Smith JL, Pathak VK. Identification of specific determinants of human APOBEC3F, APOBEC3C, and APOBEC3DE and African green monkey APOBEC3F that interact with HIV-1 Vif. J Virol. 2010;84:12599–608.
36. Zhang W, Chen G, Niewiadomska AM, Xu R, Yu XF. Distinct determinants in HIV-1 Vif and human APOBEC3 proteins are required for the suppression of diverse host anti-viral proteins. PLoS ONE. 2008;3:e3963.
37. Refsland EW, Stenglein MD, Shindo K, Albin JS, Brown WL, Harris RS. Quantitative profiling of the full *APOBEC3* mRNA repertoire in lymphocytes and tissues: implications for HIV-1 restriction. Nucleic Acids Res. 2010;38:4274–84.
38. Albin JS, Brown WL, Harris RS. Catalytic activity of APOBEC3F is required for efficient restriction of Vif-deficient human immunodeficiency virus. Virology. 2014;450–451:49–54.
39. Haché G, Shindo K, Albin JS, Harris RS. Evolution of HIV-1 isolates that use a novel Vif-independent mechanism to resist restriction by human APOBEC3G. Curr Biol. 2008;18:819–24.
40. Haché G, Abbink TE, Berkhout B, Harris RS. Optimal translation initiation enables Vif-deficient human immunodeficiency virus type 1 to escape restriction by APOBEC3G. J Virol. 2009;83:5956–60.
41. Sanjana NE, Shalem O, Zhang F. Improved vectors and genome-wide libraries for CRISPR screening. Nat Methods. 2014;11:783–4.
42. Ikeda T, Symeonides M, Albin JS, Li M, Thali M, Harris RS. HIV-1 adaptation studies reveal a novel Env-mediated homeostasis mechanism for evading lethal hypermutation by APOBEC3G. PLoS Pathog. 2018;14:e1007010.
43. Rose PP, Korber BT. Detecting hypermutations in viral sequences with an emphasis on G→A hypermutation. Bioinformatics. 2000;16:400–1.

8

$CD8^+$ T cells specific for conserved, cross-reactive Gag epitopes with strong ability to suppress HIV-1 replication

Hayato Murakoshi[1†], Chengcheng Zou[1†], Nozomi Kuse[1], Tomohiro Akahoshi[1], Takayuki Chikata[1], Hiroyuki Gatanaga[1,3], Shinichi Oka[1,3], Tomáš Hanke[2,4] and Masafumi Takiguchi[1*]

Abstract

Background: Development of AIDS vaccines for effective prevention of circulating HIV-1 is required, but no trial has demonstrated definitive effects on the prevention. Several recent T-cell vaccine trials showed no protection against HIV-1 acquisition although the vaccines induced HIV-1-specific T-cell responses, suggesting that the vaccine-induced T cells have insufficient capacities to suppress HIV-1 replication and/or cross-recognize circulating HIV-1. Therefore, it is necessary to develop T-cell vaccines that elicit T cells recognizing shared protective epitopes with strong ability to suppress HIV-1. We recently designed T-cell mosaic vaccine immunogens tHIVconsvX composed of 6 conserved Gag and Pol regions and demonstrated that the T-cell responses to peptides derived from the vaccine immunogens were significantly associated with lower plasma viral load (pVL) and higher $CD4^+$ T-cell count (CD4 count) in HIV-1-infected, treatment-naive Japanese individuals. However, it remains unknown T cells of which specificities have the ability to suppress HIV-1 replication. In the present study, we sought to identify more T cells specific for protective Gag epitopes in the vaccine immunogens, and analyze their abilities to suppress HIV-1 replication and recognize epitope variants in circulating HIV-1.

Results: We determined 17 optimal Gag epitopes and their HLA restriction, and found that T-cell responses to 9 were associated significantly with lower pVL and/or higher CD4 count. T-cells recognizing 5 of these Gag peptides remained associated with good clinical outcome in 221 HIV-1-infected individuals even when comparing responders and non-responders with the same restricting HLA alleles. Although it was known previously that T cells specific for 3 of these protective epitopes had strong abilities to suppress HIV-1 replication in vivo, here we demonstrated equivalent abilities for the 2 novel epitopes. Furthermore, T cells against all 5 Gag epitopes cross-recognized variants in majority of circulating HIV-1.

Conclusions: We demonstrated that T cells specific for 5 Gag conserved epitopes in the tHIVconsvX have ability to suppress replication of circulating HIV-1 in HIV-1-infected individuals. Therefore, the tHIVconsvX vaccines have the right specificity to contribute to prevention of HIV-1 infection and eradication of latently infected cells following HIV-1 reactivation.

Keywords: HIV-1, Gag, CTL, Vaccine, Conserved epitope

*Correspondence: masafumi@kumamoto-u.ac.jp
†Hayato Murakoshi and Chengcheng Zou contributed equally to this work
[1] Center for AIDS Research, Kumamoto University, 2-2-1 Honjo, Chuo-ku, Kumamoto 860-0811, Japan
Full list of author information is available at the end of the article

Background

Development of effective vaccines against HIV-1 is the best hope for controlling the AIDS epidemic, but no trials has yet showed definitive effect on prevention of HIV-1 infection. Although the RV144 trial in Thailand showed weak protection against HIV-1 most likely through generation of non-neutralizing antibodies [1–7], the outcome of this vaccine trial remains to be reproduced [8, 9]. The STEP study of a candidate T-cell vaccine induced low frequency responses in 77% of vaccine recipients and showed no protection against HIV-1 acquisition [10, 11]. Although a sieve effect of break-through viruses was described, overall, the vaccine-elicited CD8$^+$ T cells had insufficient capacity to suppress HIV-1 replication. There is no simple functional or phenotypic T-cell marker consistently associated with HIV-1 control. The protective capacity of CD8$^+$ T cells likely comes from multiple attributes including specificity, breadth, quality, quantity and being at the right time at the right place. Targeting protective epitopes is one of the key traits.

Our strategy for induction of effective responses is to focus the CD8$^+$ T cells on the highly functionally conserved regions of the HIV-1 proteome [12, 13]. To further improve their efficacy, vaccine immunogens tHIVconsvX consist of 6 protein regions from Gag and Pol with high coverage of known protective epitopes and employ a bivalent mosaic (two versions of each region, which differ in approximately 1/10 amino acids) to maximize the match of the vaccine to the global circulating viruses. Indeed, using overlapping 15-mer peptides derived from the tHIVconsvX immunogens, we demonstrated the correlation of both the total magnitude and breadth of the tHIVconsvX-specific T-cell responses to lower plasma viral load (pVL) and higher CD4$^+$ T-cell count (CD4 count) in a cohort of 120 treatment-naïve, HIV-1-positive patients in Japan [14].

Numerous studies have showed that CD8$^+$ T cell targeting of HIV-1 Gag linked to viral load or disease outcome in HIV-1 infection [15–17]. The mechanism of this effect may involve that the viral genome is delivered into the cell in a ribonucleoprotein complex composed largely of Gag after infection, so that Gag epitopes can be processed, presented on the cell surface and finally recognized by Gag-specific CD8$^+$ T cells within a few hours of infection before Nef-mediated down-regulation takes place [18–20].

We previously showed association of CD8$^+$ T cells specific for several Gag and Pol epitopes with significantly lower pVL and higher CD4 count in chronically HIV-1-infected Japanese patients [21]. The majority of these CD8$^+$ T cells recognized conserved or cross-recognized mutated epitopes mostly on the Gag protein. Half of these T cells were restricted by 2 protective alleles, HLA-B*52:01 or HLA-B*67:01, while the remaining recognized peptides presented by HLA-A*02:06, HLA-B*40:02, or HLA-B*40:06.

In the present study, we determined fine specificities and HLA-restriction of CD8$^+$ T-cell responses specific for the two Gag conserved regions of the candidate tHIVconsvX vaccine and extended the protective correlations to clinical outcome in 221 treatment-naïve HIV-1-positive patients. The study identified additional CD8$^+$ T cells specific for Gag conserved epitopes with strong ability to suppress HIV-1 replication. Thus, Gag-specific CD8$^+$ T cells induced by the tHIVconsvX vaccine have the potential to significantly contribute to prevention of HIV-1 infection and eradication of latently infected cells.

Results

T-cell responses to the conserved regions of Gag were protective

We previously showed in 120 treatment-naive HIV-1$^+$ Japanese patients that CD8$^+$ T-cell responses specific for the conserved regions of the tHIVconsvX immunogen were protective [14]. Here, we focused on the two vaccine conserved regions of Gag, since the responses to Gag had the strongest effect on the suppression of HIV-1 replication in most previous studies [15–17]. We analyzed additional 80 treatment-naive HIV-1-positive patients and reanalyzed the data together from 200 individuals for the magnitude and breadth of CD8$^+$ T-cell responses to 3 pools of the Gag 15-mer peptides (Pools 1, 2, and 3). The magnitude and breadth of the T-cell responses to Pools 2 and 3, and particularly responses to Pool 3, were significantly correlated with high CD4 count and low pVL, whereas those to Pool 1 were not (Fig. 1 and Additional file 1: Fig. S1). These results suggest that T cells specific for peptides within Pools 2 and 3 contribute to the suppression of HIV-1 replication.

Mapping the CD8$^+$ T-cell Gag-specificity to optimal epitopes

We found that 89 and 59 individuals recognized 15-mer peptides in Pools 2 and 3, respectively. Of these, we selected 50 and 53 based on sufficient PBMC sample availability for the fine definition of optimal epitopes. Pools 2 and 3 included peptide pairs derived from the bivalent mosaic and one single peptide common between the two mosaics. By using an IFN-γ ELISPOT assay, we found T-cell responses to 12 peptide pairs in Pool 2 (Fig. 2a) and 10 peptide pairs and one common peptide in Pool 3 (Fig. 3a) in at least one individual. These 15-mer peptides contained sequences of previously reported epitopes; 13 epitopes in Pool 2 and 10 in Pool 3 (Figs. 2b and 3b). Upon inspection of the subjects' HLA, most of the responders had the reported restricting

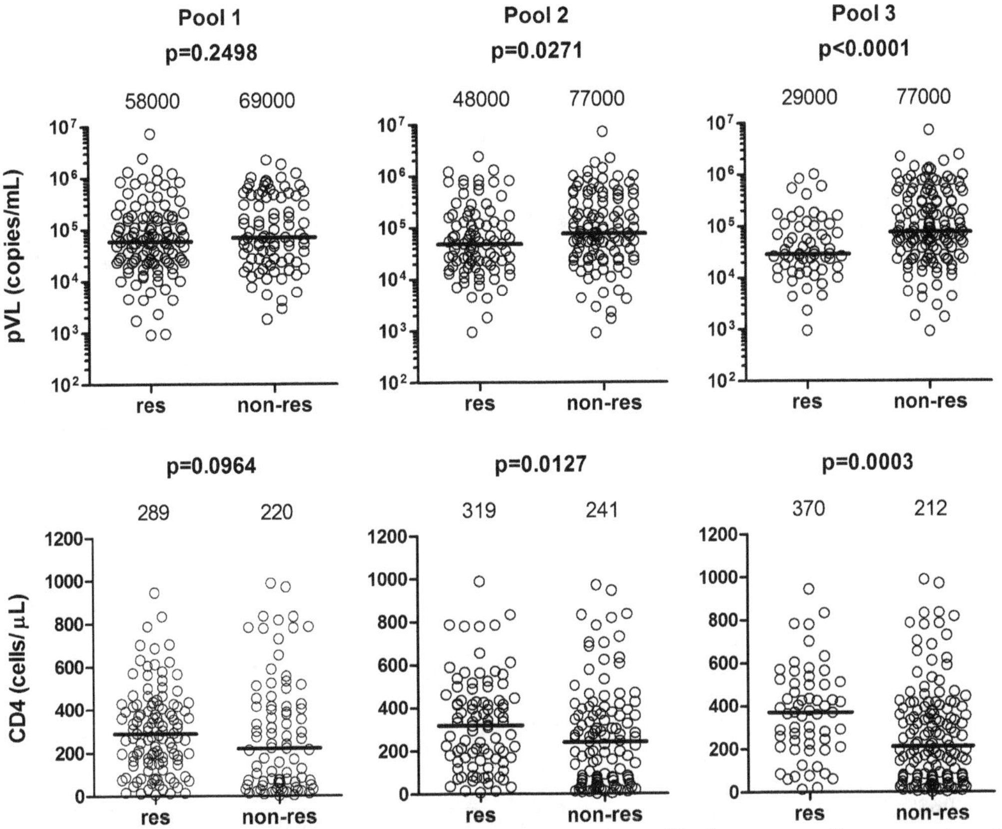

Fig. 1 Correlation of T-cell responses to Gag conserved peptides of tHIVconsvX. Correlation of T-cell responses to Gag conserved peptides of tHIVconsvX with pVL and CD4 count. T-cell responses to Gag peptide Pools 1, 2 and 3 derived from vaccine immunogen tHIVconsvX were enumerated using an IFN-γ ELISPOT assay in 200 HIV-1-infected Japanese individuals. Comparison of pVL and CD4 count between responders and non-responders to the Gag peptides was statistically analyzed using the Mann-Whitney test. The value in each figure represents the median of pVL and CD4 count

alleles (Figs. 2b and 3b). However, responders to 15-mer peptides C48/49, C052/053, C054/055, C113/114, or C125/126 did not have the matching restricting HLA alleles, indicating that their CD8$^+$ T cells may recognize novel, previously unreported epitopes.

Next, we sought to define these novel epitopes and their HLA restriction. We did not pursue the epitope in C48/49 since only one responder to this peptide pair was detected in the 50 tested individuals. The responder KI-1020 had T cells specific for the C052/053 and C054/055 peptide pairs, while KI-1102 and KI-1114 had responses to C125/126 and C113/114 peptides, respectively (Fig. 4a). To define the HLA restriction, subjects' PBMCs were first expanded with each peptide pair for 12–14 days and these STCL were tested in ICS assay using either C1R or 721.221 cells stably transfected with all subject's HLA class I molecules and pulsed with the peptides. The results showed that both T-cell responses to the C052/053 and C054/055 15-mer peptide pairs were restricted by HLA-A*26:02 and responses to the C113/114 and C125/126 peptides were restricted by HLA-A*33:03 and HLA-A*02:06, respectively (Fig. 4b).

We next narrowed the optimal epitopes by using overlapping 11-mer and further truncated peptides. Both the C052/053 STCL and C054/055 STCL recognized both Gag 11-121 and Gag 11-122 peptides (Fig. 4c), suggesting that all the STCL were specific for the same epitope. Analysis using truncated peptides showed that the optimal peptide was TLQEQIGWM (TM9) (Fig. 4d) restricted by HLA-A*26:02. This epitope has not been previously reported. The C113/114 STCL and C125/126 STCL recognized Gag 11-200 and Gag 11-214/215, respectively (Fig. 4c). Again using truncated peptides, we identified previously unreported epitopes HIAKNCRAPR (HR10) and RQANFLGKI (RI9) as the optimal peptides presented by HLA-A*33:03 and HLA-A*02:06, respectively (Fig. 4d). Thus, we identified three novel candidate peptides for the Los Alamos National Laboratory HIV Sequence Database (LANL-HSD; www.hiv.lanl.gov) 'A' list of well-defined epitopes.

a

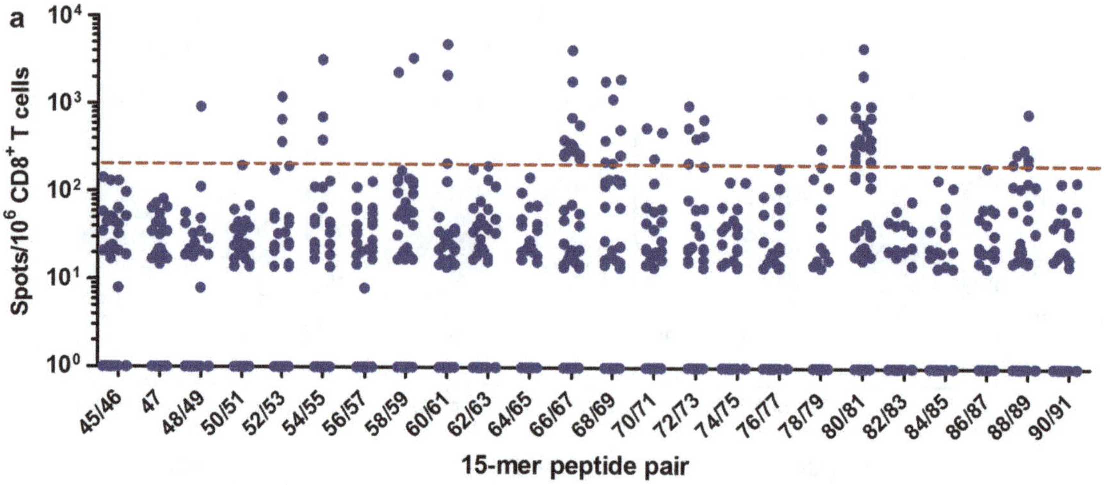

b

15-mer peptide pairs	Frequency of responders	Reported epitopes		
		HLA Allele	Epitope	HLA+responders / all responders
C048/049	1/50	not reported		
C052/053	3/50	B*5801	TSTLQEQIGW	1/3
C054/055	3/50	not reported		
C058/059	2/50	B*3501	NPPIPVGEIY	2/2
C060/061	3/50	B*3501	NPPIPVGEIY	3/3
C066/067	15/50	B*1501	GLNKIVRMY	14/15
		A*2	IILGLNKIV	6/15
C068/069	8/50	B*1501	GLNKIVRMY	8/8
		A*2402	MYSPTSIL	7/8
C070/071	3/50	B*5201	RMYSPTSI	1/3
		A*2402	MYSPTSIL	1/3
		C*0102	YSPTSILDI	3/3
C072/073	6/50	C*0102	YSPTSILDI	6/6
		A*1101	SILDIKQGPK	2/6
C078/079	2/50	A*2402	DYVDRFYKT	2/2
C080/081	16/50	A*3303	DYVDRFYKTLR	5/16
		A*0207	YVDRFYKTL	3/16
		A*2402	DYVDRFYKT	9/16
		C*03	YVDRFYKTL	9/16
C088/089	6/50	B*5201	WMTETLLV	6/6

Fig. 2 T-cell responses to 15-mer peptide pairs in Gag Pool 2. **a** The responses to individual pairs of 15-mer peptides in Gag Pool 2. Responses to peptide pairs in 50 responders to Pool 2 were analyzed by an IFN-γ ELISPOT assay. The dotted line at 200 SFU/10^6 $CD8^+$ T cells indicates a threshold for a positive response. **b** Summary of responders and peptide pairs. Reported epitopes present in the pair of the 15-mer peptides according to the LANL-HSD. HLA^+ responders are those with the matching restricting HLA allele for the reported epitope

$CD8^+$ T cells recognizing Gag conserved epitopes are protective in vivo

As described in above (Figs. 2b and 3b), T cells specific for 23 reported and 3 novel HIV-1 epitopes recognize tHIVconsvX-derived Gag peptides in HIV-1-infected patients in Japan. To investigate further these T-cell responses, we selected 221 individuals whose PBMCs were available for this analysis. Thus, T cells specific for epitopes SK10/HLA-A*11:01 and KR9/HLA-A*31:01 were not detected in any patients. $CD8^+$ T-cells responses recognizing epitopes DT9/HLA-A*24:02, ML8/HLA-A*24:02, NY10/HLA-B*35:01, NL11/HLA-B*67:01, IV9/HLA-A*02, YL9/HLA-A*02:07 and TW10/HLA-B*58:01 were only detected in 1 or 2 patients

a

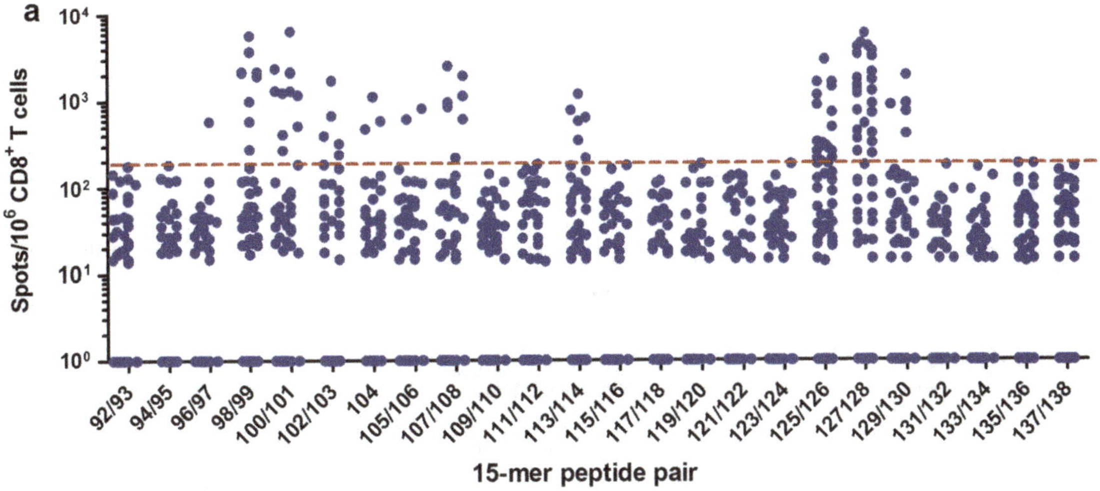

b

15-mer peptide pairs	Frequency of responders	Reported epitopes		
		HLA Allele	Epitope	HLA+responders / all responders
C096/097	1/53	B*6701	NPDCKTILRAL	1/1
C098/099	9/53	C*03	KALGPAATL	4/9
		C*08	KALGPAATL	7/9
C100/101	11/53	C*03	KALGPAATL	4/11
		C*08	KALGPAATL	8/11
C102/103	5/53	A*0206	ATLEEMMTA	5/5
C104	3/53	A*0206	ATLEEMMTA	3/3
C105/106	2/53	A*1101	ACQGVGGPGHK	2/2
C107/108	7/53	A*1101	ACQGVGGPGHK	3/7
		B*0702	GPGHKARVL	4/7
C113/114	6/53	A*3101	IAKNCRAPRK	5/6
C125/126	17/53	B*4002	TERQANFL	16/17
C127/128	24/53	B*4002	TERQANFL	22/24
		A*0201	FLGKIWPS	8/24
C129/130	5/53	A*0201	FLGKIWPS	5/5
		A*3101	KIWPSHKGR	1/5

Fig. 3 T-cell responses to 15-mer peptide pairs in Gag Pool 3. **a** The responses to individual pairs of 15-mer peptides in Gag Pool 3. Responses to peptide pairs in 53 responders to Pool 3 were analyzed by an IFN-γ ELISPOT assay. The dotted line at 200 SFU/10^6 $CD8^+$ T cells indicates a threshold for a positive response. **b** Summary of responders and peptide pairs. Reported epitopes included in the pair of the 15-mer peptides are shown according to the LANL-HSD. HLA^+ responders are those with the matching restricting HLA allele for the reported epitope

(Table 1) and were excluded from further statistical analysis because of the small number of responders. The analysis of $CD8^+$ T-cell responses to 9 epitopes with sufficient number of responders showed a significant association with lower pVL and/or higher CD4 count relative to non-responders to these epitopes. T cells specific for the WV8/HLA-B*52:01, RI8/HLA-B*52:01 and AA9/HLA-A*02:06 epitopes were previously shown to inhibit HIV-1 in vivo [21, 22]. Therefore, the present study in treatment-naïve, HIV-1-positive patients newly identified T cells specific for 6 epitopes FS8/HLA-A*02:01, TL8/HLA-B*40:02, KL9/HLA-C*08, RI9/HLA-A*02:06, TM9/HLA-A*26:02 and HR10/HLA-A*33:03, that could effectively suppress HIV-1 replication (Table 1).

We further analyzed the association of responses to these 9 epitopes with lower pVL and/or higher CD4 count in individuals having the epitopes' restricting HLA molecules. We again confirmed that responders to the WV8/HLA-B*52:01, RI8/HLA-B*52:01 or AA9/HLA-A*02:06 epitopes had significantly lower

a

KI#	HLA-A		HLA-B		HLA-C		C052 C053	C054 C055	C113 C114	C125 C126
KI-1020	2601	2602	1501	3901	0303	0702	**663**	**3072**	0	0
KI-1102	0206	2402	0702	3901	0702	-	0	0	0	**216**
KI-1114	3303	-	4403	5801	0302	1403	0	0	**583**	0

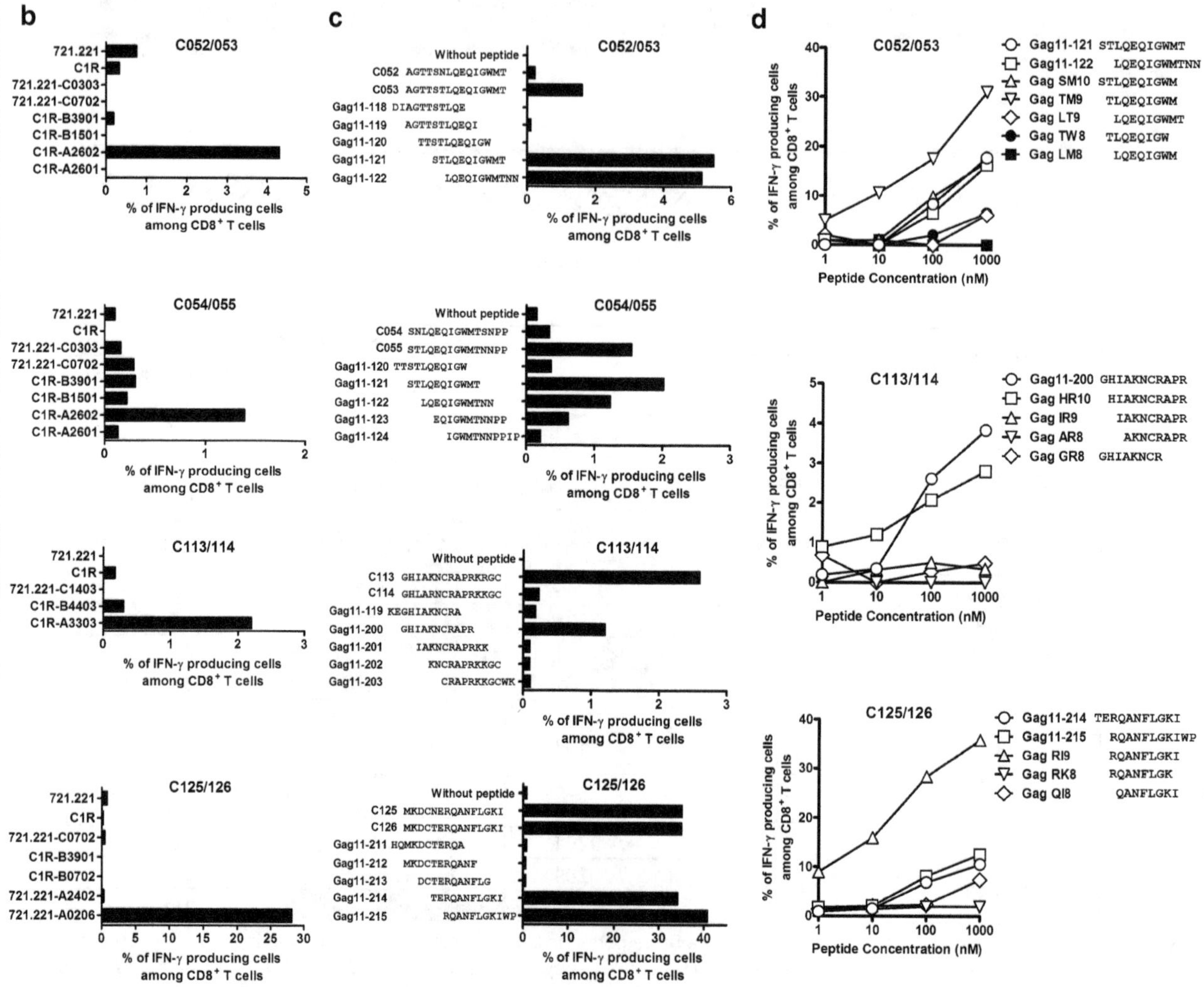

Fig. 4 Identification of novel T-cell Gag epitopes. **a** T-cell responses to peptide pairs in individuals KI-1020, KI-1102, and KI-1114, who did not have matching HLA alleles for previously reported epitopes. **b** HLA restriction. Responses by STCL stimulated with C1R or 721.221 cells expressing individual HLA molecule shared by the responders and pulsed with the peptide pair were analyzed by ICS assay. C052/053 and C054/055 peptides were analyzed by using the T cells derived from KI-1020, while C113/114 and C125/126 by using the T cells from KI-1114 and KI-1102, respectively. **c** Identification of 11-mer HIV-1 clade B overlapping peptides recognized by the HLA-A*26:02-, HLA-A*33:03-, or HLA-A*02:06-restricted T cells. The responses by STCL expanded with C052/053, C054/055, C113/114, or C125/126 peptide pairs to the corresponding stimulator cells pre-pulsed with 11-mer HIV-1 clade B-derived overlapping peptides across the parental 15-mer were analyzed in ICS assay. **d** Identification of optimal peptides. The STCL responses stimulated with peptide pairs C052/053, C113/114, or C125/126 to the corresponding stimulator cells C1R-A2602, C1R-A3303, or 721.221-A0206 pre-pulsed with individual truncated peptides were analyzed by ICS assay

pVL than non-responders with the same HLA alleles, while responders to WV8 and RI8 also had significantly higher CD4 count than the non-responders (Additional file 2: Fig. S2). Responders to HR10/HLA-A*33:03 and TL8/HLA-B*40:02 had significantly both lower pVL and higher CD4 count than the non-responders (Fig. 5a). Thus overall, responses to epitopes WV8, RI8, AA9, HR10, and TL8 exhibited signs of better clinical outcome.

Table 1 Association of CTL responses to Gag epitopes with pVL and CD4 count

Epitope	Sequence	HLA	Frequency		Median of pVL		Median of CD4		P value[b]	
			Res	Non-res	Res	Non-res	Res	Non-res	pVL	CD4
IV9	IILGLNKIV	A*02	2	219	–	–	–	–	–	–
FS8	FLGKIWPS	A*02:01	3	218	910,000	55,500	629	274	0.9882	*0.012*
AA9	ATLEEMMTA	A*02:06	16	205	22,500	58,000	388	269	*0.0057*	0.1788
YL9	YVDRFYKTL	A*02:07	2	219	–	–	–	–	–	–
SK10	SILDIKQGPK	A*11:01	0	221	–	–	–	–	–	–
AK11	ACQGVGGPGHK	A*11:01	4	217	36,500	56,000	450	274	0.3586	0.1277
DT9	DYVDRFYKT	A*24:02	1	220	–	–	–	–	–	–
ML8	MYSPTSIL	A*24:02	1	220	–	–	–	–	–	–
KR9	KIWPSHKGR	A*31:01	0	221	–	–	–	–	–	–
IK10	IAKNCRAPRK	A*31:01	9	212	67,000	55,500	325	275	0.4332	0.7564
DR11	DYVDRFYKTLR	A*33:03	5	216	55,000	56,000	185	279	0.5951	0.3866
GL9	GPGHKARVL	B*07:02	3	218	38,000	56,500	646	274	0.402	0.0741
GY9	GLNKIVRMY	B*15:01	13	208	100,000	55,000	269	285	0.6736	0.8588
NY10	NPPIPVGEIY	B*35:01	1	220	–	–	–	–	–	–
TL8	TERQANFL	B*40:02	23	198	45,000	59,500	416	275	0.1029	*0.0392*
WV8	WMTETLLV	B*52:01	22	199	39,500	58,000	405	269	*0.0481*	*0.0006*
RI8	RMYSPTSI	B*52:01	31	190	43,000	59,500	389	264	*0.0414*	*0.0005*
TW10	TSTLQEQIGW	B*58:01	2	219	–	–	–	–	–	–
NL11	NPDCKTILRAL	B*67:01	1	220	–	–	–	–	–	–
YI9	YSPTSILDI	C*01:02	10	211	123,000	55,000	214	281	0.0919	0.159
KL9	KALGPAATL	C*03	9	212	24,000	56,000	307	275	0.368	0.7865
YL9	YVDRFYKTL	C*03	5	216	76,000	55,500	224	285	0.7639	0.6464
KL9	KALGPAATL	C*08	13	208	24,000	58,000	320	274	*0.0228*	0.64
RI9[a]	RQANFLGKI	A*02:06	4	217	15,000	57,000	319	274	*0.009*	0.9741
TM9[a]	TLQEQIGWM	A*26:02	7	214	39,000	57,500	410	274	*0.032*	0.0921
HR10[a]	HIAKNCRAPR	A*33:03	9	212	4900	58,000	416	273	*0.0002*	*0.0196*

[a] New epitope

[b] Statistically analyzed differences in pVL or CD4 count between responders and non-responders by Mann–Whitney test. Italics indicates that differences were statistically significant

We also analyzed the impact of the total breadth (number of recognized epitopes) and magnitude of the T-cell responses to 5 protective epitopes AA9, TL8, WV8, RI8, and HR10 on clinical outcome and found their very strong negative and positive association with respective pVL (breadth: $p<1\times10^{-4}$, r = −0.37; magnitude: $p<1\times10^{-4}$, r = −0.40) and CD4 count (breadth: $p<1\times10^{-4}$, r=0.44; magnitude: $p<1\times10^{-4}$, r=0.44), respectively (Fig. 5b and Additional file 3: Fig. S3).

TL8- and HR10-specific T cells suppress HIV-1 replication in vitro

We next investigated the ability of T cells specific for the reported TL8/HLA-B*40:02 [23, 24] and novel HR10/A*33:03 epitopes to suppress HIV-1 replication *in vitro*. We established T-cell lines specific for TL8 and HR10 from PBMCs of individuals KI-1391 (HLA-B*40:02$^+$) and KI-1320 (HLA-A*33:03$^+$), respectively, by FACS sorting using the HLA/peptide tetrameric complexes (Fig. 6a). TL8- and HR10-specific T-cell lines were effectively stimulated by peptide-pulsed and HIV-1-infected target cells 721.221/HLA-B*40:02 and 721.221/HLA-A*33:03, respectively, but not uninfected or HLA-untransfected 721.221 cells (Fig. 6b). These T-cell lines efficiently suppressed HIV-1 replication in vitro (Fig. 6c). This HIV-1-inhibitory activity concurs with their improved clinical outcome.

Gag-specific CD8$^+$ T cells cross-recognize epitope variants

Our previous study showed that CD8$^+$ T cells specific for conserved epitopes WV8/HLA-B*52:01 and AA9/HLA-A*02:06 recognized 97 and 90%, respectively, of the circulating viruses in Japan. In contrast, RI8/HLA-B*52:01-specific T cells selected escape mutations, which were present in approximately 27% of the circulating HIV-1 variants [21]. Consensus sequences of HR10/

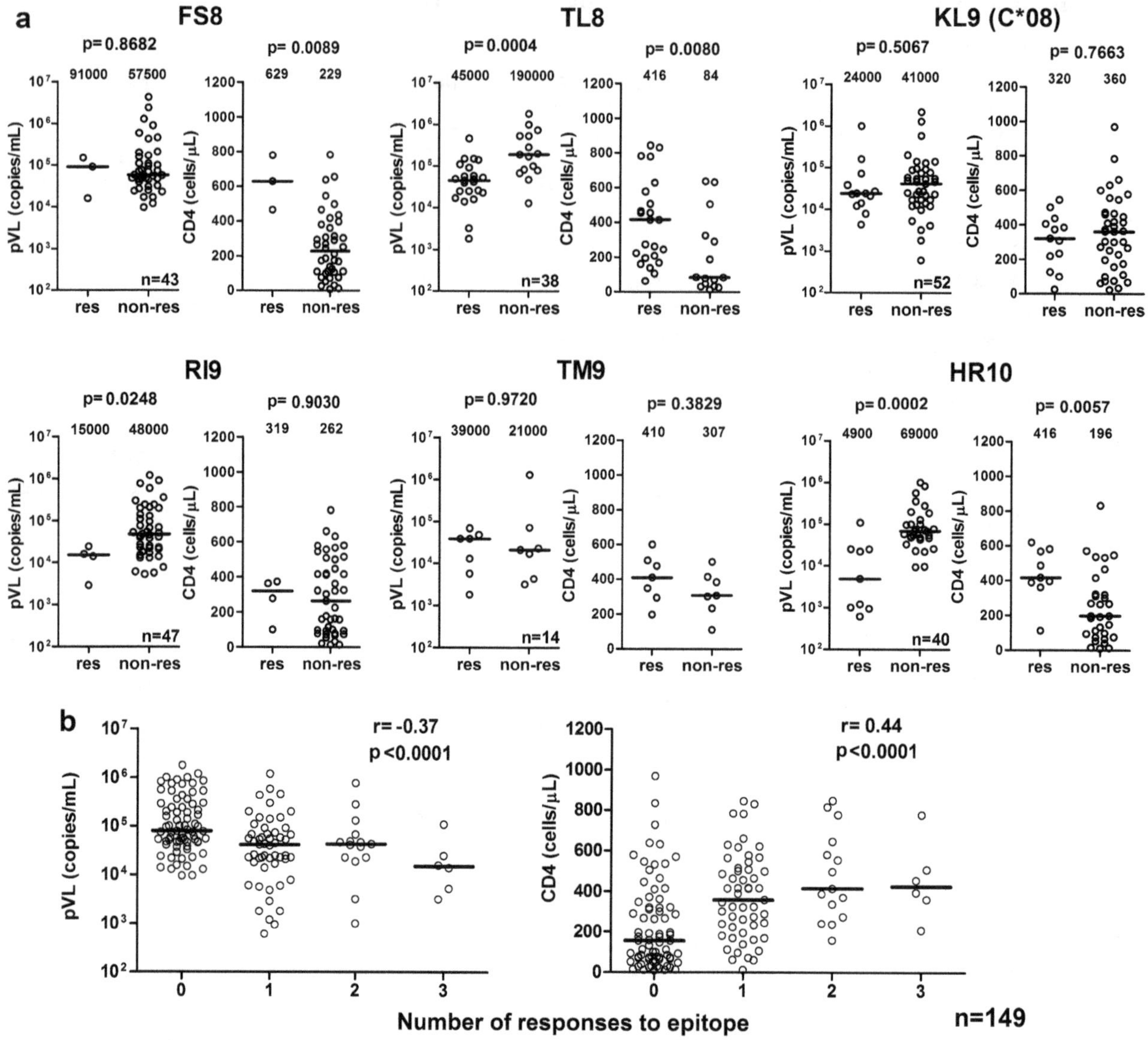

Fig. 5 Association of epitope-specific T-cell responses with clinical parameters in HIV-1-infected individuals with the restricting HLA. **a** T-cell responses to 6 epitope peptides were analyzed by using an IFN-γ ELISPOT assay. The differences in pVL or CD4 count between responders and non-responders to each epitope peptide in the individuals having HLA restriction molecules for the epitopes were statistically analyzed by using the Mann-Whitney test. **b** Correlation between a breadth of T-cell responses to 5 epitopes (AA9, TL8, WV8, RI8, and HR10) and pVL and CD4 count in 149 individuals carrying the HLA restriction molecules. The correlation was statistically analyzed using Spearman rank test

HLA-A*33:03 and TL8/HLA-B*40:02 were found in approximately 60% of circulating viruses in the present Japanese cohort. We therefore analyzed cross-recognition of mutant epitopes by $CD8^+$ T cells recognizing the two conserved Gag epitopes, HR10 and TL8. Three mutants for each of these epitopes were detected in >5% of circulating viruses (Fig. 7a). We therefore assessed the recognition of these mutant epitope peptides by using the IFN-γ ELISPOT assay. Three TL8 mutant peptides were cross-recognized by T cells in 5 HIV-1-infected HLA-B*40:02$^+$ patients with the TL8-specific responses, although the recognition of the 1D mutation was reduced (Fig. 7b). Indeed in a dose response analysis, the stimulation of the TL8-specific T-cell line by the 1D mutant peptide was severely reduced (Fig. 7c). The HLA-B*40:02-associated mutation in Gag T427N ($p=1.0\times10^{-9}$, $q=3.2\times10^{-6}$) was found within the TL8/HLA-B*40:02 epitope [25], suggesting that this mutation represented an escape from $CD8^+$ T-cell recognition. Nevertheless, the TL8-1N mutated peptide was cross-recognized in all of the 5 tested individuals including 2 having the TL8-1N mutant viruses (Additional file 4:

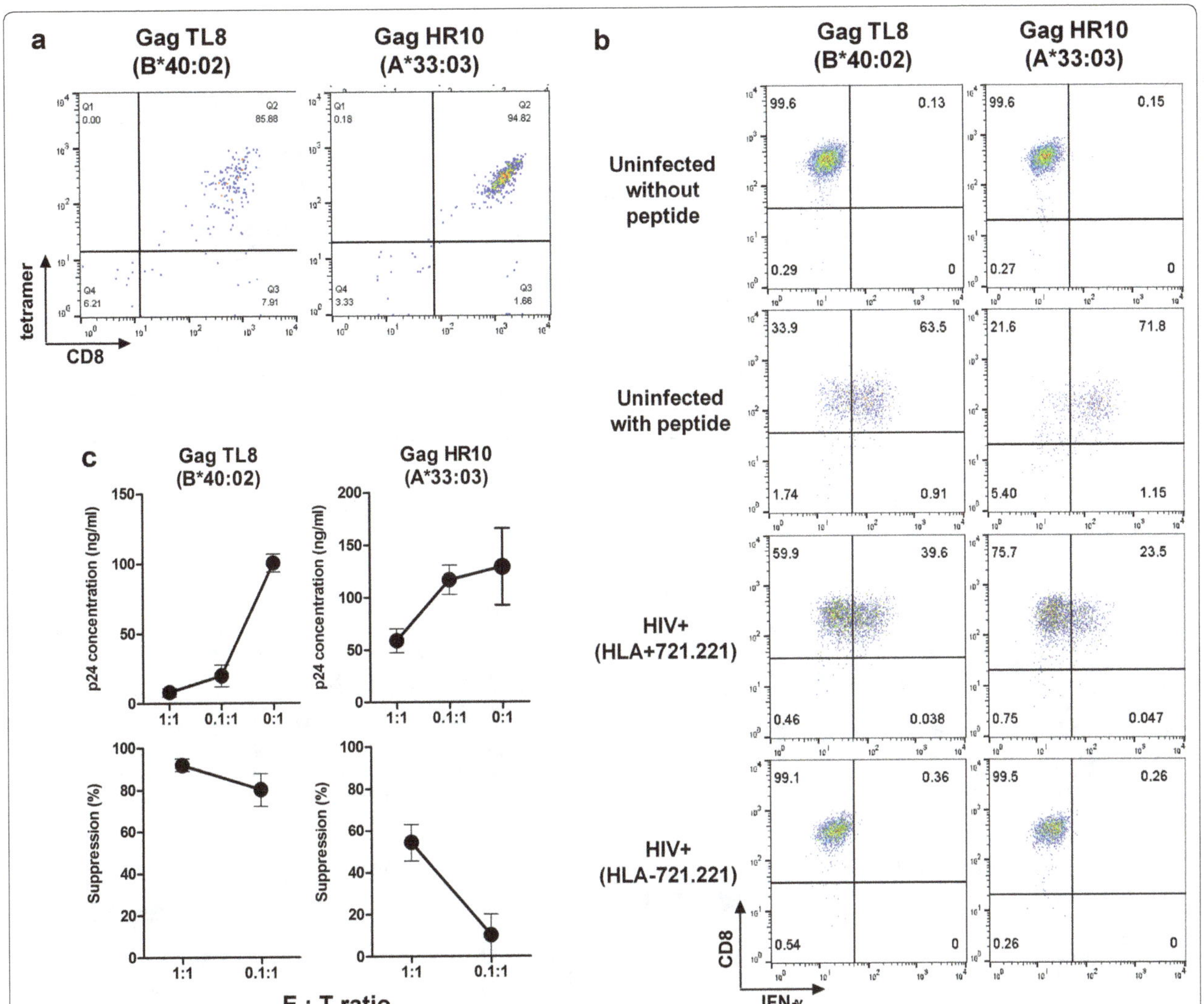

Fig. 6 Ability of CTLs to recognize HIV-1-infected cells and to suppress HIV-1 replication in vitro. **a** Generation of the T-cell lines specific for TL8 or HR10 from PBMCs of HLA-B*40:02^{+} individual (KI-1391) or HLA-A*33:03^{+} one (KI-1320). The T-cell lines were established as shown in Materials and Methods. The T-cell lines established were stained with the specific tetramers. **b** Recognition of HIV-1-infected cells by CTLs specific for TL8 or HR10 epitopes. The T-cell lines stimulated with HIV-1 (NL4-3)-infected 721.221 cells (HIV+) expressing the corresponding HLAs or HLA-negative 721.221 cells, and IFN-γ production from the T cells was measured by the ICS assay. The proportions of 721.221-B4002, -A3303 and HLA-negative 721.221 cells infected with NL4-3 were 56.0, 59.7, and 64.1%, respectively. **c** Suppression of HIV-1 replication by the T-cell lines specific for TL8 or HR10. Primary CD4^{+} T-cells from healthy donors carrying the corresponding HLA alleles were infected with NL4-3, and then co-cultured with epitope-specific T cells at E:T ratios of 1:1 and 0.1:1. The concentration of p24 Ag in the culture supernatant was measured by using an enzyme-linked immunosorbent assay. The percentage of suppression was calculated as follows: (concentration of p24 without CTLs − concentration of p24 with CTLs) / concentration of p24 without CTLs × 100. The data are presented as the mean and SD (n = 3)

Fig. S4). Similarly, the HR10-2L mutant peptide was cross-recognized by T cells in all 5 HIV-1-infected HLA-A*33:03^{+} individuals who were not infected with HR10-2L or HR10-4R mutant viruses (Additional file 4: Fig. S4), whereas HR10-2L4R and/or HR10 4R mutant peptides were cross-recognized by T cells in patients KI-1002 and KI-1320. HR10-specific T cells from KI-1320 confirmed cross-recognition of these mutant peptides (Fig. 7c). Thus, CD8^{+} T-cells responses specific for TL8 and HR10 epitopes can recognize 70–75% of the circulating HIV-1 isolates in Japan.

Discussion

Unique vaccine immunogens tHIVconsvX were assembled, which use 2 Gag and 4 Pol conserved regions to focus T-cell responses on the most vulnerable regions (not full-size proteins and not epitopes) of HIV-1 with the bi-valent mosaic design providing the vaccine with a

a

HLA	Epitope	Sequence	Frequency (%)
B*40:02	TL8	TERQANFL	194/338 (57.4)
	TL8-1N	N-------	47/338 (13.9)
	TL8-1D	D-------	19/338 (5.6)
	TL8-1I	I-------	17/338 (5.0)
A*33:03	HR10	HIAKNCRAPR	206/335 (61.5)
	HR10-2L	-L--------	34/335 (10.1)
	HR10-2L4R	-L-R------	31/335 (9.3)
	HR10-4R	---R------	30/335 (9.0)

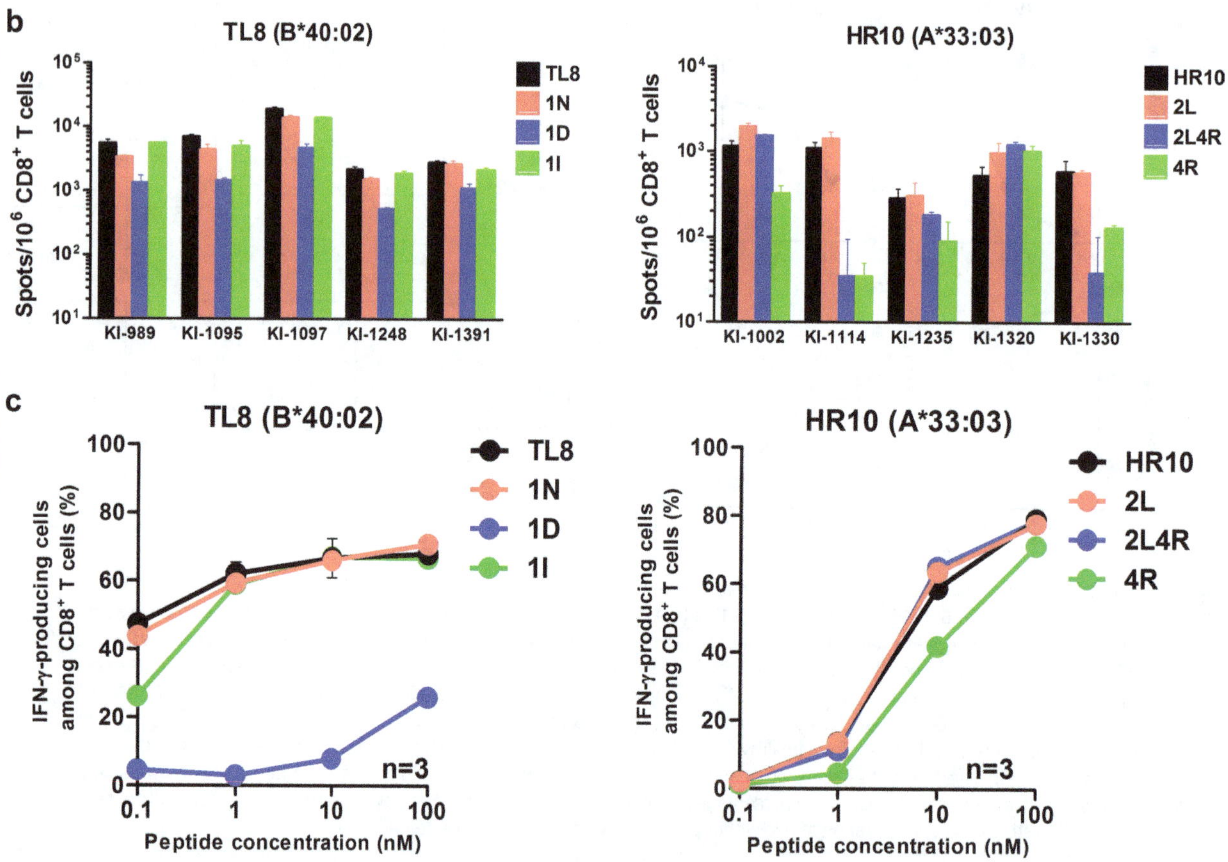

Fig. 7 Recognition of mutant epitope peptides by T cells in HIV-1-infected individuals. **a** Frequencies of TL8 and HR10 mutant epitopes in Japanese individuals. The frequencies of mutant epitopes were investigated from 430 chronically HIV-1-infected Japanese individuals. **b** The $CD8^+$ T-cell responses to the index and mutant epitope peptides in HIV-1-infected Japanese individuals were analyzed by using an IFN-γ ELISPOT assay. The results are shown as mean and SD (n = 3). **c** The responses of the T cell-lines specific for TL8 or HR10 to wild-type or mutant peptide-prepulsed 721.221 cells expressing the corresponding HLA alleles were analyzed by using the ICS assay. The results are shown as mean and SD (n = 3)

high match to global HIV-1 variants. To date, the immunogenicity of these regions was demonstrated in preclinical studies [14, 26]. Importantly, we were previously able to show that $CD8^+$ T-cell responses to 10 peptide pools derived from the two tHIVconsvX mosaic immunogens correlated significantly with low pVL and high CD4 count in 120 treatment-naïve, HIV-1-infected Japanese patients [14]. In the present study, we focused on

Gag-specific CD8$^+$ T-cells because of their strong contribution to favorable clinical outcome. We defined a number of optimal epitopes and their HLA restriction, which were recognized by CD8$^+$ T cells specific for the Gag conserved regions of tHIVconsvX. Furthermore, we demonstrated that 9 Gag-specific CD8$^+$ T-cell responses were significantly associated with low pVL and/or high CD4 count in an extended 221-patient cohort of treatment-naïve patients, implying that T cells specific for these 9 epitopes may contribute to suppression of HIV-1 replication in these individuals. To increase the analysis sensitivity and further endorse the ability of these CD8$^+$ T-cell responses to suppress HIV-1 replication, we confirmed the association of these T-cells with good clinical parameters among the individuals carrying the appropriate HLA restriction molecules. Finally, we identified 5 Gag epitope-specific protective CD8$^+$ T-cell responses and showed a strong association of the magnitude and the breadth of the responses to the 5 epitopes with favorable pVL or CD4 count, indicating again that the responses to 5 Gag epitopes are beneficial. Although escape mutants were selected by the RI8-specific T cells and these HIV-1 mutants accumulated in the Japanese population [21], 70% of circulating HIV-1 still contain variants of the RI8 epitope recognized by RI8-elicited T cells.

It has now been well documented that not all CD8$^+$ T-cell responses contribute to the suppression of HIV-1 replication equally [14, 15, 21, 27]. Mothe et al. showed that the T-cell responses to forty-eight 18-mer overlapping peptides associated with low viral load in South Africa, Peru, and Spain cohorts [27]. These 18-mer overlapping peptides covering Gag, Pol, Nef, and Vif regions include 59 optimal defined epitopes in the LANL-HSD. The tHIVconsvX immunogen contains most of these; 19 in Gag and 14 in Pol. In the present study, we identified CD8$^+$ T cells specific for 5 Gag conserved epitopes with the ability to suppress HIV-1 replication in Japanese individuals. These CD8$^+$ T cells are restricted by HLA-B*52:01, HLA-A*02:06, HLA-A*33:03, and HLA-B*40:02, and at least one of these four alleles are shared in approximately 70% of Japanese population [28]. Since at least one of these alleles is also found in 45% of Chinese [29], 35% of Vietnamese [30], 40% of Thai [31], 50% of Indian [32], and 65% of Korean [33], we hypothesize that if indeed effective, the tHIVconsvX vaccine could be deployed in all these countries; the vaccine would be universal.

To confirm whether the 5 Gag epitopes were recognized by specific T cells as conserved epitopes, we analyzed cross-recognitions of mutant epitopes among the subtype B using the IFN-γ ELISPOT assay. Notably, the circulating viruses carry epitopes with a perfect 9/9 amino acid match or cross-recognized variations in 70–97% isolates from the Japanese population and 46–97% of subtype B viruses (LANL-HSD) and were recognized by T cells specific for 4 epitopes WV8, AA9, TL8, and HR10, though escape mutants selected by the RI8-specific T cells were detected in 27% of the circulating virus in the Japanese cohort [21]. These results suggest that the 4 epitopes are functionally conserved at least among the subtype B viruses. We further analyzed HLA-associated mutations within these six epitopes and found HLA-B*40:02-associated Gag mutation T427N within the HLA-B*40:02/TL8 epitope [25], suggesting that this mutant is accumulated in the HLA-B*40:02$^+$ Japanese individuals. However, since mutant Gag epitope TL8-1N was cross-recognized by CD8$^+$ T cells in 5 individuals with HLA-B*40:02 (Fig. 7b), TL8-1N may elicit a new T-cell repertoire specific for the 1N mutant or cross-recognizing this mutant. CD8$^+$ T-cells responses specific for TL8 and HR10 epitopes can recognize 70–75% of the circulating HIV-1 isolates in Japan while those specific for WV8, AA9, and RI8 did recognized > 70% of them.

We previously demonstrated that three HLA-B*52:01-, two HLA-B*67:01-, and one HLA-B*02:06-restricted CD8$^+$ T cells specific for Gag epitopes had a strong ability to suppress HIV-1 replication in the Japanese population [21]. The present study demonstrated that CD8$^+$ T cells specific for additional 2 Gag epitopes also contribute to the suppression of HIV-1. Although contained in the tHIVconsvX vaccines, T cells specific for MI8 restricted by HLA-B*52:01 [21], and TL9 and NL11 restricted by HLA-B*67:01 [21] were not identified in this study. This may be explained by the fact that MI8 and TL9 are included in Pool 1 and that the HLA-B*67:01 is a rare allele found in only 4 out of 221 patients involved in the present study. These findings suggest that Japanese individuals vaccinated with the tHIVconsvX vaccines may respond to 8 Gag epitopes and suppress better HIV-1 replication by killer T cells (Additional file 5: Fig. S5).

The present work supports the hypothesis that the CD8$^+$ T-cell responses, which the tHIVconsvX vaccines aim to elicit, have a real chance to contribute to the suppression of HIV-1 replication *in vivo* in Japan and other Asian populations both in the context of prevention of HIV-1 infection complementing neutralizing antibodies and in HIV cure by eradicating latently infected cells after HIV-1 reactivation. These results warrantee timely testing of this target immunogen strategy in the clinic.

Methods

Subjects

All treatment-naïve Japanese individuals chronically infected with HIV-1 subtype B were recruited from the National Center for Global Health and Medicine. This study was approved by the ethics committees of Kumamoto University and the National Center for Global

Health and Medicine. Informed consent was obtained from all individuals according to the Declaration of Helsinki. Plasma and PBMCs were separated from whole blood. HLA types of the individuals were determined by standard sequence-based genotyping.

Peptides

The mosaic proteins maximize the coverage of potential T-cell epitopes for the global circulating viruses. We generated three pools containing pairs of 15-mer Gag peptides overlapped by 11 amino acids covering two mosaic regions in the tHIVconsvX. Each pool contains 17 to 23 pairs of the 15-mer peptides. Pool 1, 2, and 3 cover Gag 133-231, Gag 221-327, and Gag 317-363/391-459, respectively [14]. The sequences of the 15-mer peptides in each Pool are shown in Additional file 6: Fig. S6. The 15-mer peptides derived from the tHIVconsvX mosaics were generously provided by the International AIDS Vaccine Initiative. Shorter mapping peptides were synthesized by utilizing an automated multiple peptide synthesizer and purified by high-performance liquid chromatography (HPLC). Purity of all peptides (>90%) was examined by HPLC and mass spectrometry.

Cell lines

C1R cells expressing HLA-A*26:01 (C1R-A2601), HLA-A*26:02 (C1R-A2602), HLA-A*33:03 (C1R-A3303), HLA-B*07:02 (C1R-B0702), HLA-B*15:01 (C1R-B1501), HLA-B*39:01 (C1R-B3901), or HLA-B*44:03 (C1R-B4403) were previously generated by transfecting these genes into C1R cell lines [21, 34–39]. 721.221 cells expressing CD4 molecules and HLA-C*03:03 (721.221-C0303) were generated by transfecting the genes into the 721.221 cell line. 721.221 cells expressing CD4 molecules and HLA-A*02:06 (721.221-A0206), HLA-A*24:02 (721.221-A2402), HLA-A*33:03 (721.221-A3303), HLA-B*40:02 (721.221-B4002), HLA-C*14:03 (721.221-C1403), or HLA-C*07:02 (721.221-C0702) were previously generated [21, 40–43]. All cell lines were cultured in RPMI 1640 medium containing 10% FCS medium (R10) with 0.15 mg/ml hygromycin B.

Expansion of HIV-1-specific T cells from HIV-1-infected individuals

PBMCs from KI-1020, KI-1102, and KI-1114 were incubated with 1 μM 15-mer peptide pairs (C052/053 or C054/055, C113/114, and C125/126, respectively) and cultured for 12-14 days to induce peptide-specific short-term cell lines (STCL).

Intracellular cytokine staining (ICS) assay

C1R and 721.221 cells prepulsed with each peptide or 721.221 cells infected with the HIV-1, strain NL4-3, were added to the effector STCLs in a 96-well plate and incubated for 2 h at 37 °C. Brefeldin A (10 μg/ml) was then added and the cells were incubated further for 4 h, fixed with 4% paraformaldehyde and incubated in permeabilization buffer [0.1% saponin–10% FBS–phosphate-buffered saline (PBS)] after staining with allophycocyanin (APC)-labeled anti-CD8 monoclonal antibody (mAb) (Dako, Glostrup, Denmark). Thereafter, the cells were stained with fluorescein isothiocyanate (FITC)-labeled anti-interferon γ (IFN-γ) mAb (BD Bioscience, CA). The percentage of IFN-γ-producing cells among the $CD8^+$ T-cell population was determined by flow cytometry.

IFN-γ enzyme-linked immunospot (ELISPOT) assay

1×10^5 PBMCs from HIV-1-positive individuals and peptides were added to 96-well polyvinylidene plates (Millipore, Bedford, MA) that had been precoated with 5 μg/ml anti-IFN-γ mAb; 1-D1K (Mabtech, Stockholm, Sweden). Peptide pools or 15-mer peptide pairs were used at a concentration of 1 μM whereas optimal epitope peptides at a concentration of 100 nM in this assay. The plates were then incubated for 16 h at 37 °C before the addition of biotinylated anti-IFN-γ mAb (Mabtech) at 1 μg/ml at room temperature for 90 min, streptavidin-conjugated alkaline phosphatase (Mabtech) for at room temperature 60 min. Individual cytokine-producing cells were visualized as dark spots after a 20-min reaction with 5-bromo-4-chloro-3-indolyl phosphate and nitro blue tetrazolium in the presence of an alkaline phosphatase-conjugated substrate (Bio-Rad, Richmond, CA, USA). The spots were counted with an Eliphoto-Counter (Minerva Teck, Tokyo, Japan). The frequencies of the responding cells were represented as spot-forming units (SFU)/10^6 $CD8^+$ T cells by measuring frequency of $CD8^+$ T cells using a flow cytometry. A mean+5 SD of the SFUs of samples (N=3) from 12 HIV-1-naïve individuals for the peptide pool was 115 SFU/10^6 $CD8^+$ T cells. Therefore, we defined a positive ELISPOT response as larger than 200 SFU/10^6 $CD8^+$ T cells to exclude false positive.

Establishment of T-cell lines specific for TL8 and HR10 peptides using HLA/peptide tetramer complexes

To establish T-cell lines specific for TL8 and HR10 peptides, HLA-B*40:02/TL8 or HLA-A*33:03/HR10 tetrameric complexes (tetramers) were synthesized as previously described [44]. Briefly, PBMCs of HLA-B*40:02$^+$ individual (KI-1391) and HLA-A*33:03$^+$ one (KI-1320) were stained with PE-conjugated specific tetramers at a concentration of 100 nM at 37 °C for 30 min. The cells were then washed twice with R10, followed by staining with FITC-conjugated anti-CD8 mAb and 7-AAD at 4 °C for 30 min. The $CD8^+$ T cells specific for the TL8 and HR10 peptides were then sorted by the

FACSAria. The sorted cells were stimulated with 100 nM of the corresponding epitope peptides and then cultured for 12–14 days to induce T cell lines specific for TL8 and HR10. To confirm purities of the specific T cells, the cell lines were analyzed by using the specific tetramers.

In vitro virus inhibition assay

The ability of HIV-1-specific CTLs to suppress HIV-1 replication was examined as previously described [45, 46]. $CD4^+$ T cells isolated from PBMCs of healthy donors carrying HLA-B*40:02 or -A*33:03 were infected with NL4-3 and then the infected cells were co-cultured with epitope-specific T-cell lines at E:T ratios of 1:1 and 0.1:1. On day 3–4 post infection, the concentration of p24 Ag in the culture supernatant was measured by using an enzyme-linked immunosorbent assay.

Statistical analyses

Two-tailed Mann–Whitney's test was performed for comparison of two groups. Correlations between magnitudes and breadths of T cell responses and pVL or CD4 count were statistically analyzed using Spearman rank test. P values < 0.05 were considered to be statistically significant.

Additional files

Additional file 1: Fig. S1. Correlation of the magnitudes of the Gag responses with pVL and CD4 count. T-cell responses to Gag peptide Pools 1, 2 and 3 derived from vaccine immunogen tHIVconsvX were enumerated using an IFN-γ ELISPOT assay in 200 HIV-1-infected Japanese individuals. Correlation coefficients (r) and p-values were determined by using the Spearman rank correlation test.

Additional file 2: Fig. S2. Association of the T-cell responses to AA9, WV8, or RI8 with pVL or CD4 count. T-cell responses to the 3 epitope peptides were analyzed by using the IFN-γ ELISPOT assay. The differences in pVL or CD4 count between responders and non-responders to each epitope peptide in the individuals having HLA restriction molecules for the epitopes were statistically analyzed by using the Mann-Whitney test.

Additional file 3: Fig. S3. Correlation between a total magnitude of T-cell responses to 5 epitopes and pVL and CD4 count. T-cell responses to 5 epitope peptides (AA9, TL8, WV8, RI8, and HR10) were analyzed in 149 individuals carrying the HLA restriction molecules by using the IFN-γ ELISPOT assay. Correlation coefficients (r) and p-values were determined by using the Spearman rank correlation test.

Additional file 4: Fig. S4. HIV-1 sequences within Gag TL8 and Gag HR10 epitopes in HIV-1-infected individuals. HIV-1 sequences within Gag TL8 and Gag HR10 were analyzed in HIV-1-infected individuals tested in Figure 7b. Mutant positions are highlighted in red.

Additional file 5: Fig. S5. Location of the 8 Gag CTL epitopes in the tHIVconsvX. The tHIVconsvX vaccine is composed of 2 Gag and 4 Pol conserved fragments. The two complementing mosaic immunogens corresponding to the 6 conserved regions are used in this vaccine. HLA-B*67:01-restricted TL9-specific, HLA-B*52:01-restricted MI8-specific, and HLA-B*67:01-restricted NL11-specific CTLs also have strong abilities to suppress HIV-1 replication in vivo (highlighted in green, Murakoshi et al., 2015).

Additional file 6: Fig. S6. List of 15-mer overlapping peptide pairs in Pools 1-3. Pool 1, 2, and 3 cover Gag133-231, Gag221-327, and Gag317-363 / 391-459, respectively.

Authors' contributions

HM wrote the manuscript, performed assays, and analyzed assay results. CZ performed assays and analyzed assay results. NK, TA and TC performed assays. HG and SO collected samples and clinical data from patients. TH provided the overlapping 15-mer peptides derived from the tHIVconsvX immunogens and contributed to writing the manuscript. MT designed the study, oversaw the experiments, analyzed all of the data, and wrote and edited the manuscript. All authors read and approved the final manuscript.

Author details

[1] Center for AIDS Research, Kumamoto University, 2-2-1 Honjo, Chuo-ku, Kumamoto 860-0811, Japan. [2] International Research Center of Medical Sciences, Kumamoto University, Kumamoto, Japan. [3] AIDS Clinical Center, National Center for Global Health and Medicine, Tokyo, Japan. [4] The Jenner Institute, University of Oxford, Old Road Campus Research Building, Roosevelt Drive, Oxford, UK.

Acknowledgements

Not applicable.

Competing interests

The authors declare that they have no competing interests.

Funding

This research was supported by grants-in-aid (15fk0410019h0001, 16fk0410202h0002, and17fk0410302h0003) for AIDS Research from AMED and by grants-in-aid (26293240, 17K10021) for scientific research from the Ministry of Education, Science, Sports, and Culture, Japan. C.Z. was supported by the China Scholarship Council (CSC) scholarship. The funders had no role in study design, data collection and interpretation, or the decision to submit the work for publication.

References

1. Rerks-Ngarm S, Pitisuttithum P, Nitayaphan S, Kaewkungwal J, Chiu J, Paris R, et al. Vaccination with ALVAC and AIDSVAX to prevent HIV-1 infection in Thailand. N Engl J Med. 2009;361:2209–20.
2. Haynes BF, Gilbert PB, McElrath MJ, Zolla-Pazner S, Tomaras GD, Alam SM, et al. Immune-correlates analysis of an HIV-1 vaccine efficacy trial. N Engl J Med. 2012;366:1275–86.
3. Rolland M, Edlefsen PT, Larsen BB, Tovanabutra S, Sanders-Buell E, Hertz T, et al. Increased HIV-1 vaccine efficacy against viruses with genetic signatures in Env V2. Nature. 2012;490:417–20.
4. Liao HX, Bonsignori M, Alam SM, McLellan JS, Tomaras GD, Moody MA, et al. Vaccine induction of antibodies against a structurally heterogeneous site of immune pressure within HIV-1 envelope protein variable regions 1 and 2. Immunity. 2013;38:176–86.
5. Chung AW, Kumar MP, Arnold KB, Yu WH, Schoen MK, Dunphy LJ, et al. Dissecting polyclonal vaccine-induced humoral immunity against HIV using systems serology. Cell. 2015;163:988–98.
6. Yates NL, Liao HX, Fong Y, deCamp A, Vandergrift NA, Williams WT, et al. Vaccine-induced Env V1-V2 IgG3 correlates with lower HIV-1 infection risk and declines soon after vaccination. Sci Transl Med. 2014;6:228ra239.
7. McMichael AJ, Haynes BF. Lessons learned from HIV-1 vaccine trials: new priorities and directions. Nat Immunol. 2012;13:423–7.
8. Gray GE, Allen M, Moodie Z, Churchyard G, Bekker LG, Nchabeleng M, et al. Safety and efficacy of the HVTN 503/Phambili study of a clade-B-based HIV-1 vaccine in South Africa: a double-blind, randomised, placebo-controlled test-of-concept phase 2b study. Lancet Infect Dis. 2011;11:507–15.
9. Hammer SM, Sobieszczyk ME, Janes H, Karuna ST, Mulligan MJ, Grove D, et al. Efficacy trial of a DNA/rAd5 HIV-1 preventive vaccine. N Engl J Med. 2013;369:2083–92.
10. Buchbinder SP, Mehrotra DV, Duerr A, Fitzgerald DW, Mogg R, Li D, et al. Efficacy assessment of a cell-mediated immunity HIV-1 vaccine (the Step Study): a double-blind, randomised, placebo-controlled, test-of-concept trial. Lancet. 2008;372:1881–93.

11. McElrath MJ, De Rosa SC, Moodie Z, Dubey S, Kierstead L, Janes H, et al. HIV-1 vaccine-induced immunity in the test-of-concept Step Study: a case-cohort analysis. Lancet. 2008;372:1894–905.
12. Hanke T. Conserved immunogens in prime-boost strategies for the next-generation HIV-1 vaccines. Expert Opin Biol Ther. 2014;14:601–16.
13. Letourneau S, Im EJ, Mashishi T, Brereton C, Bridgeman A, Yang H, et al. Design and pre-clinical evaluation of a universal HIV-1 vaccine. PLoS ONE. 2007;2:e984.
14. Ondondo B, Murakoshi H, Clutton G, Abdul-Jawad S, Wee EG, Gatanaga H, et al. Novel conserved-region T-cell mosaic vaccine with high global HIV-1 coverage is recognized by protective responses in untreated infection. Mol Ther. 2016;24:83242.
15. Kiepiela P, Ngumbela K, Thobakgale C, Ramduth D, Honeyborne I, Moodley E, et al. CD8+ T-cell responses to different HIV proteins have discordant associations with viral load. Nat Med. 2007;13:46–53.
16. Klein MR, van Baalen CA, Holwerda AM, Kerkhof Garde SR, Bende RJ, Keet IP, et al. Kinetics of Gag-specific cytotoxic T lymphocyte responses during the clinical course of HIV-1 infection: a longitudinal analysis of rapid progressors and long-term asymptomatics. J Exp Med. 1995;181:1365–72.
17. Ogg GS, Jin X, Bonhoeffer S, Dunbar PR, Nowak MA, Monard S, et al. Quantitation of HIV-1-specific cytotoxic T lymphocytes and plasma load of viral RNA. Science. 1998;279:2103–6.
18. Chen DY, Balamurugan A, Ng HL, Cumberland WG, Yang OO. Epitope targeting and viral inoculum are determinants of Nef-mediated immune evasion of HIV-1 from cytotoxic T lymphocytes. Blood. 2012;120:100–11.
19. Kloverpris HN, Payne RP, Sacha JB, Rasaiyaah JT, Chen F, Takiguchi M, et al. Early antigen presentation of protective HIV-1 KF11Gag and KK10Gag epitopes from incoming viral particles facilitates rapid recognition of infected cells by specific CD8+ T cells. J Virol. 2013;87:2628–38.
20. Sacha JB, Chung C, Rakasz EG, Spencer SP, Jonas AK, Bean AT, et al. Gag-specific CD8+ T lymphocytes recognize infected cells before AIDS-virus integration and viral protein expression. J Immunol. 2007;178:2746–54.
21. Murakoshi H, Akahoshi T, Koyanagi M, Chikata T, Naruto T, Maruyama R, et al. Clinical control of HIV-1 by cytotoxic T cells specific for multiple conserved epitopes. J Virol. 2015;89:5330–9.
22. Chikata T, Murakoshi H, Koyanagi M, Honda K, Gatanaga H, Oka S, et al. Control of HIV-1 by an HLA-B*52:01-C*12:02 Protective Haplotype. J Infect Dis. 2017;216:1415–24.
23. Sabbaj S, Bansal A, Ritter GD, Perkins C, Edwards BH, Gough E, et al. Cross-reactive CD8+ T cell epitopes identified in US adolescent minorities. J Acquir Immune Defic Syndr. 2003;33:426–38.
24. Borthwick N, Lin Z, Akahoshi T, Llano A, Silva-Arrieta S, Ahmed T, et al. Novel, in-natural-infection subdominant HIV-1 CD8+ T-cell epitopes revealed in human recipients of conserved-region T-cell vaccines. PLoS ONE. 2017;12:e0176418.
25. Chikata T, Carlson JM, Tamura Y, Borghan MA, Naruto T, Hashimoto M, et al. Host-specific adaptation of HIV-1 subtype B in the Japanese population. J Virol. 2014;88:4764–75.
26. Wee EG, Ondondo B, Berglund P, Archer J, McMichael AJ, Baltimore D, et al. HIV-1 conserved mosaics delivered by regimens with integration-deficient DC-Targeting lentiviral vector induce robust T cells. Mol Ther. 2017;25:494–503.
27. Mothe B, Llano A, Ibarrondo J, Daniels M, Miranda C, Zamarreno J, et al. Definition of the viral targets of protective HIV-1-specific T cell responses. J Transl Med. 2011;9:208.
28. Ikeda N, Kojima H, Nishikawa M, Hayashi K, Futagami T, Tsujino T, et al. Determination of HLA-A, -C, -B, -DRB1 allele and haplotype frequency in Japanese population based on family study. Tissue Antigens. 2015;85:252–9.
29. Qin Qin P, Su F, Xiao Yan W, Xing Z, Meng P, Chengya W, et al. Distribution of human leucocyte antigen-A, -B and -DR alleles and haplotypes at high resolution in the population from Jiangsu province of China. Int J Immunogenet. 2011;38:475–81.
30. Hoa BK, Hang NT, Kashiwase K, Ohashi J, Lien LT, Horie T, et al. HLA-A, -B, -C, -DRB1 and -DQB1 alleles and haplotypes in the Kinh population in Vietnam. Tissue Antigens. 2008;71:127–34.
31. Kupatawintu P, Pheancharoen S, Srisuddee A, Tanaka H, Tadokoro K, Nathalang O. HLA-A, -B, -DR haplotype frequencies in the Thai Stem Cell Donor Registry. Tissue Antigens. 2010;75:730–6.
32. Rani R, Marcos C, Lazaro AM, Zhang Y, Stastny P. Molecular diversity of HLA-A, -B and -C alleles in a North Indian population as determined by PCR-SSOP. Int J Immunogenet. 2007;34:201–8.
33. Huh JY, Yi DY, Eo SH, Cho H, Park MH, Kang MS. HLA-A, -B and -DRB1 polymorphism in Koreans defined by sequence-based typing of 4128 cord blood units. Int J Immunogenet. 2013;40:515–23.
34. Satoh M, Takamiya Y, Oka S, Tokunaga K, Takiguchi M. Identification and characterization of HIV-1-specific CD8+ T cell epitopes presented by HLA-A*2601. Vaccine. 2005;23:3783–90.
35. Kawashima Y, Satoh M, Oka S, Shirasaka T, Takiguchi M. Different immunodominance of HIV-1-specific CTL epitopes among three subtypes of HLA-A*26 associated with slow progression to AIDS. Biochem Biophys Res Commun. 2008;366:612–6.
36. Hossain MS, Tomiyama H, Inagawa T, Ida S, Oka S, Takiguchi M. Identification and characterization of HLA-A*3303-restricted, HIV type 1 Pol- and Gag-derived cytotoxic T cell epitopes. AIDS Res Hum Retroviruses. 2003;19:503–10.
37. Falk K, Rotzschke O, Takiguchi M, Gnau V, Stevanovic S, Jung G, et al. Peptide motifs of HLA-B58, B60, B61, and B62 molecules. Immunogenetics. 1995;41:165–8.
38. Ogawa A, Tokunaga K, Nakajima F, Kikuchi A, Karaki S, Kashiwase K, et al. Identification of the gene encoding a novel HLA-B39 subtype. Two amino acid substitutions on the beta-sheet out of the peptide-binding floor form a novel serological epitope. Hum Immunol. 1994;41:241–7.
39. Kato N, Kikuchi A, Kano K, Egawa K, Takiguchi M. Molecular analysis of a novel HLA-A33 subtype associated with HLA-B44. Tissue Antigens. 1993;41:211–3.
40. Watanabe K, Murakoshi H, Tamura Y, Koyanagi M, Chikata T, Gatanaga H, et al. Identification of cross-clade CTL epitopes in HIV-1 clade A/E-infected individuals by using the clade B overlapping peptides. Microbes Infect. 2013;15:874–86.
41. Koizumi H, Iwatani T, Tanuma J, Fujiwara M, Izumi T, Oka S, et al. Escape mutation selected by Gag28-36-specific cytotoxic T cells in HLA-A*2402-positive HIV-1-infected donors. Microbes Infect. 2009;11:198–204.
42. Lin Z, Kuroki K, Kuse N, Sun X, Akahoshi T, Qi Y, et al. HIV-1 Control by NK Cells via Reduced Interaction between KIR2DL2 and HLA-C *12:02/C *14:03. Cell Rep. 2016;17:2210–20.
43. Watanabe T, Murakoshi H, Gatanaga H, Koyanagi M, Oka S, Takiguchi M. Effective recognition of HIV-1-infected cells by HIV-1 integrase-specific HLA-B *4002-restricted T cells. Microbes Infect. 2011;13:160–6.
44. Akahoshi T, Chikata T, Tamura Y, Gatanaga H, Oka S, Takiguchi M. Selection and accumulation of an HIV-1 escape mutant by three types of HIV-1-specific cytotoxic T lymphocytes recognizing wild-type and/or escape mutant epitopes. J Virol. 2012;86:1971–81.
45. Tomiyama H, Akari H, Adachi A, Takiguchi M. Different effects of Nef-mediated HLA class I down-regulation on human immunodeficiency virus type 1-specific CD8(+) T-cell cytolytic activity and cytokine production. J Virol. 2002;76:7535–43.
46. Tomiyama H, Fujiwara M, Oka S, Takiguchi M. Cutting Edge: Epitope-dependent effect of Nef-mediated HLA class I down-regulation on ability of HIV-1-specific CTLs to suppress HIV-1 replication. J Immunol. 2005;174:36–40.

9

Lipid levels, insulin resistance and cardiovascular risk over 96 weeks of antiretroviral therapy: a randomised controlled trial comparing low-dose stavudine and tenofovir

Alinda G. Vos[1,2,3*], Matthew F. Chersich[1], Kerstin Klipstein-Grobusch[2,4], Peter Zuithoff[2], Michelle A. Moorhouse[1], Samanta T. Lalla-Edward[1], Andrew Kambugu[5], N. Kumarasamy[6], Diederick E. Grobbee[2], Roos E. Barth[3] and Willem D. Venter[1]

Abstract

Background: HIV infection and antiretroviral treatment are associated with changes in lipid levels, insulin resistance and risk of cardiovascular disease (CVD). We investigated these changes in the first 96 weeks of treatment with low-dose stavudine or tenofovir regimens.

Methods: This is a secondary analysis of a double blind, randomised controlled trial performed in South-Africa, Uganda and India comparing low-dose stavudine (20 mg twice daily) with tenofovir in combination with efavirenz and lamivudine in antiretroviral-naïve adults ($n = 1067$) (Clinicaltrials.gov, NCT02670772). Over 96 weeks, data were collected on fasting lipids, glucose and insulin. Insulin resistance was assessed with the HOMA-IR index and 10-year CVD risk with the Framingham risk score (FRS). A generalized linear mixed model was used to estimate trends over time.

Results: Participants were on average 35.3 years old, 57.6% female and 91.8% Black African. All lipid levels increased following treatment initiation, with the sharpest increase in the first 24 weeks of treatment. The increase in all lipid subcomponents over 96 weeks was higher among those in the stavudine than the tenofovir group. Insulin resistance increased steadily with no difference detected between study groups. FRS rose from 1.90% (1.84–1.98%) at baseline to 2.06 (1.98–2.15%) at week 96 for the total group, with no difference between treatment arms ($p = 0.144$). Lipid changes were more marked in Indian than African participants.

Conclusion: Lipid levels increased in both groups, with low-dose stavudine resulting in a worse lipid profile compared to tenofovir. Insulin resistance increased, with no difference between regimens. CVD risk increased over time and tended to increase more in the group on stavudine. The low CVD risk across both arms argues against routine lipid and glucose monitoring in the absence of other CVD risk factors. In high risk patients, monitoring may only be appropriate at least a year after treatment initiation.

Keywords: HIV, South Africa, Uganda, India, Stavudine, Tenofovir, Cardiovascular disease risk

*Correspondence: avos@wrhi.ac.za

[1] Wits Reproductive Health and HIV Institute, Faculty of Health Sciences, University of Witwatersrand, Johannesburg, South Africa

Full list of author information is available at the end of the article

Introduction

Globally, cardiovascular disease (CVD) is the leading cause of mortality [1]. Low- and middle-income countries (LMICs) share this burden: the leading cause of death has changed from infectious diseases to ischaemic heart disease over the last two decades [1]. Timely recognition and treatment of cardiovascular risk factors are key for reducing the burden of CVD. Human immunodeficiency virus (HIV) infection and treatment with antiretroviral therapy (ART) affects risk factors for CVD [2–5] and some studies indicate that HIV infection increases the risk of myocardial infarction or stroke by up to 50% [6, 7].

The contribution of ART to CVD risk is less clear, and risk profiles vary by antiretroviral drug [8]. Protease inhibitors and efavirenz, a non-nucleoside reverse transcriptase inhibitor, are well known for their adverse effects on lipid and glucose metabolism [9–11]. Stavudine, a nucleoside reverse transcriptase inhibitor (NRTI), has considerable metabolic side effects. It is associated with mitochondrial toxicity, resulting in lipodystrophy, disturbances in lipid levels and an increase in insulin resistance [12–15]. Stavudine was part of first-line ART for many years globally, as it was well tolerated in the first few months of treatment, cheap, widely available in twice-daily fixed-dose formulations and effective in suppressing viral load [16]. When the extent of mitochondrial toxicity became apparent, use of the drug declined rapidly. Between 2012 and 2016, a clinical trial was initiated, comparing a lower dose of stavudine to the current most commonly used first-line drug, tenofovir disoproxil fumarate ('tenofovir'). The overall conclusion was that low-dose stavudine was equally effective as tenofovir in reducing viral load after 48 weeks, but that lipoatrophy occurred more often in the low-dose stavudine group [17]. Extensive metabolic and toxicity monitoring allow us to conduct an in-depth analysis of the effects of ART initiation with low-dose stavudine or tenofovir on lipid levels, insulin resistance and CVD risk, an important analysis as the vast majority of people on ART are taking regimens containing tenofovir. We present these results in this paper.

Methods

A randomised 1:1 double blind placebo-controlled trial was conducted in Johannesburg, South Africa, Kampala, Uganda and Chennai, India to assess the efficacy and safety of treatment with either low dose stavudine (20 mg twice a day) or tenofovir (300 mg daily) tablets administered in combination with lamivudine (150 mg BD) and efavirenz (600 mg daily) over 96 weeks (Clinicaltrials.gov, NCT02670772). The methods including quality control and safety evaluation are described in detail elsewhere [17]. In brief, data were captured in an electronic data system. Computerised and manual procedures as well as regular site visits were implemented to optimise data quality. Monitoring and support was undertaken by Pharmaceutical Product Development (PPD, Wilmington, USA). 1067 HIV-positive, ART-naive participants aged 18 years and over with unsuppressed HIV viral load were included from clinical trial sites. Recruitment in India was stopped early due to a regulatory change, and it was decided to raise the number of participants recruited in the remaining sites so as to reach the target number timeously. Exclusion criteria were age above 65 years for the Indian site, pregnancy, CD4+ > 350 cells/μL, hepatitis B antigen positivity, or an estimated glomerular filtration rate (eGFR) < 60 mL/min, calculated using the Cockcroft-Gault equation. Visits took place at baseline, 1 month, 3 months and every 3 months thereafter until 96 weeks. Data collected at baseline included demographic information, medical history including use of medication. In the South African site only, additional data on employment, marital status, having children, as well as data on current smoking, alcohol and drug use were collected.

At each visit, weight, heart rate and blood pressure were measured and blood samples taken. Blood pressure was measured once in a seated position after a 5-min rest. Blood sampling was performed after an overnight fast and included lipids, glucose, insulin, HIV viral load and CD4+ count. Blood samples were analysed by contracted and accredited laboratory services (see Additional file 1: Table S1 for more details). Insulin resistance was quantified with the HOMA-IR, using the following formula: fasting glucose (mmol/L) * fasting insulin (mU/L))/22.5 [18]. Framingham risk score (FRS) was calculated using the 10-year CVD risk score equation, and only available for South Africa, as the risk score includes smoking history [19]. The study was approved by the Human Ethics Research Committee of the University of the Witwatersrand, ethics clearance number 111112, the Research and Ethics Committee of the Joint Clinical Research Centre, Uganda, and the Uganda National Council for Science and Technology clearance number HS 1219, and the IRB & Ethics Research Committee of YRGCARE. All participants provided written informed consent prior to participation. All participants in the stavudine arm were switched to a tenofovir containing regimen at the end of follow as per the national HIV treatment guidelines.

Statistical analysis

Demographics were reported as means and standard deviations, or medians with interquartile ranges as appropriate. Non-normal outcomes were log transformed to obtain a normal distribution. In line with logistic regression models, we used a logit transformation for the predicted probabilities based on the Framingham risk model. Lipid levels, glucose, insulin, HOMA-IR and FRS measurements over the 96 week period were analysed with a linear mixed model (estimated with restricted maximum likelihood) with a random intercept and a random effect for time. We included a quadratic term for time (i.e. time^2), considering that outcomes are expected to level off at a certain point in time. Treatment arm, time, time^2 and the interaction between treatment arm and both time and time^2 were included as fixed effects together with a correction for age, sex, site of inclusion (South-Africa, Uganda or India), body mass index and viral load at baseline. p values for the interactions between treatment and both time and time^2 were estimated with likelihood ratio tests. Likelihood ratio tests were chosen, as it allows estimation of one p value (per outcome) for both interactions. For these tests, models were refitted with maximum likelihood estimation [20]. The estimated mean values at week 0, 24 weeks, 48 weeks, 72 weeks and 96 weeks for all outcomes were displayed graphically (transformed back to the original scale when applicable). The trend over time (per month) per treatment arm was calculated for all outcomes. Results are presented as regression coefficients (β) with 95% confidence intervals. Statistical analysis was done with SPSS version 24 (IBM SPSS Statistics for Windows, Version 24.0. Armonk, NY: IBM Corp.). Validity of the models was evaluated with residual analyses with SAS software version 9.4 (SAS Institute Inc., Cary, NC, USA).

Results

The mean age of participants was 35.3 years, the majority were female (57.6%), 24.2% was married, 76.0% was employed and most were of African descent (91.8%). The median nadir CD4+ cell count was 206 cells/μL (IQR 124–272). There were no differences in demographics, CVD risk factors or HIV-related factors between the study arms or study sites at baseline (Table 1). Participant retention was similar in each arm, with an overall median follow-up time of 95.4 weeks (Interquartile range (IQR) 95.0–95.7 weeks). Viral suppression (< 50 copies/mL) was reached in 71.1% in the stavudine arm and 76.7% in the tenofovir arm at the end of follow up. In total, 12.2% had a follow-up time of less than 48 weeks. Alcohol use was the most frequently reported substance used, followed by smoking and illegal drug use.

Figure 1 shows the trend in lipid levels following treatment initiation (Fig. 1). Total-C, HDL-C and LDL-C increased over the course of 96 weeks with a sharper increase in the group receiving stavudine than in the group receiving tenofovir, $p < 0.001$ for Total-C and HDL-C, and $p = 0.036$ for LDL-C (Table 2, see Additional file 2: Table S2). TG levels decreased from week 24 in the group on tenofovir and the estimated mean at week 96 was almost identical to the estimated mean at baseline (1.148 mmol/L (95% CI 1.102–1.197) at baseline versus 1.154 mmol/L (95% CI 1.097–1.212) at week 96).

Glucose values were similar in both groups, with a steep increase in the first 48 weeks (from 4.63 mmol/L (95% CI 4.56–4.69) at baseline to 5.13 mmol/L (95% CI 5.06–5.20) at week 48) and values tended to level off from week 48 to reach 5.12 mmol/L (95% CI 5.05–5.31) at week 96 (see Fig. 2 for trend per treatment arm and Additional file 2: Table S2). Insulin increased over 96 weeks, with no difference detected between arms ($p = 0.677$, Table 2). The increase in insulin resistance per month (with HOMA-IR) was not different for participants on stavudine versus tenofovir ($p = 0.698$; Table 2).

Framingham risk score rose from 1.90% (1.84–1.98%) at baseline to 2.06 (1.98–2.15%) at week 96 for the total group (see Fig. 3 for trend per treatment arm and Additional file 2: Table S2). FRS in the stavudine group was significantly lower at baseline than in the tenofovir group, 1.82% (95% CI 1.73–1.92) and 2.00% (95% CI 1.89–2.10) respectively, $p = 0.013$. The risk in the group on stavudine went up to 2.11% (95% CI 1.99–2.24) at week 96. The risk in the group on tenofovir decreased initially to 1.94% (95% CI 1.85–2.05) at week 48 to increase to 2.01% (95% CI 1.90–2.13) at week 96. The overall trend between the groups was not significantly different, $p = 0.144$.

Age, sex and BMI were associated with all lipid subcomponents, glucose, insulin, HOMA-IR and FRS (Additional file 3: Table S3a–h). The 10-year CVD risk according to the FRS was almost double for males compared to females at week 96, 2.88% (95% CI 2.68–3.00) for males versus 1.50% (95% CI 1.42–1.58) for females, $p < 0.001$. Log viral load at baseline was associated with all lipid subcomponents (with a higher log viral load linked to a lower Total-C, HDL-C and LDL-C) but not to glucose, insulin, HOMA-IR or FRS. Indian participants had a worse lipid profile and glucose homeostasis compared to their African counterparts. For example, at baseline Total-C was 4.38 mmol/L (95% CI 4.21–4.55) for the Indian site and 3.72 mmol/L (95% CI 3.66–3.79) for the South African site, and fasting glucose was

Table 1 Demographics, HIV disease and cardiovascular disease risks of patients at baseline

	All ($n = 1067$)	Stavudine ($n = 533$)	Tenofovir ($n = 534$)
Sex, female (*n*, %)	615 (57.6)	324 (60.8)	291 (54.5)
Age in years	35.3 (8.2)	35.5 (8.4)	35.0 (8.1)
Race (*n*, %)			
African	979 (91.8)	489 (91.7)	490 (91.8)
Indian	86 (8.1)	43 (8.1)	43 (8.1)
Other	2 (0.2)	1 (0.2)	1 (0.2)
Site (*n*, %)			
South Africa	600 (56.2)	300 (56.3)	300 (56.2)
Uganda	381 (35.7)	190 (35.6)	191 (35.8)
India	86 (8.1)	43 (8.1)	43 (8.1)
Married[a] (*n*, %)	139 (24.2)	69 (24.1)	70 (24.2)
Having children[a] (*n*, %)	492 (85.9)	243 (84.1)	249 (87.7)
Employed[a] (*n*, %)	434 (76.0)	220 (76.1)	214 (75.9)
CD4+ cell count (cells/μL) (median, IQR)	206 (124–272)	206 (128–274)	206 (123–270)
Log HIV viral load (copies/mL)	11.26 (1.61)	11.34 (1.64)	11.18 (1.58)
History of hypertension[b] (*n*, %)	59 (5.5)	28 (5.3)	31 (5.8)
History of diabetes mellitus[b] (*n*, %)	6 (0.6)	4 (0.8)	2 (0.4)
History of dyslipidemia[b] (*n*, %)	2 (0.2)	1 (0.2)	1 (0.2)
Current smoking[c] (*n*, %)	86 (15)	44 (15.3)	42 (14.7)
Current alcohol use[c] (*n*, %)	203 (35.4)	101 (35.1)	102 (35.8)
Current drug use[c] (*n*, %)	17 (3.0)	8 (2.8)	9 (3.2)
Systolic blood pressure (mmHg)	118.3 (13.2)	117.5 (13.5)	119.1 (12.8)
Diastolic blood pressure (mmHg)	72.9 (10.2)	72.5 (10.6)	73.4 (9.9)
Body mass index (kg/m^2)	23.5 (4.4)	23.6 (4.6)	23.5 (4.2)
Follow-up time, median months, IQR	95.4 (95.0–95.7)	95.4 (95.0–95.7)	95.4 (95.1–95.8)
Total-C (mmol/L)	4.30 (1.02)	4.39 (1.11)	4.22 (0.93)
HDL-C (mmol/L)	1.32 (0.49)	1.37 (0.54)	1.27 (0.44)
LDL-C (mmol/L)	2.47 (0.86)	2.50 (0.93)	2.44 (0.79)
TG (mmol/L) (median, IQR)	1.00 (0.74–1.37)	1.00 (0.76–1.40)	0.98 (0.72–1.33)
Fasting glucose (mmol/L)	4.77 (0.97)	4.72 (0.64)	4.81 (1.20)
Fasting insulin (pmol/L) (median, IQR)	38.20 (25.70–55.56)	40.28 (27.09–57.64)	36.81 (24.31–54.17)
HOMA-IR (median, IQR)	1.13 (0.74–1.75)	1.18 (0.78–1.78)	1.09 (0.71–1.70)
Framingham risk score (%) (median, IQR)	1.86 (1.02–3.45)	1.90 (1.05–3.44)	1.82 (0.98–3.46)

Values in mean with standard deviation unless otherwise specified

HDL-C high density lipoprotein cholesterol, *IQR* interquartile range, *LDL-C* low density lipoprotein cholesterol, *TG* triglycerides; *Total-C* total cholesterol

[a] Data only available for the South African site, n=575 for marital status, $n = 571$ for employment status, $n = 573$ for having children

[b] History or use of medication

[c] Data only available for the South African site, n=574 for current smoking, $n = 573$ for current alcohol use, $n = 562$ for current drug use

5.05 mmol/L (95% CI 4.90–5.22) for the Indian site versus 4.45 mmol/L (95% CI 4.38–4.51) for the South African site, $p < 0.001$ for both comparisons. In addition to these differences in baseline levels, Indian participants had a sharper increase in total-C, LDL-C and TG, and a larger decrease in HDL-C with treatment (See Additional file 3: Table S3a–d). The effect of stavudine and tenofovir on lipid components, as described above, was similar for all three populations (South Africa, Uganda and India).

Discussion

Initiation of low-dose stavudine and tenofovir with a backbone of lamivudine and efavirenz resulted in a rise in lipid levels with a worse lipid profile for stavudine

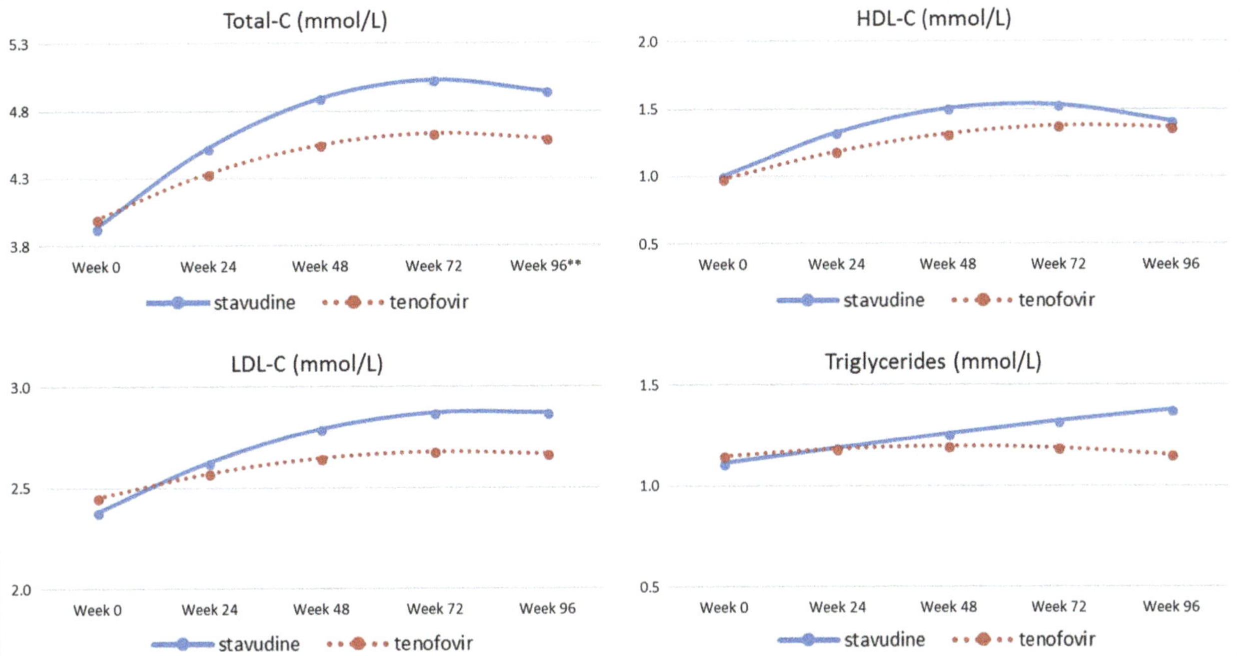

Fig. 1 Trends in estimated marginal mean lipid levels in each study group. *HDL-C* high density lipoprotein cholesterol, *LDL-C* low densiy lipoprotein cholesterol, *Total-C* total cholesterol

compared to tenofovir. Insulin resistance went up but there was no difference in trend between both groups. CVD risk was low in general and with no statistically significant difference between the groups over time.

Untreated HIV infection is characterized by an increase in TG and a decrease in both LDL-C and HDL-C [8, 21, 22]. The disturbance in lipid levels is likely a result of chronic inflammation that accompanies HIV infection [23]. Chronic inflammation influences the way lipids are metabolized and activated. Oxidized LDL and HDL may directly induce monocyte and endothelial cell activation, and this is related to the development of CVD [24].

Studies on initiation of ART have attributed both improvement, as well as deterioration, in lipid levels to ART [8, 22, 24–26]. A meta-analysis of cardiovascular risk factors in relation to HIV and ART in sub-Saharan Africa found that ART was associated with an increase in HDL-C and a decrease in TG, but also with an increase in LDL-C [2]. Liu et al. investigated changes in lipid levels during 3 years following ART initiation in Tanzania. They found an increase in LDL and HDL, and a decrease in TGs in the first 6 months. Between 6 months and 3 years HDL levels topped off and TG levels continued to increase [25]. Our results are broadly in line with these findings. We did not see a decrease in TG between treatment initiation and week 96, but the TG level in the group receiving tenofovir at week 96 was approximately the same as the level before treatment initiation. Our results underline the importance of taking treatment duration into account.

There is no agreed threshold to define insulin resistance in an urban African population [27] but based on a study in an urban Ghanaian population, a HOMA-IR cut-off of 2.3 seems appropriate [28]. According to this definition, neither low-dose stavudine nor tenofovir resulted in significant insulin resistance. Stavudine's mitochondrial toxicity is likely dose dependent [29, 30], and the low dose used in this study may account for the lack of impact on glucose metabolism. We can, however, not exclude the possibility that the drug's deleterious effects could still occur with more extended durations of treatment.

Lipid levels and insulin resistance are increased in both treatment arms. This may partly rely on the effects of efavirenz, as this drug is known to alter lipid levels and glucose metabolism [11, 15]. Our findings underline the need for more 'lipid-friendly' ART, of which dolutegravir is a promising alternative [31].

CVD risk according to the 10-year FRS increased in the total group with 0.16% between baseline and week 96. This was statistically significant, but it's questionable whether this slight increase over almost 2 years is clinically relevant. There was a trend towards a sharper increase in CVD risk in the group receiving

Table 2 Estimated changes per month in lipid levels, insulin resistance and Framingham risk score

Variable	Increase per month Regression coefficient (95% CI)#	Increase per month² Regression coefficient (95% CI)#	*p*
Total-C (mmol/L)			
Stavudine	0.1275 (0.1035–0.1514)	− 0.0037 (− 0.0047 to − 0.0027)	
Tenofovir	0.0719 (0.0621–0.0818)	− 0.0020 (− 0.0025 to − 0.0016)	< 0.001
HDL-C (mmol/L)			
Stavudine	0.0721 (0.0572–0.0870)	− 0.0024 (− 0.0031 to − 0.0018)	
Tenofovir	0.0433 (0.0372–0.0495)	− 0.0012 (− 0.0014 to − 0.0009)	< 0.001
LDL-C (mmol/L)			
Stavudine	0.0516 (0.0324–0.0709)	− 0.0013 (− 0.0021 to − 0.0005)	
Tenofovir	0.0254 (0.0175–0.0334)	− 0.0007 (− 0.0010 to − 0.0004)	0.036
Log TG (mmol/L)			
Stavudine	0.0125 (− 0.0017–0.0268)	− 0.0001 (− 0.0007–0.0005)	
Tenofovir	0.0073 (0.0014–0.0132)	− 0.0003 (− 0.0006 to − 0.0001)	0.058
Glucose (mmol/L)			
Stavudine	0.0689 (0.0393–0.0980)	− 0.0021 (− 0.0033 to − 0.0009)	
Tenofovir	0.0679 (0.0558–0.0800)	− 0.0020 (− 0.0026 to − 0.0015)	0.991
Insulin (pmol/L)			
Stavudine	0.3941 (− 1.3102–2.0984)	0.0045 (− 0.0659–0.0749)	
Tenofovir	0.0042 (− 0.7001–0.7086)	0.0231 (− 0.0059–0.0521)	0.677
Log HOMA-IR			
Stavudine	0.0242 (0.0006–0.0479)	− 0.0004 (− 0.0014–0.0006)	
Tenofovir	0.0158 (0.0060–0.0255)	− 0.0002 (− 0.0006–0.0003)	0.756
Logit Framingham risk score*			
Stavudine	0.0100 (− 0.0050–0.0251)	− 0.0001 (− 0.0008–0.0005)	
Tenofovir	− 0.0047 (− 0.0109–0.0016)	0.0002 (0.0000–0.0005)	0.144

CI confidence interval, *HDL-C* high density lipoprotein cholesterol, *LDL-C* low density lipoprotein cholesterol, *OR* odds ratio, *TG* triglycerides, *Total-C* total cholesterol

*Data for South African site only

The model includes adjustment for age, gender, site, body mass index and viral load

Month², time was used as a quadratic term to take non-linearity into account. *p* values were estimated with a likelihood ratio test for both interactions. See "Methods" for details

stavudine, most likely reflecting the steeper increase in LDL-C and TG in this group compared to the group on tenofovir. At the same time, it would be important to consider the relative contribution of ART to well-known modifiable cardiovascular risk factors such as obesity, smoking, diet, and physical inactivity.

Indian participants had substantially worse lipid profiles and glucose homeostasis compared to the African participants. This is a well-known phenomenon that is likely dependent on genetic factors and lifestyle factors such as being underweight during infancy, diet and physical activity [32–34].

Strengths of the study are the large, representative sample including participants from three countries, the randomized design, the standardized and regular assessment of lipids, glucose and insulin, and the follow up duration of almost 2 years. However, some limitations need to be recognized. Information on socio-demographics, alcohol use, drug use and smoking was only collected for the South African site, relied on patient reporting as opposed to biochemical measures, and only at baseline. No information on family history of CVD was collected, so we could not quantify CVD risk with other CVD risk prediction scores such as the Data collection on Adverse events of Anti-HIV Drugs (D:A:D) score, which is considered to be a more accurate risk prediction tool for HIV-infected populations than the FRS [35]. Also, we could not examine effects of the metabolic syndrome as waist circumference had not been measured. In line with an intention-to-treat approach for clinical trials, we only corrected for viral load at baseline, even though viral load at follow-up was available in the data. Both adherence and the proportion of participants with suppressed viral load between treatment groups were approximately equal. We therefore assume that any bias is small. Finally, the study did not include an HIV-negative control group

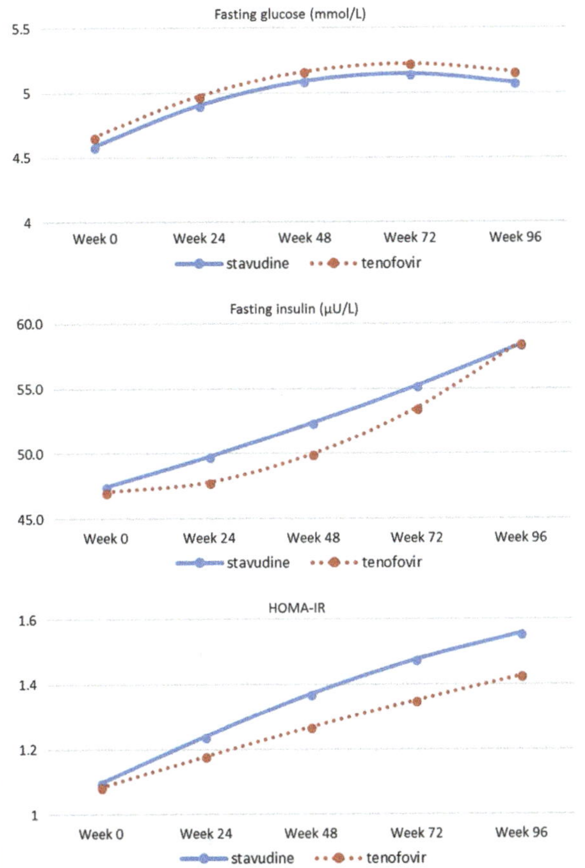

Fig. 2 Trends in estimated marginal mean levels of indicators of insulin resistance

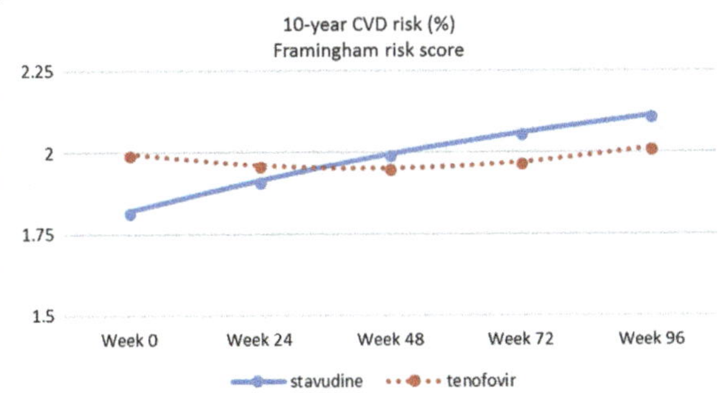

Fig. 3 Trend in estimated marginal mean levels of 10-year CVD risk

to compare our findings to age-related, HIV-unrelated changes in lipid levels and insulin resistance over time.

Conclusion

This study showed that low-dose stavudine has more deleterious effects on lipids than tenofovir. However, the impact of stavudine on this short term appears to be small, and it's questionable whether the increase in lipid levels is clinical relevant in this relatively young population with a low CVD risk. The need to monitor lipid and glucose levels has to be determined using CVD risk calculators that take other risk factors like age and hypertension into account. There should be extra awareness for CVD risk prevention in the Indian population. We recommend to measure lipids and glucose at the earliest a year after treatment initiation, as levels change substantially following treatment initiation.

Additional files

Additional file 1: Table S1. Title: Laboratory methods. Description of data: this table includes details on the measurement methods of lipids, glucose, insulin, HIV viral load and CD4+ cell count.

Additional file 2: Table S2. Title: Estimated means with 95% confidence intervals per outcome. Description of data: This table presents estimates of the means for each treatment group at 4 follow-up measurements based on a linear mixed model with correction for age, sex, site of inclusion (South-Africa, Uganda or India), body mass index and viral load at baseline (see "Methods" for details).

Additional file 3: Table S3. Title: Generalized linear mixed models. Description of data: generalized linear mixed models for the outcomes total cholesterol, HDL cholesterol, LDL cholesterol, triglycerides, glucose, insulin, HOMA-IR and Framingham risk score.

Authors' contributions

MM, AK, NK and FV were involved in data collection. AV, MC and FV designed this analysis, AV, MC and PZ analyzed and interpreted the patient data. AV and MM wrote the manuscript. All co-authors contributed to the final manuscript. All authors read and approved the final manuscript.

Author details

[1] Wits Reproductive Health and HIV Institute, Faculty of Health Sciences, University of Witwatersrand, Johannesburg, South Africa. [2] Julius Global Health, Julius Center for Health Sciences and Primary Care, University Medical Center Utrecht, Utrecht University, Utrecht, The Netherlands. [3] Department of Internal Medicine and Infectious Diseases, University Medical Center Utrecht, Utrecht University, Utrecht, The Netherlands. [4] Division of Epidemiology and Biostatistics, School of Public Health, Faculty of Health Sciences, University of the Witwatersrand, Johannesburg, South Africa. [5] Infectious Diseases Institute, Makerere University, Kampala, Uganda. [6] YRGCARE Medical Centre, CART Clinical Research Site, Voluntary Health Services, Chennai, India.

Acknowledgements

Not applicable.

Competing interests

The authors declare that they have no competing interests.

Consent for publication

Not applicable.

Funding

Bill and Melinda Gates Foundation. The foundation was in not involved in the design of this secondary analysis, interpretation of the data nor in writing the manuscript.

References

1. GBD 2016 Causes of Death Collaborators. Global, regional, and national age-sex specific mortality for 264 causes of death, 1980–2016: a sys-

tematic analysis for the Global Burden of Disease Study 2016. Lancet. 2017;390(10100):1151–210.

2. Dillon DG, Gurdasani D, Riha J, Ekoru K, Asiki G, Mayanja BN, et al. Association of HIV and ART with cardiometabolic traits in sub-Saharan Africa: a systematic review and meta-analysis. Int J Epidemiol. 2013;42(6):1754–71.
3. Kingery JR, Alfred Y, Smart LR, Nash E, Todd J, Naguib MR, et al. Short-term and long-term cardiovascular risk, metabolic syndrome and HIV in Tanzania. Heart. 2016;102(15):1200–5.
4. Nduka CU, Stranges S, Sarki AM, Kimani PK, Uthman OA. Evidence of increased blood pressure and hypertension risk among people living with HIV on antiretroviral therapy: a systematic review with meta-analysis. J Hum Hypertens. 2016;30(6):355–62.
5. Nou E, Lo J, Grinspoon SK. Inflammation, immune activation, and cardiovascular disease in HIV. AIDS. 2016;30(10):1495–509.
6. Freiberg MS, Chang CC, Kuller LH, Skanderson M, Lowy E, Kraemer KL, et al. HIV infection and the risk of acute myocardial infarction. JAMA Internal Med. 2013;173(8):614–22.
7. Zanni MV, Schouten J, Grinspoon SK, Reiss P. Risk of coronary heart disease in patients with HIV infection. Nat Rev Cardiol. 2014;11(12):728–41.
8. Dube MP, Cadden JJ. Lipid metabolism in treated HIV Infection. Best Pract Res Clin Endocrinol Metab. 2011;25(3):429–42.
9. Clotet B, Negredo E. HIV protease inhibitors and dyslipidemia. AIDS Rev. 2003;5(1):19–24.
10. Stein JH. Dyslipidemia in the era of HIV protease inhibitors. Prog Cardiovasc Dis. 2003;45(4):293–304.
11. Sinxadi PZ, McIlleron HM, Dave JA, Smith PJ, Levitt NS, Haas DW, et al. Plasma efavirenz concentrations are associated with lipid and glucose concentrations. Medicine (Baltimore). 2016;95(2):e2385.
12. Fleischman A, Johnsen S, Systrom DM, Hrovat M, Farrar CT, Frontera W, et al. Effects of a nucleoside reverse transcriptase inhibitor, stavudine, on glucose disposal and mitochondrial function in muscle of healthy adults. Am J Physiol Endocrinol Metab. 2007;292(6):E1666–73.
13. Gallant JE, Staszewski S, Pozniak AL, DeJesus E, Suleiman JM, Miller MD, et al. Efficacy and safety of tenofovir DF vs stavudine in combination therapy in antiretroviral-naive patients: a 3-year randomized trial. JAMA. 2004;292(2):191–201.
14. George JA, Venter WD, Van Deventer HE, Crowther NJ. A longitudinal study of the changes in body fat and metabolic parameters in a South African population of HIV-positive patients receiving an antiretroviral therapeutic regimen containing stavudine. AIDS Res Hum Retroviruses. 2009;25(8):771–81.
15. Mouton JP, Cohen K, Maartens G. Key toxicity issues with the WHO-recommended first-line antiretroviral therapy regimen. Expert Rev Clin Pharmacol. 2016;9(11):1493–503.
16. World HO. Antiretroviral therapy for HIV infection in adults and adolescents. Recommendations for a public health approach. 2010 revision. 2010.
17. Venter WDF, Kambugu A, Chersich MF, Becker S, Hill A, et al. Efficacy and safety of tenofovir disoproxil fumarate versus low-dose stavudine over 96 weeks: a multi-country randomised, non-inferiority trial. J Acquir Immune Defic Syndr. 2018. https://doi.org/10.1097/QAI.0000000000001908.
18. Wallace TM, Levy JC, Matthews DR. Use and abuse of HOMA modeling. Diabetes Care. 2004;27(6):1487–95.
19. Framingham Heart Study10-year CVD risk score. https://www.framinghamheartstudy.org/fhs-risk-functions/cardiovascular-disease-10-year-risk/. Accessed 27 April 2018.
20. Roy J. Applied Longitudinal Analysis, Second Edition, by G. M. Fitzmaurice, N. M. Laird, and J. H. Ware. J Biopharm Stat. 2013;23(2):491–2.
21. Grunfeld C, Pang M, Doerrler W, Shigenaga JK, Jensen P, Feingold KR. Lipids, lipoproteins, triglyceride clearance, and cytokines in human immunodeficiency virus infection and the acquired immunodeficiency syndrome. J Clin Endocrinol Metab. 1992;74(5):1045–52.
22. Dave JA, Levitt NS, Ross IL, Lacerda M, Maartens G, Blom D. Anti-retroviral therapy increases the prevalence of dyslipidemia in South African HIV-infected patients. PLoS ONE. 2016;11(3):e0151911.
23. Miedema F, Hazenberg MD, Tesselaar K, van Baarle D, de Boer RJ, Borghans JA. Immune activation and collateral damage in AIDS pathogenesis. Front Immunol. 2013;26(4):298.
24. Funderburg NT, van Mehta NN. Lipid abnormalities and inflammation in HIV inflection. Curr HIV/AIDS Rep. 2016;13(4):218–25.
25. Liu E, Armstrong C, Spiegelman D, Chalamilla G, Njelekela M, Hawkins C, et al. First-line antiretroviral therapy and changes in lipid levels over 3 years among HIV-infected adults in Tanzania. Clin Infect Dis. 2013;56(12):1820–8.
26. Nduka C, Sarki A, Uthman O, Stranges S. Impact of antiretroviral therapy on serum lipoprotein levels and dyslipidemias: a systematic review and meta-analysis. Int J Cardiol. 2015;15(199):307–18.
27. Tang Q, Li X, Song P, Xu L. Optimal cut-off values for the homeostasis model assessment of insulin resistance (HOMA-IR) and pre-diabetes screening: developments in research and prospects for the future. Drug Discov Ther. 2015;9(6):380–5.
28. Meeks KA, Stronks K, Adeyemo A, Addo J, Bahendeka S, Beune E, et al. Peripheral insulin resistance rather than beta cell dysfunction accounts for geographical differences in impaired fasting blood glucose among sub-Saharan African individuals: findings from the RODAM study. Diabetologia. 2017;60(5):854–64.
29. Hill A, Ruxrungtham K, Hanvanich M, Katlama C, Wolf E, Soriano V, et al. Systematic review of clinical trials evaluating low doses of stavudine as part of antiretroviral treatment. Expert Opin Pharmacother. 2007;8(5):679–88.
30. McComsey GA, Lo Re V 3rd, O'Riordan M, Walker UA, Lebrecht D, Baron E, et al. Effect of reducing the dose of stavudine on body composition, bone density, and markers of mitochondrial toxicity in HIV-infected subjects: a randomized, controlled study. Clin Infect Dis. 2008;46(8):1290–6.
31. Quercia R, Roberts J, Martin-Carpenter L, Zala C. Comparative changes of lipid levels in treatment-naive, HIV-1-infected adults treated with dolutegravir vs. efavirenz, raltegravir, and ritonavir-boosted darunavir-based regimens over 48 weeks. Clin Drug Investig. 2015;35(3):211–9.
32. Anand SS, Yusuf S, Vuksan V, Devanesen S, Teo KK, Montague PA, et al. Differences in risk factors, atherosclerosis and cardiovascular disease between ethnic groups in Canada: the study of health assessment and risk in ethnic groups (SHARE). Indian Heart J. 2000;52(7 Suppl):S35–43.
33. Frank AT, Zhao B, Jose PO, Azar KM, Fortmann SP, Palaniappan LP. Racial/ethnic differences in dyslipidemia patterns. Circulation. 2014;129(5):570–9.
34. Patel SA, Shivashankar R, Ali MK, Anjana RM, Deepa M, Kapoor D, et al. Is the "South Asian phenotype" unique to South Asians?: comparing cardiometabolic risk factors in the CARRS and NHANES studies. Glob Heart. 2016;11(1):89-96.e3.
35. Friis-Moller N, Ryom L, Smith C, Weber R, Reiss P, Dabis F, et al. An updated prediction model of the global risk of cardiovascular disease in HIV-positive persons: the Data-collection on Adverse Effects of Anti-HIV Drugs (D:A:D) study. Eur J Prev Cardiol. 2016;23(2):214–23.

10

Six-helix bundle completion in the distal C-terminal heptad repeat region of gp41 is required for efficient human immunodeficiency virus type 1 infection

Dehua Liu[1,5], Hongyun Wang[1], Mizuki Yamamoto[2,4], Jiping Song[1], Rui Zhang[1], Qingling Du[1], Yasushi Kawaguchi[2,3], Jun-ichiro Inoue[4] and Zene Matsuda[1,2*]

Abstract

Background: The native pre-fusion structure of gp120/gp41 complex of human immunodeficiency virus type 1 was recently revealed. In the model, the helices of gp41 (α6, α7, α8, and α9) form a four-helix collar underneath trimeric gp120. Gp41 is a class I fusion protein and mediates membrane fusion by forming a post-fusion structure called the six-helix bundle (6HB). The comparison of the pre- and post-fusion structures revealed the large conformational changes in gp41 during the antiparallel packing of the N- and C-terminal heptad repeats (NHRs and CHRs) in membrane fusion. Several mutagenesis studies of gp41 performed in the past were interpreted based on 6HB, the only available structure at that time. To obtain an insight about the current pre-fusion structural model and conformational changes during membrane fusion, alanine insertion mutagenesis of the NHR, CHR and connecting loop regions of HXB2 gp41 was performed. The effects of mutations on biosynthesis and membrane fusion were analyzed by immunoblotting and fusion assays, respectively. The extent of membrane fusion was evaluated by split luciferase-based pore formation and syncytia formation assays, respectively.

Results: Consistent with the current structural model, drastic negative effects of mutations on biosynthesis and membrane fusion were observed for NHR, loop, and proximal regions of CHR (up to amino acid position 643). The insertions in α9 after it leaves the four-helix collar were tolerable for biosynthesis. These CHR mutants showed varying effects on membrane fusion. Insertion at position 644 or 645 resulted in poor pore and syncytia formation. Efficient pore and syncytia formation almost similar to that of the wild type was observed for insertion at position 647, 648 or 649. However, recovery of virus infectivity was only observed for the insertions beyond position 648.

Conclusions: The mutagenesis data for HXB2 gp41 is in agreement with the recent pre-fusion structure model. The virus infection data suggested that fusion pores sufficiently large enough for the release of the virus genome complex are formed after the completion of 6HB beyond position 648.

Keywords: Human immunodeficiency virus type 1, Envelope protein, Membrane fusion, Fusion pore, Heptad repeat, Six-helix bundle, Split luciferase, Split green fluorescent protein

*Correspondence: zmatsuda@ims.u-tokyo.ac.jp
[2] Research Center for Asian Infectious Diseases, Institute of Medical Science, The University of Tokyo, 4-6-1 Shirokanedai, Minato-ku, Tokyo 108-8639, Japan
Full list of author information is available at the end of the article

Background

The envelope protein of human immunodeficiency virus type 1 (HIV-1) is composed of gp120 and gp41. The gp120 subunit recognizes the receptor and co-receptor expressed on target cells and determines the host range of HIV-1 [1–3]. The gp41 subunit is a transmembrane protein and mediates membrane fusion between the viral and cellular membranes, an essential initial step of HIV-1 infection [4–6]. Recent structural analyses of BG505 SOSIP HIV-1 Env revealed the native pre-fusion

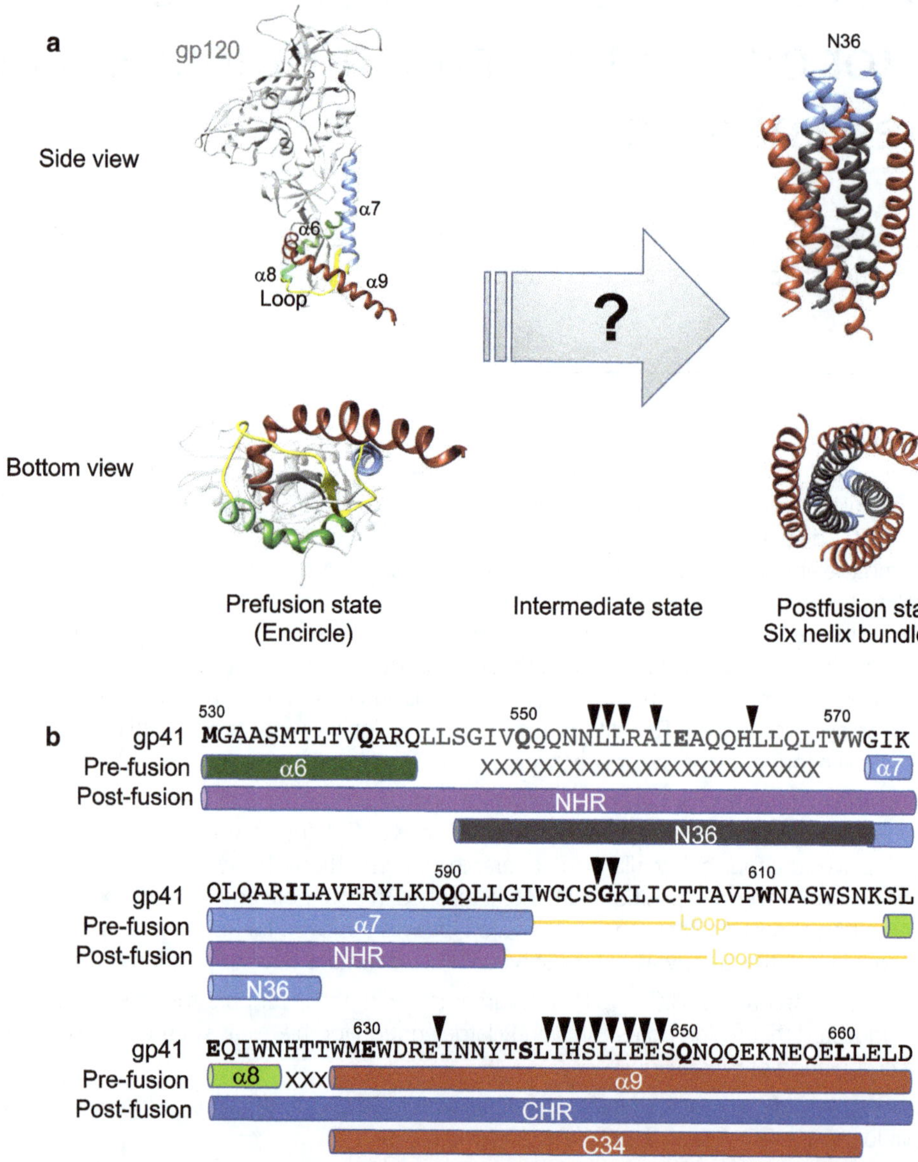

Fig. 1 Conformational changes in Env during membrane fusion. **a** Pre-fusion and post-fusion structures of gp41. Left panel: Schematic representation of the prefusion structure of Env based on the structure of BG505 SOSIP.664, as determined by CryoEM (PDB 4TVP). The upper and lower images depict the side and bottom (from the viral membrane) views, respectively. The main chain of the gp120 subunit is shown in gray, and the gp41 subunit is shown in various colors corresponding to different α-helices. The loop between the NHR and CHR is shown in yellow. Note that only a monomer of Env is shown in the left panel, and bound antibodies PGT122 and 35O22 are removed for clarity. The N- and C- terminus of gp120 are wrapped in the center of the complex formed by gp41 [7]. Right panel: the post-fusion structures of the six-helix bundle (6HB) of gp41 (PDB 1AIK). NHR and CHR are shown in the same color assignment as in the left panel. The side and bottom views are shown. The images were generated by the University of California, San Francisco Chimera program. **b** The alanine insertion mutants used in this study. The insertion sites for alanine are indicated by arrowheads above the sequence of gp41. The sequence and secondary structures of HIV-1 gp41 are indicated. Both pre-fusion and post-fusion secondary structures are colored corresponding to **a**. Cylinders represent α-helices, and the disordered regions are indicated by "x". The regions corresponding to N36 and C34 are also shown. Mutants are named by the position of the inserted alanine residue (such as 645+A). The numbering is based on HXB2 Env

state of gp120/gp41 complex [7, 8]. In this structure, the four helices of gp41 (α6, α7, α8 and α9) form a collar underneath trimeric gp120. Trimeric gp41 non-covalently associates with a gp120 trimer by encircling the N- and C-terminus of gp120 (Fig. 1a). Thus, gp120 and gp41 have a large interface between them and this intimate structural arrangement is believed to be responsible for the transmission of conformational changes of gp120 and gp41 during membrane fusion. The comparison of the pre-fusion structure with that of post-fusion gp41, called the six-helix bundle (6HB), revealed the presence of dynamic conformational rearrangements during membrane fusion such as loop to helix conversions, similar to those of influenza HA2 [7–12]. However, the details of the conformational changes remain elusive because the intermediate forms of gp41 remain unknown.

The gp41 subunit is a member of the class I fusion proteins, and the progress of membrane fusion is tightly linked with the formation of 6HB. After the interaction of gp120 and the receptor/co-receptor complex, changes of the gp120–gp41 interaction are expected to lead to exposure of the N-terminal heptad repeat (NHR) region of gp41. The C-terminal heptad repeat (CHR) bends over onto the grooves formed by the exposed trimeric NHRs to form 6HB [10, 11, 13, 14]. This packing brings the viral and cellular membranes closer, leading to the initial fusion pore formation via a hemifusion intermediate [15–19]. The current data suggested that the initial fusion pore is generated before completion of 6HB formation [4, 13, 20]. The stabilization and enlargement of the initial labile fusion pore to achieve full fusion require energy released by the completion of 6HB [4, 20–22].

In the past, extensive mutagenesis of gp41 such as alanine scanning mutagenesis [23–26] has been performed to reveal the structure–function relationship of gp41. The results were mainly interpreted based on the then-available structure of gp41, i.e., the post-fusion state structure (6HB) [10, 11], as the pre-fusion structure of gp41 was unknown. Because the structure model of gp41 in the native (pre-fusion) state complexed with gp120 became available recently (Fig. 1a), we revisited the issue of the structure–function relationship of HXB2 gp41. Currently, structural models are available for R5 tropic Env, and the structure of X4 tropic HXB2 Env has not been determined experimentally. However, the relatively high sequence homology of different gp41 variants suggests that the structure of HXB2 gp41 is similar to that of the currently available models.

We employed alanine insertion mutagenesis rather than alanine substitution. Insertional mutagenesis is expected to introduce more drastic structural alterations than point mutations, but alanine insertion can be accommodated to some extent dependent on the protein structure [27]. Thus, an insertional approach may probe the structural flexibility more efficiently than the single alanine substitution approach. Furthermore, if an insertion is tolerated and does not disturb the biosynthesis of HIV-1 Env extensively, the mutant may provide an insight into the critical boundary of 6HB formation required for membrane fusion because an insertion will result in the deregistration of α-helices in 6HB beyond the insertion point. We chose an alanine residue for insertion because it can be well-tolerated in an α-helix, which is a hallmark structure of class I fusion proteins such as gp41. We subjected the NHR, CHR, and connecting loop regions of gp41 to our mutagenesis protocol.

To obtain information on the progression of membrane fusion, particularly during the early phase, we employed our split luciferase-based fusion assay, called the dual split protein (DSP) assay [28–32]. The DSP assay can quantitatively evaluate the extent of fusion pore formation and content mixing by measuring signals from re-associated split *Renilla* luciferase (RL) pre-expressed in the effector and target cells prior to membrane fusion. Application of the DSP assay to the analysis of membrane fusion of herpes simplex virus successfully revealed the difference in the early phase of membrane fusion in the mutants [33]. We also used a classic syncytia formation assay that could provide information on the whole process of membrane fusion, including fusion pore formation, enlargement of fusion pores, and the merging of the involved cells. By comparing the results of these two assays, we could evaluate the effects of a mutation on different stages of membrane fusion, i.e., from initial formation to subsequent growth of fusion pores in a relatively simple experimental setting. We also evaluated the effect of mutation on virus infectivity by one round of pseudo-typed HIV-1 infection.

Consistent with the current pre-fusion gp120/gp41 structure model, most insertions in NHR and the loop region negatively affected biosynthesis of Env. For the relationship of fusion pore growth and the generation of 6HB, our results are consistent with previous results: (i) the fusion pore is formed before the completion of 6HB formation [13], and (ii) the progress of 6HB formation toward its C-terminus is necessary for the enlargement of the fusion pore [4]. Our data for virus infection suggest that the zipping of CHR (α9) in 6HB beyond position 648 is necessary to generate a pore sufficiently large to allow the release of the virus genome complex.

Results

Insertion of an alanine residue in the NHR and loop portions more negatively affected the biosynthesis of Env than insertions in CHR

To probe the structure–function relationship of gp41, we generated alanine insertion mutants of gp41 by introducing one alanine residue at a time in the coding sequences of NHR, CHR, and the connecting loop between them (Fig. 1b). The mutants were named by the position of the inserted alanine residue. For example, the mutant 645+A had one alanine inserted between residues 644 and 645 (numbering is based on HXB2 Env). Given the more intimate interaction of the NHR and loop region of gp41 with the inner domain of gp120 in the current pre-fusion structural model, we expected to generate more biosynthesis-defective mutants in these regions of gp41, as the alanine insertion is more detrimental than the alanine substitution. Therefore, we made fewer mutations in the NHR and loop regions. For the CHR region, we focused on the region of α9 near its exit from the four-helix collar (residue 642–649). We generated five NHR mutants, two loop mutants, and nine CHR mutants (Fig. 1b).

Because the proper processing of gp160 into gp120 and gp41 is a marker of proper biosynthesis of HIV-1 Env and a prerequisite for fusion activity, we analyzed the protein profiles of the mutants by immunoblotting using anti-gp120 and anti-gp41 antibodies (Fig. 2). For the loop mutants (600+A and 601+A), the expression level of gp160 was reduced. Furthermore, these mutants also showed decreased levels of gp41, which indicates defective processing of gp160 into gp120 and gp41. For insertion mutants in the NHR region (555+A, 556+A, 557+A, 559+A and 565+A), the expression level of gp160 was relatively retained, whereas the processing of gp160 was severely impaired (Fig. 2a and b).

For insertion mutants in the CHR region, the expression level of gp160 was similar to that of the wild-type protein for most of the mutants. However, the processing of the 635+A, 642+A, 643+A, and 646+A mutants that locate in the four-helix collar was significantly reduced. Other insertions closer to the exit of α9 from the four-helix collar, such as 644+A, 645+A, 647+A, 648+A and 649+A, showed comparable gp160 processing to that of the wild-type protein. Taken together, our results showed that the insertion in the NHR, loop, and proximal half of the CHR regions (up to 643) that have a closer interaction with gp120 affected the biosynthesis of HIV-1 Env more severely than the insertion in the distal portion of the CHR region. These aspects seem to be well accounted for by the current pre-fusion structural model.

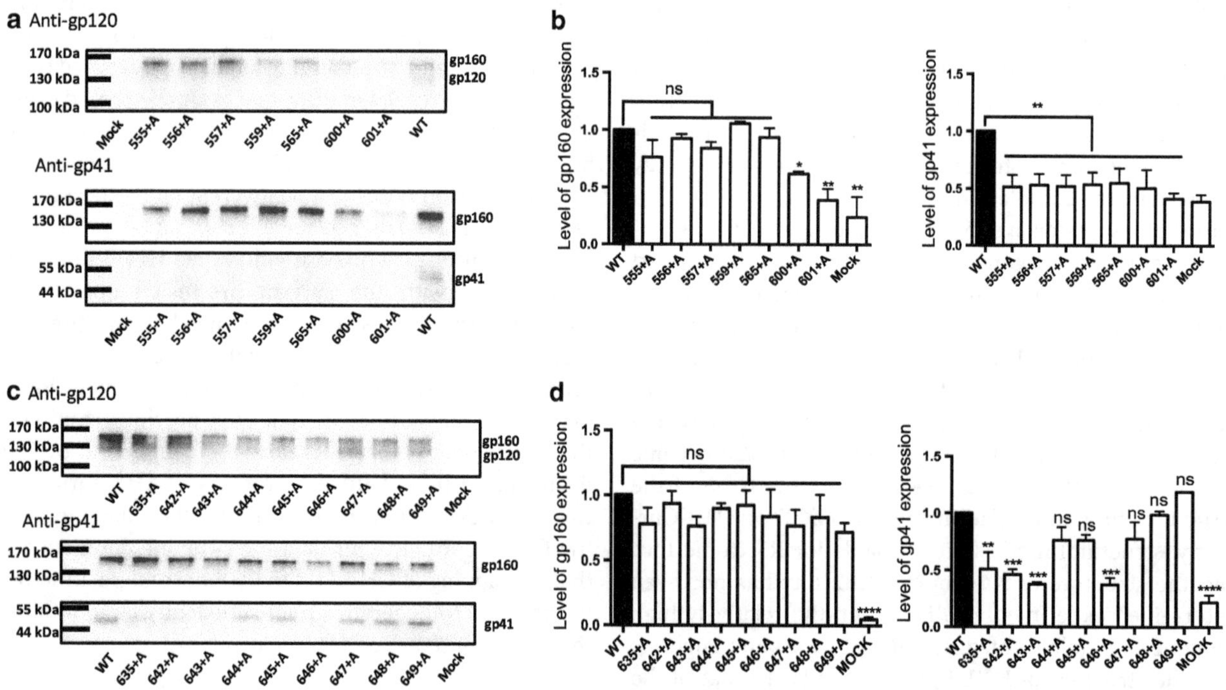

Fig. 2 The protein profiles of Env mutants. The expression and processing of Env were analyzed by immunoblotting. Representative results of immunoblotting analysis of Env with NHR and loop mutants (**a**) and CHR mutants (**c**) expressed in the transfected 293FT cells probed with goat anti-gp120 or with Chessie 8 anti-gp41 antibodies are shown. The gp160 and gp41 bands were then quantified (**b** and **d**) by measuring the intensity of bands on the anti-gp41 immunoblot. The expression levels were normalized to that of the wild-type protein. Statistical analysis was performed using one way ANOVA. ****$p < 0.0001$, ***$p < 0.001$, **$p < 0.01$, *$p < 0.05$; ns indicates no statistically significant difference when compared with the WT protein

Mutants with insertion of an alanine residue in the distal portion of CHR of gp41 retained fusion activity

Next, each mutant was evaluated for its function in cell–cell fusion assays to evaluate the impact of the alanine insertion. First, a fusion assay using co-cultures of the effector and target cells expressing the split luciferase-based reporter protein, i.e., the DSP, was employed to evaluate the rate of fusion pore formation and context mixing between the effector and target cells by measuring the activities of re-associated split RL [28–30, 34]. All the DSP activities were normalized to the surface expression level of Env determined by CELISA. As expected and consistent with the results of the immunoblotting analysis (Fig. 2a and b), most of the insertions in the loop and NHR regions resulted in severely impaired recovery of RL activity, indicating severe impairment of fusion pore formation in these mutants (Fig. 3a).

For mutations in the CHR region, insertion in the proximal portion of CHR (635+A, 642+A and 643+A) and one mutation in the distal portion of CHR (646+A) resulted in severely impaired DSP activity (Fig. 3b). Again, this phenotype was consistent with the protein profile of poor processing of gp160 into gp120 and gp41 (Fig. 2c and d). Mutants 644+A and 645+A showed decreased activity of RL in the DSP assay, indicating poor fusion pore formation. The 647+A, 648+A and 649+A mutants showed relatively high RL activity, but they were still lower than that of the wild-type protein (Fig. 3b). There was no difference in the normalized DSP activity between 647+A and 648+A.

Because the presence of mutants with no defects in fusion pore formation, but defective enlargement of the fusion pores, has been demonstrated [35], we next performed syncytia formation assays to further investigate the entire process of membrane fusion in each mutant. Transient transfection of the expression vector for each HIV-1 Env mutant in 293CD4 cells was performed as previously described [34]. The degree of syncytia formation was measured by dividing the number of nuclei contained in syncytia by the total number of nuclei in the examined fields [36, 37]. The results are shown in Fig. 3c. Overall, there was a good correlation between the results of the DSP activity assay and degree of syncytia formation. For example, consistent with the null DSP activity, cells transfected with NHR and loop region mutants showed no syncytia formation (Fig. 3c). For the CHR mutants, insertions in the proximal portion of CHR tended to have more negative effects on syncytia formation activity than insertion in the distal portion of CHR. The 644+A and 645+A mutants with lower DSP activity showed a decreased level of syncytia formation. The mutants that showed comparable DSP activity to that of the wild-type protein (647+A, 648+A and 649+A) had a degree of syncytia formation similar to that of the wild-type protein (Fig. 3c).

To examine the effect of the mutations on virus infection, we performed a one-round pseudotyped virus infection assay using some CHR mutant Envs. The representative results are shown in Fig. 4a. Only 648+A and 649+A Env showed a low level of infectivity, but the other tested mutants failed to achieve efficient infection. Notably, mutant 647+A that showed pore formation and syncytia formation similar to that of the wild type failed to infect the target cells. This failure of infection by 644+A, 645+A, 646+A and 647+A was not due to poor incorporation of Env into virions, because a similar amount of gp41 as the wild type was observed for these mutants (see the gp41/p24 ratio in Fig. 4b).

Discussion

We performed alanine insertion mutagenesis to probe the structure–function relationship of HXB2 gp41. Our assumption is that structure of HXB2 Env may be similar, if not identical, to the structure of BG505SOSIP. Similar to the previous studies [4, 25], mutations in the NHR and loop regions of gp41 tend to be defective in the biosynthesis of Env. An alanine residue insertion not only changed the amino acid residue at the insertion point to an alanine residue, but also induced a shift in the primary sequence and could introduce a rotational shift of 100° around the axis in the α-helix of the protein. This type of structural alterations seems to not be tolerable due to the recent structural model of the native gp120/gp41 complex [7–9], in which a trimeric coiled coil of NHRs forms a central hydrophobic core structure that interacts with the inner domain of gp120. The loop region of gp41 also has intimate hydrophobic interactions with gp120 by forming the base of a four-helix collar of gp41 (Fig. 5a). Alanine insertion is also not tolerable in the loop region.

The results showed that alanine insertion mutations in the distal portion of CHR were less damaging to biosynthesis. In the current structural model of the pre-fusion status, α9 leaves the bottom of the four-helix collar (formed by α6, α7, α8, and α9 of gp41) that wraps around the N- and C-terminus ends of the gp120 at position 647; therefore, the α9 residues beyond position 648 have less interaction with trimeric gp120 (Fig. 5a). This structural arrangement may account for the tolerance of mutations in the distal region of CHR (α9) for biosynthesis. Similar tolerance against mutations was observed in previous studies [4, 26].

It has been shown that the initial pore is formed through partial formation of 6HB [20]; these pores are labile and cannot progress to the stage of enlargement until the zipping proceeds towards the distal ends [4, 16, 19, 20]. The results of cell–cell fusion assays performed

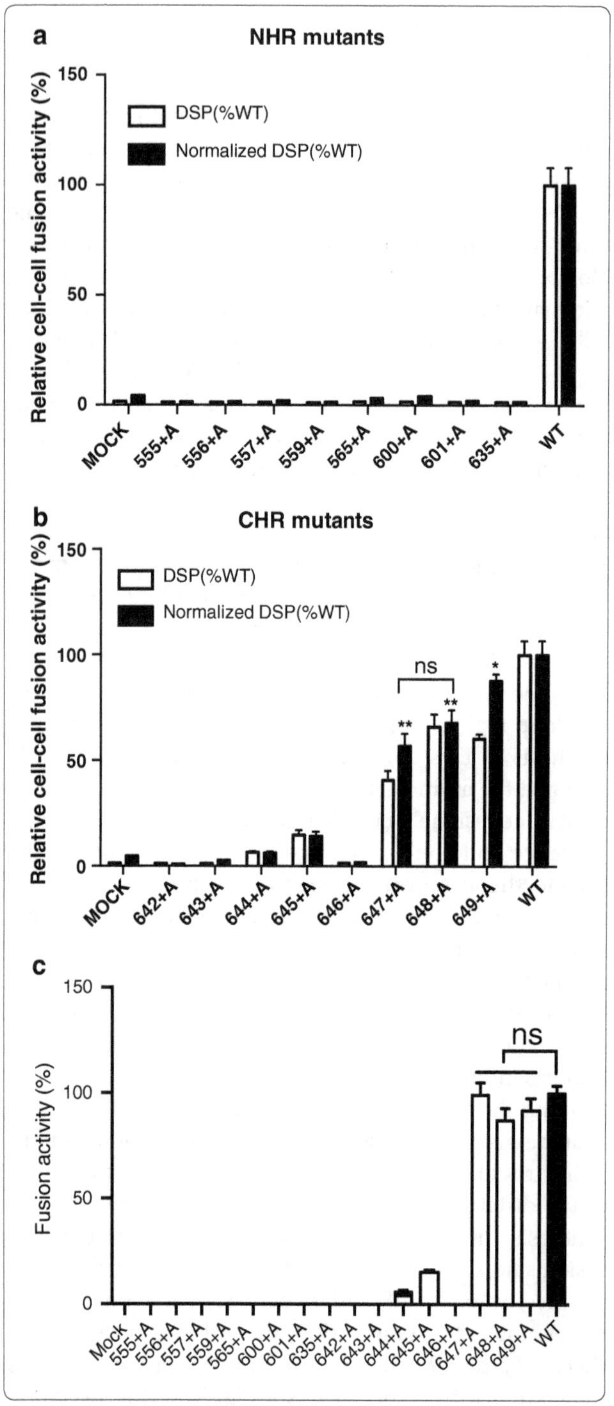

Fig. 3 Fusion assay of alanine insertion mutants of gp41. **a** and **b** The DSP assay was performed at 2 h after coculture to examine the fusion activities of mutants. Surface expression levels of mutant Envs were determined by CELISA, and the fusion activity measured by the DSP assay (open bar) was normalized to the surface expression level (dark bar). Error bars represent the standard deviation of the results of triplicate experiments. The results for NHR and loop mutants are shown in **a**. **b** The results for CHR mutants. Statistical analysis was performed using *t* test. $^{**}p < 0.01$ and $^{*}p < 0.05$ when comparison of the indicated mutant with wild type was done. ns indicates no significant differences ($p > 0.05$) when 647+A and 648+A were compared. **c** Syncytia formation assays were used to evaluate the fusion activities of the mutants. The number of nuclei in the syncytia was divided by the total number of the nuclei in the field. Five randomly chosen fields were used to measure the value. The degree of syncytia formation is shown as a percentage (the wild-type was set as 100%). Statistical analysis was performed using *t* test. ns indicates no significant differences ($p > 0.05$) when compared with the value of the wild type

by several groups with alanine substitutions are available. For example, mutation studies of the **a** and **d** positions of HR2 by Markosyan et al. suggested that the pore enlargement gradually progresses during the formation of 6HB toward C-terminus end of CHR [4]. They observed slight gain of fusion activity for mutant 645, and larger recovery was observed for mutant 649. Diaz-Aguilar et al. observed good recovery of fusion after substitution of position 647, though with some fluctuation in fusion activities in position 643 [26]. We detected slight recovery of DSP activities for 644+A and good recovery for 647+A. Although the reporters used in these studies were different (dye, viral proteins, or split protein), the obtained results are similar. Our results for the CHR mutants showed that the fusion pore can be formed by partial zipping of 6HB until residue 644 and more efficient pore formation occurred when zipping progressed towards the C-terminus direction of CHR.

Different from the above cell–cell fusion assays, a previous virus infectivity assay of the alanine substitution mutants did not reveal such a clear boundary of change in infectivity [26]. The analysis of insertion mutants in our virus infection assay, however, showed that the critical boundary for infectivity was located between position 647 and 648; we found that the 647+A mutant failed to establish infection, whereas the 648+A mutant virus was able to infect target cells. The difference in the two studies may arise from the differences in the types of mutations. An alanine substitution may be more tolerable by only altering the local structure of 6HB and may not prevent the zipping beyond the mutation site.

Our alanine insertion, however, seems to have a more drastic impact on the zipping of 6HB because the CD analysis using the 647+A version of C34 with N36 failed to show the typical α-helical CD profile (Additional file 1: Fig. S1). Interestingly, the infectivity of the virus did not correlate well with the pore formation activities estimated by our DSP assay. This was probably because the size of our split reporter proteins was much smaller than the virus genome complex and could pass through the pores. Therefore, the fusion pores in 647+A mutant were of sufficient size to allow reporter proteins, but not the virus genome complex, to pass. However, since our

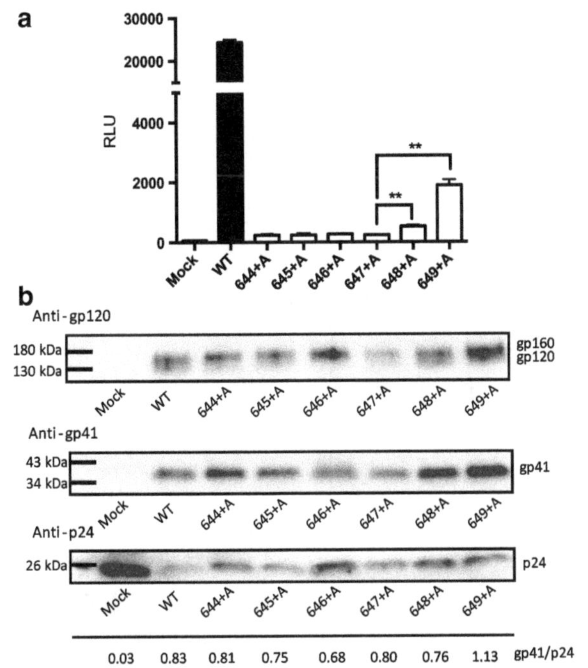

Fig. 4 Pseudotype virus infection assay. **a** Infectivity of pseudotype virus bearing mutant Env was evaluated by measuring the reporter (n-luc) activity after 48 h of infection. The CHR mutants, 644+A, 645+A, 646+A, 647+A, 648+A, 649+A, were tested together with the wild type. Error bars represent the standard deviation of the results of triplicate experiments. Statistical analysis was performed using t test. $^{**}p < 0.01$ when compared with 647+A. **b** The protein profiles of VLPs of CHR mutants were analyzed by immunoblotting. Anti-gp120, gp41 and p24 antibodies were used to detect the respective bands. The Env expression vector was omitted for mock transfection. The ratio of the band intensities of gp41 and p24 was indicated under the immunoblot

assay could not directly evaluate the size of the formed fusion pores, alternative possibilities such as short half-life of the fusion pores made by mutant Env or the impact of the differences in the stability of mutant Env cannot be excluded.

In addition to the difference in the required pore size for the reporter proteins and viral genome, the limited number of available Env molecules on virions compared with the cell surface may be the reason for the more severe defect in the virus infection assay. Different from a previous study [26], we did not observe any recovery or compensation of defective cell–cell fusion activities in our virus assay. This could be due to the greater deterioration with our insertion mutation compared with a point mutation as discussed above. It will be also interesting to determine whether our findings in X4 tropic HXB2 can be replicated in other X4 strains or in R5 tropic Env, as one previous study indicated potential differences between the HXB2 and JRFL strains [26].

Based on our data, a schematic diagram of the progress of membrane fusion during virus infection is shown in Fig. 5b. When zipping proceeds beyond position 647 of CHR, the energy released becomes sufficient to produce larger fusion pores for the release of the virus genome. After further zipping of 6HB proceeds, membrane fusion proceeds, eventually leading to full fusion. This position of critical zipping coincidentally mapped to position 647 where a9 leaves the four-helix collar of gp41.

The exact nature or structure of the intermediate gp41 and its relationship with the lipid bilayer are still unclear; however, 644+A or 645+A mutant may be a useful model to investigate the elusive intermediate status of gp41. In addition to the structure of 6HB, when the pre-fusion structure of the gp120/gp41 is considered, the region around residue 647 may be important for regulation of the initial conformational change of gp41. For example, the 647 E residue may be involved in the interaction with residues within α7, such as 592, 595, and 596 (Fig. 5a). The insertion at position 647 may affect these intra-gp41 interactions and the initial conformational changes in gp41 [19]. Further mutations in α7 and α9 in future may shed light on this aspect.

Our analysis using the DSP reporter is relatively simple and does not require a very sophisticated laboratory setup. The analysis of the early stage of membrane fusion will provide useful information to further characterize the detailed mechanisms of membrane fusion mediated by HIV-1 Env. These analyses may lead to the design of rational fusion inhibitors that act on HIV-1 infection at its earliest stage.

Conclusions

Analyses of the effects of alanine-insertion mutants of HXB2 gp41 for their biosynthesis and fusion activities in both cell–cell and virus-cell fusion assays were performed. The results suggested that HXB2 gp41 may have a structural arrangement similar to the currently available structural model based on R5 tropic BG505 Env. Our data implies that the fusion pore can be generated by partial 6HB completion. However, completion of 6HB beyond position 647 may be a prerequisite for fusion pore enlargement sufficient for the virus to acquire infectivity.

Methods

Plasmid construction

The HXB2 envelope gene sequence used in this study was codon-optimized for mammalian cell expression [34]. A QuikChange site-directed mutagenesis kit (Stratagene, La Jolla, CA, USA) was used to generate the mutants in this study. Briefly, the plasmid pHIV-env-opt HXB2, which contains the 2.6-kb *Sal*I-*Eco*RI fragment corresponding to HXB2 Env, was used as a template.

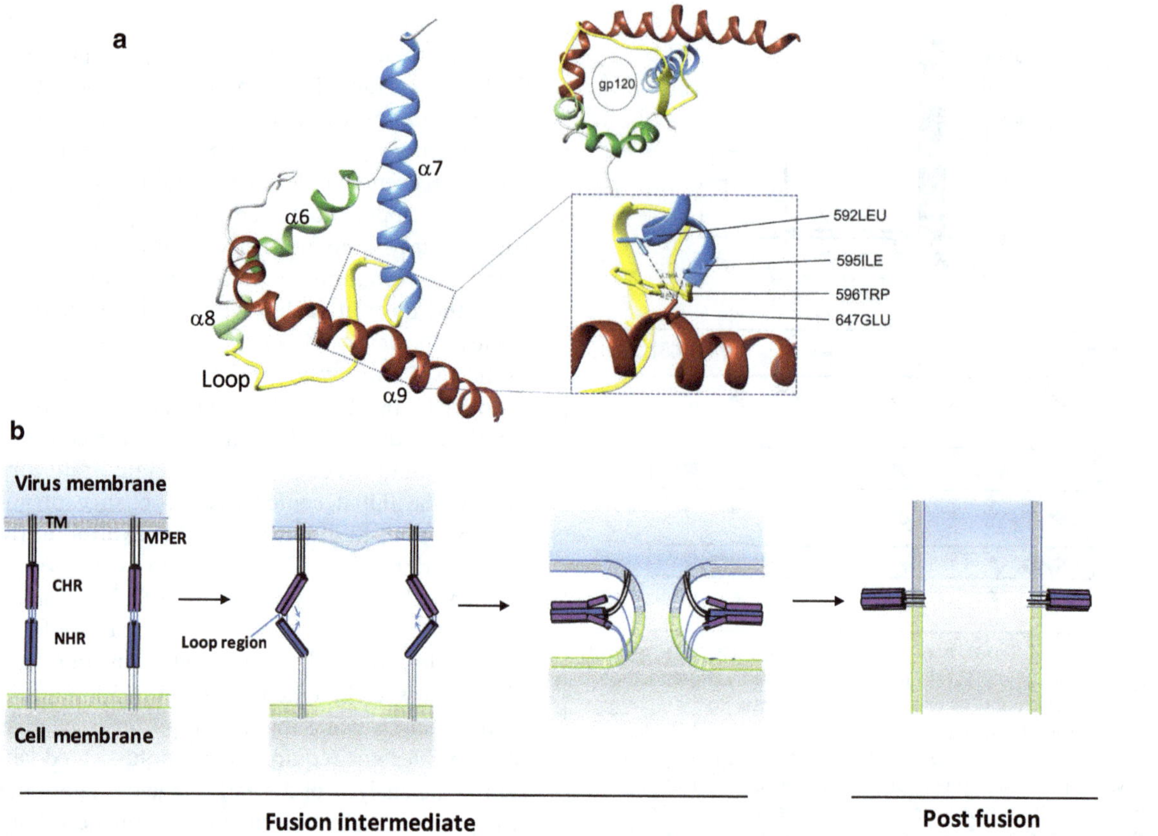

Fig. 5 Schematic representation of the virus and cell fusion. **a** Prefusion structure of gp41. Side (left panel) and bottom (right upper panel) views of gp41 are shown. The four-helix collar (α6–α9) of gp41 encircled gp120 (gp120 is indicated as a dotted circle). Close-up view of gp41 around residue 647 is shown in the right lower panel. The distances between residues 647 and residues 592 (4.786 Å), 595 (3.199 Å), and 596 (4.829 Å) are indicated. **b** Schematic description of conformational changes in gp41 during membrane fusion based on our data. After the fusion peptide (FP) reached the host-cell membrane, packing of CHR into the NHR grooves was initiated near the connecting loop region between the CHR and NHR. Pairing of residues in the CHR located before residue 647 may be sufficient for generation of initial unstable fusion pores. Residues beyond 647 of the CHR needed to interact with the NHR to achieve pore enlargement. The cytoplasmic (CT) domain, transmembrane (TM) domains, fusion peptide (FP), membrane proximal external region (MPER), NHR, CHR, and loop region are indicated

Complementary oligonucleotide pairs containing an insertion codon for the alanine residue, GCA, were used. The mutant identified by sequencing was cloned back into the pHIV-env-opt HXB2 as the *BsrG*I and *BamH*I fragment. After cloning, the entire *Sal*I-*EcoR*I region including the vector junction was verified by sequencing. Polymerase chain reaction (PCR) was performed using Pfu Turbo (Stratagene).

Cell culture and transfection

293FT cells (Invitrogen, Carlsbad, CA, USA) or 293CD4 cells (293 cells constitutively expressing human CD4) [37] were grown in Dulbecco's modified Eagle's medium (DMEM) supplemented with 10% fetal bovine serum (FBS). Cells were cultured in a 5% CO_2 in a humidified incubator (Sanyo, Japan) and then transferred to 6- or 96-well plates 1 day before transfection, and Fugene HD (Fugene HD [μL]: DNA [μg]: DMEM [μL] = 5:2:200; Promega, Madison, WI, USA) was used for transient transfection. The transfection mix was incubated for 15 min at room temperature prior to addition to the cell culture in a drop-wise manner (10 μL/well). After indicated time after the transfection, the transfected cells were subjected to further analyses, as indicated below.

DSP assay and CELISA

To quantify cell–cell fusion, we used the DSP assay, as described previously [28–30, 34, 36]. Briefly, the stable cell line 293FT expressing DSP_{1-7} (293FT/DSP_{1-7}), was plated in a 12-well plate (1.5×10^5 cells/well) (BD Falcon, San Jose, CA, USA) with DMEM (Sigma, St. Louis, MO, USA) supplemented with 10% FBS (Hyclone Labs, Logan, UT, USA). 293FT/DSP_{1-7} cells were then transfected with expression vectors of interest in triplicate to generate effector cells. To prepare the target cells, 293CD4/DSP$_{8-11}$ cells, a stable cell line expressing CD4 and DSP$_{8-11}$, were seeded in 6-cm dishes (7.5 × 10^5/dish) (BD Falcon, San Jose, CA, USA). The DSP assay was performed in triplicate 20 h post-transfection by mixing the effector and target cells at 37 °C in fresh medium containing 60 μM membrane-permeable substrate for RL (Enduren Live Cell Substrate; Promega) added 1 h prior to mixing. *Renilla* activity was measured at 2 h after coculture using a GloMax-Multi Plus Detection System (Promega). Half of the same effector cells were subjected to CELISA to measure the surface expression level of Env. CELISA was performed in a 96-well ViewPlate (PerkinElmer Life Sciences). Briefly, cells were fixed in PBS with 4% paraformaldehyde at 25 °C for 5 min and washed with PBS, then treated with 3% hydrogen peroxide in PBS at 25 °C for 10 min to deactivate endogenous peroxidases. Fixed cells were incubated with 3% ECL Prime Blocking Reagent (GE Healthcare, NY, USA) in PBS at 25 °C for 30 min. After blocking, cells were incubated with a saturating amount of 2G12 (13.1 μg/mL, 1:1000 dilution) at 25 °C for 1 h and then incubated with anti-human IgG-HRP at 1:3000 dilution (Santa Cruz Biotechnology) at 25 °C for 1 h. Between each procedure, cells were washed with PBS three times. Antibodies were diluted in 1.5% ECL Prime Blocking Reagent. Luminescence was detected using ECL Prime Western Blotting Detection Reagent (GE Healthcare) and a GloMax 96 microplate luminometer (Promega). Before calculation of the relative expression levels, the average background signal from blank control wells was subtracted.

Syncytia formation assay and virus infection assay

Fusion activity was further evaluated by syncytia formation assays. Expression vectors for Env were transfected into 293CD4 cells using FuGENE HD. Unless otherwise indicated, visible syncytia formation was evaluated at 16–24 h post-transfection. To visualize the nuclei of the cells in syncytia formation assays, Hoechst 33,342 (0.2 mg/mL; Invitrogen) was applied at 37 °C for 15 min. After labeling, the cells were washed three times with 200 μL of prewarmed DMEM/FBS, and images were captured using a confocal microscope (Olympus FluoView FV1000). The number of nuclei in syncytia was counted and divided by the total number of the nuclei in the field to estimate the degree of syncytia formation. Five randomly chosen fields were examined.

Pseudotyped virus was generated by transfecting the following plasmids in the ratio of 2:1:1; pLenti-CMV-GFPnluc (reporter expressing GFP-nluc fusion protein), psPAX2 (gag-pol expression), and envelope expression vectors for several gp41 mutants and the wild-type. For the infection assay, the plasmids were transfected into 293FT cells by the calcium phosphate precipitation method. One day after transfection, the medium was replaced and virus-like particles were harvested 2 days after transfection. The virus amount was measured by the p24 ELISA kit (Zeptrometrix, Buffalo, NY), and the same amount of filtered (0.45-μm filter) virus (4.5 ng of p24) was applied to NP2-CD4-CXCR4-DSP1 cells [32] in a 96 well plate (1 × 10^4 cells/well) in the presence of polybrene (10 μg/mL). The medium was replaced 24 h after infection, and the infectivity of the virus was evaluated at 48 h after infection by measuring n-luc reporter activity. The measurements were taken after incubating infected cells for 1 h in the medium containing 6 μM EnduRen substrate (Promega, WI) using the luminometer (CentroXS3, Berthold Technologies, Bad Wildbad, Germany).

Synthesis of peptides and CD analysis

The following peptides were synthesized: N36, SGIVQQQNNLLRAIEAQQHLLQLTVWGIKQLQARIL; C34, WMEWDREINNYTSLIHSLIEESQNQQEKNEQELL; and C34 (647+A), WMEWDREINNYTSLIHSLIAEESQNQQEKNEQELL. The peptides were synthetized by GL Biochem (Shanghai) Limited (Shanghai, China) with a purity of more than 95%. CD spectra were acquired on a Chirascan Plus CD spectrometer (Applied Photophysics Limited, Surry, UK) with a thermoelectric sample temperature controller. Samples for wavelength spectra were prepared as 10 μM peptide in phosphate-buffered saline (PBS). The cuvette had a 0.1 cm path length. The wavelength dependence of molar ellipticity, [θ], was monitored at 17 °C as the average of three scans, using a 2.5-s integration time at 1.0-nm wavelength increments. Spectra were baseline-corrected against the value for the cuvette with buffer alone. Thermal stability was determined in the same buffer by measuring $[\theta]_{222}$ as a function of temperature. A cell with a path length of 0.1 cm was used with continuous stirring. Thermal melts were monitored in 2 °C intervals with a 2-min equilibration at the desired temperature and an integration time of 30 s. Protein concentrations were determined by measuring the absorbance at 280 nm in the same buffer.

Immunoblotting

293FT cells (2 × 10^5) were transiently transfected with Env expression vector using FuGENE HD in a 6-well culture plate. Cells were lysed with 60 μL of RIPA lysis buffer (Thermofisher Scientific, MA, USA) at 48 h after transfection. After centrifugation at 20,000×*g* for 30 min at 4 °C, the supernatant was collected, and the protein concentration was determined by Pierce BCA protein assay (Pierce Biotechnology, Rockford, IL, USA). Samples containing approximately 50 μg of protein were loaded into each well, separated by sodium dodecyl sulfate polyacrylamide gel electrophoresis on 10% gels

(Bio-Rad Ready Gel J), and transferred to polyvinylidene fluoride membranes (Immobilon-PSQ; Millipore, MA, USA). The blots were probed with goat anti-gp120 polyclonal antibodies (Fitzgerald, Concord, MA, USA) or the monoclonal antibody Chessie 8 [38]. Donkey anti-goat IgG-horseradish peroxidase (HRP; Santa Cruz Biotechnology, Santa Cruz, CA, USA) or goat anti-mouse IgG-HRP (Santa Cruz Biotechnology) was used as the secondary antibody. The blot was further treated with an ECL Western Blot Kit (CWBIO, Beijing, China). Images were obtained with an LAS3000 system (Fujifilm, Tokyo, Japan). Band intensities were analyzed using ImageJ software (NIH, MD, USA).

For the protein profile analysis of VLPs, the pseudotyped virus was generated by Fugene HD- mediated transfection of 293FT cells with the following plasmids: pLenti-CMV-GFPnluc, psPAX2, and envelope expression vectors (in the ratio of 2:1.5:1). The protein profile of the released VLPs was analyzed by immunoblotting. The VLPs were collected by pelleting down from the cleared and filtered (0.45-μm) culture supernatant of the transfected 293FT cells by ultracentrifugation (20,000 rpm, SW32 rotor, Beckman) at 4 °C for 2 h. The VLPs were lysed in RIPA lysis buffer (Thermofisher Scientific, MA, USA) and subjected to SDS PAGE. The immunoblotting procedure was similar to that for cell lysates as described above. For VLPs, an anti-p24 monoclonal antibody (Santa Cruz Biotechnology, Dallas, TX), was used in addition to anti-Env (gp120 and gp41).

Statistical analysis

Statistical analyses were performed using Prism software (GraphPad, USA). *T* test was used to compare groups of normalized DSP activity of CHR mutants, syncytia formation assay, and pseudotype virus infection assay. One-way ANOVA test were used to assess the expression level of HIV-1 envelop proteins. Statistical significance was set as $p < 0.05$.

Abbreviations

Env: envelope glycoprotein; HIV-1: human immunodeficiency virus type 1; CHR: C-terminal heptad repeat; NHR: N-terminal heptad repeat; 6HB: six-helix bundle; GFP: green fluorescent protein; RL: *Renilla* luciferase; RLU: relative light units; DSP: dual split protein; CD: circular dichroism; MSD: membrane-spanning domain; CT: cytoplasmic; TM: transmembrane; FP: fusion peptide; FBS: fetal bovine serum; PBS: phosphate-buffered saline.

Authors' contributions

The study was conceived by ZM. ZM supervised the entire work. ZM, DL, HW, JS, RZ, and QD performed the experiments. Mutagenesis was performed by DL and QD. The protein profiles of the mutants were determined by DL. The DSP assay and syncytia formation analysis were performed by DL, HW, JS, and HW. The work regarding CELISA was performed by HW, JS, and QD. DL and RZ performed the CD analysis. The virus infection assay was performed by MY. The protein profile of VLPs was examined by ZM, RZ, and HW. DL and ZM wrote the manuscript. YK and JI joined the discussion. All authors read and approved the final manuscript.

Author details

[1] Laboratory of Structural Virology and Immunology, Institute of Biophysics, Chinese Academy of Sciences, Beijing, China. [2] Research Center for Asian Infectious Diseases, Institute of Medical Science, The University of Tokyo, 4-6-1 Shirokanedai, Minato-ku, Tokyo 108-8639, Japan. [3] Division of Molecular Virology, Department of Microbiology and Immunology, Institute of Medical Science, The University of Tokyo, Tokyo, Japan. [4] Division of Cellular and Molecular Biology, Institute of Medical Science, The University of Tokyo, Tokyo, Japan. [5] Present Address: The Chinese University of Hong Kong, Sha Tin, Hong Kong, China.

Acknowledgements

We thank Dr. Gregory Melikian (Department of Pediatrics, Infectious Diseases, Emory University, Georgia, USA) for providing the Chessie 8 antibody. We acknowledge Promega as the source of NanoLuc® Technology and the EnduRen™-branded LARs, and have executed an agreement with Promega to use EnduRen™-branded LARs with NanoLuc® Technology outside the normal terms and conditions of sale described in Promega's LULL. We also thank Dr. Kunito Yoshiike for a critical reading of the manuscript. We would like to thank Editage for English language editing.

Competing interests

The authors declare that they have no competing interests.

Funding

This work was supported by a contract research fund from the Ministry of Education, Culture, Sports, Science, and Technology for Program of Japan Initiative for Global Research Network on Infectious Diseases (J-GRID). This research was also partially supported by the Research Program on Emerging and Re-emerging Infectious Diseases and the Program of J-GRID from the Japan Agency for Medical Research and Development, AMED. The funders had no role in study design, data collection and analysis, decision to publish, or preparation of the manuscript.

References

1. Wilen CB, Tilton JC, Doms RW. HIV: cell binding and entry. Cold Spring Harb Perspect Med. 2012;2:a006866.
2. White JM, Delos SE, Brecher M, et al. Structures and mechanisms of viral membrane fusion proteins: multiple variations on a common theme. Crit Rev Biochem Mol Biol. 2008;43:189–219.
3. Harrison SC. Viral membrane fusion. Nat Struct Mol Biol. 2008;15:690–8.
4. Markosyan RM, Leung MY, Cohen FS. The six-helix bundle of human immunodeficiency virus Env controls pore formation and enlargement and is initiated at residues proximal to the hairpin turn. J Virol. 2009;83:10048–57.
5. Ashkenazi A, Shai Y. Insights into the mechanism of HIV-1 envelope induced membrane fusion as revealed by its inhibitory peptides. Eur Biophys J. 2011;40:349–57.
6. Hu X, Saha P, Chen X, et al. Cell surface assembly of HIV gp41 six-helix bundles for facile, quantitative measurements of hetero-oligomeric interactions. J Am Chem Soc. 2012;134:14642–5.
7. Pancera M, Zhou T, Druz A, et al. Structure and immune recognition of trimeric pre-fusion HIV-1 Env. Nature. 2014;514:455–61.
8. Lee JH, Ozorowski G, Ward AB. Cryo-EM structure of a native, fully glycosylated, cleaved HIV-1 envelope trimer. Science. 2016;351:1043–8.
9. Matsuda Z. Recent advance in the structural analysis of HIV-1 envelope protein. Sci China Life Sci. 2015;58:420–4.
10. Chan DC, Fass D, Berger JM, et al. Core structure of gp41 from the HIV envelope glycoprotein. Cell. 1997;89:263–73.
11. Winfried W, Dessen A, Harrison SC, et al. Atomic structure of the ectodomain from HIV-1 gp41. Nature. 1997;387:426–30.

12. Ward AB, Wilson IA. Insights into the trimeric HIV-1 envelope glycoprotein structure. Trends Biochem Sci. 2015;40:101–7.
13. Melikyan GB, Markosyan RM, Hemmati H, et al. Evidence that the transition of HIV-1 gp41 into a six-helix bundle, not the bundle configuration, induces membrane fusion. J Cell Biol. 2000;151:413–23.
14. Gallo SA, Puri A, Blumenthal R. HIV-1 gp41 six-helix bundle formation occurs rapidly after the engagement of gp120 by CXCR4 in the HIV-1 Env-mediated fusion Process. Biochemistry. 2001;40:12231–6.
15. Blanco J, Barretina J, Ferri KF, et al. Cell-surface-expressed HIV-1 envelope induces the death of CD4 T cells during GP41-mediated hemifusion-like events. Virology. 2003;305:318–29.
16. Bar S, Alizon M. Role of the ectodomain of the gp41 transmembrane envelope protein of human immunodeficiency virus type 1 in late steps of thememembrane fusion process. J Virol. 2003;78:811–20.
17. Garg H, Joshi A, Freed EO, et al. Site-specific mutations in HIV-1 gp41 reveal a correlation between HIV-1-mediated bystander apoptosis and fusion/hemifusion. J Biol Chem. 2007;282:16899–906.
18. Markosyan RM, Cohen FS, Melikyan GB. Time-resolved imaging of HIV-1 Env-mediated lipid and content mixing between a single virion and cell membrane. Mol Biol Cell. 2005;16:5502–13.
19. Cohen FS, Melikyan GB. The energetics of membrane fusion from binding, through hemifusion, pore formation, and pore enlargement. J Membr Biol. 2004;199:1–4.
20. Markosyan RM, Cohen FS, Melikyan GB. HIV-1 envelope proteins complete their folding into six-helix bundles immediately after fusion pore formation. Mol Biol Cell. 2003;14:926–38.
21. Eckert DM, Kim PS. Mechanisms of viral membrane fusion and its inhibition. Annu Rev Biochem. 2001;70:777–810.
22. Hughson FM. Enveloped viruses: A common mode of membrane fusion? Curr Biol. 1997;7:565–9.
23. Jacobs A, Sen J, Rong L, et al. Alanine scanning mutants of the HIV gp41 loop. J Biol Chem. 2005;280:27284–8.
24. Lu M, Stoller MO, Wang S, et al. Structural and functional analysis of interhelical interactions in the human immunodeficiency virus type 1 gp41 envelope glycoprotein by alanine-scanning mutagenesis. J Virol. 2001;75:11146–56.
25. Sen J, Yan T, Wang J, et al. Alanine scanning mutagenesis of HIV-1 gp41 heptad repeat 1: insight into the gp120–gp41 interaction. Biochemistry. 2010;49:5057–65.
26. Diaz-Aguilar B, Dewispelaere K, Yi HA, et al. Significant differences in cell–cell fusion and viral entry between strains revealed by scanning mutagenesis of the C-heptad repeat of HIV gp41. Biochemistry. 2013;52:3552–63.
27. Vetter IR, Baase WA, Heinz DW, et al. Protein structural plasticity exemplified by insertion and deletion mutants in T4 lysozyme. Protein Sci. 1996;5:2399–415.
28. Ishikawa H, Meng F, Kondo N, et al. Generation of a dual-functional split-reporter protein for monitoring membrane fusion using self-associating split GFP. Protein Eng Des Sel. 2012;25:813–20.
29. Kondo N, Miyauchi K, Meng F, et al. Conformational changes of the HIV-1 envelope protein during membrane fusion are inhibited by the replacement of its membrane-spanning domain. J Biol Chem. 2010;285:14681–8.
30. Nakane S, Matsuda Z. Dual split protein (DSP) assay to monitor cell–cell membrane fusion. Methods Mol Biol. 2015;1313:229–36.
31. Nakane S, Iwamoto A, Matsuda Z. The V4 and V5 variable loops of HIV-1 envelope glycoprotein are tolerant to insertion of green fluorescent protein and are useful targets for labeling. J Biol Chem. 2015;290:15279–91.
32. Teeranaipong P, Hosoya N, Kawana-Tachikawa A, et al. Development of a rapid cell-fusion-based phenotypic HIV-1 tropism assay. J Int AIDS Soc. 2013;16:18723.
33. Atanasiu D, Saw WT, Gallagher JR, et al. Dual split protein-based fusion assay reveals that mutations to herpes simplex virus (HSV) glycoprotein gB alter the kinetics of cell-cell fusion induced by HSV entry glycoproteins. J Virol. 2013;87:11332–45.
34. Wang H, Li X, Nakane S, et al. Co-expression of foreign proteins tethered to HIV-1 envelope glycoprotein on the cell surface by introducing an intervening second membrane-spanning domain. PLoS ONE. 2014;9:e96790.
35. Munoz-Barroso I, Salzwedel K, Hunter E, et al. Role of the membrane-proximal domain in the initial stages of human immunodeficiency virus type 1 envelope glycoprotein-mediated membrane fusion. J Virol. 1999;73:6089–92.
36. Long Y, Meng F, Kondo N, et al. Conserved arginine residue in the membrane-spanning domain of HIV-1 gp41 is required for efficient membrane fusion. Protein Cell. 2011;2:369–76.
37. Miyauchi K, Komano J, Yokomaku Y, et al. Role of the specific amino acid sequence of the membrane-spanning domain of human immunodeficiency virus type 1 in membrane fusion. J Virol. 2005;79:4720–9.
38. Abacioglu Y, Fouts T, Laman J, et al. Epitope mapping and topology of baculovirus-expressed HIV-1 gp160 determined with a panel of murine monoclonal antibodies. AIDS Res Hum Retrovir. 1994;10:371–81.

11

The HIV-1 integrase-LEDGF allosteric inhibitor MUT-A: resistance profile, impairment of virus maturation and infectivity but without influence on RNA packaging or virus immunoreactivity

Céline Amadori[1,2,3,4†], Yme Ubeles van der Velden[5†], Damien Bonnard[1], Igor Orlov[6], Nikki van Bel[5], Erwann Le Rouzic[1], Laia Miralles[7], Julie Brias[1], Francis Chevreuil[1], Daniele Spehner[6], Sophie Chasset[1], Benoit Ledoussal[1], Luzia Mayr[8], François Moreau[1], Felipe García[7], José Gatell[7], Alessia Zamborlini[9], Stéphane Emiliani[2,3,4], Marc Ruff[6], Bruno P. Klaholz[6], Christiane Moog[8], Ben Berkhout[5*], Montserrat Plana[7*] and Richard Benarous[1,10*]

Abstract

Background: HIV-1 Integrase (IN) interacts with the cellular co-factor LEDGF/p75 and tethers the HIV preintegration complex to the host genome enabling integration. Recently a new class of IN inhibitors was described, the IN-LEDGF allosteric inhibitors (INLAIs). Designed to interfere with the IN-LEDGF interaction during integration, the major impact of these inhibitors was surprisingly found on virus maturation, causing a reverse transcription defect in target cells.

Results: Here we describe the MUT-A compound as a genuine INLAI with an original chemical structure based on a new type of scaffold, a thiophene ring. MUT-A has all characteristics of INLAI compounds such as inhibition of IN-LEDGF/p75 interaction, IN multimerization, dual antiretroviral (ARV) activities, normal packaging of genomic viral RNA and complete Gag protein maturation. MUT-A has more potent ARV activity compared to other INLAIs previously reported, but similar profile of resistance mutations and absence of ARV activity on SIV. HIV-1 virions produced in the presence of MUT-A were non-infectious with the formation of eccentric condensates outside of the core. In studying the immunoreactivity of these non-infectious virions, we found that inactivated HIV-1 particles were captured by anti-HIV-specific neutralizing and non-neutralizing antibodies (b12, 2G12, PGT121, 4D4, 10-1074, 10E8, VRC01) with efficiencies comparable to non-treated virus. Autologous CD4$^+$ T lymphocyte proliferation and cytokine induction by monocyte-derived dendritic cells (MDDC) pulsed either with MUT-A-inactivated HIV or non-treated HIV were also comparable.

*Correspondence: b.berkhout@amc.uva.nl; mMPLANA@clinic.ub.es; MPLANA@clinic.cat; benarous.r@wanadoo.fr
†Céline Amadori and Yme Ubeles van der Velden contributed equally to this work
[1] Biodim Mutabilis, 93230 Romainville, France
[5] Laboratory of Experimental Virology, Department of Medical Microbiology, Center for Infection and Immunity Amsterdam (CINIMA), Academic Medical Center, University of Amsterdam, Amsterdam, The Netherlands
[7] AIDS Research Group, IDIBAPS, Hospital Clinic, Barcelona, Spain
Full list of author information is available at the end of the article

Conclusions: Although strongly defective in infectivity, HIV-1 virions produced in the presence of the MUT-A INLAI have a normal protein and genomic RNA content as well as B and T cell immunoreactivities comparable to non-treated HIV-1. These inactivated viruses might form an attractive new approach in vaccine research in an attempt to study if this new type of immunogen could elicit an immune response against HIV-1 in animal models.

Keywords: HIV-1, Integrase, LEDGF, Allosteric integrase inhibitor, LEDGIN, INLAI, Immunoreactivity

Background

The integration of a DNA copy of the HIV RNA genome into host chromatin is a crucial step of HIV replication [1]. The HIV-1 pre-integration complex is tethered to the host chromosome via the cellular co-factor lens epithelium-derived growth factor (LEDGF/p75) [2], together with the involvement of the capsid binding protein CPSF6 [3]. LEDGF/p75 is a chromatin-bound protein that interacts with HIV-1 Integrase (IN) via its C-terminal IN binding domain (IBD) [4, 5]. A new class of IN-inhibitors was designed that prevents this IN-LEDGF/p75 interaction, named first LEDGINs [6], then ALLINIs [7] for Allosteric IN inhibitors, NCINIs [8–10] for non catalytic IN inhibitors, MINIs for Multimerization Integrase Inhibitors [11] or INLAIs for Integrase-LEDGF allosteric inhibitors [12]. Since their first description by the group of Zeger Debyser [6], there is not yet a consensus name or acronym for this new class of IN inhibitors; we chose in this report the acronym INLAI as a generic name for these inhibitors, which has the advantage to recall the dual mechanism of action of these inhibitors: inhibition of the IN-LEDGF/p75 interaction and induction of an allosteric conformational change and multimerization of IN.

INLAIs are allosteric IN inhibitors that bind to the LEDGF binding pocket of IN and are fully active on HIV-1 resistant to INSTIs [6–10, 12–14]. From a chemical point of view, all INLAIs described up to date share a common motif composed of a tert-butylether and a carboxylic acid group that can be linked to different scaffolds, quinoline, naphthyl, phenyl or pyrimidine [6–10, 12–14]. INLAIs have a dual antiretroviral (ARV) activity at two different steps of the HIV-1 replication cycle: Inhibition of the LEDGF/p75-IN interaction accounts for an "early" block of HIV-1 replication at integration, but the major impact of INLAIs is during virus maturation or the "late" phase, leading to the production of normal CA-p24 amounts of non-infectious virus. This late effect on virus maturation is linked to INLAI-promoted IN multimerization [9, 12–15].

HIV-1 virions produced in the presence of INLAIs are non-infectious and contain eccentric condensates outside of the cores as shown by electron microscopy [9, 14, 15]. However, using HIV-1 produced in the presence of the quinoline INLAI compound BI-D (developed by Boehringer Ingelheim), we recently described that a wild-type level of HIV-1 genomic RNA is packaged in these virions in a dimeric state, and the $tRNA^{lys3}$ primer for reverse transcription was properly placed on the genomic RNA and could be extended ex vivo. In addition, RT enzyme extracted from these virions was fully active although these virions were unable to complete reverse transcription in target cells [9]. Fontana et al. [16] found that INLAIs block ribonucleoprotein complex packaging inside viral cores leading to the formation of "eccentric condensates" with high Nucleocapsid (NC) content outside the core. Kessl et al. [17] showed recently that IN directly binds the viral RNA genome in virions and that ALLINIs impair IN binding to viral RNA in wild-type virions. These INLAI-inactivated virions were able to infect target cells, but the subsequent reverse transcription step in target cells was blocked [9, 15]. Vranckx et al. [18] have shown that LEDGF/p75 depletion hampers HIV-1 reactivation in cell culture, and they demonstrated that LEDGINs relocate and retarget HIV integration, resulting in a HIV reservoir that is refractory to reactivation by different latency-reversing agents.

Several attempts have been made to produce inactivated SIV vaccines by treatment with denaturing and inactivating agents such as zinc chelators (2,2′-dithiobisbenzamide (DIBA)), heat, UV or cross-linking agents such as psoralen in non-human primate studies. But none of these attempts has shown significant efficacy in preventing SIV infection or in controlling virus replication and delaying progression of disease (for review [19]). HIV-1 inactivated by similar denaturing methods has also been used in dendritic cell-based therapeutic vaccination, but did not yet yield convincing results [20, 21]. INLAI-inactivated HIV-1 particles have remarkably conserved many properties of "native" non-treated HIV-1 particles, but are based on a new ARV drug class under preclinical development and not on a denaturing chemical. In front of the persisting bottlenecks in identifying viable HIV vaccine candidates, we explored the main B and T cell immunoreactivities of INLAI inactivated virus particles to study whether they could be considered as a new type of immunogen able to elicit immune response against HIV-1 in animal models.

Here we describe the new INLAI compound MUT-A. MUT-A shares with all previously described INLAIs a tert-butylether motif linked to a carboxylic acid group, but it is composed of an original scaffold, a 5-atom thiophene ring (Fig. 1). We studied the ARV activity of MUT-A, its resistance profile, its influence on viral RNA packaging and the B-cell and T-cell immunoreactivities of MUT-A-inactivated HIV-1.

Methods

Cell culture

SupT1 or MT4 T cells were cultured in advanced RPMI 1640 medium (Gibco) supplemented with 10% (v/v) heat-inactivated fetal bovine serum (FBS, Gibco), 2 mM L-glutamine (Gibco), 15 μg/mL streptomycin and 15 units/mL penicillin at 37 °C and 5% CO_2. Human embryonic kidney (HEK) 293T cells were grown in DMEM (Gibco) supplemented with 10% (v/v) heat-inactivated FBS (Gibco) and 1× minimum essential medium non-essential amino acids (MEM NEAA, Gibco) at 37 °C and 5% CO_2.

Virus strains and recombinant molecular clones

HIV-1 NL4-3, HxB2 molecular clones were obtained from the NIH AIDS Research and Reference Reagent Program and SIVmac239 was a gift of R. Desrosiers. The SpeI-SalI fragment from pNL4-3 containing the full *pol* gene was cloned into the pUC18 plasmid. In vitro mutagenesis was performed with the PfuTurbo (Agilent) and specific sets of primers to engineer the single mutants in *integrase* (Y99H, A128T, L102F, H171Q/T, T174I, N222K). The mutated fragment was validated by sequencing (Eurofins MWG Operon) and cloned back into pNL4-3 to generate a HIV-1 mutant molecular clone.

Compound

Optimization of our small molecule series led to compound MUT-A, one of the most active molecules. MUT-A was prepared as described in details in Additional file 1 and in a patent application [22], according to the example 11. The chemical identification of MUT-A

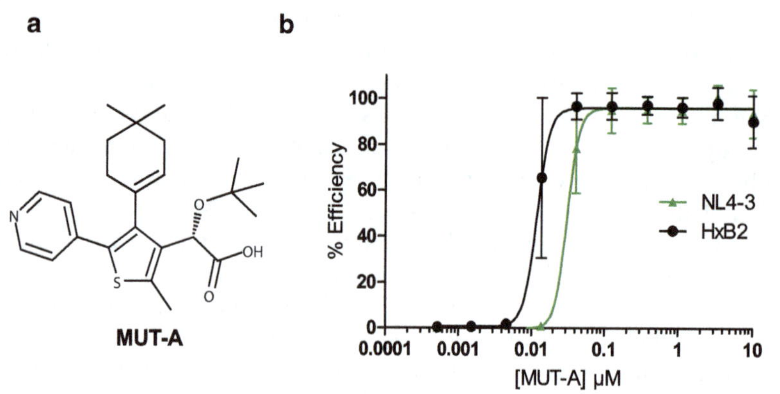

c

IN-LEDGF p75 IC_{50}= 95 nM
IN multimerization AC_{50}= 54 nM
IN multimerization Plateau= 600%

d

	MT4 infection EC_{50} (nM)	
Compound	HIV-1 NL4-3	HIV-1 HxB2
MUT-A	32 ± 9	12 ± 5
BI-D (rac)	190 ± 60	42 ± 14
BI-224436	46 ± 16	19 ± 5
DTG	2.6 ± 1.3	1.9 ± 1.0

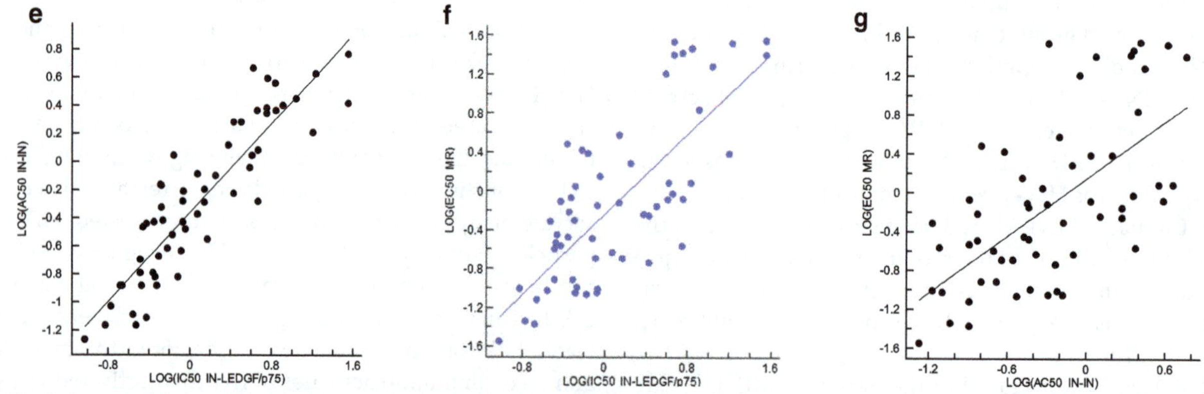

Fig. 1 Structure, biochemical and antiretroviral activities of MUT-A. **a** Chemical structure of MUT-A. **b** Titration of MUT-A ARV activity on HIV-1 NL4-3 and HxB2 strains in MT4 cell infection assay. **c** Biochemical activities of MUT-A on IN-LEDGF/p75 interaction (IC_{50}) and of IN multimerization (activation concentration AC_{50} and maximum signal increase plateau). **d** EC_{50} of ARV activities of MUT-A and references HIV-1 NL4-3 and HxB2 strains in MT4 cell infection assay. DTG: INSTI Dolutegravir. **e–g** Correlations between IN multimerization (IN–IN) AC_{50} and IN-LEDGF/p75 IC_{50} (**e**), ARV EC_{50} and IN-LEDGF IC_{50} (**f**), ARV EC_{50}s and IN–IN AC_{50}s (**g**) of MUT-A compound series

and its purity were assessed by NMR spectrum (shown in Additional file 1: Fig. S1). MUT-A was dissolved in DMSO to generate a 10 mM stock solution and was added to the culture medium at a final concentration as indicated. The equivalent volume of DMSO was added to control cultures.

HTRF®-based IN-LEDGF interaction assay

The IN-LEDGF HTRF® assay was performed using Flag-tagged IN NL4-3 and His-tagged LEDGF/p75 as described in [12], and detailed in Additional file 1.

HTRF®-based IN multimerization assay

The IN–IN HTRF® assay was performed using Flag-tagged and His-tagged IN from NL4-3, as described in [12] and detailed in Additional file 1.

Virus production and replication

The experiments with HIV-1 isolates LAI or NL4-3 were performed as described in [23] and the data presented here were collected simultaneously with those reported previously in [23] using the INLAI BI-D together with the same untreated controls. 293T cells were seeded in T75 culture flasks, cultured to 50–70% confluency and transfected with 20 μg pLAI or pNL4-3 DNA plasmid that encodes the wt HIV-1 LAI or HIV-1 NL4-3 isolates respectively using Lipofectamine2000 (Invitrogen). MUT-A was added 6 h after transfection. The culture supernatant was harvested 48 h after transfection and used as virus stock or for viral RNA isolation. The CA-p24 level was measured by enzyme-linked immunosorbent assay (ELISA) as described previously [23]. SupT1 T cells (5×10^6 cells in 5 mL) were infected with the HIV-1 LAI virus stocks (equivalent of 1 ng CA-p24). Similarly, MT4 cells (5×10^6 cells in 5 mL) were infected with the HIV-1 NL4-3 virus stocks (equivalent of 1 ng CA-p24). When indicated, the culture was split and MUT-A or DMSO was added. Viral spread was monitored by measuring the CA-p24 level in the virus culture medium every 2 days. HIV-1 NL4-3 virus produced and inactivated in the presence of AT-2 was prepared as described in [24].

HIV-1 and SIV antiviral assays in MT4 cells

MT4 cells growing exponentially at the density of 10^6/mL were infected for 2 h with HIV-1 strain NL4-3, HxB2 or SIVmac239 (viral stock produced according to [25]) at a multiplicity of infection (MOI) of 0.006 and 0.01, respectively. The cells were washed with PBS and aliquoted, using 100 μL fresh complete RPMI, into 96-well white plates (Corning) in the presence of different concentrations of compounds. The effective concentration of compound required to inhibit 50% (EC_{50}) of viral replication was determined, after 5 or 7 days for HIV-1 and SIV respectively, using the CellTiter-Glo® luminescent reagent (Promega) to quantify cell viability.

Viral resistance selection

Resistance selection was carried out by serial passages of infection at suboptimal compound concentrations. MT4 cells were seeded in 96-well culture plates at a density of 1×10^5 cells/well in 200 μL of culture medium. HIV-1 NL4-3 was used at 0.005 MOI. Compounds were added at final concentrations corresponding to ½ EC_{50}, EC_{50} and 2× EC_{50} and the plates were done in triplicate. Every 3–4 days, measurement of the antiviral activity allowed selecting between the three successive concentrations which one to choose to go on with the next passage. The cytopathic effect induced by HIV-1 was used to follow the progression of the infection. Briefly, 50 μL of the cell culture from the passage plate were mixed with 50 μL CellTiter-Glo® reagent to quantify cell viability. The concentration corresponding to the lower protection (0–25%) at the highest concentration was chosen to proceed. All the content from the chosen well was transferred to a 1.5 mL microtube and centrifuged for 3 min at 3000 rpm at room temperature. 30 μL of the supernatant were then used to infect three wells seeded with fresh cells and twofold serial dilutions of the compound. Successive viral passages were obtained by repeating this procedure.

Clonal sequencing of viral DNA

For clonal sequencing of viral passages, total DNA was extracted from infected cells using the QIAamp DNA blood minikit (Qiagen). A 980-bp viral DNA fragment containing the entire IN orf, spanning from nucleotide 4176 in the *pol* gene to nucleotide 5154 in the *vif* gene, was amplified by PCR (nucleotide numbering according to pNL4-3 molecular clone sequence (Accession number AF324493)). A 610-bp viral DNA fragment spanning from nucleotide 2588–3199 in the *pol* gene was also amplified by PCR. This fragment contains the region of RT sensitive to mutation under the pressure of selection. DNA extracted from non-infected cells was used as a negative control in PCR experiments to ensure there was no cross contamination. PCR products were purified using EZ-10 Spin Column PCR Purification kit (BioBasic) and cloned into the pCR2.1® using the TA cloning kit (Invitrogen). The IN or RT gene from 30 to 40 single clones was sequenced by Eurofins sequencing service (Eurofins MWG Operon) using specific primers. The sequence was compared to the wild-type NL4-3 region using NCBI BLAST alignment engine and amino acid substitutions were identified.

B-cell immunoreactivity

B-cell immunoreactivity of HIV-1 NL4-3 virus produced in the absence or presence of MUT-A was studied by virus immunocapture assays as described earlier [26], using the panel of anti-HIV-1 neutralizing and non-neutralizing antibodies [27–35] listed in Fig. 6b. This assay measures the concentration of whole native virus particles captured by antibodies (Abs) coated onto 96-well plates (Maxisorp, Nunc) as previously described [26, 36, 37]. Briefly, HIV particles were incubated in Ab-coated ELISA plates for 1 h. Unbound virus was removed by washing with PBS containing 10% fetal calf serum. Virus captured by Abs was lysed with 10% NP-40 and quantified by dosage of CA-p24 by ELISA (Innogenetics/Ingen).

Patients and samples

Samples of EDTA-anticoagulated venous blood samples were obtained from chronic asymptomatic HIV-1-infected patients with baseline $CD4^+$ T-cell counts > 450 cells/mm^3, and plasma viral load (pVL) < 50 HIV-1 RNA copies/mL, who were on antiretroviral therapy (ART). All the subjects participating in the study were recruited at the Service of Infectious Diseases and AIDS Unit of the Hospital Clinic from Barcelona (Spain). All the individuals gave informed written consent and this study was reviewed and approved by the Institutional Ethical Committee board of Hospital Clinic (Barcelona, Spain).

Generation of monocyte-derived dendritic cells

Generation of monocyte-derived dendritic cells (MDDC) was performed as previously described [38]. Essentially, peripheral blood mononuclear cells (PBMC) were isolated immediately after venous extraction by using a standard Ficoll gradient. Cells were processed immediately after isolation. To obtain human monocytes, PBMC were incubated in plastic plates (2 h at 37 °C) in a humidified atmosphere with 5% CO_2 in MDDC medium (serum-free XVIVO-15 medium, Lonza) supplemented with 1% autologous serum, gentamicin (Braun Medical) and fungizone (amphotericin B, Bristol-Myers Squibb) and 1 μM zidovudine (Retrovir from GlaxoSmithKline) to avoid HIV replication. To obtain immature MDDC, adherent cells were washed four times with pre-warmed MDDC medium and then cultured for 5 days in the presence of 1000 U/mL each of recombinant human IL-4 (Strathmann Biotec AG) and recombinant human GM-CSF (Peprotech) on day 0 and 2. Immature MDDC in fresh MDDC medium, IL-4 and GM-CSF (1000 U/mL each) were exposed to NL4-3 virus either treated with DMSO, inactivated with MUT-A at 1 μM (1 or 5 μg/mL of HIV Gag CA-p24), inactivated with AT-2 1 μg/mL, and controls (SEA (Staphylococcus Enterotoxin A) 100 pg/mL). To obtain mature MDDC, a cocktail of recombinant human cytokines containing TNF-α, IL-6 (1000 IU/mL each, Strathmann Biotec AG), IL-1β (300 IU/mL, Strathmann Biotec AG) and PGE2 (1 mg/mL, Pfizer) was added at 2 h post-exposure, and the mixture was incubated for 48 h.

Autologous co-cultures

As a source of enriched T cells we employed autologous fresh PBMC depleted of monocytes after adherence to plastic as indicated above for the generation of MDDC. These monocyte-depleted lymphocytes were washed and resuspended in serum-free XVIVO-10 medium and labelled with CFSE following the instructions of the manufacturer (CellTrace CFSE cell proliferation kit, Molecular Probes). Autologous virus-exposed and matured MDDC were washed and resuspended in XVIVO-10 and co-cultured with autologous fresh CFSE-labelled lymphocytes in a final volume of 0.2 mL in XVIVO-10 medium supplemented with 1 μM zidovudine to impede possible replication of endogenous HIV-1. The contribution of MDDC alone and monocyte-depleted PBMC (lymphocytes) alone was determined as negative controls of proliferation and cytokine secretion. The co-cultures were done in triplicates at 37 °C in a humidified atmosphere of air with 5% CO_2.

Assessment of T cell proliferation and flow cytometry

After 6–7 days, proliferating $CD3^+CD4^+$ and $CD3^+CD8^+$ T cells were determined by direct staining with mAbs conjugated with a-CD3-Per-CP and a-CD8-PE. Mouse Ig isotypes mAbs (from BD Biosciences) conjugated with PerCP or PE were used as negative control mAbs. The stained cells were analyzed on a FACSCalibur flow cytometer (BD Biosciences). T cell populations were selected by forward and side light-scatter parameters and sub-gated for CD4 or CD8 expression. Cells that proliferated after the co-culture had lower intensity of CFSE ($CFSE^{low}$) in comparison with basal conditions. T cell specific proliferation was expressed as the percentage of $CFSE^{low}$ cells after co-culture with MDDC exposed to different virus minus the percentage of (mock-treated) $CFSE^{low}$ cells.

Cytokine and chemokine secretion by autologous MDDC-T cells co-cultures

The secretion of cytokines and chemokines induced during autologous MDDC-T cells co-cultures was measured in the culture supernatant using the Luminex technology (Cytokine Human 25-Plex Planel, Invitrogen), following the manufacturer's instructions. The following 25 mediators were tested: Eotaxin, GM-CSF, IL-1β, IL-1RA, IL-2, IL-2R, IL-4, IL-5, IL-6, IL-7, IL-8, IL-10, IL-12p40/p70,

IL-13, IL-15, IL-17, IFN-α, IFN-γ, IP-10, MCP-1, MIG, MIP-1α, MIP-1β, RANTES and TNF-α. Data were analyzed using the GraphPad Prism Software version 5.00 (San Diego, CA, USA). Comparisons for the different parameters were performed by Student's test. For the analysis a level of significance was set at $p < 0.05$.

Results

Biochemical and antiretroviral activity of MUT-A

We previously reported the identification of an INLAI, Mut101 of the aryl or heteroaryl-tertbutoxy-acetic acid family [12]. In an effort to diversify our INLAI library and to obtain compounds with more potent ARV activity, we produced a library of 58 original INLAI compounds. These compounds share a common motif composed of a tert-butylether and a carboxylic acid group but linked to a heterocyclic scaffold different from those previously reported in this compound family. As shown in Fig. 1a, MUT-A is an original INLAI compound of 413 g/mol molecular weight, with a 5-membered heterocyclic thiophene core substituted by a pyridine group, a cyclohexen moiety and the sterically bulky tertbutoxy-ether key side chain. MUT-A has potent anti-HIV activity with an EC_{50} of 32 ± 9 nM or 12 ± 6 nM in MT4 cells infected by NL4-3 or HxB2 HIV-1 strains, respectively (Fig. 1b). As expected for all inhibitors of the INLAI family, MUT-A had a much weaker early ARV activity at integration, with an EC_{50} of 2.3 μM, measured in single-cycle infection of MT4 cells by NL4-3/Env VSVg-pseudotyped non-replicative virus (not shown). Cellular toxicity of MUT-A was low with CC_{50} of 42 ± 9 μM in MT4 cells, yielding a selectivity index of 1355. MUT-A inhibited IN (NL4-3)-LEDGF/p75 interaction with an IC_{50} of 92 ± 5 nM and promoted IN multimerization with an activation constant AC_{50} of 55 ± 2 nM and a maximum multimerization signal plateau of 600% compared to the background IN–IN signal obtained in the absence of MUT-A (Fig. 1c). The ARV activity of MUT-A was compared to that of previously described INLAIs (BI-D racemate and BI-224436) or to that of the second generation INSTI, Dolutegravir (DTG). MUT-A was found 4–6 times more potent than racemic BI-D and 1.4–1.7 times more potent than BI-224436, but 6–12 times less potent than DTG (Fig. 1d). Structure activity relationship of all compounds in this series demonstrates on the one hand, a tight correlation between IN-LEDGF/p75 inhibition and IN multimerization (Fig. 1e), and on the other hand, a good correlation between ARV activity and IN-LEDGF/p75 inhibition (Fig. 1f) or IN multimerization (Fig. 1g). As expected the correlation between ARV activity and in vitro biochemical activities on IN-LEDGF/p75 inhibition or IN multimerization was less tight than between the two different biochemical activities. Indeed, a parameter such as cellular permeability is critical for ARV activity but not relevant for the biochemical activities of these molecules. Altogether these results show that the MUT-A series are genuine INLAI compounds.

Characterization of HIV-1 virions produced in the presence of MUT-A

We infected SupT1 T cells with the HIV-1 LAI strain and cultured the cells with or without MUT-A (160 nM, 5× EC_{50}). In this experiment we produced in parallel HIV-1 LAI treated with another reference INLAI compound, BI-D, together with a control of virus produced in the absence of any compound that was previously published in [23]. Viral spread was monitored by measuring the CA-p24 level in the culture supernatant. Whereas efficient virus replication resulted in a rapid increase in CA-p24 level in the control culture, HIV-1 LAI was efficiently blocked by MUT-A (Additional file 1: Fig. S2A) similar to the antiviral effect of BI-D [23].

To test the effect of MUT-A on virus production, we transfected HEK 293T cells with the HIV-1 encoding plasmid pLAI [39] and cultured the cells in the presence or absence of MUT-A (Additional file 1: Fig. S2B) or BI-D [23]. The virus-containing supernatant was harvested 48 h later and the CA-p24 level was determined by ELISA. As previously shown for BI-D treatment, virus production on HEK 293T cells was not affected by MUT-A (Additional file 1: Fig. S2C). Western blot analysis of virions showed a similar protein content as untreated virions, while virions treated with Saquinavir (SQV) were clearly defective for protein maturation (Additional file 1: Fig. S2D). Next, we tested whether virus produced in the presence of MUT-A can promote a spreading infection in SupT1 T cells. No additional MUT-A was added during SupT1 culturing. Although we did not wash away MUT-A that is present in the virus stock, the high dilution factor (~ 2500×) and frequent passaging of the cell cultures (every 3–4 days), makes a sustained antiviral effect unlikely. Efficient viral spreading was observed upon infection of the cells with virus produced in the absence of MUT-A. In contrast, the presence of MUT-A or of BI-D during virus production severely impaired virus infectivity such that no spreading infection could be established (Additional file 1: Fig. S2E) as previously shown [23]. Similar results were obtained for the NL4-3 strain produced in the presence of MUT-A (data not shown). These effects have been reported for other INLAIs as well [7, 8, 12–14]. As previously described for BI-D [23], MUT-A treatment does not affect HIV-1 RNA packaging thermal stability of RNA dimers, reverse transcriptase activity and initiation of reverse transcription in vitro by $tRNA^{lys3}$ primer (Additional file 1: Fig. S3).

One documented effect of INLAIs at the late phase of HIV-1 virion assembly and maturation is mislocalization of the ribonucleoprotein outside of the capsid core [9, 12, 14, 15]. We performed cryo-electron microscopy (cryo-EM) analysis on virus that were exposed or not to MUT-A treatment during virus production by 293T cells transfected with pNL4-3. The aim was to determine whether MUT-A, as previously described for other INLAIs [9, 15, 16], induced any architectural changes to the virus particle. Figure S4 illustrates the morphology of intact virus particles observed by cryo-EM, produced in the presence of MUT-A (Additional file 1: Fig. S4A) or its absence (Additional file 1: Fig. S4B). Variation in virus particle size can be observed in both panels, which is rather typical for HIV [40]. Particles were distinguished into three distinct phenotypes: (a) normal mature conical cores found mostly in virus produced in the absence of MUT-A (Additional file 1: Fig. S4B, blue arrow), (b) non-conical cores found both in virus produced in the presence of MUT-A (Additional file 1: Fig. S4A) and less frequently in the absence of MUT-A (Additional file 1: Fig. S4B), (c), virions with eccentric condensate, frequently found in virus produced in the presence of MUT-A (Additional file 1: Fig. S4A) and only very rarely observed for untreated virus. As previously indicated, these eccentric condensates likely represent the viral ribonucleoprotein (RNP) complex mislocalized outside the empty core. Indeed, these cryo-EM observations confirm that MUT-A treatment during virus production induces core abnormalities similar to those previously described for other integrase allosteric inhibitors, demonstrating that MUT-A displays all the characteristics attributed to allosteric inhibitors of integrase, such as IN multimerization, inhibition of IN-LEDGF/p75 interaction and induction of virion abnormalities such as cores with eccentric condensates.

Resistance profile and spectrum of activity of MUT-A

Several mutations that induce resistance to INLAIs have been identified [6–13]. Since MUT-A is a new INLAI with an original scaffold never described previously, we wanted to check whether its potent ARV activity was sensitive to INLAI resistant mutations. First we constructed NL4-3 HIV-1 harboring IN single site mutations such as Y99H, A128T, L102F, H171Q, H171T, T174I, N222K, and checked the ARV activity of MUT-A and other INLAIs used as controls, BI-D and BI-224436, on each of these mutants. As shown on Fig. 2a, MUT-A unfortunately is impacted by all resistant mutants we tested, although at variable levels. The impact of these mutations was evaluated by their fold change in EC_{50}. The most detrimental mutation is T174I with a considerable effect with a fold shift in EC_{50} as high as 1600. Then, mutations L102F, H171T, Y99H, A128T had a significant impact with fold changes of 39, 25, 16 and 11 respectively. Mutation H171Q had a lower impact with a fold change of 4. The other INLAIs used as controls, BI-D and BI-224436 had a similar profile of resistance with the T174I mutation having the strongest impact (EC_{50} fold change of 226 and 544 respectively) followed by L102F, H171T, Y99H or A128T. All these mutations are in the catalytic core domain (CCD) of IN, close to the INLAI binding site. On the contrary, mutation N222K is localized in the C-terminal domain (CTD) of IN at a distance of the binding site. Its impact is generally lower than the other mutations (data not shown for MUT-A, but see the effect on BI224436 in Fig. 2).

In addition to these experiments based on the fold change determination of MUT-A EC_{50} on purified mutant viruses, we sought to identify which resistant mutations are selected in serial passage experiment by progressively increasing MUT-A concentrations on MT4 cells infected with HIV-1 NL4-3, as described in the material and method section. Infected cells were passaged, starting at concentration below MUT-A EC_{50} (16 nM $= EC_{50}/2$) to favor the occurrence of resistant mutations, and ending after 67 days selection at a MUT-A concentration 1000 fold higher (16 μM). Kinetics of MUT-A resistance was compared to resistance to Nevirapine (NVP, NRTI) and Raltegravir (RAL, INSTI). As shown in Fig. 2a, the kinetics of resistance development to MUT-A was parallel to that of NVP and slightly faster than that of RAL. At three different passages, day 23, day 43 and day 67, IN sequence analysis was performed by PCR in order to identify the resistant mutations that were selected and their frequency and kinetics of occurrence. As shown in Fig. 2b, c, the most abundant mutations that were detected as soon as day 23 are A128T and Y99H. These mutations remain the most frequent and accumulate with other minor mutations up to the end of the experiment. At day 67, A128T occurred in almost all sequences (99%) and Y99H in 53% of the sequences. Interestingly the most detrimental mutation, T174I, occurred only at the end with a frequency as low as 12%. The other strong resistant mutations, L102F and H171T were not detected at any time point. Some mutations such as H171Q which has a moderate impact (EC_{50} fold change of 4) were detected early at day 23 and then at day 43, but were not detectable at the end of the experiment.

These kinetics and frequency of occurrence of resistant mutations during serial passage experiment indicate that some mutations like A128T or Y99H arise preferentially compared to more detrimental mutations like T174I, L102F and H171T. One possible explanation of such phenomenon is that the resistance mutations that are favored correspond to virus resistant mutants that have the better replication capacity. We verified that this was indeed

a

	NL4-3 EC$_{50}$ (nM)	NL4-3 with resistance mutation in IN EC$_{50}$ (Fold change vs. WT)						
Compound	NL4-3	A128T	Y99H	L102F	H171Q	H171T	T174I	N222K
EVG	2	1	1	1	1	1	1	1
MUT-A	31	11	16	39	4	25		ND
BI-D (rac)	190	1	4	35	ND	25		ND
BI-224436	45	5	2	41	4	38		2

Fold shift color code	0-4	4-10	10-20	20-50	

b

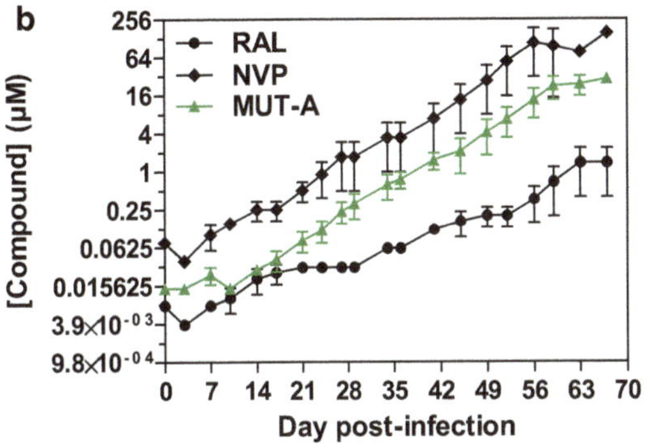

c

Frequency of mutations in IN sequences		
Day 23	Day 43	Day 67
Y99H (18%)	**Y99H** (38%)	**Y99H** (53%)
	T125S (3%)	T125K (11%)
A128T (42%)	**A128T** (91%)	**A128T** (99%)
E170K (5%)	E170K/G (12%/18%)	E170G/D (22%/15%)
H171Q (6%)	H171Q (5%)	H171R (13%)
		T174I (12%)

Fig. 2 Resistance profile of MUT-A. **a** Comparative resistance profile of MUT-A and references (EVG, BI-D and BI-224436). The color code subdivides the fold change in resistance into five levels of magnitude. **b** Long-term culture of infected MT4 cells with escalating concentrations of drugs. Cells were initially infected at 0.005 MOI with HIV-1 and were passaged twice weekly in the presence of RAL (filled circle), NVP (filled diamond), or MUT-A (filled upward triangle). The serial passage was carried out over 67 days. The highest concentration of compounds at which the virus could replicate in culture are shown. Error bars represent the variation obtained in one experiment done in triplicate. **c** Frequency of mutations in IN sequences. Cloned sequences were analyzed at day 23, 43 and 67 from 30 to 40 individual clones (% of total clones analyzed)

the case. As shown in Additional file 1: Fig. S5, viruses harboring Y99H or A128T resistance mutations have replication capacity similar or close to wt NL4-3, while viruses harboring H171T, T174I or L102F mutations had lower replication capacity. Presumably, during the resistance selection process, mutations that are favored are those which preserve the best virus fitness.

We then wanted to know if MUT-A was able or not to inhibit replication of other lentiviruses such as SIV. We compared the ARV activity of MUT-A, BI-D to that of the IN strand transfer inhibitor RAL used as control, on SIVmac239 and HIV-1 NL4-3. As shown in Fig. 3, while RAL had comparable efficiency on HIV-1 and SIVmac239 (reproducibly slightly better on SIV), neither MUT-A, nor BI-D had significant activity on SIV despite their potent activity on HIV-1. Similarly, MUT-A and BI-D had no significant activity on HIV-2 (data not shown).

Immunoreactivity of MUT-A-inactivated HIV-1 against HIV-1 Env specific antibodies

We analyzed HIV-1 NL4-3 virions produced in the absence or presence of MUT-A for their immunoreactivity against a panel of polyclonal and monoclonal neutralizing and non-neutralizing anti-HIV-1 Env antibodies. Viral particles were captured by the antibodies coated on a 96-well ELISA plate and, after removal of unbound virus, we quantitated captured virus by detection of CA-p24 by ELISA after virus lysis (Fig. 4a). This assay allows the detection of capture of complete well-preserved virus particles [36]. The neutralizing and non-neutralizing antibodies used in these virus capture assays are listed in Fig. 4b together with their binding sites. This panel comprises two polyclonal antibodies: one HIV-1 specific neutralizing antibody (F6 Gri/Ly) and one non-specific IgG antibody as negative control, and seven anti-HIV Env monoclonal antibodies (mAbs) including two neutralizing mAbs directed against the CD4 binding site (b12 and VRC01), two neutralizing mAbs against the MPER gp41 epitope (2G12 and 10E8), a non-neutralizing mAb against gp41 (4D4), two neutralizing mAbs against the V3 Envelope epitope (PGT121 and 10/1074), and one irrelevant non-specific mAb (Synagis) as an additional negative control. Three antibody concentrations were used for each capture assay. As shown in Fig. 4c, NL4-3

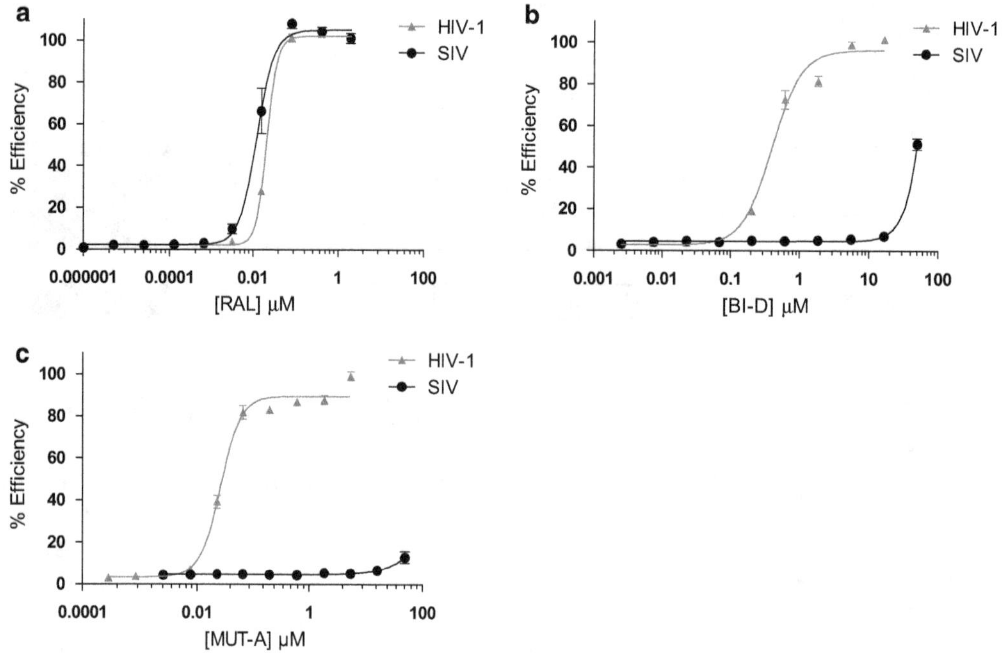

Fig. 3 Antiviral activity in infection assays with HIV-1 and SIV. Dose–response curves for RAL (**a**), BI-D (**b**) and MUT-A (**c**) activity in MT4 cells infected with either HIV-1 NL4-3 (filled upward triangle) or SIVmac239 (filled circle). Error bars represent the variation obtained from three independent experiments done in duplicate

MUT-A-inactivated particles were captured by the HIV Env specific antibodies with comparable efficiency as the non-treated NL4-3 particles. As expected, the two non-specific antibodies were negative in these capture assays. Except for the 10E8 and the 10/1074 mAbs which captured a comparable low amount of non-treated or inactivated virus, all specific antibodies were quite efficient in these virus capture assays. These results indicate that the MUT-A-inactivated NL4-3 virus particles were similarly immunoreactive as non-treated virus, indicating the preservation of the Envelope glycoprotein structure following inactivation of the virus particles. We next compared the immunoreactivity of HIV-1 virions produced in the presence of MUT-A with HIV-1 virions inactivated by other means such as AT-2 or Saquinavir treatment. As shown in Additional file 1: Table S1, Fig. S6 the immunoreactivities of these inactivated virions were comparable and similar to untreated virions.

T-cell immunoreactivity of DCs from HIV$^+$ patients loaded with MUT-A-inactivated HIV-1 NL4-3 co-cultured with autologous T lymphocytes

We next investigated the T-cell immune responses induced by MDDC pulsed with non-treated and MUT-A-inactivated HIV-1 NL4-3, as well as AT-2-inactivated NL4-3 and the SEA super antigen as positive controls. After 6-days of co-culture, T cell proliferation induced by virus-loaded MDDC or MDCC pulsed with the SEA super antigen was estimated by loss of CFSE labelling (Fig. 5). Proliferation was observed essentially on CD4$^+$ T cells (Fig. 5). Interestingly, the extend of CD4$^+$ T lymphocyte proliferation observed with MDDCs loaded with MUT-A-inactivated NL4-3 was comparable or even higher than that obtained with MDDCs loaded with non-treated NL4-3 as shown in flow cytometry dot plots in Fig. 6a (28.9% loss of CFSE labelling for MUT-A inactivated NL4-3 versus 16.8% for non-treated NL4-3) or from quantitation of three independent experiments (Fig. 6b). This CD4$^+$ T cell proliferation observed with MDDCs loaded with MUT-A-inactivated NL4-3 was also comparable to that obtained with AT-2-inactivated virus (29% loss in Fig. 6a), although smaller amount of AT-2-inactivated virion was used to obtain a similar level of CD4 proliferation. We also observed a small loss of CFSE labelling in CD8$^+$ T cells. However, it seemed that CD8$^+$ T lymphocyte reactivity was poor under the used experimental conditions, as shown by the weak loss of CFSE labelling obtained also with CD8 + T cells using MDDCs loaded with the super antigen SEA (2.7% as shown in Fig. 6a or about 5% on average in three independent experiments (Fig. 6c)).

To further characterize the effects of the MDDC exposure on MUT-A-inactivated-HIV-1 virus on the T-cell induced immune responses, the levels of secreted

a

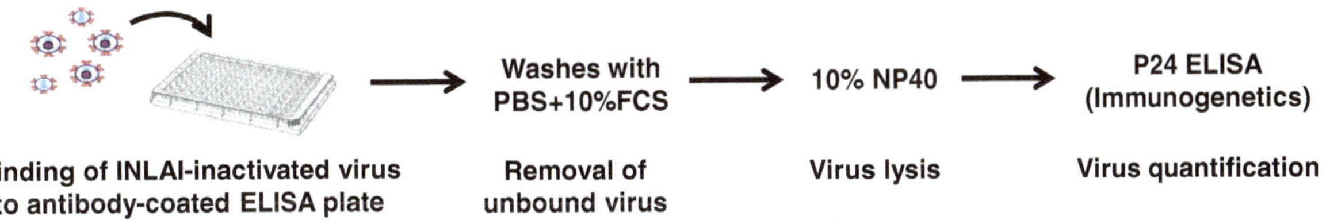

b

	Name	Binding site	Neutralizing activity
Polyclonal	Purified IgG HIV+ (F6 Gri/Ly)		+
	Purified IgG HIV-	Negative control	Irrelevant
Monoclonal	b12	CD4 bs	+
	2G12	MPER (gp41)	+
	4D4	gp41	-
	PGT121	V3	+
	10/1074	V3	+
	10E8	MPER (gp41)	+
	VRC01	CD4 bs	+
	Syn (Synagis)	Negative control	Irrelevant

c

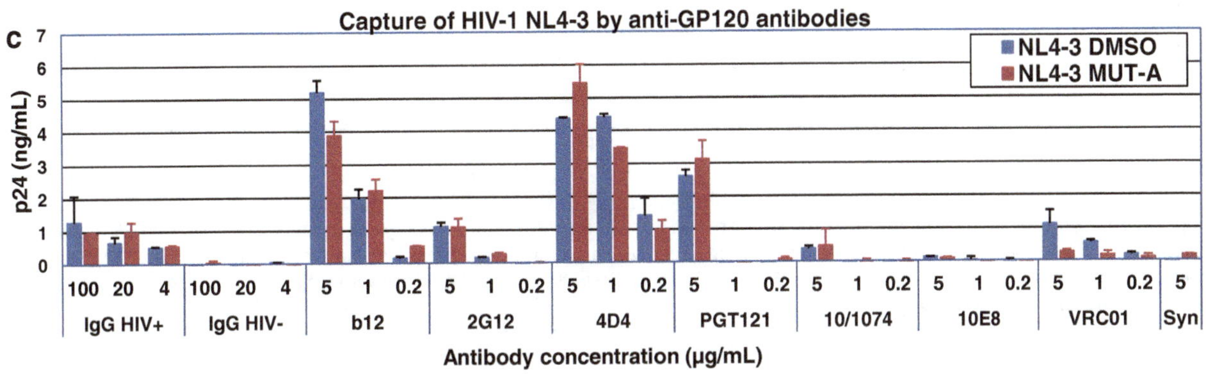

Fig. 4 Immunoreactivity of HIV-1 produced in the presence or absence of MUT-A. **a** Schematic of the antibody capture assay of native HIV-1 particles produced. The capacity of the different antibodies to capture native HIV-1 particles was assessed by ELISA. HIV-1 particles retained by the antibodies were lysed and quantified by CA-p24 detection by ELISA. **b** The different monoclonal and polyclonal antibodies used in virus-capture assays are listed together with their specific target and their ability to neutralize HIV-1. Nonspecific antibodies (polyclonal IgG HIV- and mAb Syn = Synagis) were used as negative controls. **c** Quantitation by CA-p24 ELISA of native HIV-1 particles produced in the presence (red bars) or absence (blue bars) of MUT-A that have been captured with the different antibodies used at three different concentrations

cytokines and chemokines in the culture supernatants of autologous pulsed MDDC and T cell co-cultures were assessed by Luminex technology. Results shown in Figs. 7 and 8 indicated that MDDC pulsed with MUT-A-inactivated NL4-3 induced secretion of IL-12 pro-inflammatory cytokine, Th1 cytokines (IFN-γ and IL-2R) as well as chemoattractant and antiviral chemokine (MIP-1β) (Fig. 7a–d), IL-10, IL-6, IL-13, MIP-1α, MCP-1, IL-5 and IP10 (IFN-γ-inducible protein 10). This pattern of cytokine/chemokine secretion reinforces the results obtained with $CD4^+$ T cells proliferation assays and illustrates the significant induction of HIV-specific T cell immune response promoted by MUT-A-inactivated HIV-1 loaded on DCs isolated from HIV-1 infected subjects.

Discussion

Here, we describe MUT-A as a new type of IN-LEDGF allosteric inhibitor. This compound consists of a 5-atom thiophene scaffold linked to the common carboxylic acid and tert-butylether moieties present on all INLAIs reported to date. As illustrated in Fig. 1, MUT-A as well as other compounds of the MUT-A series displayed biochemical activities specific of genuine INLAIs, such as inhibition of IN-LEDGF/p75 interaction and activation of IN multimerization. Furthermore, these two activities were tightly correlated, and both correlated with the antiretroviral activity of the MUT-A compounds (Fig. 1d–f). INLAIs that disrupt the IN-LEDGF/p75 interaction do not only affect the 'early' process of HIV-1

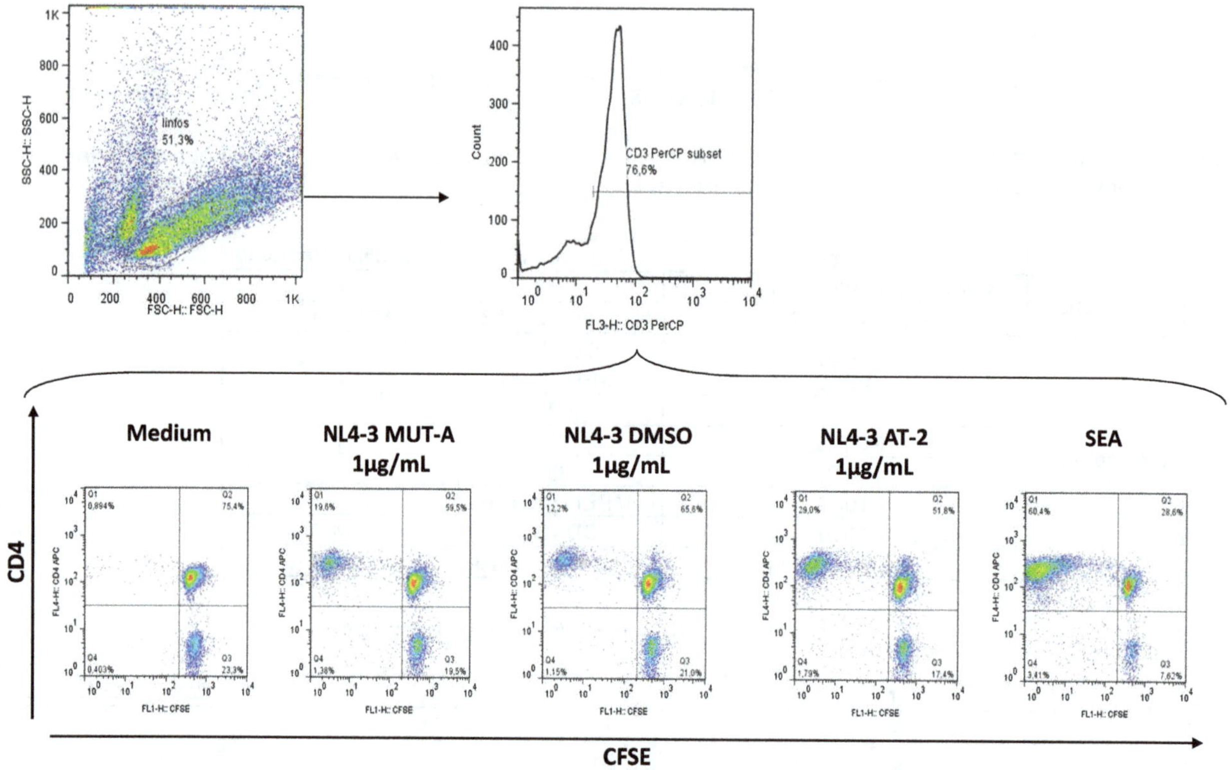

Fig. 5 T-cell proliferation induced by autologous MDDCs exposed to MUT-A-inactivated HIV-1 assessed by CFSE proliferation assay. Proliferation in response to autologous MDDCs exposed to wt HIV-1 (NL4-3 DMSO) or MUT-A or AT-2 inactivated HIV-1 in a 6-day co-culture was assessed in triplicates using the CFSE proliferation assay. MDDCs were pulsed with virus particles corresponding to 1 µg/mL of HIV-1 Gag CA-p24. A representative (out of 3 independent experiments) flow cytometry dot plots showing lymphocyte gating strategy by FSC/SSC and CD3$^+$ positive staining, as well as dot plots showing CFSE dilution in gated CD3$^+$ CD4$^+$ T lymphocytes after in vitro stimulation with the above mentioned virus or negative (medium) and positive SEA controls

integration but, unexpectedly, also the 'late' process of virus maturation, triggering a strong infectivity defect when HIV-1 is produced in the presence of these inhibitors. This infectivity defect does not correlate with any detectable change in the protein content of these non-infectious virions and the maturation state of the structural Gag and Gag-Pol proteins as indicated by western blot analysis of MUT-A inactivated virus (Additional file 1: Fig. S2D). However, the MUT-A-induced infectivity defect correlated with the appearance of numerous capsid abnormalities in virus produced in the presence of MUT-A, such as virions that contain eccentric condensates outside of empty capsids as previously described [9, 15, 16], reconfirming that MUT-A is a genuine INLAI compound. We found that MUT-A, like other INLAIs, renders the progeny virus non-infectious when present during virus production. However, like the previous report on BI-D-treated viruses [23], we did not observe any effect on HIV-1 genomic RNA packaging thermal stability of RNA dimers, tRNA^{lys3} placement on the viral RNA and activity of the co-packaged RT enzyme. These results confirm and generalize the finding that the viral RNA genome and the initiation of reverse transcription is not a target of INLAIs. These combined results imply that RNA packaging and dimerization are not affected by the aberrant localization outside the virus core in INLAI-treated virions (as visible in the cryo-EM images, Additional file 1: Fig. S4), and suggest that RNA dimer maturation does not require a correct virus core conformation. Cell entry is not affected for virus produced in the presence of INLAIs, but there is a reverse transcription block in these target cells [9, 14, 15]. We show that the packaged RT enzyme is fully active on exogenous templates and that the tRNA^{lys3} primer is correctly placed on the genomic RNA for cDNA synthesis. Thus, all basic factors required for reverse transcription are functionally present in INLAI-treated virions, which nevertheless are severely defective in reverse transcription. Several studies indicated that the first step of reverse transcription (initiation and production of strong-stop cDNA) is inhibited [9, 14, 15]. This may indicate that the components, although present and active in INLAI-treated virions, are

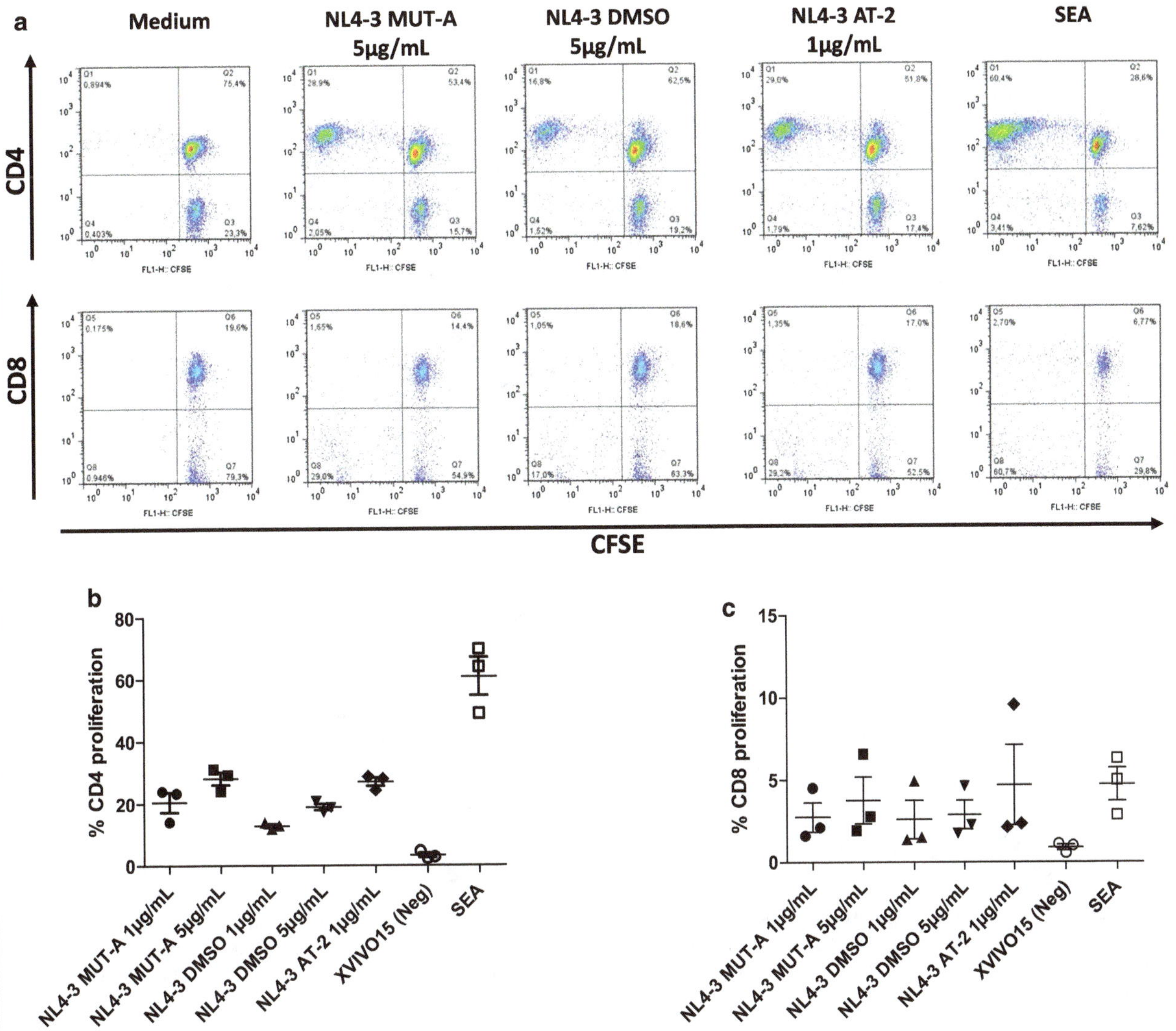

Fig. 6 $CD4^+$ and $CD8^+$ T-cell proliferation induced by autologous MDDCs exposed to MUT-A-inactivated HIV-1. **a** Proliferation in response to autologous MDDC exposed to wt HIV-1 (NL4-3 DMSO) or MUT-A inactivated HIV-1 or AT-2-inactivated HIV-1 in a 6-day co-culture was assessed in triplicates using the CFSE proliferation assay. MDDCs were pulsed with either 5 µg/mL HIV-1 (NL4-3 DMSO or MUT-A) or 1 µg/mL HIV-1 (NL4-3 AT-2) of HIV-1 gag CA-p24. A representative (out of three independent experiments) flow cytometry dot plots showing CFSE dilution in gated $CD3^+$ $CD4^+$ and $CD3^+$ $CD8^+$ T lymphocytes after in vitro stimulation with the above mentioned viruses or negative (medium) and positive SEA controls. **b, c** Quantitation of T $CD4^+$ (B) and $CD8^+$ (C) proliferation in response to autologous MDDC exposed to wt HIV virus (NL4-3 DMSO) or inactivated HIV virus (MUT-A or NL4-3 AT-2) in a 6-day co-culture was assessed in triplicates using the CFSE proliferation assay. In all viruses used, MDDCs were pulsed with 1 µg/mL of HIV gag CA-p24. Graph showing mean ± SD of percentages of CFSE low $CD3^+$ $CD4^+$ (**b**) and $CD3^+$ $CD8^+$ (**c**) T cells from three different co-cultures are represented

not located at the right position at the right time to execute the complex reverse transcription reaction.

Although MUT-A harbors an original 5-membered thiophene scaffold, it is nevertheless as sensitive as the other INLAIs to several point resistance mutations in the IN gene. These mutations are mainly located close to the INLAI binding site in the CCD, with the exception of the N222K mutation localized in the CTD of IN. Presumably mutations in the CCD could strongly diminish the binding affinity of MUT-A and other INLAIs for IN, disrupting their interaction with IN and consequently decrease strongly or cancel their ARV activity. The impact of these mutations on compound binding is due to impaired interaction with the conserved common carboxylic acid and tert-butylether moieties that are found in all INLAIs rather than with the scaffold moiety.

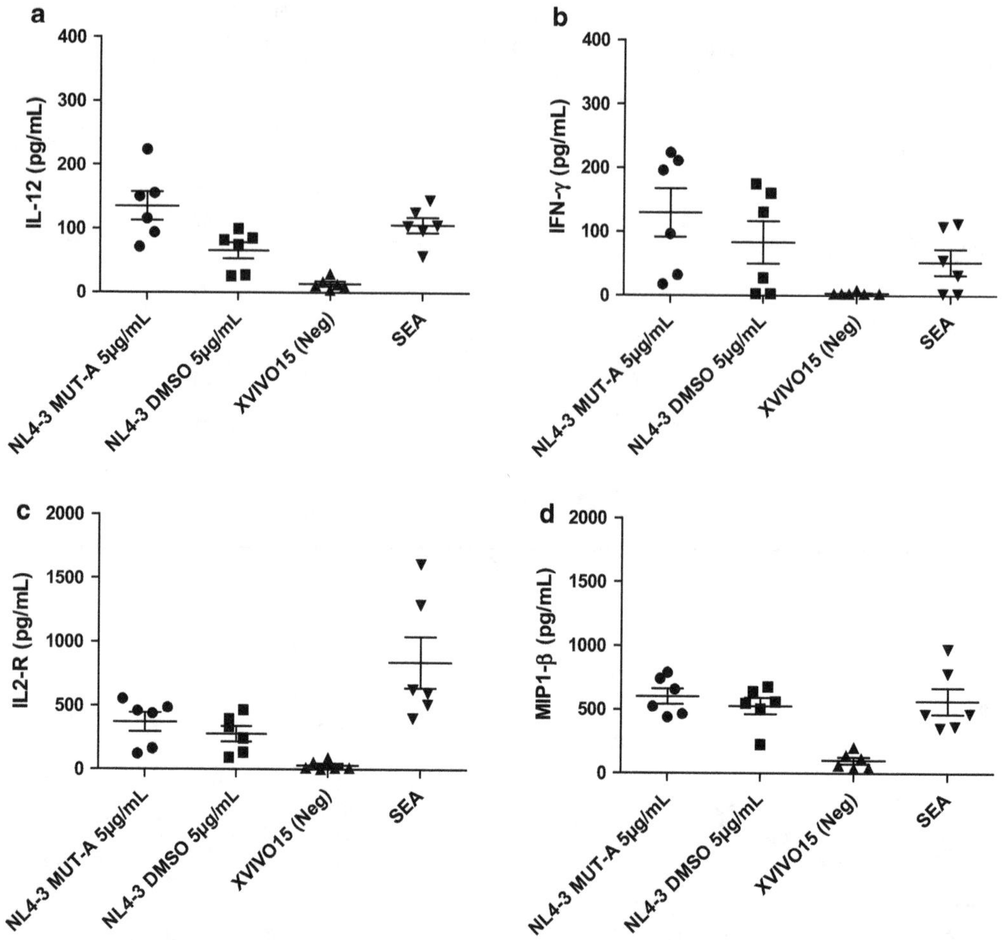

Fig. 7 Secretion of IL-12, IFN-γ, IL2-R and MIP-1β induced by autologous MDDCs exposed to MUT-A-inactivated HIV-1. Cytokine and chemokine secretion in response to autologous MDDCs exposed to wt HIV-1 (NL4-3 DMSO) or MUT-A-inactivated HIV-1 (MUT-A 1 µM) in supernatants from a 6-day co-culture was assessed in duplicates by Luminex assay. Graph representing mean ± SD concentrations (pg/mL) of IL-12 (**a**), IFN-γ (**b**), IL2-R (**c**), and MIP-1β (**d**) analyzed after three different co-cultures

On the contrary, the N222K resistant mutation that is localized in the IN CTD, at a distance from the compound binding site, could impact the conformational change induced by INLAIs for IN multimerization rather than the compound binding affinity. The resistance profile of MUT-A, BI-D and BI-224436 indicate that the genetic barrier of these compounds is weak, similar to that of NVP, since they are very sensitive to point resistant mutants. Fortunately, on the one hand, INSTIs like Elvitegravir (EVG) have fully conserved activity on these mutants, and on the other hand INLAIs conserved full ARV activity on mutants resistant to INSTIs [12]. These results allow us to consider a future combination of INLAIs and INSTIs as fully efficient. The lack of activity of INLAIs on SIV or HIV-2 compared to the fully conserved activity of INSTIs on HIV-1, HIV-2 and SIV is reminiscent of the inefficiency of NNRTIs on SIV and HIV-2. This seems to be a general property of allosteric inhibitors in contrast with catalytic inhibitors of both IN and RT enzymes.

Since we could not observe any detectable change in the protein content of MUT-A-inactivated HIV-1 neither in the packaging of viral RNA nor in exogenous RT activity, we completed this analysis by inspecting further the immunoreactivity of these MUT-A-inactivated HIV-1 particles. We indeed found that these inactivated viruses exhibit conserved B and T cell immunoreactivity, comparable with that of non-treated HIV-1. MUT-A-inactivated viruses were captured by multiple neutralizing and non-neutralizing polyclonal and monoclonal anti-HIV Env antibodies with efficiency comparable to that of non-treated HIV-1. These results indicate that MUT-A-inactivated HIV-1 particles conserve various important native epitopes for a B-cell immune response such as the

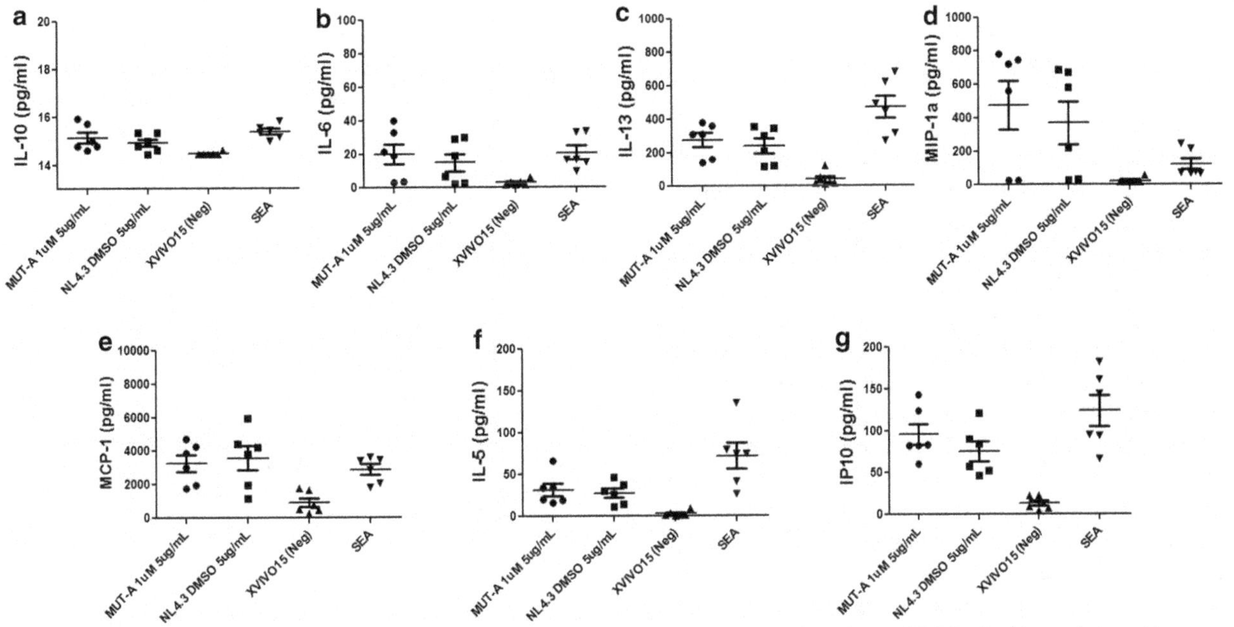

Fig. 8 IL-5, IL-6, IL-10, IL-13, MIP-1α, MCP-1 and IP-10 secretion induced by MUT-A-HIV-1-exposed autologous MDDCs. Cytokine and chemokine secretion in response to autologous MDDCs exposed to wt HIV-1 (NL4-3 DMSO) or MUT-A-inactivated HIV-1 (MUT-A 1 μM) in supernatants from a 6-day co-culture was assessed in duplicates by Luminex assay. Graph representing mean ± SD concentrations (pg/mL) of IL-10 (**a**), IL-6 (**b**), IL-13 (**c**), MIP-1α (**d**), MCP-1 (**e**), IL-5 (**f**) and IP-10 (**g**) analyzed after three different co-cultures

CD4 binding site and V3 epitope on the gp120 Env surface subunit or the major proximal epitope of the gp41 trans-membrane subunit. We also found that MUT-A-inactivated virus had B-cell immunoreactivity toward this panel of antibodies comparable to that of HIV-1 NL4-3 inactivated by other means such as Protease inhibitor or AT-2 treatment (Additional file 1: Fig. S6).

In addition to the exploration of the immunoreactivity of MUT-A-inactivated HIV-1 with regard to this panel of neutralizing and non-neutralizing anti-Env antibodies, we studied if these inactivated viruses were suitable to elicit a $CD4^+$ and $CD8^+$ T-cells immune response when loaded ex vivo on MDDCs isolated from subjects infected by HIV-1. We measured a strong $CD4^+$ T helper cell proliferative response, comparable or even higher than that induced by non-treated viruses. This indicates that MUT-A-inactivated viruses are well recognized by professional antigen presenting cells. In contrast, the $CD8^+$ T cell proliferative response was much weaker, but the meaning of this was not obvious since we also observed a very weak CD8 response against the SAE super antigen that was used as positive control.

These results suggest that MUT-A treatment during virus production did not alter the conformation of epitopes at the surface of the virus. In fact, previous work has indicated that, for MDDC, small amounts of antigen are sufficient for presentation on MHC molecules and activation of T cells. Some of these approaches have been used with antigens from non-replicating viruses, including aldrithiol-2 (AT-2)-inactivated HIV-1 [41–43] or defective virus [44]. As we describe in this report, MUT-A-inactivated virus conserves conformational and functionally intact surface envelope proteins that can interact with T cells as well as with surface receptors present on circulating dendritic cells and monocytes. One of the advantages of MUT-A inactivation over other modes of inactivation such as AT-2 treatment is that MUT-A is a non-toxic ARV compound in development compared to AT-2 which is a toxic and denaturing agent. The potential use of AT-2-treated biological would be certainly prohibited in human. Also, the well documented observation that INLAI-inactivated virus enters target cells as wt virus and could accumulate and be degraded before the reverse transcription block might also confer an advantage in term of cellular immune response in vivo. Compared to HIV-1 inactivation by protease inhibitors, INLAI-inactivated viruses did not display any Gag maturation defect (Additional file 1: Fig. S2). These combined results make INLAI-treated virus an interesting reagent that could be exploited for vaccine research in animal models of HIV infection. Unfortunately, the lack of ARV activity of INLAIs against SIV precludes the use of the macaque model infected with SIV. The only possible model that could be used today for such studies would be the humanized mouse model infected with HIV-1, although the immune B cell response is somehow limited in the current humanized mouse models.

Conclusions

In conclusion, this report describes the characteristics of a new type of INLAI drug MUT-A, which consists of an original scaffold, a 5-atom thiophene ring, linked to carboxylic acid and tert-butylether moieties common to most INLAIs. We demonstrate that HIV-1 inactivated by MUT-A displays a strong defect in infectivity. However, this strong infectivity defect did not correlate with detectable change in the viral protein content, Gag maturation, viral RNA packaging and exogenous RT activity, but correlated with the formation of numerous viral particles containing eccentric condensates with empty cores. We described resistance mutations that strongly reduce the ARV activity of this class of compounds. In addition, from the point of view of both the B cell immunoreactivity against a panel of neutralizing and non-neutralizing anti-HIV-1 Env antibodies, as well as the T $CD4^+$ helper cell proliferative response, the MUT-A-inactivated HIV-1 behaves like a strong and efficient antigen capable to promote large B and T cells immune responses comparable or even higher than non-treated virus. Up to date the only structural difference that has been observed between INLAI-inactivated HIV-1 and non-treated virus is the formation of an eccentric core and misplacement of the genetic RNA material outside the viral core. The work reported here concerning the B and T cell immunoreactivities displayed by MUT-A inactivated HIV-1 suggests that the surface of such inactivated virions, which plays an essential role in eliciting B and T cell immune responses, seems to be unaltered. However, all this work has been performed with two selected HIV-1 isolates, the primary X4-using LAI isolate and the X4-using lab-adapted NL4-3 virus. It remains to be seen whether these observations could be extended to multiple HIV-1 primary isolates. If this is the case, it would be worth checking whether the use of such INLAI-inactivated viruses as immunogen in some adapted animal models suffices to raise a protective neutralizing antibody response against HIV and could be used as therapeutic adjuvant to ART that could be tested in a small animal model of HIV-1 infection. Interestingly, while this manuscript was in preparation, Vranckx and colleagues [18] published a paper showing that LEDGINs can relocate and retarget HIV integration in a HIV reservoir that is refractory to reactivation by different latency-reversing agents, supporting the notion that this INLAI class of compounds could reduce reactivation of residual latent HIV. In a commentary on this work, Mesplede and Wainberg [45] raise the intriguing possibility that LEDGINs could play a role in a cure for HIV infection. The work presented here, together with that of Vranckx et al. [18] suggests that it is worthwhile to explore these novel possibilities and to test whether INLAIs could be exploited as adjuvant for HIV cure.

Additional file

Additional file 1. Supplementary methods. Chemical synthesis process of MUT-A,HTRF®-based IN-LEDGF interaction assay, HTRF®-based IN multimerization assay, Viral RNA isolation, Northern blot analyses, primer extension assays, RT activity assays, Cryo-electron microscopy of HIV particles, Determination of HIV-1 replication capacity. **Figure S1:** 1H NMR spectrum of MUT-A. **Figure S2:** Impact of MUT-A on HIV-1 replication and production. **Figure S3:** Analysis of genomic RNA, Reverse transcriptase and tRNALys3 primer in MUT-A-treated HIV-1. A-D. HIV-1 RNA packaging and thermal stability of HIV-1 RNA dimers. **Figure S4:** HIV-1 NL4-3 particles observed by cryo-EM. A. **Figure S5:** Replicative capacity of HIV-1 NL4-3 viruses bearing resistance mutations to INLAIs or INSTIs. **Table S1:** Immunoreactivity of HIV-1 NL4-3 produced in the presence of MUT-A, or Saquinavir, or after AT2 treatment, or in the presence of DMSO. **Figure S6:** Results of virus immunocapture from Table S1 represented in bar graphs.

Authors' contributions

CA, YUVDV, NVB, performed the experiments and analyzed the data; DB, ELR, LMi, LMa, JB, FC, made substantial contributions to acquisition, analysis and interpretation of data; ELR, DB and CA performed experiments in MUT-A resistance profile, YUVDV and NVB performed viral RNA studies, CA, L Mi and L Ma performed immunology experiments; IO, DS, MR and BPK performed and analyzed the cryo-EM results, SC, BL, FM, FG, JG, AZ, SE, CM, have been involved in the design of the experiments, interpretation of data and revising the manuscript critically; BB, MP and RB designed and coordinated the full study and drafted the manuscript. All authors read and approved the final manuscript.

Author details

[1] Biodim Mutabilis, 93230 Romainville, France. [2] INSERM, U1016, Institut Cochin, Paris, France. [3] CNRS, UMR8104, Paris, France. [4] University Paris Descartes, Sorbonne Paris Cité, Paris, France. [5] Laboratory of Experimental Virology, Department of Medical Microbiology, Center for Infection and Immunity Amsterdam (CINIMA), Academic Medical Center, University of Amsterdam, Amsterdam, The Netherlands. [6] Centre for Integrative Biology, IGBMC, CNRS, INSERM, University of Strasbourg, Strasbourg, France. [7] AIDS Research Group, IDIBAPS, Hospital Clinic, Barcelona, Spain. [8] INSERM U1109, Strasbourg, France. [9] CNRS, UMR7212, INSERM U944, Université Paris Diderot, Conservatoire National des Arts et Métiers, Paris, France. [10] Present Address: 19 rue de Croulebarbe, 75013 Paris, France.

Acknowledgements

The authors thank Pr. R Desrosiers for the gift of SIVmac 239, Stéphane Huguet for support, Juliette Nguyen, Roxane Beauvoir and Armelle Walton for assistance in virology, and the NIH AIDS research repository for reagents. Celine Amadori is a recipient of a CIFRE doctoral fellowship from the Association National Recherche Technologie (ANRT). We also thank all the patients who gave their informed consent for this study.

Competing interests

CA, DB, ELR, JB, FC, SC, BL, FM, RB are or were employees of Biodim Mutabilis. YUVDV, IO, NVB, LMi, DS, LMa, FG, JG, AZ, SE, MR, BPK, CM, BB and MP declare that they have no competing interests.

Funding

This work was supported by Biodim Mutabilis under the Authorization Number DUO 2145, assigned by the French Ministry of Research for work with genetically modified organisms, and by Grants from EU FP7 to Biodim, BB, SE and AZ under the HIVINNOV consortium, Grant Agreement No. 305137, from the Eurostar Grant ResistAids to Biodim and BB (Grant Agreement No. E!10239) and by the French Infrastructure for Integrated Structural Biology (FRISBI) ANR-10-INSB-05-01, and Instruct as part of the European Strategy Forum on Research Infrastructures (ESFRI). Felipe García team was supported by Grants SAF2015-66193-R, SAF2012-39075, Montserrat Plana by Grant FIS PI15/00480 and Nikki van Bel by NWO-CW (TOP Grant).

References

1. Lesbats P, Engelman AN, Cherepanov P. Retroviral DNA integration. Chem Rev. 2016;116:12730–57.
2. Engelman A, Cherepanov P. The lentiviral integrase binding protein LEDGF/p75 and HIV-1 replication. PLoS Pathog. 2008;4:e1000046.
3. Sowd GA, Serrao E, Wang H, Wang W, Fadel HJ, Poeschla EM, et al. A critical role for alternative polyadenylation factor CPSF6 in targeting HIV-1 integration to transcriptionally active chromatin. Proc Natl Acad Sci. 2016;113:E1054–63.
4. Cherepanov P, Maertens G, Proost P, Devreese B, Jozef VB, Engelborghs Y, et al. HIV-1 integrase forms stable tetramers and associates with LEDGF/p75 protein in human cells. J Biol Chem. 2003;278:372–81.
5. Emiliani S, Mousnier A, Busschots K, Maroun M, Van Maele B, Tempé D, et al. Integrase mutants defective for interaction with LEDGF/p75 are impaired in chromosome tethering and HIV-1 replication. J Biol Chem. 2005;280:25517–23.
6. Christ F, Voet A, Marchand A, Nicolet S, Desimmie BA, Marchand D, et al. Rational design of small-molecule inhibitors of the LEDGF/p75-integrase interaction and HIV replication. Nat Chem Biol. 2010;6:442–8.
7. Kessl JJ, Jena N, Koh Y, Taskent-Sezgin H, Slaughter A, Feng L, et al. Multimode, cooperative mechanism of action of allosteric HIV-1 integrase inhibitors. J Biol Chem. 2012;287:16801–11.
8. Tsiang M, Jones GS, Niedziela-Majka A, Kan E, Lansdon EB, Huang W, et al. New class of HIV-1 integrase (IN) inhibitors with a dual mode of action. J Biol Chem. 2012;287:21189–203.
9. Balakrishnan M, Yant SR, Tsai L, O'Sullivan C, Bam RA, Tsai A, et al. Non-catalytic site HIV-1 integrase inhibitors disrupt core maturation and induce a reverse transcription block in target cells. PLoS ONE. 2013;8:e74163.
10. Fenwick C, Amad M, Bailey MD, Bethell R, Bös M, Bonneau P, et al. Preclinical profile of BI 224436, a novel HIV-1 non-catalytic-site integrase inhibitor. Antimicrob Agents Chemother. 2014;58:3233–44.
11. Sharma A, Slaughter A, Jena N, Feng L, Kessl JJ, Fadel HJ, et al. A new class of multimerization selective inhibitors of HIV-1 integrase. PLoS Pathog. 2014;10:e1004171.
12. Le Rouzic E, Bonnard D, Chasset S, Bruneau J-M, Chevreuil F, Le Strat F, et al. Dual inhibition of HIV-1 replication by integrase-LEDGF allosteric inhibitors is predominant at the post-integration stage. Retrovirology. 2013;10:144.
13. Christ F, Shaw S, Demeulemeester J, Desimmie BA, Marchand A, Butler S, et al. Small-molecule inhibitors of the LEDGF/p75 binding site of integrase block HIV replication and modulate integrase multimerization. Antimicrob Agents Chemother. 2012;56:4365–74.
14. Desimmie BA, Schrijvers R, Demeulemeester J, Borrenberghs D, Weydert C, Thys W, et al. LEDGINs inhibit late stage HIV-1 replication by modulating integrase multimerization in the virions. Retrovirology. 2013;10:57.
15. Jurado KA, Wang H, Slaughter A, Feng L, Kessl JJ, Koh Y, et al. Allosteric integrase inhibitor potency is determined through the inhibition of HIV-1 particle maturation. Proc Natl Acad Sci USA. 2013;110:8690–5.
16. Fontana J, Jurado KA, Cheng N, Ly NL, Fuchs JR, Gorelick RJ, et al. Distribution and redistribution of HIV-1 nucleocapsid protein in immature, mature, and integrase-inhibited virions: a role for integrase in maturation. J Virol. 2015;89:9765–80.
17. Kessl JJ, Kutluay SB, Townsend D, Rebensburg S, Slaughter A, Larue RC, et al. HIV-1 integrase binds the viral RNA genome and is essential during virion morphogenesis. Cell. 2016;166(1257–1268):e12.
18. Vranckx LS, Demeulemeester J, Saleh S, Boll A, Vansant G, Schrijvers R, et al. LEDGIN-mediated inhibition of Integrase–LEDGF/p75 interaction reduces reactivation of residual latent HIV. EBioMedicine. 2016;8:248–64.
19. Girard MP, Osmanov S, Assossou OM, Kieny M-P. Human immunodeficiency virus (HIV) immunopathogenesis and vaccine development: a review. Vaccine. 2011;29:6191–218.
20. Garcia F, Routy J-P. Challenges in dendritic cells-based therapeutic vaccination in HIV-1 infection. Vaccine. 2011;29:6454–63.
21. Garcia F, Climent N, Guardo AC, Gil C, Leon A, Autran B, et al. A dendritic cell-based vaccine elicits T cell responses associated with control of HIV-1 replication. Sci Transl Med. 2013;5:166ra2.
22. Chasset S, Chevreuil F, Ledoussal B, Le Strat F, Benarous R. Inhibitors of viral replication, their process of preparation and their therapeutical uses. WO2014/053666A1. 2014.
23. van Bel N, van der Velden Y, Bonnard D, Le Rouzic E, Das AT, Benarous R, et al. The allosteric HIV-1 Integrase inhibitor BI-D affects virion maturation but does not influence packaging of a functional RNA genome. PLoS ONE. 2014;9:e103552.
24. Rutebemberwa A, Bess JW, Brown B, Arroyo M, Eller M, Slike B, et al. Evaluation of aldrithiol-2-inactivated preparations of HIV type 1 subtypes A, B, and D as reagents to monitor T cell responses. AIDS Res Hum Retroviruses. 2007;23:532–42.
25. Kodama T, Wooley DP, Naidu YM, Kestler HW, Daniel MD, Li Y, et al. Significance of premature stop codons in env of simian immunodeficiency virus. J Virol. 1989;63:4709–14.
26. Moog C, Dereuddre-Bosquet N, Teillaud J-L, Biedma ME, Holl V, Van Ham G, et al. Protective effect of vaginal application of neutralizing and nonneutralizing inhibitory antibodies against vaginal SHIV challenge in macaques. Mucosal Immunol. 2014;7:46–56.
27. Wu X, Yang Z-Y, Li Y, Hogerkorp C-M, Schief WR, Seaman MS, et al. Rational design of envelope identifies broadly neutralizing human monoclonal antibodies to HIV-1. Science. 2010;329:856–61.
28. Walker LM, Phogat SK, Chan-Hui P-Y, Wagner D, Phung P, Goss JL, et al. Broad and potent neutralizing antibodies from an African donor reveal a new HIV-1 vaccine target. Science. 2009;326:285–9.
29. Zhou T, Xu L, Dey B, Hessell AJ, Van Ryk D, Xiang S-H, et al. Structural definition of a conserved neutralization epitope on HIV-1 gp120. Nature. 2007;445:732–7.
30. Trkola A, Purtscher M, Muster T, Ballaun C, Buchacher A, Sullivan N, et al. Human monoclonal antibody 2G12 defines a distinctive neutralization epitope on the gp120 glycoprotein of human immunodeficiency virus type 1. J Virol. 1996;70:1100–8.
31. Muster T, Steindl F, Purtscher M, Trkola A, Klima A, Himmler G, et al. A conserved neutralizing epitope on gp41 of human immunodeficiency virus type 1. J Virol. 1993;67:6642–7.
32. Stiegler G, Kunert R, Purtscher M, Wolbank S, Voglauer R, Steindl F, et al. A potent cross-clade neutralizing human monoclonal antibody against a novel epitope on gp41 of human immunodeficiency virus type 1. AIDS Res Hum Retroviruses. 2001;17:1757–65.
33. Zwick MB, Labrijn AF, Wang M, Spenlehauer C, Saphire EO, Binley JM, et al. Broadly neutralizing antibodies targeted to the membrane-proximal external region of human immunodeficiency virus type 1 glycoprotein gp41. J Virol. 2001;75:10892–905.
34. Lorin V, Mouquet H. Efficient generation of human IgA monoclonal antibodies. J Immunol Methods. 2015;422:102–10.
35. Kwong PD, Mascola JR. Human antibodies that neutralize HIV-1: identification, structures, and B cell ontogenies. Immunity. 2012;37:412–25.
36. Burrer R, Haessig-Einius S, Aubertin A-M, Moog C. Neutralizing as well as non-neutralizing polyclonal immunoglobulin (Ig)G from infected patients capture HIV-1 via antibodies directed against the principal immunodominant domain of gp41. Virology. 2005;333:102–13.
37. Nyambi PN, Burda S, Bastiani L, Williams C. A virus binding assay for studying the antigenic landscape on intact, native, primary human immunodeficiency virus-type 1. J Immunol Methods. 2001;253:253–62.

38. Climent N, Guerra S, García F, Rovira C, Miralles L, Gómez CE, et al. Dendritic cells exposed to MVA-based HIV-1 vaccine induce highly functional HIV-1-specific CD8(+) T cell responses in HIV-1-infected individuals. PLoS ONE. 2011;6:e19644.
39. Peden K, Emerman M, Montagnier L. Changes in growth properties on passage in tissue culture of viruses derived from infectious molecular clones of HIV-1LAI, HIV-1MAL, and HIV-1ELI. Virology. 1991;185:661–72.
40. Zhang W, Cao S, Martin JL, Mueller JD, Mansky LM. Morphology and ultrastructure of retrovirus particles. AIMS Biophys. 2015;2:343–69.
41. Bender A, Bui LK, Feldman MA, Larsson M, Bhardwaj N. Inactivated influenza virus, when presented on dendritic cells, elicits human $CD8^+$ cytolytic T cell responses. J Exp Med. 1995;182:1663–71.
42. Larsson M, Fonteneau J-F, Lirvall M, Haslett P, Lifson JD, Bhardwaj N. Activation of HIV-1 specific CD4 and CD8 T cells by human dendritic cells: roles for cross-presentation and non-infectious HIV-1 virus. AIDS. 2002;16:1319–29.
43. Buseyne F, Le Gall S, Boccaccio C, Abastado JP, Lifson JD, Arthur LO, et al. MHC-I-restricted presentation of HIV-1 virion antigens without viral replication. Nat Med. 2001;7:344–9.
44. Guardo AC, Alvarez-Fernández C, Arberas H, García-Pérez J, García F, Bargalló ME, et al. Use of RT-defective HIV virions: new tool to evaluate specific response in chronic asymptomatic HIV-infected individuals. PLoS ONE. 2013;8:e58927.
45. Mesplède T, Wainberg MA. Will LEDGIN molecules be able to play a role in a cure for HIV infection? EBioMedicine. 2016;8:14–5.

12

Characterization update of HIV-1 M subtypes diversity and proposal for subtypes A and D sub-subtypes reclassification

Nathalie Désiré[1,2], Lorenzo Cerutti[3], Quentin Le Hingrat[4,5], Marine Perrier[4,5], Stefan Emler[3], Vincent Calvez[1,2], Diane Descamps[4,5], Anne-Geneviève Marcelin[1,2], Stéphane Hué[6] and Benoit Visseaux[4,5*]

Abstract

Background: The large and constantly evolving HIV-1 pandemic has led to an increasingly complex diversity. Because of some taxonomic difficulties among the most diverse HIV-1 subtypes, and taking advantage of the large amount of sequence data generated in the recent years, we investigated novel lineage patterns among the main HIV-1 subtypes.

Results: All HIV full-length genomes available in public databases were analysed (n = 2017). Maximum likelihood phylogenies and pairwise genetic distance were obtained. Clustering patterns and mean distributions of genetic distances were compared within and across the current groups, subtypes and sub-subtypes of HIV-1 to detect and analyse any divergent lineages within previously defined HIV lineages. The level of genetic similarity observed between most HIV clades was deeply consistent with the current classification. However, both subtypes A and D showed evidence of further intra-subtype diversification not fully described by the nomenclature system at the time and could be divided into several distinct sub-subtypes.

Conclusions: With this work, we propose an updated nomenclature of sub-types A and D better reflecting their current genetic diversity and evolutionary patterns. Allowing a more accurate nomenclature and classification system is a necessary step for easier subtyping of HIV strains and a better detection or follow-up of viral epidemiology shifts.

Keywords: HIV, Nomenclature, Viral diversity, Phylogenetic, Evolution

Background

HIV-1 presents an extraordinary degree of genetic diversity and has been classified, based on phylogenetic clustering, into groups, subtypes and sub-subtypes. While groups correspond to distinct lineages independently introduced into the human population from non-human apes, subtypes and sub-subtypes results from post-introduction founder events and further diversification. These distinctions have led to the formal recognition of 4 groups (M, N, O and P), 9 group M subtypes (A, B, C, D, F, G, H, J and K) and several sub-subtypes (A1, A2 for subtype A and F1, F2 for subtype F) [1].

HIV classification is used routinely by all medical laboratories performing genotyping resistance testing to characterize patient's strain, to detect reinfection and to analyse viral epidemiology shifts [2, 3]. HIV subtypes can also present different antiretroviral drug or vaccine response [4–6] [7, 8], disease progression [9–11] or transmission rates [12–14]. All of these observations rely on an up-to-date and representative classification system, reflecting as best as possible the true and constantly evolving diversity of HIV-1 strains. Thus, since HIV-1 is rapidly and continuously evolving, it is important to keep track of the diversification of the virus on a regular basis.

Due to some routinely experienced classification difficulties, particularly with subtype A for which only sub-subtypes A1 and A2 were formally retained before this work, we reanalysed the global HIV-1 group M diversity

*Correspondence: benoit.visseaux@aphp.fr
[4] IAME, UMR 1137, Université Paris Diderot, INSERM, Paris, France
Full list of author information is available at the end of the article

among all subtypes, at exclusion of recombinant forms, using all available near full genome sequences and combining a systematic phylogenetic distance analysis with maximum likelihood phylogenetic analyses. We identified significant divergence patterns not fully included in the classification at the time among two subtypes, A and D. Thus, we propose a new sub-subtypes classification for these two subtypes to better describe their diversity and provide a more accurate description of their currently observed genetic diversity.

Methods

Sequence data

All HIV-1 groups M, N, O and P near full-genome sequences with available subtype information were downloaded from the Los Alamos National Laboratory (LANL) database on June 2016. All LANL partial *gag, pol* and *env* gene sequences of subtypes A and D were also collected in specific datasets. CRF (circulating recombinant form) sequences as well as clonal or iterative sequences were excluded from the analysis. As the A3 and A4 sub-subtype were not formally retained into the LANL classification at the time of this analysis, all subtype A sequences were manually re-annotated according to the corresponding publications [15, 16]. The country of sampling was also retrieved for each included sequence.

Alignments

After removing LTR regions, sequence datasets were aligned using Mafft version 7 with default settings [17] and manually edited using Aliview [18]. Potential recombinants were identified using the package RDP4 [19] and excluded. All genome regions that could not be unequivocally aligned (e.g. highly variable *env* regions) were removed from the alignments. All sequences retained in the final are depicted in Additional file 1: Table S1.

Pairwise genetic distance calculations

Pairwise genetic distances were calculated using HyPhy 2.2.4 with the GTR model of nucleotide substitutions and a gamma distribution with 4 parameters [20]. This model was chosen using jModelTest 2 [21]. The distributions of near full genome pairwise genetic distance within and between known groups, subtypes and sub-subtypes were plotted using R 3.3.2 and ggplot2 package [22] to describe genetic distance ranges and net genetic divergence observed for intergroup, intersuptypes, intersub-subtypes and intrasub-subtypes comparisons.

Phylogenetic analyses

In case of large genetic distribution for a particular subtype or sub-subtype, compatible with the existence of various sub-subtypes not retained in the current HIV nomenclature, neighbour joining and maximum likelihood phylogenies were reconstructed using PhyML 3.0 [23] using the GTR model of nucleotide substitutions with a gamma distribution with 4 parameters (GTR-G). Maximum likelihood phylogenies were reconstructed using PhyML 3.0 [23] with GTR-G. An initial NJ tree calculated by default by PhyML was used with both NNI (Next Neighbour Inversion) and SPR (Subtree Pruning and Regrafting) for the tree-space exploration. Branch support was estimated by bootstrap analysis with 1000 replicates. Observed clades strongly supported (i.e. above 90%) and presenting genetic distance in the range of sub-subtype area according to our analysis were used to propose a better classification of these subtypes.

According to current guidelines, a new lineage must include three non-directly linked viral isolates and at least two must be a full genome sequence to be retained as a new subtype or sub-subtype [1]. Thus, when a single near full genome sequence was suggestive of a previously non-described sub-subtype, phylogenies of all available *gag, pol* and *env* sequences were used to identify if sequences from other isolates of the same clade were available.

Results

A total of 2017 near full genome sequences were included in the study, representing 204, 1185, 488, 66, 29, 39, 4, 3 and 2 sequences for A, B, C, D, F, G, H, J and K subtypes, respectively.

Near full genome pairwise genetic distance distributions showed patterns consistent with the classification at the time, delimiting ranges of distances within which inter-group, inter-subtypes, inter-sub-subtypes and intra-sub-subtypes genetic diversities almost consistently fall (Fig. 1a–d). Indeed, using these ranges observed in our dataset, two clades, A1 and D, were identified as presenting intra-clade genetic distances inconsistent with the range expected from their respective classification. This highlighted their large diversity and the existence of distinct sub-subtypes within both of them (Fig. 1e).

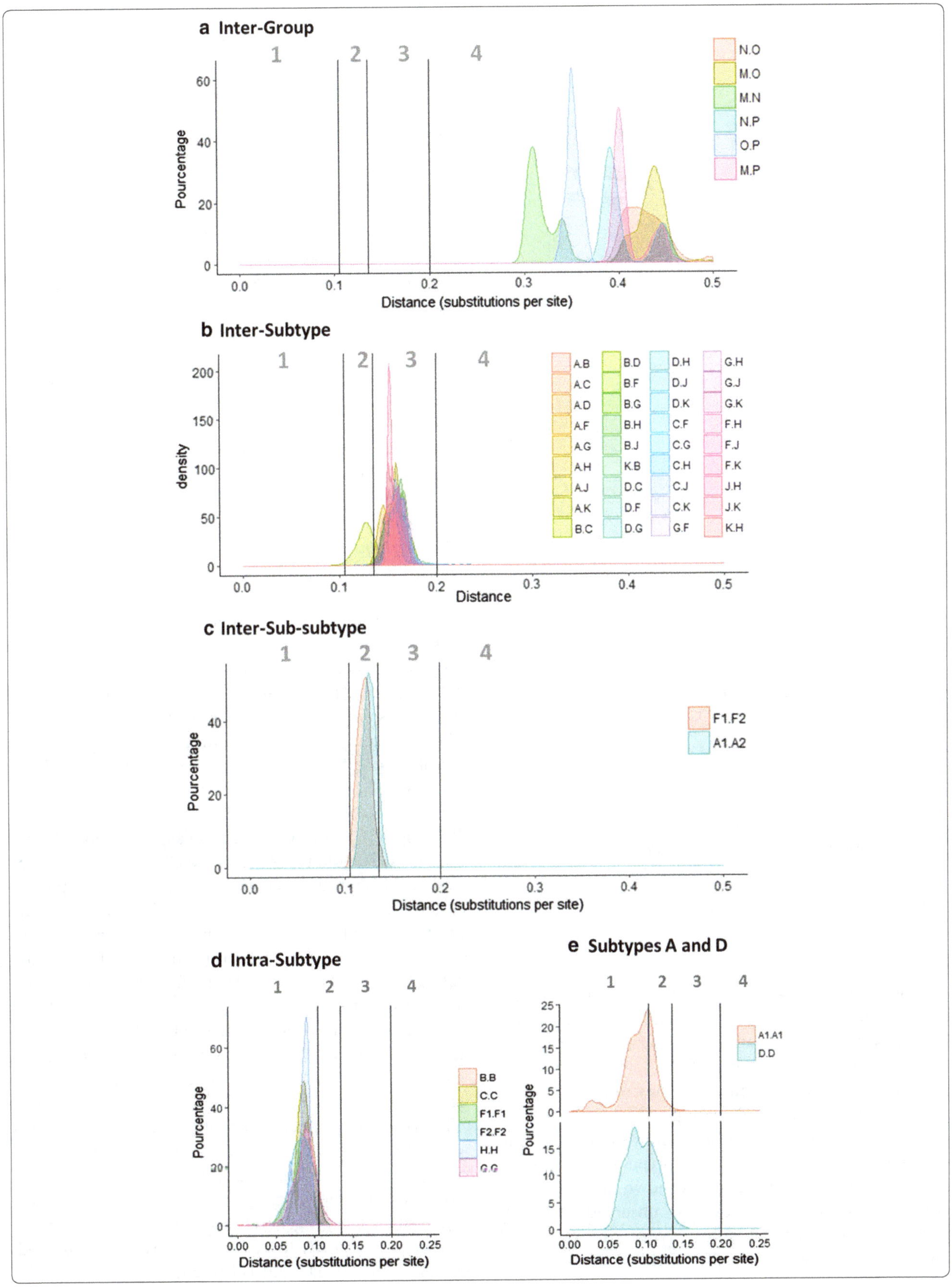
a Inter-Group
1
2
3
4
N.O
M.O
M.N
N.P
O.P
M.P
Pourcentage
Distance (substitutions per site)
b Inter-Subtype
A.B
A.C
A.D
A.F
A.G
A.H
A.J
A.K
B.C
B.D
B.F
B.G
B.H
B.J
K.B
D.C
D.F
D.G
D.H
D.J
D.K
C.F
C.G
C.H
C.J
C.K
G.F
G.H
G.J
G.K
F.H
F.J
F.K
J.H
J.K
K.H
density
Distance
c Inter-Sub-subtype
F1.F2
A1.A2
Pourcentage
Distance (substitutions per site)
d Intra-Subtype
B.B
C.C
F1.F1
F2.F2
H.H
G.G
Pourcentage
Distance (substitutions per site)
e Subtypes A and D
A1.A1
D.D
Pourcentage
Distance (substitutions per site)

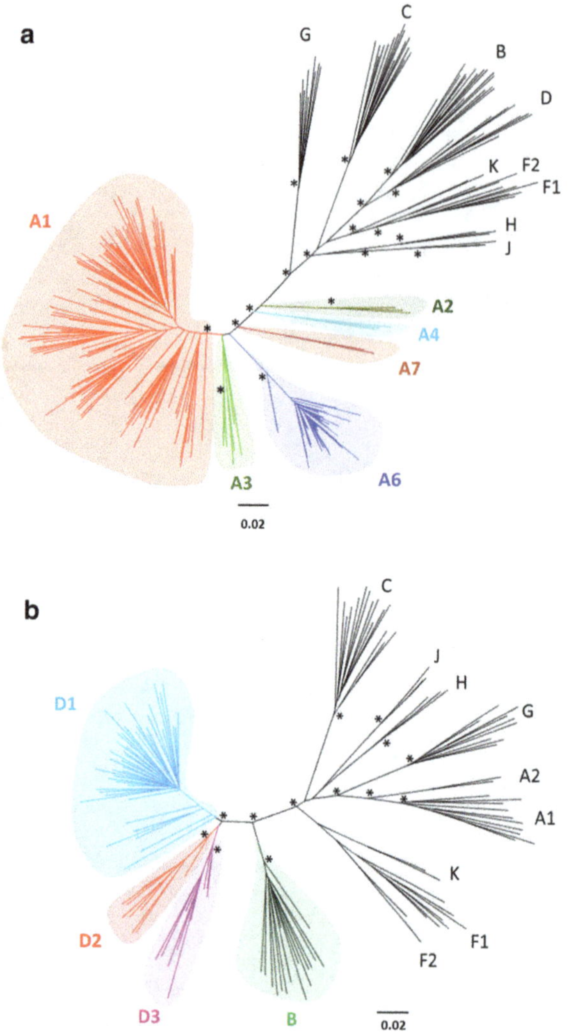

Fig. 2 Subtypes A (**a**) and D (**b**) near full genome sequences maximum likelihood phylogenetic trees. Names of all sub-subtypes are indicated according to the new classification proposal. For all identified sub-subtypes and the other major clades, i.e. subtypes, nodes presenting a branch support > 95% (bootstrap analysis using 1000 replicates) are indicated by an asterisk

To note, and as previously described, genetic distances observed between subtypes B and D were smaller than those observed between all others inter-subtype comparisons but felt within the range of inter-sub-subtype comparisons (Fig. 1b), confirming that they should represent two sub-subtypes of a single subtype.

To assess possible classification improvements within subtype A, a phylogeny of all subtype A full genome sequences was constructed and is shown in Fig. 2a. Several sub-subtypes candidates were observed: the well-established A1 and A2 sub-subtypes, the described but previously not formally retained A3 and A4 sub-subtypes, the former FSU-cluster (herein called A6) and a well separated cluster of 2 sequences (herein called A7). All these clades were supported by a strong branch support at 100%. Four sequences also clustered with the A3 sub-subtypes but were poorly supported (branch support of 3%) and could not be unambiguously included into A3 nor any of the observed clades in the near full genome nor *gag*, *pol* and *env* specific phylogenies. The same tree topology was observed with both neighbour-joining phylogenetic and maximum likelihood reconstruction. To assess if A3 and A4 not formally retained sub-subtypes, as well as FSU-cluster (herein renamed A6) and the newly described A7 clade, all felt in the range of formally retained sub-subtypes, near full genome pairwise genetic distances were calculated. Their distribution supported all these A sub-subtype as candidates with the exception of the A3 sub-subtype presenting a genetic distance distribution with A1 slightly lower than other sub-subtypes (Additional file 1: Fig. S1, Tables S2 and S3). The same patterns of phylogenetic tree topology and genetic distance distributions were also observed when analysing the *gag*, *pol* and *env* genes separately.

Concerning the A7 clade, as only two near full genome sequences are currently available and as at least three sequences, including two full genome sequences, are required to describe a new sub-subtype [1], all available partial *pol* (1329 bp), *gag* (1466 bp) and *env* (2566 bp) genes from subtype A were retrieved from the LANL database (n = 244, 387 and 399 sequences, respectively) and corresponding phylogenies were constructed. Four additional sequences were identified as belonging to the A7 clade, one from the *pol* tree (DQ273915; Additional file 1: Fig. S2) and three from the *gag* tree (Accession number JF683739, EU673432, EU673431; Additional file 1: Fig. S3), allowing to retain A7 as a new sub-subtype.

All these sub-subtypes presented slightly different geographical dispersal as A1 strains were mostly sampled in Eastern Africa from Kenya, Uganda, Tanzania, Rwanda and Cyprus (113/132); A2 in Cyprus, Eastern and Central Africa(3/3); A3 in Senegal (3/3), A4 in Democratic Republic of Congo (3/3); A6 in former Soviet Union countries (Russia, Ukraine, Uzbekistan, etc.) (44/52); A7 in Nigeria for strains identified with the complete genome analysis (2/2), in Cyprus for strains identified with gag gene (3/3) and in Nigeria for strains identified with pol gene (1/1).

To assess if distinct sub-subtypes can be identified within the subtype D, the phylogenetic tree obtained for subtype D near full-length genome sequences was constructed and is shown in Fig. 2b. Within subtype D, characterised by a high genetic diversity in the same range than for subtype A, three main clades were observed, herein called D1, D2 and D3 (branch support for 100%

for all of them). The proposed D1 clade is composed of a large clade of sequences with a few outsider groups of sequences located nearer of the most recent common ancestor of the D subtype. The same tree topology was observed with both neighbour-joining phylogenetic and maximum likelihood reconstruction. Most of these outsider sequences were sampled in Uganda as for the major clade of D1 and, thus, were also retained into the proposed D1 clade for genetic distance analysis. Genetic distances observed between D1, D2 and D3 sub-subtype candidates showed large distributions, largely compatible with inter-sub-subtype range but also overlapping the intra-sub-subtype range (Additional file 1: Fig. S4 and Tables S2, S3). This large distribution is explained by the few outsider sequences located nearer to the D common ancestor. No other classification choices provided better genetic distance distribution and outsider sequences always presented very close genetic distance to the main groups, preventing to process them as separate sub-subtypes. Despite these outsider sequences, and because of the large genetic distance observed between the three D sub-subtype candidates, we propose to include D1, D2 and D3 into the HIV nomenclature. The same patterns of phylogenetic tree topology and genetic distance distributions were also observed when analysing the *gag*, *pol* and *env* genes separately.

These three sub-subtype candidates presented distinct geographic distribution as D1 strains were sampled in Eastern Africa, mostly in Uganda (38/46); D2 was sampled in Democratic Republic of Congo, South Africa and Brazil (10/10); D3 was mostly sampled in Cameroun, Chad and Republic Democratic of Congo (7/10).

All these observations have also been confirmed by separate *gag*, *pol* and *env* genes for both subtypes A and D (Additional file 1: Figs. S5, S6, S7). However, genetic distance observed for *pol* with the subtype D, if within the range of inter-sub-subtypes comparisons, were not as well separated than with the other genes or full genome analyses (Additional file 1: Fig. S8).

Discussion

The large HIV genetic diversity is a challenging reflection of HIV complex and changing epidemiology leading to an evolving nomenclature and sometime unresolved issues. Given the amount of sequence data regularly and increasingly generated in the recent years, we re-investigated HIV-1 genetic diversity of all subtypes initially to resolve some issues within subtype A. We propose to slightly modify the subtype A classification by dividing it into 6 sub-subtypes, called A1, A2, A3, A4, A6 and A7. We also propose to divide the highly diverse subtype D into 3 sub-subtypes named D1 to D3.

Thus, at the time of this work, subtype A was formally divided into only two sub-subtypes A1 and A2 and two other sub-subtypes, A3 and A4, were also proposed but not formally retained in the Los Alamos National Laboratory nomenclature. A partially described A5 sub-subtype has also been previously described based on analysis of recombinant viruses but no prototype of this strain has been identified to date [24]. In our analysis, A3 and A4 confirmed to have a degree of diversity compatible with sub-subtypes definition. Interestingly, the Former Soviet Union (FSU) cluster, resulting from the introduction of HIV-1 subtype A in intravenous drug users population of the former Soviet Union countries in the 80s [25], also depicted a high degree of diversity compatible with sub-subtype and, thus, is proposed in this work as a separate sub-subtype called A6. Before the presentation of this part of the current study at the International AIDS Society conference 2017, A2, A3 and the FSU cluster were only identified as sub-subtype A1 by available subtyping tools. Since then, A3, A4 and A6 sub-subtypes have been formally implemented into the Los Alamos National Laboratory HIV database and allow a quick and more accurate description of the subtype A diversity. In this work we also added the description of the A7 sub-subtype that was not previously described. We also enlarged the analysis to all other HIV-1 subtypes and identified evidence of further diversification consistent with distinct sub-subtypes among the subtype D. Thus, 3 sub-subtypes are proposed for this latter subtype, called D1 to D3, allowing a better description of D diversity despite the existence of a few outsider sequences. Their classification may have to continue to evolve in the future depending of the potential acquisition of new sequences.

Due to the constant evolution of HIV and regular addition of new sequences in public databases, uncertainties about HIV-1 classification have regularly emerged. For example, the previous classification proposal did no resolve whether subtypes F and K represented two different subtypes or sub-subtypes [1]. Our results support the notion that they represent two different subtypes, since they exhibit genetic distances within in the range of other inter-subtype comparisons. A few previous studies also reported large genetic diversity among existing subtypes, as for subtype C [26, 27]. However, in our current work we did not find enough genetic variations within any other subtypes than subtypes A and D to define specific clades according to the current classification rules.

The classification changes proposed in the current work (Fig. 3) help to provide an up-to-date description of HIV-1 group M heterogeneity and, if some challenges remains, will help to more accurate descriptions of

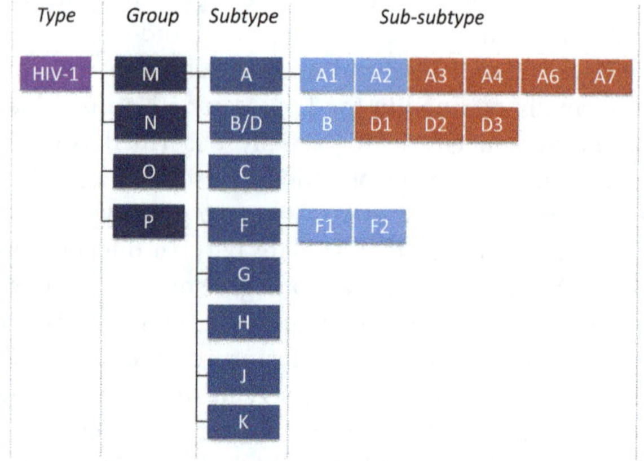

Fig. 3 HIV-1 updated classification proposal. HIV-1 remains classified into groups, subtypes and sub-subtypes, with changes from the previous classification colored in red

HIV-1 diversification and to keep a better track of HIV-1 upcoming evolutions.

Additional file

Additional file 1: Fig. S1. Full genome genetic distance comparisons between HIV-1 subtype A sub-subtypes according to our classification proposal. X-axis scale lines indicate genetic distance thresholds allowing, in our alignment and model conditions, for group, subtype and sub-subtype identification. **Fig. S2.** Phylogenetic tree of sub-type A obtained with *pol* gene. Sequences from A1, A2, A3, A4 and A6 clades have been collapsed for readability. One *pol* sequence was identified in the A7 clade in addition to the two full genome sequences and is highlighted by an arrow. The tree has been obtained with PhyML 3.0, using GTR-G nucleotide substitution model and branch support obtained by bootstrap method is given for each node. Several sequences, depicted in black, clustered outside the defined clades but cannot be retained in the classification proposal because of poor branch support values or absence of available full genome sequences. **Fig. S3.** Phylogenetic tree of sub-type A obtained with *gag* gene. Sequences from A1, A2, A3, A4 and A6 clades have been collapsed for readability. One pol sequence was identified in the A7 clade in addition to the two full genome sequences and is highlighted by an arrow. The tree has been obtained with PhyML 3.0, using GTR-G nucleotide substitution model and branch support obtained by bootstrap method is given for each node. Several sequences, depicted in black, clustered outside the previously defined clades but cannot be retained in the classification proposal because of poor branch support values or absence of available full genome sequences. **Fig. S4.** Genetic distance comparisons between HIV-1 subtype D sub-subtypes according to our new classification proposal. X-axis scale lines indicate genetic distance thresholds allowing, in our alignment and model conditions, for group, subtype and sub-subtype identification. **Fig. S5.** Genetic distance comparisons between HIV-1 groups, subtypes and sub-subtypes using *pol* sequences. X-axis scale lines indicate genetic distance thresholds allowing, in our alignment and model conditions, for group, subtype and sub-subtype identification. Genetic distance ranges compatible with intra-sub-subtype, inter-sub-subtype, inter-subtype and inter-group comparisons are indicated by the numbers 1, 2, 3 and 4, respectively. **Fig. S6.** Subtypes A (A) and D (B) *pol* sequences maximum likelihood phylogenetic trees. Names of all sub-subtypes are indicated according to the new classification proposal. **Fig. S7.** *Pol* gene genetic distance comparisons between HIV-1 subtype A sub-subtypes according to our classification proposal. X-axis scale lines indicate genetic distance thresholds allowing, in our alignment and model conditions, for group, subtype and sub-subtype identification. **Fig. S8.** *Pol* gene genetic distance comparisons between HIV-1 subtype D sub-subtypes according to our classification proposal. X-axis scale lines indicate genetic distance thresholds allowing, in our alignment and model conditions, for group, subtype and sub-subtype identification. **Table S1.** Full genome sequence used for our analysis and the corresponding clade in our classification proposal. **Table S2.** Full genome genetic distance distributions observed within each corresponding clade. **Table S3.** Net genetic divergence between each identified sub-subtypes. The "net genetic divergence" between each identified sub-subtypes within corresponding subtypes, which also takes into account the within-sub-subtype diversity, was calculated as follow: if $d_{x,y}$ is the average genetic divergence between two sub-subtypes, x and y, and d_x and d_y are the genetic diversities within populations x and y, respectively, net divergence, $D_{x,y}$, is given by the expression $D_{x,y} = d_{x,y} - (d_x + d_y)/2$.

Authors' contributions

SH, BV, ND, LC and SE participated to the conception of this study. ND, LC, QLH and MP participated to acquisition and analysis of data. SH, BV, VC and DD participated to the interpretation of the results. ND, SH and BV participated to the drafting of the manuscript. All authors read and approved the final manuscript.

Author details

[1] INSERM, Institut Pierre Louis d'Epidémiologie et de Santé Publique (iPLESP), Sorbonne Université, 75013 Paris, France. [2] AP-HP, Hôpital Pitié-Salpêtrière, Virologie, 75013 Paris, France. [3] SmartGene Services, EPFL Innovation Park, 1015 Lausanne, Switzerland. [4] IAME, UMR 1137, Université Paris Diderot, INSERM, Paris, France. [5] AP-HP, Hôpital Bichat, Virologie, 75018 Paris, France. [6] Department of Epidemiology & Population Health, London School of Hygiene and Tropical Medicine, London WC1E 7HT, UK.

Competing interests

The authors have no competing interest that could bias this work.

Funding

This study did not receive any specific funding.

References

1. Robertson DL, Anderson JP, Bradac JA, Carr JK, Foley B, Funkhouser RK, et al. HIV-1 nomenclature proposal. Science. 2000;288:55.
2. Ambrosioni J, Sued O, Nicolas D, Parera M, López-Diéguez M, Romero A, et al. Trends in transmission of drug resistance and prevalence of non-B subtypes in patients with acute or recent HIV-1 infection in Barcelona in the last 16 Years (1997–2012). PLoS ONE. 2015;10:e0125837.
3. Descamps D, Assoumou L, Chaix M-L, Chaillon A, Pakianather S, de Rougemont A, et al. National sentinel surveillance of transmitted drug resistance in antiretroviral-naive chronically HIV-infected patients in France over a decade: 2001–2011. J Antimicrob Chemother. 2013;68:2626–31.
4. Abecasis AB, Deforche K, Snoeck J, Bacheler LT, McKenna P, Carvalho AP, et al. Protease mutation M89I/V is linked to therapy failure in patients infected with the HIV-1 non-B subtypes C, F or G. AIDS Lond Engl. 2005;19:1799–806.
5. Abecasis AB, Deforche K, Bacheler LT, McKenna P, Carvalho AP, Gomes P, et al. Investigation of baseline susceptibility to protease inhibitors in HIV-1 subtypes C, F, G and CRF02_AG. Antivir Ther. 2006;11:581–9.
6. Brenner B, Turner D, Oliveira M, Moisi D, Detorio M, Carobene M, et al. A V106 M mutation in HIV-1 clade C viruses exposed to efavirenz confers cross-resistance to non-nucleoside reverse transcriptase inhibitors. AIDS Lond Engl. 2003;17:F1–5.
7. Weaver EA, Lu Z, Camacho ZT, Moukdar F, Liao H-X, Ma B-J, et al. Cross-subtype T-cell immune responses induced by a human immunodeficiency virus type 1 group m consensus env immunogen. J Virol. 2006;80:6745–56.

8. Gaschen B, Taylor J, Yusim K, Foley B, Gao F, Lang D, et al. Diversity considerations in HIV-1 vaccine selection. Science. 2002;296:2354–60.
9. Baeten JM, Chohan B, Lavreys L, Chohan V, McClelland RS, Certain L, et al. HIV-1 subtype D infection is associated with faster disease progression than subtype A in spite of similar plasma HIV-1 loads. J Infect Dis. 2007;195:1177–80.
10. Kiwanuka N, Laeyendecker O, Robb M, Kigozi G, Arroyo M, McCutchan F, et al. Effect of human immunodeficiency virus Type 1 (HIV-1) subtype on disease progression in persons from Rakai, Uganda, with incident HIV-1 infection. J Infect Dis. 2008;197:707–13.
11. Kouri V, Khouri R, Alemán Y, Abrahantes Y, Vercauteren J, Pineda-Peña A-C, et al. CRF19_cpx is an evolutionary fit HIV-1 variant strongly associated with rapid progression to AIDS in Cuba. EBioMedicine. 2015;2:244–54.
12. Renjifo B, Gilbert P, Chaplin B, Msamanga G, Mwakagile D, Fawzi W, et al. Preferential in utero transmission of HIV-1 subtype C as compared to HIV-1 subtype A or D. AIDS Lond Engl. 2004;18:1629–36.
13. Yang C, Li M, Newman RD, Shi Y-P, Ayisi J, van Eijk AM, et al. Genetic diversity of HIV-1 in western Kenya: subtype-specific differences in mother-to-child transmission. AIDS Lond Engl. 2003;17:1667–74.
14. John-Stewart GC, Nduati RW, Rousseau CM, Mbori-Ngacha DA, Richardson BA, Rainwater S, et al. Subtype C Is associated with increased vaginal shedding of HIV-1. J Infect Dis. 2005;192:492–6.
15. Vidal N, Mulanga C, Bazepeo SE, Lepira F, Delaporte E, Peeters M. Identification and molecular characterization of subsubtype A4 in central Africa. AIDS Res Hum Retroviruses. 2006;22:182–7.
16. Meloni ST, Kim B, Sankale J-L, Hamel DJ, Tovanabutra S, Mboup S, et al. Distinct human immunodeficiency virus type 1 subtype A virus circulating in West Africa: sub-subtype A3. J Virol. 2004;78:12438–45.
17. Katoh K, Standley DM. MAFFT multiple sequence alignment software version 7: improvements in performance and usability. Mol Biol Evol. 2013;30:772–80.
18. Larsson A. AliView: a fast and lightweight alignment viewer and editor for large datasets. Bioinforma Oxf Engl. 2014;30:3276–8.
19. Martin DP, Murrell B, Golden M, Khoosal A, Muhire B. RDP4: detection and analysis of recombination patterns in virus genomes. Virus Evol. 2015;1:vev003.
20. Pond SLK, Frost SDW, Muse SV. HyPhy: hypothesis testing using phylogenies. Bioinforma Oxf Engl. 2005;21:676–9.
21. Darriba D, Taboada GL, Doallo R, Posada D. jModelTest 2: more models, new heuristics and parallel computing. Nat Methods. 2012;9:772.
22. R Core Team. R: a language and environment for statistical computing. 2015; http://www.R-project.org/. Accessed 15 Sept 2017.
23. Guindon S, Dufayard J-F, Lefort V, Anisimova M, Hordijk W, Gascuel O. New algorithms and methods to estimate maximum-likelihood phylogenies: assessing the performance of PhyML 3.0. Syst Biol. 2010;59:307–21.
24. Vidal N, Bazepeo SE, Mulanga C, Delaporte E, Peeters M. Genetic characterization of eight full-length HIV type 1 genomes from the Democratic Republic of Congo (DRC) reveal a new subsubtype, A5, in the A radiation that predominates in the recombinant structure of CRF26_A5U. AIDS Res Hum Retroviruses. 2009;25:823–32.
25. Thomson MM, de Parga EV, Vinogradova A, Sierra M, Yakovlev A, Rakhmanova A, et al. New insights into the origin of the HIV type 1 subtype A epidemic in former Soviet Union's countries derived from sequence analyses of preepidemically transmitted viruses. AIDS Res Hum Retroviruses. 2007;23:1599–604.
26. Engelbrecht S, de Villiers T, Sampson CC, zur Megede J, Barnett SW, van Rensburg EJ. Genetic analysis of the complete gag and env genes of HIV type 1 subtype C primary isolates from South Africa. AIDS Res Hum Retroviruses. 2001;17:1533–47.
27. Khan IF, Vajpayee M, Prasad VVSP, Seth P. Genetic diversity of HIV type 1 subtype C env gene sequences from India. AIDS Res Hum Retroviruses. 2007;23:934–40.

13

Reconstruction of a replication-competent ancestral murine endogenous retrovirus-L

Daniel Blanco-Melo[1,3], Robert J. Gifford[2] and Paul D. Bieniasz[1*]

Abstract

Background: About 10% of the mouse genome is composed of endogenous retroviruses (ERVs) that represent a molecular fossil record of past retroviral infections. One such retrovirus, murine ERV-L (MuERV-L) is an *env*-deficient ERV that has undergone episodic proliferation, with the most recent amplification occurring ~2 million years ago. MuERV-L related sequences have been co-opted by mice for antiretroviral defense, and possibly as promoters for some genes that regulate totipotency in early mouse embryos. However, MuERV-L sequences present in modern mouse genomes have not been observed to replicate.

Results: Here, we describe the reconstruction of an ancestral MuERV-L (ancML) sequence through paleovirological analyses of MuERV-L elements in the modern mouse genome. The resulting MuERV-L (ancML) sequence was synthesized and a reporter gene embedded. The reconstructed MuERV-L (ancML) could replicate in a manner that is dependent on reverse transcription and generated de novo integrants. Notably, MuERV-L (ancML) exhibited a narrow host range. Interferon-α could reduce MuERV-L (ancML) replication, suggesting the existence of interferon-inducible genes that could inhibit MuERV-L replication. While mouse APOBEC3 was able to restrict the replication of MuERV-L (ancML), inspection of endogenous MuERV-L sequences suggested that the impact of APOBEC3 mediated hypermutation on MuERV-L has been minimal.

Conclusion: The reconstruction of an ancestral MuERV-L sequence highlights the potential for the retroviral fossil record to illuminate ancient events and enable studies of the impact of retroviral elements on animal evolution.

Background

Uniquely among animal viruses, retroviruses integrate into the genome of the host cell as an obligate step in their replication cycle. Because the target cells of some retroviruses can include cells of the germ line, proviruses can occasionally become vertically inherited [1]. A subset of these inherited proviruses can become fixed in the population through genetic drift, or sometimes by providing an evolutionary advantage to the host. Inherited proviruses are termed endogenous retroviruses (ERVs) and are present in all animal species that have been examined, accounting for approximately 8 and 10% of the human and mouse genomes, respectively [2]. In nearly every case, however, fixed proviruses have inactivating mutations that prevent their further spread.

The vast array of ERVs represent an extensive viral fossil record that provides an opportunity to study the biology of ancient or extinct retroviruses, and the effects that these viruses have had on the evolution of their hosts [3]. Previously, we and others have reconstructed a full-length infectious human ERV (HERV-K) [4, 5], functional capsid proteins of endogenous chimpanzee gammaretroviruses (CERV 1 and 2) [6, 7], and lentiviruses, PSIV and RELIK [8], as well as functional envelope proteins from CERV2 and HERV-T [9, 10]. These reconstruction experiments have enabled the identification of ancient virus receptors [9, 10] and demonstrated the ancient origin of cyclophilin A-lentiviral capsid interactions [8].

*Correspondence: pbieniasz@rockefeller.edu
[1] Laboratory of Retrovirology and Howard Hughes Medical Institute, The Rockefeller University, New York, NY, USA
Full list of author information is available at the end of the article

Additionally, these studies have shown that the replication of HERV-K, CERV1 and CERV2 was affected by the APOBEC3 cytidine deaminases [5, 6, 11]. Overall these "paleovirological" studies have provided previously inaccessible insights into the co-evolution of viruses and hosts.

Murine ERV-L (MuERV-L) is an abundant mouse ERV that is transcriptionally active at an early (2-cell) stage of the mouse embryo [12–15]. Previous analyses have established that MuERV-L underwent two amplification bursts, one after the divergence of the *Mus* and *Rattus* genera around ~10 million years ago (MYA), and a more recent and prolific burst about 2 MYA, which is distinguished by the presence of a 33nt in-frame deletion in the 5′ half of the *gag* ORF (Fig. 1a) [16]. These amplifications led to the deposition of thousands of MuERV-L derived sequences in the mouse genome. Moreover, MuERV-L belongs to a larger family of ERV-L elements that have been active throughout the evolution of mammals [17–19]. In contrast to their human counterparts, many MuERV-L elements have complete coding potential, encoding open reading frames (ORFs) for *gag* and *pol* (Fig. 1a) [20]. However, like other ERV-L elements, MuERV-L is characterized by the complete absence of an *env* gene that, coupled with its highly restricted early transcription profile, suggests an entirely intracellular retrotransposon-like replication cycle [21]. Indeed, MuERV-L transcripts are able to give rise to intracellular viral-like particles that accumulate in the endoplasmic reticulum [13] but are not thought to be replication competent.

MuERV-L related or derived sequences appear to have been co-opted for two distinct biological activities in the mouse. The antiretroviral restriction factor Fv1, which inhibits infection by MuLV and certain other retroviruses is derived from MuERV-L-like Gag sequences and appeared in the mouse genome at least 5 MYA [22–24]. Additionally, recent studies have suggested that the propensity of MuERV-L to be transcriptionally active only at the two-cell stage of mouse embryogenesis may have led to the co-option of its long terminal repeats (LTRs), as promoters of genes involved in the zygotic genome activation [14, 15]. Transcriptional activity of MuERV-L LTRs in two-cell mouse embryos may drive the expression of hundreds of genes that contribute to the totipotency of the blastomeres, but also results in the expression of MuERV-L Gag–Pol polyprotein and the formation of intracellular viral-like particles [13–15]. As development progresses, MuERV-L LTRs appear to be silenced [14, 25, 26]. The expression of MuERV-L at the two-cell stage does not induce an increase in their copy numbers, suggesting that the expressed proviruses do not have the potential to re-integrate into the genome [25].

In fact, no extant copy of MuERV-L has been demonstrated to be capable of completing a replication cycle. Therefore, we set out to derive a replication-competent MuERV-L, based on the premise that an ancestral reconstruction would deliver a sequence that most closely resembles that of a functional ancestor. Herein, we describe the analysis of MuERV-L elements in the mouse genome, a successful reconstruction of a ~2MY old replication-competent ancestral sequence and an analysis of its replication and its interaction with the components of the host intrinsic/innate immune systems.

Methods

Bioinformatic analyses and ancestral reconstruction

Screening for MuERV-L elements was performed using amino acid and nucleotide sequences from the MuERV-L reference sequence (MuERV-L^{ref}, GenBank: Y12713) [20] as probes for tBLASTn (*gag* and *pol*) and BLASTn (LTR) [27] searches of two mouse genome assemblies: Mm_Celera (NCBI: GCF_000002165.2) [28] and GRCm38/mm10 (UCSC: mm10) [29]. To avoid the identification of sequences from related but distinct retroviruses, BLAST hits with an e-value ≤ 1e−10 were used as probes for a second round of BLASTx or BLASTn searches against a previously constructed database of endogenous and exogenous class III retroviral sequences. Results from these BLAST searches were imported into a relational database to facilitate the management of the screening process and analysis of hits. Reciprocal hits to MuERV-L^{ref} were first ordered by chromosome and orientation and then adjacent or overlapping hits were assembled into proviral loci by comparison with the MuERV-L^{ref} sequence, allowing for insertions no longer than 10,000 nucleotides. The resulting MuERV-L loci were annotated as genomic features of GRCm38 (downloaded from Ensembl [30] using BioMart) by comparison of their chromosome location using in-house Perl scripts.

Dates of integration for MuERV-L elements were estimated by determining the divergence (K) to a consensus sequence (for solo LTRs) or between paired LTRs (for provirus-containing loci) using PAUP* [31], divided by 2× the mouse neutral substitution rate (r) of 4.5×10^{-9} substitutions per site per year [29] (K/2r) [32, 33]. The mean sequence identity between the consensus LTR sequence and each of the solo LTRs was 95.43%, with very few outliers (1.5% of solo LTRs showed less than 80% identity), suggesting that the consensus sequence is adequate to perform these estimations. Nevertheless, the estimated ages derived using this method represent approximations to the integration dates and should be treated as such.

Statistical analyses of the positions of MuERV-L elements relative to mouse genomic features were

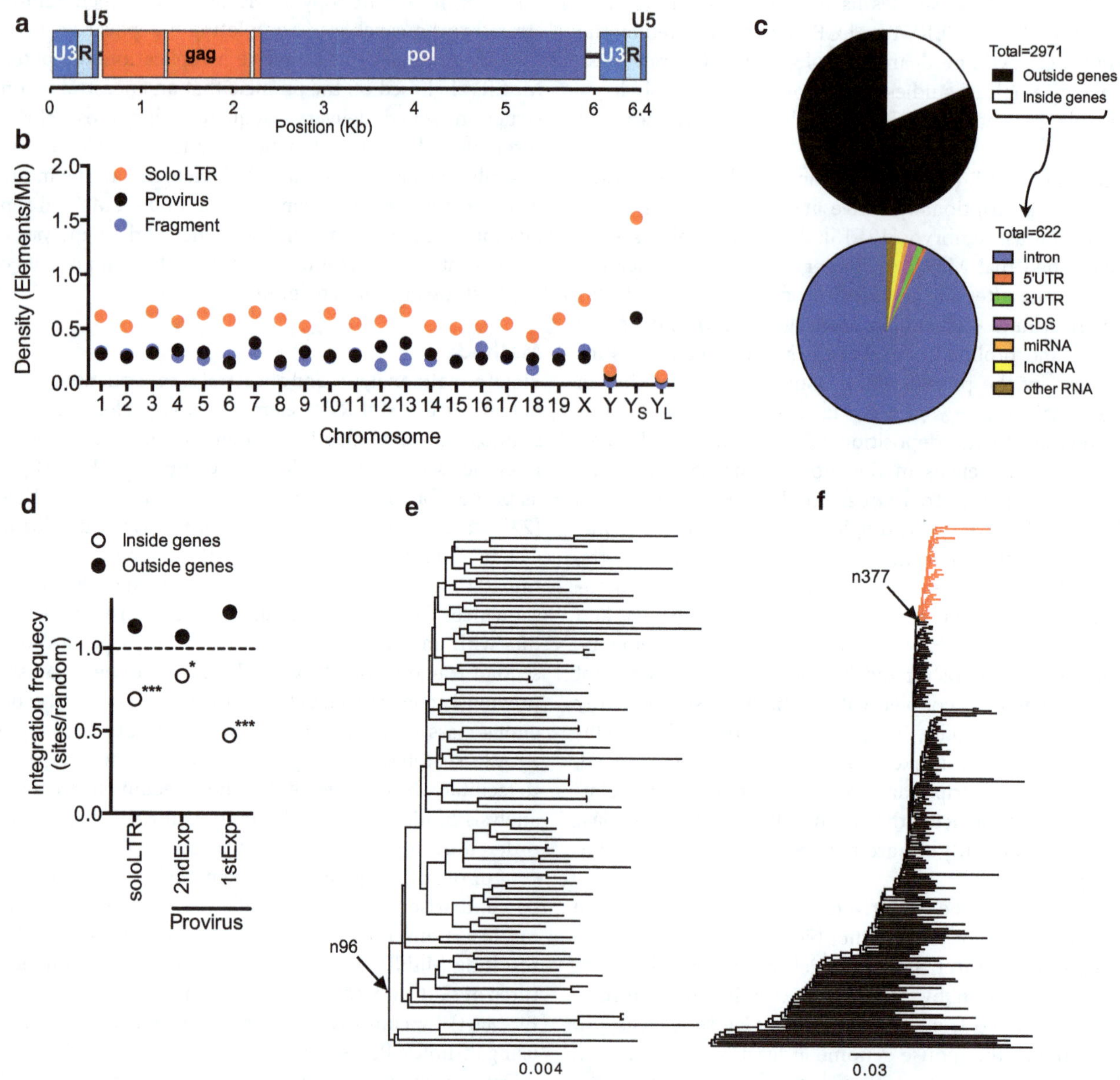

Fig. 1 Structure and distribution of MuERV-L elements in the mouse genome and reconstruction of an ancestral ~2MY old MuERV-L sequence. **a** Schematic representation of the structure of MuERV-L elements. White boxes in *gag* represent the 33 and 39nt deletions in *gag* at nucleotide positions 671 and 1597, respectively. **b** Distribution of distinct MuERV-L structures among mouse chromosomes. Y_S = Y chromosome short arm, Y_L = Y chromosome long arm. **c** Distribution of MuERV-L elements in mouse genomic features (GRCm38/mm10). The fraction of the 2971 elements inside and outside genes is depicted (top chart), as the fraction of elements present in each type of gene or gene feature (bottom chart). **d** Distribution of MuERV-L elements in genic or intergenic regions relative to random controls. The measured value indicates the percentage of MuERV-L elements in each population divided by that of the random controls (see "Methods"). The horizontal dotted line indicates no difference between the ratio of MuERV-L elements in each population and that of the controls. (*) p value < 0.05. (***) p value < 0.001. P values are based on Chi-squared goodness-of-fit or contingency table tests. **e** Maximum likelihood phylogenetic tree of 95 LTR–*gag*–*pol*–LTR MuERV-L elements in the mouse genome. Arrow denotes the ancestral node reconstructed by baseml (*pol* and LTR, node 96). **f** Maximum likelihood phylogenetic tree of 230 gag–pol containing MuERV-L elements in the mouse genome. The monophyletic red clade contains only elements with a 33nt deletion in *gag* at position 671 with or without an additional 39nt deletion in *gag* at position 1597. Arrow denotes the ancestral node reconstructed by baseml (*gag*, node 377)

performed using the Pearson's Chi-squared test for count data (chisq.test) implemented in R. As controls, two sets of random coordinates in the mouse genome were computationally generated using in-house Perl scripts and the runif function implemented in R. For solo LTR comparisons, 10,000 random mouse sequences of 500 nucleotides in length were generated. For proviral loci comparisons, 5000 random mouse sequences of 6500 nucleotides in length were generated. The coordinates of both MuERV-L integrations and random sequences were then mapped to the GRCm38 genome assembly to determine overlapping genomic features (intergenic regions, genes, common repeats and others) using in-house Perl scripts. For ancML integration comparisons, in-house Perl scripts were used to generate controls, consisting of 1000 randomly selected EcoRV-containing fragments (of 1000 nucleotides in length) from the Chinese hamster genome. Each ancML integration site was matched with three of these genomic sequences such that control sites were equidistant from an EcoRV site as was each ancML integration site, as described in [34]. These sites were then mapped to the Chinese hamster genome (criGri1) [35] to determine the overlapping genomic features.

Ancestral reconstruction of MuERV-L was performed using two distinct sequence sets. For the reconstruction of the ancestral *pol* gene and the LTRs, we used a set of 95 complete proviral sequences (LTR–*gag*–*pol*–LTR) identified by default-parameter based BLASTn searches of GRCm38 using MuERV-L^{ref}. Each proviral sequence was individually aligned to MuERV-L^{ref} using MUSCLE [36] and a multiple sequence alignment (MSA) was generated using the profile alignment function of MUSCLE. Insertions relative to MuERV-L^{ref} were eliminated from the MSA, except a 6nt insertion at position 298 and 6249 in both LTRs that was shared between 25% of the sequences. The MSA was used to construct a maximum likelihood (ML) phylogenetic tree using raxML [37] with the following parameters: rapid bootstrap analysis with 1000 replicates under GTRCAT followed by a ML search under GTRGAMMA to evaluate the final tree topology (-m GTRCAT -# 1000 -x 13 -k -f a). Thereafter, the tree was midpoint rooted. The MSA together with the phylogenetic tree were used to guide an ML ancestral reconstruction using baseml from the PAML package [38] (model: REV, initial values of alpha and kappa were calculated on the MSA by jmodeltest [39], branch lengths were used as initial values). For the ancestral reconstruction of the *gag* ORF we first aligned and constructed a phylogenetic tree using 230 *gag*–*pol* containing sequences (identified by our screening of the mouse genome described above). We determined the presence or absence of the 33 and 39nt deletions in the *gag* ORF relative to MuERV-L^{ref} (that does not show any deletion) and identified a monophyletic clade of 40 sequences that had the 33nt deletion in *gag* (irrespective of the status of the 39nt deletion). Thereafter, reconstruction of the ancestral *gag* sequence corresponding to the internal node for this monophyletic clade was performed as described above. A correction for the effect of methylation-induced mutations at CpG dinucleotides was applied on both strands of all three sequences (*gag*, *pol* and LTR) as described in [8]. Specifically, if a particular site where the ancestral reconstruction estimated a TG dinucleotide but at least 10% of the sequences in the MSA encoded a CG at that position, the TG state was considered to be the result of methylation-induced mutation and the sequence at this position was assigned as CG. The resulting sequences were combined to produce the sequence from which ancML was derived.

Hypermutation analysis and statistics were performed using Hypermut 2.0 [40] on the set of 230 gag–pol containing sequences used to reconstruct an ancestral *gag*, using either default parameters or with exclusion of sites with a 5′ C next to the mutated G.

Plasmid construction

To construct ancML, sequences from the U3 region of the MuERV-L 5′LTR 5′ to the TATA box were substituted with corresponding CMV promoter sequences. We also added an extra 12nt containing two MluI sites immediately 3′ to the *pol* stop codon to facilitate the insertion of a reporter gene. The modified ancML sequence was synthesized and inserted into pUC57 (Genewiz, NJ). The replication dependent LINE-1 element (L1.3 plasmid) [41] was kindly provided by Dr. John V. Moran. The replication dependent *neo* cassette (a *neo* gene controlled by a SV40 promoter and interrupted by an intron) was amplified by PCR from the L1.3 plasmid and inserted into ancML using the MluI sites at the 3′ end of *pol*. A separate pCR3.1 based plasmid, expressing GFP (from a CMV promoter) and a *neo* gene (NEO, expressed from a SV40 promoter) was used as a control.

The ancML-RTmut construct was created by using overlapping PCR and primers that annealed to the RT active site with four nucleotide mismatches, the PCR fragment was inserted into ancML using unique surrounding BstZ17I and NheI restriction sites contained in the outmost primers, generating an ancML with a mutated RT active site (YIDD to AIAA).

The ancMLΔGAAGT construct was generated using PCR and a reverse primer that annealed to the 5′ end of the PBS and the 3′ end of the U5 region of the 5′LTR, and lacked the intervening 5nt linker sequence (GAAGT). The PCR fragment was inserted into ancML using unique AgeI and KpnI restriction sites contained in the forward and reverse primers, respectively.

A plasmid expressing mouse APOBEC3 (mA3, C57BL/6J strain) was kindly provided by Rachel Liberatore (unpublished). A C-termini HA-tagged version of mA3 was produced by PCR using primers containing two HA tags and a 15nt linker sequence, following previously published functional human HA-tagged APOBEC3 proteins [42, 43]. This construct was introduced into the retroviral expression plasmid LBCX using unique SfiI sites. A retroviral vector (LBCX) expressing $Fv1^{bbn}$ was kindly provided by Dr. Theodora Hatziioannou [44].

Cell culture

Cell lines (except CHO-K1 and pgsA cells) were maintained in Dulbecco's Modified Eagle Medium (DMEM), Eagle's Minimum Essential Medium (EMEM) or Roswell Park Memorial Institute medium (RPMI) supplemented with 10% FBS and gentamycin (2 µg/ml, Gibco) according to ATCC instructions. CHO-K1 and pgsA cells were maintained in Ham's F-12 media supplemented with 10% FBS, 1 mM of L-glutamine and 2 µg/ml of gentamycin. All cells were incubated at 37 °C, except DF-1 cells that were incubated at 39 °C.

Generation of CHO cell lines expressing murine APOBEC3

293T cells were transfected (using polyethylenimine) with plasmids expressing MuLV gag–pol, and VSV-G, along with an LBCX based retroviral vector expressing HA-tagged mA3 or $Fv1^{bbn}$. Viral stocks were harvested and filtered (0.22 µm) 2 days after transfection, and were used to transduce CHO-K1 cells (seeded in 24 well plates). Transduced cells were expanded in 10 cm dishes with media supplemented with 5 µg/ml of blasticidin (Thermo Fisher Scientific Inc.). Single cell clones expressing mA3 were isolated by seeding blasticidin resistant cells at 0.5 cells per well in a 96 well plate. Three distinct single clones that expressed mA3 in 100% of the cells (tested by immunofluorescence) were used in ancML replication assays.

Immunofluorescence assay

Individual clones of CHO cells expressing murine APOBEC3 were fixed with 4% paraformaldehyde (PFA) for 30 min followed by treatment with 10 mM glycine (diluted in PBS) for another 30 min. Cells were permeabilized with a buffer containing 0.1% of Triton X-100 and 5% goat serum (diluted in PBS) for 15 min. Cells were then washed 2 times with PBS before being treated with mouse monoclonal anti-HA antibody (Covance) diluted in a buffer containing 0.1% Tween-20 and 5% goat serum (diluted in PBS) for 2 h at room temperature. Cells were washed three times with PBS before being treated with goat anti-mouse secondary antibody (Alexa Fluor 488 dye, ThermoFisher) diluted in a buffer containing 0.1% Tween-20 and 5% goat serum (diluted in PBS) for 1 h at room temperature. Cells were washed three more times with PBS and fluorescent microscopy images were analyzed using the EVOS FL Cell Imaging System.

MuLV infection assay

293T cells were transfected (using polyethylenimine) with plasmids expressing N-tropic or B-tropic MuLV gag–pol, and VSV-G, along with a CNCG based retroviral vector expressing GFP [45]. Viral stocks were harvested 2 days after transfection, filtered (0.22 µm) and were used to infect control or $Fv1^{bbn}$-expressing CHO cells. Two days post infection the percentage of GFP positive population was quantified using the Guava EasyCyte flow cytometer (Millipore).

MuERV-L(ancML) replication assays

The cell lines listed in Table 2 were seeded in 12 well plates 1 day before being transfected with 700 ng of plasmids containing L1.3, ancML or a plasmid expressing *gfp* and a *neo* gene, using 4 µl of Lipofectamine 2000 (Thermo Fisher Scientific Inc.) according to manufacturer instructions. Two days after transfection, cells were plated in 6-well plates with G418 selection media (containing concentrations of G418 that were previously calibrated for each cell type). Ten days later, surviving cells were fixed with 4% PFA and colonies were stained using 0.3% crystal violet in 20% ethanol for counting.

Subsequently, the ancML replication assays were routinely done using CHO-K1 cells as follows. CHO-K1 cells were seeded at 3×10^5 cells per well in a 12 well plate. One day later the cells were transfected with 1 µg of plasmid DNA, using 3 µl of Transit-CHO supplemented with 0.5 µl of CHO-mojo reagent (Mirus) diluted in Opti-MEM (Gibco). One day later, the cells were expanded on a 10 cm dish with media supplemented with or without AZT (obtained through the NIH AIDS Reagent Program, Division of AIDS, NIAID, NIH) or mouse IFNα (Pestka Biomedical Laboratories, Inc.). Two days later, cells were plated in a 15 cm dish or three 96 well plates (for analysis of single cell clones) with media supplemented with 1 µg/ml of G418. For controls, 1/1000 of the cells transfected with the NEO plasmid were plated in the 15 cm dish with selection media. Cells in 15 cm dishes were cultured under selection for 10 days before treatment with 4% PFA and colonies were stained with 0.3% crystal violet in 20% ethanol for counting. Single colonies in 96 well plates were monitored and expanded until reaching confluence in a 10 cm dish. Genomic DNA (gDNA) was extracted from 5×10^6 cells using QIAmp DNA mini kit (QIAGEN) for analysis of ancML integration (see below).

To determine the fate of the intron interrupting the *neo* gene during ancML replication, gDNA extracted from CHO pools of cells transfected with a plasmid expressing ancML, ancMLΔGAAGT or an empty vector, was used as template for PCR analysis. Forward and reverse primers were design to anneal to the extreme 5′ and 3′ ends of the *neo* gene. For all PCRs performed in this study we used Phusion High-Fidelity DNA Polymerase (Thermo Fisher Scientific Inc.).

Integration site analyses

Sites of ancML integration were determined using a universal Genome Walker kit (Clontech). Briefly, gDNA was extracted from expanded single cell clones of CHO cells following transfection with a plasmid containing ancML and selected in G418. The gDNA was digested with EcoRV (New England Biolabs) and ligated to adaptors. Nested PCRs were performed using forward primers that were designed to anneal to the R region of the 3′LTR and the reverse primers to the adaptor sequence, thereby amplifying 3′ flanking sequences. Bands from second round PCR reactions were gel purified and inserted into pCR-Blunt II-TOPO using the Zero Blunt TOPO PCR cloning Kit (Life technologies) for sequencing. To amplify and sequence the 5′ flanking site we use reverse primers specific to the 5′ LTR and designed forward primers that would specifically anneal to the predicted integration site, based on the previously sequenced 3′ flanking sequence. The resulting CHO gDNA sequences were mapped to the CHO genome (criGri1) using BLAT [46] searches on the UCSC genome browser [47]. To account for biases due to location and density of EcoRV restriction sites, we compared the distribution of the 26 ancML integration sites to matched random controls consisting of three random genomic locations that were at the same distance from an EcoRV site as the site found in the flanking CHO DNA sequence for each MuERV-L integration site [34] (see above).

Results

Bioinformatic screens for MuERV-L elements in the mouse genome

To construct a replication-competent MuERV-L sequence we first catalogued the diversity of MuERV-L related sequences in the mouse genome. Currently, there are two available complete mouse genome assemblies: the Mouse Genome Reference Consortium build 38 (GRCm38 also known as mm10) corresponding to the C57BL/6J strain [29], and the whole genome shotgun (WGS) assembly (Celera) that corresponds to a mixture of 5 strains (129X1/SvJ, 129S1/SvImJ, DBA/2J, A/J and C57BL/6J) [28]. We mined both genome assemblies using BLASTn and tBLASTn [27] searches with separate *gag*, *pol* and LTR probes from a MuERV-L reference sequence (GenBank: Y12713) [20]. The resulting hits were defragmented by merging contiguous hits that mapped to the same locus, representing individual elements.

Overall, we found nearly 3000 MuERV-L elements in the mouse genome that had three major types of structures (Table 1, Fig. 1b, Additional file 1: Table S1). One type comprised complete or near complete proviruses, consisting of an internal *gag–pol* region flanked by two LTRs. The frequency with which this structure occurred was highly discrepant in the two genome assemblies, with 220 proviruses being present in the Celera assembly and 719 in GRCm38 (Table 1). It is unclear whether this discrepancy is due to the different mouse strains or the assembly methods used in each genome project. A second type of MuERV-L elements were solo LTRs that typically arise from recombination between the LTRs flanking a provirus, resulting in the complete excision of the internal sequence. This type of element represents the single most abundant ERV type in animal genomes and accounts for >50% of the MuERV-L elements in the mouse genome (Table 1). The third type of MuERV-L structures, representing ~25% of all elements, were composed of internal sequences with or without a single associated LTR (Table 1).

Because the GRCm38 assembly was better supported by external and internal annotations we utilized this data source to analyze the distribution of MuERV-L elements in the genome (Fig. 1b–d). MuERV-L elements were roughly evenly distributed across mouse chromosomes (Fig. 1b), with the exception of the Y chromosome in which MuERV-L was underrepresented. Indeed, there were only a few elements in the Y chromosome. This finding was surprising, given that other ERVs, (e.g. HERV-K) are enriched in the human Y chromosome [48]. However, in contrast to human Y chromosome, the long arm of the mouse Y chromosome is a highly dynamic gene rich region that has frequently expanded and undergone rearrangement over the past ~3 MY [49]. The distribution of MuERV-L elements in this mouse chromosome shows a clear discrepancy between the short and long arms (Fig. 1b). This finding is likely explained by the low recombination rate of the short arm of the mouse Y chromosome, resulting in an enrichment of mobile DNA elements, while recent gene amplification and

Table 1 MuERV-L sequences identified in mouse genome assemblies

Assembly	Strain	References	Total	soloLTR	Provirus	Fragments
GRCm38	C57BL/6J	[29]	2971	1588	719	664
Mm_Celera	Mixture	[28]	2768	1775	220	773

rearrangement events in the long arm may have inhibited the fixation of MuERV-L elements.

The majority (79%) of MuERV-L elements were found in intergenic regions (Fig. 1c). About ~19.5% of elements were found inside introns (Fig. 1c), of which the majority (65%) were found in antisense orientation relative to of the corresponding gene, as previously documented for Intracisternal A particles (IAPs) [50]. The remaining elements (1.5%) were found primarily in non-coding RNA genes and untranslated exons (UTRs, Fig. 1c). Only 10 LTRs were found overlapping coding exons. This distribution differs significantly compared to randomized controls ($p < 0.001$). This enrichment of elements outside genes was also apparent if analysis was confined to solo LTRs (p value < 0.001, Fig. 1d) or if proviruses from the 2MYA or 10MYA expansions were analyzed separately (p values < 0.05, Fig. 1d). Overall, these observations are consistent with an expected selective pressure against retention of MuERV-L elements in genes, which was less evident for younger loci (2nd expansion, Fig. 1d). Interestingly, 10.13% of proviruses implicated in the 10MYA expansion have lost a recognizable ORF (either *gag* or *pol*), in contrast to integrations implicated in the 2MYA expansion where only 5.02% have lost an ORF, consistent with the notion that a modest selection to purge MuERV-L sequences is ongoing.

Reconstruction of an ancestral MuERV-L

To reconstruct the ancestral MuERV-L LTRs and the *pol* ORF, we selected 95 complete (LTR–*gag*–*pol*–LTR) proviruses that were most closely related to a reference MuERV-L sequence (defined by BLAST searches), as this sequence has retained coding potential for both ORFs, has almost identical LTRs, and contained recognizable functional motifs [20]. These sequences were used to guide a maximum likelihood (ML) reconstruction of the root node (*pol* n96) and a pair of identical LTRs (Fig. 1e).

For reconstruction of the Gag gene we took a slightly different approach. As the 2nd expansion was the most prolific and we expected that younger integrations would be less divergent from a functional ancestor than older integrations, we selected *gag* sequences that were specific to the most recent expansion (based on the presence of the 33nt in-frame deletion in *gag*). After aligning 230 *gag*–*pol* containing loci that were present in both Celera and GRCm38 assemblies to the reference sequence (that does not exhibit the deletion), we identified a monophyletic clade of 40 elements that all contained a 33nt in-frame deletion in the 5′ half of their *gag* ORFs (Fig. 1f). The *gag* sequences in this clade were selected to guide a ML reconstruction of this internal node (*gag* n377). Thereafter, the combined ancestral LTR, *gag* and *pol* ancestral sequences were corrected for probable errors derived from deamination of methylated CpG dinucleotides to create a ~2 MY ancestral MuERV-L sequence (ancML, Fig. 2 and Additional file 2) (see "Methods").

ancML is replication-competent and its replication is dependent on a functional reverse transcriptase

To assess the potential replication competence of the reconstructed ancestral MuERV-L sequence, we inserted the full-length ancML sequence into a plasmid vector, replacing the U3 region of the 5′ LTR by a CMV promoter to overcome the highly restricted promoter activity of the MuERV-L LTR (Fig. 3a). This design allows for the loss of the CMV promoter sequence after transcription, reverse transcription and integration, resulting in two identical flanking LTRs. A replication-dependent reporter gene, consisting of a *neo* gene controlled by a separate SV40 promoter and interrupted by an intron, was inserted between *pol* and the 3′ LTR (Fig. 3a). While the SV40p-*neo* cassette is in reverse orientation relative to ancML transcription, the splice donor and acceptor sites of the intron are in the same orientation as the ancML transcription. Thus, only if ancML undergoes splicing, reverse transcription and integration in the host cell genome the intron is removed and a functional Neo (G418 resistance) protein is expressed. This approach has been used previously to monitor the intracellular replication of retrotransposons [51].

We determined whether transfection of a plasmid harboring ancML could produce G418 resistant colonies in cultured cell lines. For this purpose we tested a set of 13 cell lines: 6 of primate origin, 5 from rodents, 1 from a carnivore and 1 from an avian species (Table 2). Despite efficient transfection in the majority of the cells tested, and the abundant formation of G418-resistant colonies using a control plasmid, the ancML expression plasmid was able to generate G418 resistant colonies only in Chinese hamster ovary K1 (CHO) cells, CHO-derived pgsA cells, and (to a greatly reduced extent) in Vero cells (Table 2 and Fig. 3b). Surprisingly, a control plasmid expressing a LINE1 element containing the same replication dependent *neo* resistant gene (L1.3 plasmid) [41] failed to produce G418 resistant colonies in 4 of the cells tested, including Vero (Table 2). Conversely, all the cell lines tested generated G418 resistant colonies when transfected with a control plasmid expressing an intact *neo* gene.

To determine whether the ancML-associated G418 resistant colonies had arisen as a result of ancML replication, we mutated the predicted ancML reverse transcriptase (RT) active site from YIDD to AIAA (Fig. 2). This mutation completely abolished the production of G148 resistant colonies following ancML transfection (Fig. 3c). Additionally, we found that the formation

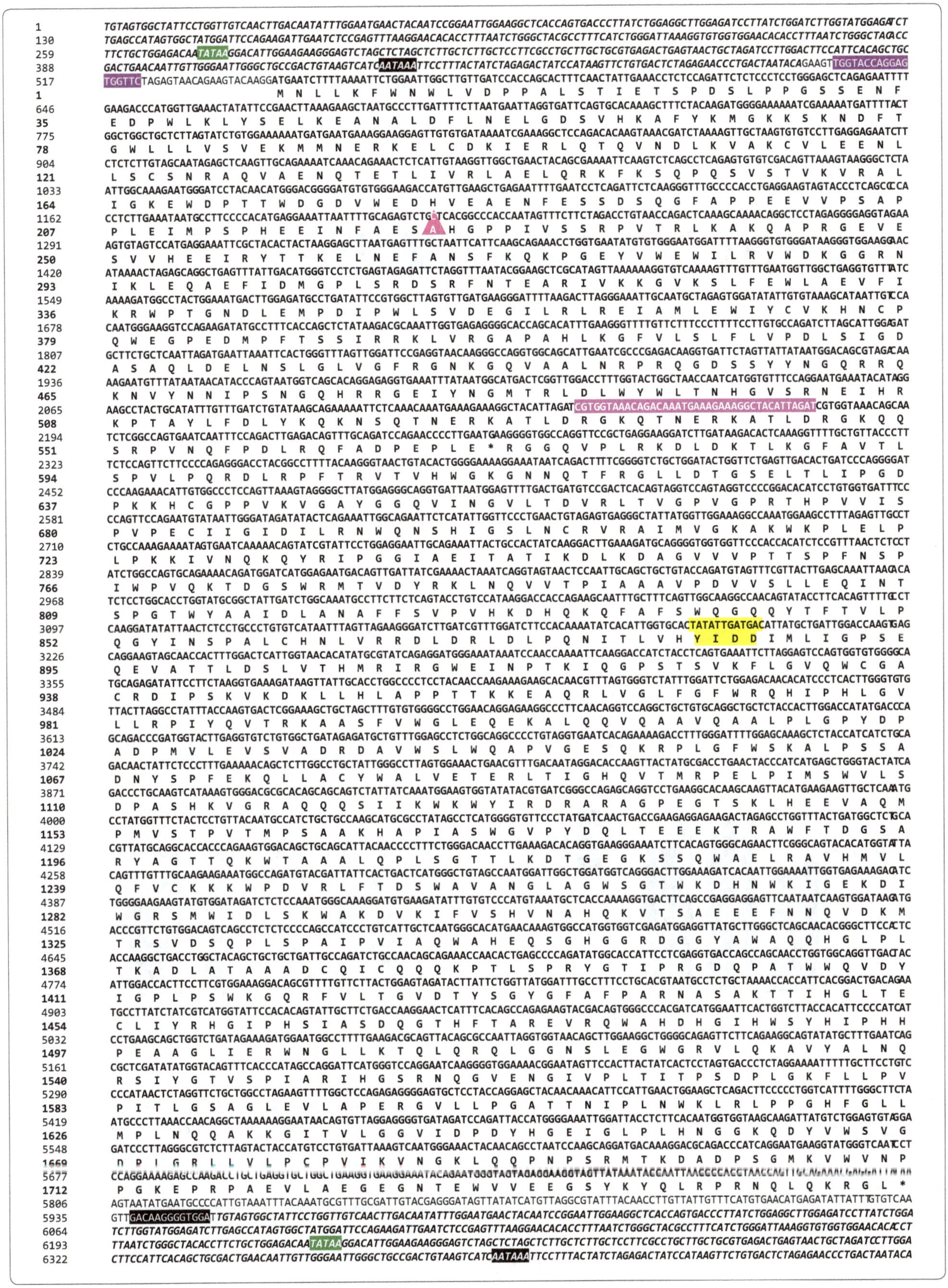

(See figure on previous page.)

Fig. 2 Nucleotide sequence and translation products of a reconstructed ancestral MuERV-L. LTR sequences are shown in bold italics. Nucleotide and protein sequence of *gag* and *pol* are indicated in red and blue, respectively (amino acid single letter code, (*) represents stop codons). The 33nt deletion in gag is shown with a magenta triangle. The position of the 39nt that are deleted in some MuERV-Ls is highlighted in magenta. The RT active site is highlighted in yellow. The PBS is indicated in violet, the polypurine tract in red, the TATA box in green and the polyadenylation site in bright blue

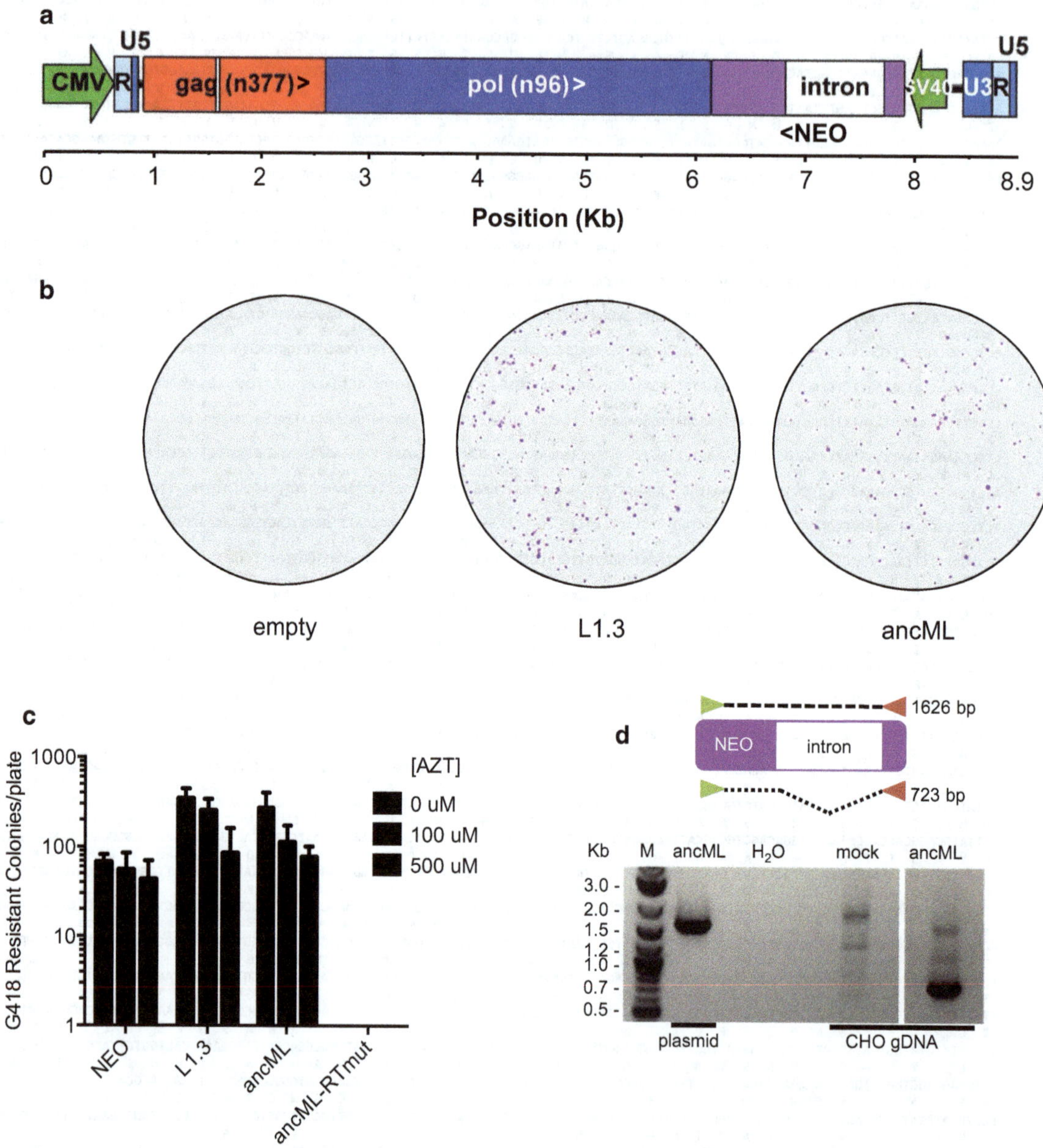

Fig. 3 Reverse transcription-dependent ancML replication. **a** Organization of the ancML construct. Green arrows indicate promoter sequences. NEO: *neo* gene in reverse orientation relative to ancML transcription. Chevrons indicate the orientation for each ORF (>: forward, <: reverse). A white box indicates a 33nt deletion in *gag* at position 671. **b** G418 resistant colonies on 15 cm cell culture plates derived from CHO cells transfected with plasmids expressing ancML, L1.3 or an empty vector. **c** Quantification of G418 resistant CHO cell colonies following transfection and treatment in the presence or absence AZT. CHO cells were transfected with plasmids expressing a *neo* gene (NEO), L1.3, ancML or an ancML construct with inactivating mutations in the RT active site (ancML-RTmut). AZT treatment was applied for 2 days before G418 selection. Data are mean ± SD from 3 independent experiments. **d** PCR amplification of the *neo* gene in genomic DNA (gDNA) extracted from CHO cells following transfection with a plasmid expressing ancML or an empty vector. A scheme of the PCR amplification strategy is shown on top. The use of template DNA from plasmid or from CHO gDNA as well as a water control is indicated. M: molecular weight ladder

Table 2 Cell lines tested for replication of ancML

Cell Line	Organism	Lineage	GFP[#]	NEO[a]	L1.3[b]	ancML[c]
DF-1	*Gallus gallus*	Aves	***	+++	++	−
CRFK	*Felis catus*	Carnivora	**	+++	+	−
CV-1	*Cercopithecus aethiops*	Primates	*	++	+	−
Vero	*Cercopithecus aethiops*	Primates	***	+++	−	+
HT1080	*Homo sapiens*	Primates	**	++	+	−
HOS	*Homo sapiens*	Primates	*	++	−	−
Huh7.5	*Homo sapiens*	Primates	*	++	−	−
HeLa	*Homo sapiens*	Primates	***	+++	++	−
pgsA745	*Cricetulus griseus*	Rodentia	**	++	+	++
CHO-K1	*Cricetulus griseus*	Rodentia	**	++	+	++
MusDunni	*Mus dunni*	Rodentia	***	+++	+	−
SC-1	*Mus musculus*	Rodentia	**	++	−	−
NIH3T3	*Mus musculus*	Rodentia	**	++	++	−

All cell lines were transfected using lipofectamine 2000

[#] A plasmid expressing GFP was utilized as a transfection control. Percentage of GFP positive cells: (−) 0%, (*) < 10%, (**) 10–50%, (***) > 50%

[a] A pCR3.1 plasmid expressing a *neo* gene from a SV40 promoter was used as a control for G418 resistant colony formation

[b] The replication dependent LINE1 (L1.3) plasmid [41] was used as a control for retrotransposition and G418 resistance from the interrupted NEO reporter cassette

[c] ancML correspond to a pUC57 plasmid containing the cassette depicted in Fig. 3a

Number of G418 resistant colonies: (−) None, (+) ≤ 10, (++) > 10, (+++) > 50

of G418 resistant colonies by the ancML and L1.3 constructs was modestly reduced in the presence of azidothymidine (AZT), a retroviral RT inhibitor (Fig. 3c), while the formation of G418 resistant colonies by CHO cells transfected with a control plasmid expressing the *neo* gene was nearly unaffected. We isolated genomic DNA (gDNA) from CHO cells that had been transfected with the ancML plasmid and selected in G418, and determined the fate of the intron in DNA forms of ancML by PCR (Fig. 3d). In CHO cells transfected with the ancML plasmid, the vast majority of the amplified DNA sequences corresponded to the properly processed *neo* gene, with the intron excised (Fig. 3d).

Overall, these results indicate that the reconstructed ancestral MuERV-L sequence is replication competent and able to undergo transcription and reverse transcription upon transfection into CHO cells.

Analysis of ancML integration in CHO cells

We next determined whether ancML underwent bona fide integration into CHO cell DNA. For this purpose we used an adapter ligation-PCR technique (Genome Walker kit, Clontech) to amplify integration sites using primers specific to the MuERV-L LTR and an adaptor sequence. The risk of amplifying CHO genomic DNA from hamster ERV-L LTR sequences was minimal as these LTR sequences are quite divergent from those of ancML (only 55.8% of sequence identity) with indels and substitutions in the annealing sites for the primer sets used. Additionally, hamster ERV-L elements exist only in moderate copy numbers in the Chinese hamster genome [52]. Therefore, DNA from G418-resistant single cell clones of CHO cells, previously transfected with ancML, was digested with restriction enzymes, ligated to linkers and subjected to nested PCR reactions. This procedure revealed bands of varying sizes, sometimes consistent with multiple integrations per cell (Fig. 4a). We cloned and sequenced some of these PCR products. Although some clearly resulted from amplification of the transfected ancML plasmid DNA, we were able to identify 26 *bona-fide* integration sites with CHO genomic DNA flanking the 3′ LTR (Fig. 4b and Additional file 3: Table S2). Amplification of sequences flanking the 5′ LTR for one of these integrants revealed a five-nucleotide target site duplication (Fig. 4b), as was observed for MuERV-L sequences present in the mouse genome.

Surprisingly, 10 of the 26 3′ integration sites included a portion of the 5′ leader sequence containing various lengths of the primer-binding site (PBS) and a five-nucleotide LTR-PBS linker sequence, flanking CHO DNA (Fig. 4b). This separation of LTR and PBS is uncommon in exogenous retroviruses and has only previously been observed in another ERV, HERV-E [53]. Elimination of these five nucleotides in the ancML sequence resulted in a ~4 fold reduction in the number of G418 resistant colonies, suggesting an enhancing, but non-essential role for the five nucleotide linker in MuERV-L replication (Fig. 4c). Intriguingly, ~6.5% of the complete

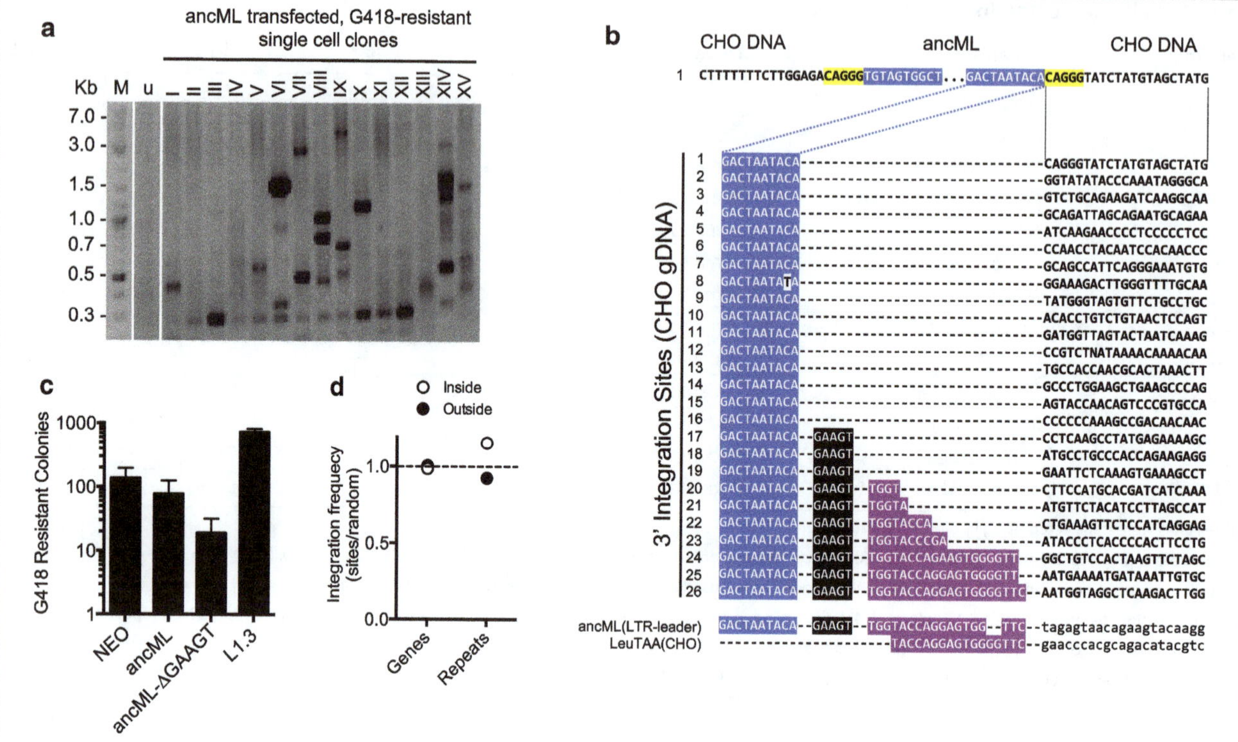

Fig. 4 Integration of ancML into CHO cell DNA. **a** Example of a genome walker experiment to determine the 3′ flanking sequence of ancML integration events in 15 single cell clones that became resistant to G418 following transfection with ancML. Nested PCR reactions were done using EcoRV digested, adapter ligated, gDNA from single G418-resistat cell clones. Forward and reverse primers were designed to anneal to the R region of the 3′LTR and to the adaptor sequence, respectively. M: molecular weight ladder. u: CHO DNA without an integrated ancML insertion. **b** Top: the sequence of an integration site with both 5′ and 3′ flanking CHO gDNA. The five-nucleotide target site duplication is indicated in yellow. Bottom: Sequences of 26 ancML integration sites in the CHO genome. Sequences of the ancML U5-PBS region as well as the Leucine (TAA) tRNA sequence are included at the bottom of the diagram. Sequence from the U5 region of the 3′ ancML LTR is indicated in blue. The 5nt linker sequence is indicated in black. The PBS sequence is indicated in purple. CHO genomic sequences are indicted in bold. Dotted lines indicate correspondence of each sequenced 3′ integration junction to the integration site at the top. **c** Enumeration of G418 resistant colonies of CHO cells transfected with plasmids expressing a *neo* gene (NEO), L1.3, ancML and an ancML construct with a deletion of the 5nt linker sequence between the 5′ LTR and the PBS (ancML ΔGAAGT). Data are mean ± SD from 3 independent experiments. **d** Distribution of ancML integration sites in genic, intergenic, or repeat regions relative to matched random controls. The measured value indicates the percentage of ancML integration sites in each population divided by that of the matched random controls (each integration site was matched to three random genomic sequences equidistant to the EcoRV site where the adaptor was ligated). The horizontal dashed line indicates no difference between the frequencies of ancML integration sites in each population compared to the matched controls

(LTR–*gag*–*pol*–LTR) proviruses in the mouse genome also contain similar sequences (5-nt linker/PBS) at the end of the 3′ LTR, thus showing that this phenomenon also occurred during ancient MuERV-L replication events (Additional file 1: Table S1). During reverse transcription, after the synthesis of the plus-strand strong-stop DNA (+sssDNA), RNase H should remove the primer tRNA, thereby exposing sequences on the +sssDNA that are complementary to the minus strand PBS which will guide the second strand transfer [54, 55]. Inefficient removal of the tRNA primer might result in the synthesis of +sssDNA that includes additional sequences 3′ to the PBS. Such a scenario might explain the unusual integration site structure that we observed for some MuERV-L and ancML insertions (Fig. 4b).

We mapped the position of the 26 ancML integration sites to the Chinese hamster genome using the UCSC genome browser (Fig. 4d) [35, 47]. The Chinese hamster genome (CriGri_1.0) is currently assembled to the scaffold level and has been annotated by distinct de novo, expression-based and homology gene prediction systems [35]. The majority of the ancML integration sites (19/26 sites) corresponded to intergenic regions, 5/26 sites corresponded to introns and one corresponded to exon 3 of *Znf462*. The single remaining site could not be classified as intergenic or in genes because it mapped to multiple scaffolds. Of the 26 integration sites, 10 were in elements corresponding to SINE (4), LINE (4) and ERV-L (2) elements. This distribution of ancML integration sites, i.e. within genes versus intergenic regions, as well as within versus outside repetitive sequences, did not differ significantly from matched randomized controls (p value = 0.97 and 0.56 respectively) (Fig. 4d). Although the distribution of the sequenced ancML integration sites and the distribution of MuERV-L elements in the mouse genome appeared different (Figs. 1d and 4b), our ancML integration site dataset was too small to establish statistical significance. Nonetheless, our results suggest that ancML integration sites are random (or close to random) in their distribution in CHO DNA, in contrast to the distribution of MuERV-L proviruses that are found in the mouse genome which have been subject to selection.

ancML is sensitive to innate host antiviral defenses

In response to exogenous microbial threats, hosts have evolved sets of genes that sense, and directly interfere with replication of pathogens. One class of such genes which are expressed in response to viral infection following induction by interferons (IFNs), cause a so-called antiviral state [56]. It is not known whether IFNs can inhibit the intracellular replication of retrotransposons. To determine whether ancML replication is affected by type-I IFNs, we transfected CHO cells with plasmids expressing ancML, L1.3 or a *neo* gene and cultured them with media containing varying amounts of murine IFNα (mIFNα) for 2 days prior to selection in G418 (Fig. 5a). The replication of L1.3 was reduced by ~4 fold upon mIFN-α treatment, and there was a larger, dose dependent effect on ancML, reaching a ~20-fold reduction in G418 resistant colony formation with 50U/ml of mIFNα (Fig. 5a). Notably, generation of G418-resistant colonies by transfected, non-replicated DNA was not affected by mIFNα. Previous studies have observed that mouse IFNα can stimulate an antiviral state in CHO cells [57–60] and promote the induction of hamster ISGs [61]. Thus, these experiments suggest that IFNα is able to inhibit one or more steps in MuERV-L replication. Interestingly pluripotent stem cells have been shown to express a subset of ISGs [62], and suppression of ERV replication may be one impetus for the acquisition of this property.

We also tested whether specific candidate innate immune effectors could inhibit ancML replication. We first tested if the murine restriction factor Fv1 (thought to have been co-opted from a MuERV-L-like element) could have had an impact on MuERV-L replication. For this we constructed CHO cells stably expressing a chimeric form of Fv1 that shows an expanded resistance to different MuLVs ($Fv1^{bbn}$) [63]. As expected, $Fv1^{bbn}$-expressing CHO cells exhibited resistance to infection by both N-tropic and B-tropic MuLV (Fig. 5b). However, $Fv1^{bbn}$-expressing CHO cells supported ancML, or L1.3 replication (Fig. 5c), at levels similar to those of control cells, indicating that ancML is insensitive to this Fv1 protein. We also tested the ability of mouse APOBEC3, that has been previously shown to inhibit endogenous and exogenous retroviruses (reviewed in [64]), to inhibit ancML replication. For this purpose, we generated CHO cell clones that stably expressing the mouse *Apobec3* in 100% of the cells (Fig. 5d). Remarkably, mouse *Apobec3* (mA3) was able to inhibit ancML replication, reducing G418 resistant colony formation by ~30-fold, but did not affect L1.3 replication (Fig. 5e). The inability of mA3 to

(See figure on next page.)

Fig. 5 MuERV-L(ancML) replication can be inhibited by innate immune effectors. **a** Enumeration of G418 resistant colonies generated in the presence of increasing amounts of mouse IFNα. CHO cells were transfected with plasmids expressing a *neo* gene (NEO), L1.3, or ancML and cultured with increasing amounts of mouse IFNα for 2 days before G418 selection. Data are mean ± SD from 3 independent experiments. **b** Infectivity of MuLV on CHO cells expressing $Fv1^{bbn}$. Percentage of MuLV infected (GFP positive) cells in CHO cells stably expressing $Fv1^{bbn}$ (red) or an empty vector (black). Circles and triangles indicate infection by N-tropic or B-tropic MuLV, respectively. **c** Enumeration of G418 resistant colonies of CHO cells expressing $Fv1^{bbn}$ or an empty vector were transfected with plasmids expressing a *neo* gene (NEO), L1.3, or ancML. Data are mean ± SD from 2 independent experiments. **d** Representative images of Immunofluorescence assays on CHO cells stably expressing an HA-tagged version of mouse APOBEC3 or an empty vector. CHO cells were fixed with 4% PFA and stained with anti-HA antibodies. **e** Enumeration of G418 resistant colonies of CHO cells expressing mouse APOBEC3. Three clones of CHO cells expressing HA-tagged mA3 or an empty vector were transfected with plasmids expressing a *neo* gene (NEO), L1.3, or ancML. Data are mean ± SD from 3 experiments with independent single cell clones. **f** and **g** Analysis of MuERV-L elements using Hypermut 2.0. Ratio of G to A mutations at preferred mA3 editing sites (RD 3′ to a G) (Y-axis) plotted against ratio of G to A mutations at disfavored mA3 editing sites (YN|RC 3′ to a G) (control ratio, X-axis). No 5′ context was imposed (**f**), or sites with a 5′ C to the mutated G were excluded (**g**). 230 gag–pol containing MuERV-L elements in the mouse genome were compared to their consensus sequence. Data points in red and orange indicate MuERV-L sequences that were statistically significantly enriched in putative mA3 induced mutations (p value < 0.05). Data points in orange represent MuERV-L elements that are statistically significantly enriched mA3-induced mutations in both analyses (p value < 0.01). **h** Profile of G to A transitions in two putatively mA3-edited MuERV-L proviral sequences compared to a consensus sequence. The profile of the reference MuERV-L sequence is shown for comparison (ML_{ref}, non significantly mA3 edited). Lines in red and cyan represent putative mA3-derived G to A transitions, not accounting for the +2 position (dinucleotide changes from GG to AG and GA to AA respectively), whereas lines in green and magenta represent non mA3-derived G to A transitions (GC to AC and GT to AT respectively). Lines in yellow indicate gaps compared to the consensus sequence

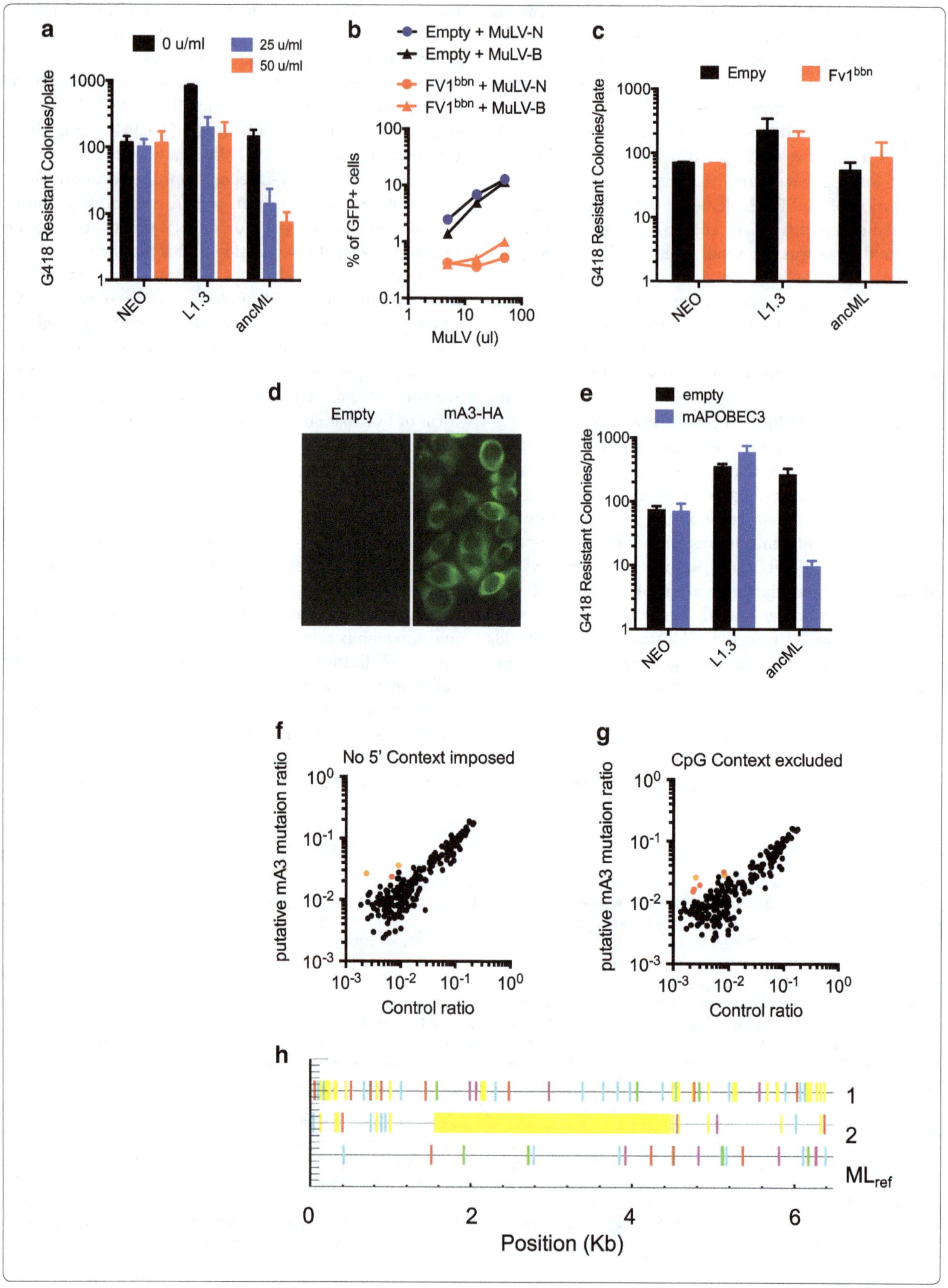
a
0 u/ml
25 u/ml
50 u/ml
G418 Resistant Colonies/plate
1000
100
10
1
NEO
L1.3
ancML
b
Empty + MuLV-N
Empty + MuLV-B
FV1bbn + MuLV-N
FV1bbn + MuLV-B
% of GFP+ cells
100
10
1
0.1
1
10
100
MuLV (ul)
c
Empy
Fv1bbn
G418 Resistant Colonies/plate
1000
100
10
1
NEO
L1.3
ancML
d
Empty
mA3-HA
e
empty
mAPOBEC3
G418 Resistant Colonies/plate
1000
100
10
1
NEO
L1.3
ancML
f
No 5' Context imposed
putative mA3 mutaion ratio
Control ratio
g
CpG Context excluded
putative mA3 mutaion ratio
Control ratio
h
1
2
MLref
0
2
4
6
Position (Kb)

restrict human L1.3 retrotransposition has been previously documented [65], while some human APOBEC3 proteins inhibit L1.3 retrotransposition [42, 66, 67], suggesting that species-dependent differences exist in the ability of APOBEC3 proteins to inhibit the replication of endogenous retroelements.

Because mA3 clearly inhibited ancML replication, and therefore might have affected MuERV-L sequence or replication in vivo, we inspected the 230 gag–pol containing MuERV-L elements that were used to derive ancML *gag* (Fig. 1f) using Hypermut 2.0 [40] (Fig. 5f–h). For each MuERV-L element we compared the number of G to A transitions in mA3-preferred motifs (5′ G(A|G)(A|G|T) 3′) with those in control sites (5′ G(C|T)N 3′ or 5′ G(A|G) C 3′) relative to a consensus sequence (Additional file 4: Table S3). Only three MuERV-L elements showed significant ($p<0.05$) evidence of mA3 dependent hypermutation when no 5′ context was enforced (Fig. 5f). Because spontaneous deamination of methylated CpG dinucleotides can also produce G to A transitions, we performed the same analysis after excluding sites containing a C nucleotide 5′ to the mutated G. When these sites were excluded, 10 MuERV-L elements showed a significant ($p<0.05$) evidence for mA3 dependent hypermutation (Fig. 5g). Only two MuERV-L elements exhibited statistically significant evidence of mA3 dependent hypermutation in both analyses (p value <0.01), and both of these elements carried a relatively low mutational burden (Fig. 5f–h). Thus, although ancML replication can be inhibited by mA3, analysis of MuERV-L proviruses in the mouse genome suggests that MuERV-L either rarely encountered mA3, or is inhibited in a manner that prevents the deposition of hypermutated proviruses.

Discussion

Here, we report the successful reconstruction of a ~2MY old replication competent ancestral MuERV-L sequence, through the analysis of a recently expanded subset of fossilized MuERV-L elements in the mouse genome. According to previous studies [16], and corroborated here, MuERV-L originated ~10 MYA, after the *Rattus–Mus* split and underwent a prolific expansion ~2 MYA. In fact, almost 65% of solo LTRs and MuERV-L proviruses identified herein have an estimated integration date of < 3 MYA. Furthermore, the estimated dates of solo LTRs follow a bimodal distribution (a major one centered ~3MYA and the other ~8MYA) consistent with the estimated times of both expansions (Additional file 1: Table S1). A combination of homology searches and defragmentation methods provided the material for the estimation of the sequence of the ~2MY old replication-competent ancestor.

Other highly abundant *env*-defective ERVs typically appear to be derived from closely related elements that possess an *env* gene. While other closely related elements do possess an *env* gene, there are no documented ERV-L elements that encode an *env*. It is likely, therefore, that an ancestral ERV-L element lacked an *env* gene. Thus, the bulk of MuERV-L replication likely occurred through entirely intracellular retrotransposon-like mechanisms [21]. Moreover, the bulk of MuERV-L replication likely occurred in early embryos, as the expression of MuERV-L elements appears to be restricted to the 2-cell embryo, although It is unknown whether this property is confined to the subset of elements that proliferated ~2MYA. It is possible that the early embryonic environment is also necessary in some other way for MuERV-L replication given its apparently restricted tropism in cell lines. In particular, it is intriguing that (and as yet unexplained why) MuERV-L only replicated with reasonable efficiency in Chinese hamster ovary cells, even when provided with a promoter that should drive its expression in nearly any cell type.

MuERV-L belongs to an ancient mammalian ERV family (which originated >100 MYA [18]) that is distantly related to spumaviruses. Therefore, modern functional viral sequences are therefore not useful for attempts to increase the replicative efficiency of ancML. Remarkably, there is a high number of MuERV-L proviruses that have retained their coding potential, and share a high degree of sequence similarity to the functional ancML (with only few coding differences and overall nucleotide identity ranging from 96.16 to 99.31%). However, currently there is no evidence that the ongoing expression of MuERV-L elements at the two-cell stage of the mouse embryo results in successful re-integration, although it is possible that MuERV-L replication and reintegration occurs in modern mouse embryos at some very low rate. Nevertheless, examination of recent *bona fide* integrations might highlight important residues that might be altered to improve ancML replication and/or integration.

We found that mouse IFNα was able to inhibit ancML replication, suggesting that interferon stimulated genes can directly inhibit MuERV-L replication, possibly leading to its recent extinction as a replication competent entity. Alternatively, early embryos may express antiviral proteins that inhibit re-integration of modern MuERV-L elements that would otherwise be intrinsically replication competent [62]. We found that mouse APOBEC3 inhibits ancML replication, but mutational profiles of MuERV-L elements in the mouse genome provide minimal evidence for mA3-dependent hypermutation as a mechanism for inhibition in vivo. During mouse development, mA3 is expressed at the two-cell stage, increasing at the four-cell stage to become one of the top 30% most highly expressed genes [68]. Thus, it is at least possible that mA3 may have acted on replicating MuERV-L elements, perhaps in part through deaminase-independent mechanisms [69].

Despite the apparently random integration pattern of ancML, the analysis of fixed MuERV-L elements showed that there has been a selective pressure to eliminate MuERV-L integrations from genes. Conversely, MuERV-L related sequences (Fv1) have clearly been positively selected to provide defense against retroviral infection [22–24] and recent studies have suggested that regulatory elements of MuERV-L LTRs may have been co-opted to regulate the expression of numerous genes during embryogenesis [14, 15]. While Fv1 arose at least ~5–7 MYA, it is unclear whether the potential exaptation of MuERV-L regulatory sequences occurred during the 10MYA expansion or the more recent ~2 MYA expansion. Nonetheless, there appears to be both a benefit (co-option for antiviral defense and regulation of embryogenesis) and cost (disruption of gene function) associated with the presence of MuERV-L elements in the mouse genome.

MuERV-L appears to be the only member of the ERV-L family that seems to have been reactivated in recent evolutionary times. It is particularly intriguing that the recent expansion is characterized by an in-frame deletion in *gag*, as it could be this deletion the responsible for releasing some MuERV-L elements from the deleterious effects of a hypothetical inhibitory factor ~2MYA. Recent studies have shown the fundamental role that some endogenous retroviral sequences may play in mammalian development and protection from exogenous retroviral infection [15, 23, 24, 70–73]. Indeed one report has suggested that knockdown of MuERV-L transcripts impacts embryonic development [74]. Nevertheless, it remains to be determined whether the current presence of MuERV-L transcripts, proteins and virus-like particles at the two-cell stage of the mouse embryo might be beneficial or deleterious to the mouse.

Conclusions

The reconstruction of an ancestral MuERV-L sequence highlights the potential for the retroviral fossil record to illuminate ancient events and represents a unique opportunity to study ERV-L biology and reactivation, the role of MuERV-L in mouse development and potentially uncover new roles for ERVs in mammalian biology.

Additional files

Additional file 1: Table S1. MuERV-L loci identified in mouse genome assemblies.

Additional file 2.ancML sequence in FASTA format.

Additional file 3: Table S2. ancML integration sites cloned from CHO gDNA.

Additional file 4: Table S3. Analysis for hypermutation in MuERV-L elements in the mouse genome.

Authors' contributions

DBM performed all the experimental work and analysis. RG supervised the computational work and PDB supervised the experimental work. All authors wrote, read and approved the final manuscript.

Author details

[1] Laboratory of Retrovirology and Howard Hughes Medical Institute, The Rockefeller University, New York, NY, USA. [2] MRC-University of Glasgow Centre for Virus Research, Glasgow, UK. [3] Present Address: Department of Microbiology, Icahn School of Medicine at Mount Sinai, New York, NY, USA.

Acknowledgements

We thank Dr. John V. Moran for kindly sharing the L1.3 plasmid (JM101), Theodora Hatziioannou for the plasmid expressing Fv1^{bbn}, and Rachel Liberatore for the plasmid expressing mouse APOBEC3. We also thank all the members of the Bieniasz lab for their help and suggestions on the project.

Competing interests

The authors declare that they have no competing interests.

Funding

This work was supported by a grant from the National Institute of Allergy and Infectious diseases (R3764003 to PDB) and a grant from the UK Medical Research Council (MC_UU_12014/10 to RJG).

References

1. Weiss RA. The discovery of endogenous retroviruses. Retrovirology. 2006;3:67.
2. Lander ES, Linton LM, Birren B, Nusbaum C, Zody MC, Baldwin J, Devon K, Dewar K, Doyle M, FitzHugh W, et al. Initial sequencing and analysis of the human genome. Nature. 2001;409(6822):860–921.
3. Emerman M, Malik HS. Paleovirology—modern consequences of ancient viruses. PLoS Biol. 2010;8(2):e1000301.
4. Dewannieux M, Harper F, Richaud A, Letzelter C, Ribet D, Pierron G, Heidmann T. Identification of an infectious progenitor for the multiple-copy HERV-K human endogenous retroelements. Genome Res. 2006;16(12):1548–56.
5. Lee YN, Bieniasz PD. Reconstitution of an infectious human endogenous retrovirus. PLoS Pathog. 2007;3(1):e10.
6. Perez-Caballero D, Soll SJ, Bieniasz PD. Evidence for restriction of ancient primate gammaretroviruses by APOBEC3 but not TRIM5alpha proteins. PLoS Pathog. 2008;4(10):e1000181.
7. Kaiser SM, Malik HS, Emerman M. Restriction of an extinct retrovirus by the human TRIM5alpha antiviral protein. Science. 2007;316(5832):1756–8.
8. Goldstone DC, Yap MW, Robertson LE, Haire LF, Taylor WR, Katzourakis A, Stoye JP, Taylor IA. Structural and functional analysis of prehistoric lentiviruses uncovers an ancient molecular interface. Cell Host Microbe. 2010;8(3):248–59.
9. Soll SJ, Neil SJ, Bieniasz PD. Identification of a receptor for an extinct virus. Proc Natl Acad Sci USA. 2010;107(45):19496–501.
10. Blanco-Melo D, Gifford RJ, Bieniasz PD. Co-option of an endogenous retrovirus envelope for host defense in hominid ancestors. Elife. 2017;6:e22519.
11. Lee YN, Malim MH, Bieniasz PD. Hypermutation of an ancient human retrovirus by APOBEC3G. J Virol. 2008;82(17):8762–70.
12. Kigami D, Minami N, Takayama H, Imai H. MuERV-L is one of the earliest transcribed genes in mouse one-cell embryos. Biol Reprod. 2003;68(2):651–4.
13. Ribet D, Louvet-Vallee S, Harper F, de Parseval N, Dewannieux M, Heidmann O, Pierron G, Maro B, Heidmann T. Murine endogenous retrovirus MuERV-L is the progenitor of the "orphan" epsilon viruslike particles of the early mouse embryo. J Virol. 2008;82(3):1622–5.
14. Macfarlan TS, Gifford WD, Agarwal S, Driscoll S, Lettieri K, Wang J, Andrews SE, Franco L, Rosenfeld MG, Ren B, et al. Endogenous retroviruses and neighboring genes are coordinately repressed by LSD1/KDM1A. Genes Dev. 2011;25(6):594–607.

15. Macfarlan TS, Gifford WD, Driscoll S, Lettieri K, Rowe HM, Bonanomi D, Firth A, Singer O, Trono D, Pfaff SL. Embryonic stem cell potency fluctuates with endogenous retrovirus activity. Nature. 2012;487(7405):57–63.
16. Costas J. Molecular characterization of the recent intragenomic spread of the murine endogenous retrovirus MuERV-L. J Mol Evol. 2003;56(2):181–6.
17. Benit L, Lallemand JB, Casella JF, Philippe H, Heidmann T. ERV-L elements: a family of endogenous retrovirus-like elements active throughout the evolution of mammals. J Virol. 1999;73(4):3301–8.
18. Lee A, Nolan A, Watson J, Tristem M. Identification of an ancient endogenous retrovirus, predating the divergence of the placental mammals. Philos Trans R Soc Lond B Biol Sci. 2013;368(1626):20120503.
19. Cordonnier A, Casella JF, Heidmann T. Isolation of novel human endogenous retrovirus-like elements with foamy virus-related pol sequence. J Virol. 1995;69(9):5890–7.
20. Benit L, De Parseval N, Casella JF, Callebaut I, Cordonnier A, Heidmann T. Cloning of a new murine endogenous retrovirus, MuERV-L, with strong similarity to the human HERV-L element and with a gag coding sequence closely related to the Fv1 restriction gene. J Virol. 1997;71(7):5652–7.
21. Magiorkinis G, Gifford RJ, Katzourakis A, De Ranter J, Belshaw R. Env-less endogenous retroviruses are genomic superspreaders. Proc Natl Acad Sci USA. 2012;109(19):7385–90.
22. Best S, Le Tissier P, Towers G, Stoye JP. Positional cloning of the mouse retrovirus restriction gene Fv1. Nature. 1996;382(6594):826–9.
23. Yan Y, Buckler-White A, Wollenberg K, Kozak CA. Origin, antiviral function and evidence for positive selection of the gammaretrovirus restriction gene Fv1 in the genus Mus. Proc Natl Acad Sci USA. 2009;106(9):3259–63.
24. Yap MW, Colbeck E, Ellis SA, Stoye JP. Evolution of the retroviral restriction gene Fv1: inhibition of non-MLV retroviruses. PLoS Pathog. 2014;10(3):e1003968.
25. Gualdar D, Perez-Palacios R, Climent M, Martinez-Abadia I, Larraga A, Fernandez-Juan M, Vallejo C, Muniesa P, Schoorlemmer J. Expression of endogenous retroviruses is negatively regulated by the pluripotency marker Rex1/Zfp42. Nucleic Acids Res. 2012;40(18):8993–9007.
26. Rowe HM, Trono D. Dynamic control of endogenous retroviruses during development. Virology. 2011;411(2):273–87.
27. Camacho C, Coulouris G, Avagyan V, Ma N, Papadopoulos J, Bealer K, Madden TL. BLAST+: architecture and applications. BMC Bioinformatics. 2009;10:421.
28. Mural RJ, Adams MD, Myers EW, Smith HO, Miklos GL, Wides R, Halpern A, Li PW, Sutton GG, Nadeau J, et al. A comparison of whole-genome shotgun-derived mouse chromosome 16 and the human genome. Science. 2002;296(5573):1661–71.
29. Mouse Genome Sequencing Consortium, Waterston RH, Lindblad-Toh K, Birney E, Rogers J, Abril JF, Agarwal P, Agarwala R, Ainscough R, Alexandersson M, et al. Initial sequencing and comparative analysis of the mouse genome. Nature. 2002;420(6915):520–62.
30. Hubbard T, Barker D, Birney E, Cameron G, Chen Y, Clark L, Cox T, Cuff J, Curwen V, Down T, et al. The Ensembl genome database project. Nucleic Acids Res. 2002;30(1):38–41.
31. Swofford DL. PAUP*. Phylogenetic analysis using parsimony (*and other methods). Version 4. Sunderland: Sinauer Associates; 2002.
32. Lebedev YB, Belonovitch OS, Zybrova NV, Khil PP, Kurdyukov SG, Vinogradova TV, Hunsmann G, Sverdlov ED. Differences in HERV-K LTR insertions in orthologous loci of humans and great apes. Gene. 2000;247(1–2):265–77.
33. Subramanian RP, Wildschutte JH, Russo C, Coffin JM. Identification, characterization, and comparative genomic distribution of the HERV-K (HML-2) group of human endogenous retroviruses. Retrovirology. 2011;8:90.
34. Marshall HM, Ronen K, Berry C, Llano M, Sutherland H, Saenz D, Bickmore W, Poeschla E, Bushman FD. Role of PSIP1/LEDGF/p75 in lentiviral infectivity and integration targeting. PLoS ONE. 2007;2(12):e1340.
35. Xu X, Nagarajan H, Lewis NE, Pan S, Cai Z, Liu X, Chen W, Xie M, Wang W, Hammond S, et al. The genomic sequence of the Chinese hamster ovary (CHO)-K1 cell line. Nat Biotechnol. 2011;29(8):735–41.
36. Edgar RC. MUSCLE: multiple sequence alignment with high accuracy and high throughput. Nucleic Acids Res. 2004;32(5):1792–7.
37. Stamatakis A. RAxML version 8: a tool for phylogenetic analysis and post-analysis of large phylogenies. Bioinformatics. 2014;30(9):1312–3.
38. Yang Z. PAML: a program package for phylogenetic analysis by maximum likelihood. Comput Appl Biosci. 1997;13(5):555–6.
39. Darriba D, Taboada GL, Doallo R, Posada D. jModelTest 2: more models, new heuristics and parallel computing. Nat Methods. 2012;9(8):772.
40. Rose PP, Korber BT. Detecting hypermutations in viral sequences with an emphasis on G → A hypermutation. Bioinformatics. 2000;16(4):400–1.
41. Moran JV, DeBerardinis RJ, Kazazian HH Jr. Exon shuffling by L1 retrotransposition. Science. 1999;283(5407):1530–4.
42. Kinomoto M, Kanno T, Shimura M, Ishizaka Y, Kojima A, Kurata T, Sata T, Tokunaga K. All APOBEC3 family proteins differentially inhibit LINE-1 retrotransposition. Nucleic Acids Res. 2007;35(9):2955–64.
43. Mariani R, Chen D, Schrofelbauer B, Navarro F, Konig R, Bollman B, Munk C, Nymark-McMahon H, Landau NR. Species-specific exclusion of APOBEC3G from HIV-1 virions by Vif. Cell. 2003;114(1):21–31.
44. Hatziioannou T, Cowan S, Bieniasz PD. Capsid-dependent and -independent postentry restriction of primate lentivirus tropism in rodent cells. J Virol. 2004;78(2):1006–11.
45. Soneoka Y, Cannon PM, Ramsdale EE, Griffiths JC, Romano G, Kingsman SM, Kingsman AJ. A transient three-plasmid expression system for the production of high titer retroviral vectors. Nucleic Acids Res. 1995;23(4):628–33.
46. Kent WJ. BLAT—the BLAST-like alignment tool. Genome Res. 2002;12(4):656–64.
47. Kent WJ, Sugnet CW, Furey TS, Roskin KM, Pringle TH, Zahler AM, Haussler D. The human genome browser at UCSC. Genome Res. 2002;12(6):996–1006.
48. Brady T, Lee YN, Ronen K, Malani N, Berry CC, Bieniasz PD, Bushman FD. Integration target site selection by a resurrected human endogenous retrovirus. Genes Dev. 2009;23(5):633–42.
49. Soh YQ, Alfoldi J, Pyntikova T, Brown LG, Graves T, Minx PJ, Fulton RS, Kremitzki C, Koutseva N, Mueller JL, et al. Sequencing the mouse Y chromosome reveals convergent gene acquisition and amplification on both sex chromosomes. Cell. 2014;159(4):800–13.
50. Qin C, Wang Z, Shang J, Bekkari K, Liu R, Pacchione S, McNulty KA, Ng A, Barnum JE, Storer RD. Intracisternal A particle genes: distribution in the mouse genome, active subtypes, and potential roles as species-specific mediators of susceptibility to cancer. Mol Carcinog. 2010;49(1):54–67.
51. Moran JV, Holmes SE, Naas TP, DeBerardinis RJ, Boeke JD, Kazazian HH Jr. High frequency retrotransposition in cultured mammalian cells. Cell. 1996;87(5):917–27.
52. Lewis NE, Liu X, Li Y, Nagarajan H, Yerganian G, O'Brien E, Bordbar A, Roth AM, Rosenbloom J, Bian C, et al. Genomic landscapes of Chinese hamster ovary cell lines as revealed by the Cricetulus griseus draft genome. Nat Biotechnol. 2013;31(8):759–65.
53. Repaske R, Steele PE, O'Neill RR, Rabson AB, Martin MA. Nucleotide sequence of a full-length human endogenous retroviral segment. J Virol. 1985;54(3):764–72.
54. Coffin JM, Hughes SH, Varmus H. Retroviruses. Plainview, NY: Cold Spring Harbor Laboratory Press; 1997.
55. Champoux JJ, Schultz SJ. Ribonuclease H: properties, substrate specificity and roles in retroviral reverse transcription. FEBS J. 2009;276(6):1506–16.
56. Schoggins JW, Wilson SJ, Panis M, Murphy MY, Jones CT, Bieniasz P, Rice CM. A diverse range of gene products are effectors of the type I interferon antiviral response. Nature. 2011;472(7344):481–5.
57. Zwarthoff EC, Bosveld IJ, Vonk WP, Trapman J. Constitutive expression of a murine interferon alpha gene in hamster cells and characterization of its protein product. J Gen Virol. 1985;66(Pt 4):685–91.
58. Van Heuvel M, Bosveld IJ, Mooren AA, Trapman J, Zwarthoff EC. Properties of natural and hybrid murine alpha interferons. J Gen Virol. 1986;67(Pt 10):2215–22.
59. Trapman J, van Heuvel M, de Jonge P, Bosveld IJ, Klaassen P, Zwarthoff EC. Structure-function analysis of mouse interferon alpha species: MuIFN-alpha 10, a subspecies with low antiviral activity. J Gen Virol. 1988;69(Pt 1):67–75.
60. Van Heuvel M, Bosveld IJ, Klaassen P, Zwarthoff EC, Trapman J. Structure-function analysis of murine interferon-alpha: antiviral properties of novel hybrid interferons. J Interferon Res. 1988;8(1):5–14.
61. van Heuvel M, Govaert-Siemerink M, Bosveld IJ, Zwarthoff EC, Trapman J. Interferon-alpha-(IFN) producing CHO cell lines are resistant to the antiproliferative activity of IFN: a correlation with gene expression. J Cell Biochem. 1988;38(4):269–78.

62. Wu X, Dao Thi VL, Huang Y, Billerbeck E, Saha D, Hoffmann HH, Wang Y, Silva LAV, Sarbanes S, Sun T, et al. Intrinsic immunity shapes viral resistance of stem cells. Cell. 2018;172(3):423–38.
63. Bock M, Bishop KN, Towers G, Stoye JP. Use of a transient assay for studying the genetic determinants of Fv1 restriction. J Virol. 2000;74(16):7422–30.
64. Rehwinkel J. Mouse knockout models for HIV-1 restriction factors. Cell Mol Life Sci. 2014;71(19):3749–66.
65. Lovsin N, Peterlin BM. APOBEC3 proteins inhibit LINE-1 retrotransposition in the absence of ORF1p binding. Ann N Y Acad Sci. 2009;1178:268–75.
66. Muckenfuss H, Hamdorf M, Held U, Perkovic M, Lower J, Cichutek K, Flory E, Schumann GG, Munk C. APOBEC3 proteins inhibit human LINE-1 retrotransposition. J Biol Chem. 2006;281(31):22161–72.
67. Stenglein MD, Harris RS. APOBEC3B and APOBEC3F inhibit L1 retrotransposition by a DNA deamination-independent mechanism. J Biol Chem. 2006;281(25):16837–41.
68. Xie D, Chen CC, Ptaszek LM, Xiao S, Cao X, Fang F, Ng HH, Lewin HA, Cowan C, Zhong S. Rewirable gene regulatory networks in the preimplantation embryonic development of three mammalian species. Genome Res. 2010;20(6):804–15.
69. MacMillan AL, Kohli RM, Ross SR. APOBEC3 inhibition of mouse mammary tumor virus infection: the role of cytidine deamination versus inhibition of reverse transcription. J Virol. 2013;87(9):4808–17.
70. Lavialle C, Cornelis G, Dupressoir A, Esnault C, Heidmann O, Vernochet C, Heidmann T. Paleovirology of 'syncytins', retroviral env genes exapted for a role in placentation. Philos Trans R Soc Lond B Biol Sci. 2013;368(1626):20120507.
71. Wang J, Xie G, Singh M, Ghanbarian AT, Rasko T, Szvetnik A, Cai H, Besser D, Prigione A, Fuchs NV, et al. Primate-specific endogenous retrovirus-driven transcription defines naive-like stem cells. Nature. 2014;516(7531):405–9.
72. Lu X, Sachs F, Ramsay L, Jacques PE, Goke J, Bourque G, Ng HH. The retrovirus HERVH is a long noncoding RNA required for human embryonic stem cell identity. Nat Struct Mol Biol. 2014;21(4):423–5.
73. Armezzani A, Varela M, Spencer TE, Palmarini M, Arnaud F. "Menage a trois": the evolutionary interplay between JSRV, enJSRVs and domestic sheep. Viruses. 2014;6(12):4926–45.
74. Huang Y, Kim JK, Do DV, Lee C, Penfold CA, Zylicz JJ, Marioni JC, Hackett JA, Surani MA. Stella modulates transcriptional and endogenous retrovirus programs during maternal-to-zygotic transition. Elife. 2017;6:e22345.

14

Impact of the HIV-1 genetic background and HIV-1 population size on the evolution of raltegravir resistance

Axel Fun[1], Thomas Leitner[2], Linos Vandekerckhove[3], Martin Däumer[4], Alexander Thielen[5], Bernd Buchholz[6], Andy I. M. Hoepelman[7], Elizabeth H. Gisolf[8], Pauline J. Schipper[1], Annemarie M. J. Wensing[1,7] and Monique Nijhuis[1*]

Abstract

Background: Emergence of resistance against integrase inhibitor raltegravir in human immunodeficiency virus type 1 (HIV-1) patients is generally associated with selection of one of three signature mutations: Y143C/R, Q148K/H/R or N155H, representing three distinct resistance pathways. The mechanisms that drive selection of a specific pathway are still poorly understood. We investigated the impact of the HIV-1 genetic background and population dynamics on the emergence of raltegravir resistance. Using deep sequencing we analyzed the integrase coding sequence (CDS) in longitudinal samples from five patients who initiated raltegravir plus optimized background therapy at viral loads > 5000 copies/ml. To investigate the role of the HIV-1 genetic background we created recombinant viruses containing the viral integrase coding region from pre-raltegravir samples from two patients in whom raltegravir resistance developed through different pathways. The in vitro selections performed with these recombinant viruses were designed to mimic natural population bottlenecks.

Results: Deep sequencing analysis of the viral integrase CDS revealed that the virological response to raltegravir containing therapy inversely correlated with the relative amount of unique sequence variants that emerged suggesting diversifying selection during drug pressure. In 4/5 patients multiple signature mutations representing different resistance pathways were observed. Interestingly, the resistant population can consist of a single resistant variant that completely dominates the population but also of multiple variants from different resistance pathways that coexist in the viral population. We also found evidence for increased diversification after stronger bottlenecks. In vitro selections with low viral titers, mimicking population bottlenecks, revealed that both recombinant viruses and HXB2 reference virus were able to select mutations from different resistance pathways, although typically only one resistance pathway emerged in each individual culture.

Conclusions: The generation of a specific raltegravir resistant variant is not predisposed in the genetic background of the viral integrase CDS. Typically, in the early phases of therapy failure the sequence space is explored and multiple resistance pathways emerge and then compete for dominance which frequently results in a switch of the dominant population over time towards the fittest variant or even multiple variants of similar fitness that can coexist in the viral population.

*Correspondence: m.nijhuis@umcutrecht.nl

[1] Department of Medical Microbiology, Virology, University Medical Center Utrecht, Heidelberglaan 100, HP G04.614, 3584 CX Utrecht, The Netherlands

Full list of author information is available at the end of the article

Background

Currently, viral replication is successfully suppressed in the majority of HIV-infected patients treated with combination antiretroviral therapy (cART) [1]. However, virological failure associated with the emergence of drug resistant viruses may still limit the success of cART. The emergence of drug resistance in HIV is a direct consequence of the high error-rate of the HIV reverse transcriptase (RT) enzyme [2–4]. The frequent incorrect nucleotide incorporations result in evolution of the viral population and generate a myriad of viral variants upon which selective forces may act. The population size and replication rate are important viral parameters that contribute to the probability that resistance emerges and to how HIV-1 drug resistance evolves [5–9].

Integrase inhibitors comprise a class of antiretroviral drugs that specifically prevent the integration of the viral genome into the human genome. Raltegravir is the first representative of a class of integrase inhibitors that target the strand transfer reaction (INSTIs) of the viral DNA into the host genome which is performed by the viral enzyme integrase. Like other INSTIs, raltegravir preferentially binds and inhibits the viral DNA-integrase complex (intasome) over unbound integrase [10–12].

It was the first integrase inhibitor used in clinical practice (since 2007) but was recently registered by the FDA for once daily dosing [13] and is very well tolerated [14] due to a low toxicity profile [15]. Resistance is commonly associated with selection of one of the signature raltegravir resistance mutations Y143C/R/H, Q148H/K/R or N155H [16–18]. Mutations at each of these three amino acids represent a distinct resistance pathway and all signature mutations are associated with reductions in viral replication [17, 19, 20]. Accumulation of secondary resistance mutations is often associated with a greater loss of drug susceptibility and/or improved viral fitness [21–23]. Different mutational combinations vary greatly in their impact on raltegravir susceptibility and viral replication. In general, substitutions at amino acid position 148 confer higher levels of resistance than substitutions at amino acid Y143 or N155. The G140S plus Q148H combination is considered the most resistant variant and has little effect on viral replication. The resistance patterns observed in HIV-1 patients on a raltegravir containing regimen are very diverse and Q148H/K/R (usually with G140A/C/S and/or E138A/K) and N155H (often together with E92Q or V151I) mutations are observed more frequently than Y143 mutations [24]. The different resistance pathways are believed to be mutually exclusive and multiple primary mutations (especially 148 + 155 mutations) are generally not observed on the same viral genome [25]. Remarkably, replacement of the dominant resistant population by a viral population with a completely different resistance pattern during continuous non-suppressive INSTI therapy has been observed [26–29].

The mechanisms that drive selection and switching of resistance pathways are inadequately understood. Understanding these mechanisms is essential in view of other INSTIs that are in clinical use (elvitegravir and dolutegravir) or in clinical trial (bictegravir and cabotegravir), since their resistance profiles partially overlap with that of raltegravir [25]. For instance, raltegravir resistance mutations Q148H/K/R and N155H show a high level of cross-resistance with elvitegravir, but elvitegravir susceptibility is unaffected by Y143 mutations [30, 31], a difference that is beautifully explained by crystal structures of the intasome in presence of raltegravir or elvitegravir [11]. The VIKING studies demonstrated dolutegravir's superior resistance profile attested by sustained activity against all raltegravir resistant variants, except for viruses with a mutation at amino acid 148 in combination with at least one secondary mutation at position 138 and/or 140 [32–34]. These observations were corroborated by in vitro analysis of resistance profiles [35–37] which also uncovered two atypical INSTI resistance mutations (G118R and F121Y) that confer pan-INSTI resistance [38, 39].

We investigated the impact of the HIV-1 genetic background and population size on the evolution of raltegravir resistance and their role in determining selection of a particular resistance pathway. Using next-generation sequencing (NGS) we analyzed patient-derived viral integrase sequences from samples taken before and during raltegravir therapy failure. Frequency and dynamics analysis of the deep sequencing data was used to evaluate intra-patient evolution of resistance.

To investigate the role of the genetic background we generated recombinant viruses containing the viral integrase CDS from pre-raltegravir samples from two patients experiencing virological failure receiving raltegravir therapy. With these recombinant viruses we conducted multiple low multiplicity of infection (MOI) in vitro selections in parallel, with the advantage that all resistant variants generated are allowed to replicate and different raltegravir resistance pathways can be identified.

Results

During raltegravir resistance development, multiple resistant variants emerge that compete to become the dominant variant

We studied five patients who initiated raltegravir therapy as part of a cART regimen and subsequently demonstrated virological failure to the raltegravir containing regimen. Of note, all 5 patients were heavily pre-treated and raltegravir was part of a (partly) compromised backbone. This may partially explain the moderate virological responses and limited therapy success in these patients. Patient (Pt) 1, Pt2, Pt3 and Pt5 were infected with HIV-1 subtype B strains, Pt4 was infected with a HIV-1 CRF02_AG strain. All patients started raltegravir therapy with viral loads > 5000 copies/ml (c/ml) of HIV-1 RNA, as measured in the last viral load test before initiation of raltegravir therapy. To investigate if the emerged resistant variants existed as minority variants before raltegravir therapy and how the resistant population evolved we analyzed longitudinal samples from these patients by NGS.

Patient 1

HIV-1 RNA initially decreased after start of raltegravir containing cART (2.2 log decrease in HIV-1 RNA), but the viral load rebounded quickly after therapy initiation (< 81 days, Fig. 1a). Population sequencing revealed presence of raltegravir resistance mutations (E138E/K + Q148Q/K/R + N155H/H) and raltegravir was discontinued from the regimen after 124 days. NGS revealed very small populations at baseline containing Q148R and E138K (0.1% of the population, Fig. 2a), but they were not present on the same genomes (Fig. 2b). 40 days after start of raltegravir, virus with Q148R had increased to 1.7% of the population and virus with E138K + Q148K had increased to 0.5% (Fig. 2a, b). These two variants dominated the population after 90 days (red nodes, Fig. 1a) but a third resistant variant appeared, N155H which already comprised 13% of the population. In the subsequent sample (153 days after start of raltegravir, black nodes Fig. 1a) the N155H variant replaced the Q148R variant and dominated the population together with the E138K + Q148K variant. Surprisingly, 4 weeks after raltegravir discontinuation, wild-type virus had not reseeded the viral population which was consisted entirely of raltegravir resistant variants (Fig. 2).

Patient 2

Pt2 was off therapy for a few weeks due to toxicity related issues but a new five-drug regimen including raltegravir and darunavir resulted in a rapid decline of the viral load but was discontinued again because of darunavir-related toxicity (Fig. 1b). Shortly thereafter, the same regimen was restarted without darunavir resulting in a further decline of the viral load to undetectable levels. After brief virological suppression (HIV-1 RNA < 50 c/ml, a 3.0 log decrease in HIV-1 RNA), viral load rebounded and population sequencing showed gradual accumulation of raltegravir resistance mutations: initially primary resistance mutation N155H appeared followed by two secondary resistance mutations, first Q95K and later V151I. NGS revealed no major raltegravir resistance mutations at baseline or mutations at positions 143 or 148 at any time-point, not even as minority variants (Fig. 2).

Patient 3

Pt3 only showed a partial virological response to the raltegravir containing therapy (a 1.0 log decrease in HIV-1 RNA was observed) and full suppression was never achieved. The viral load rebounded shortly after start of raltegravir containing cART (< 70 days, Fig. 1c) and the viral population contained N155H mutants. In a later time-point, 169 days after start of raltegravir, this population was replaced by a variant with primary mutation Y143R and several secondary mutations including L74M, T97A and G163E. NGS revealed only secondary mutations in the sample before raltegravir therapy, mostly as very low frequency variants (Fig. 2), including T97A and Y143C (both 0.2%). In a sample taken 28 days later, 77.5% of the population contained N155H, 15.4% had T97A and variants with Y143C were not detected anymore. In the subsequent sample 70 days after start of raltegravir therapy, N155H had increased to 88.6% and Y143C had reappeared in 11.8% of the reads. Y143C and N155H did not appear to be on the same genome (Fig. 2b). After 169 days the resistant population had shifted dramatically, with Y143R making up 99.7% of the population and complete absence of N155H. Mutations L74M, T97A and G163R appeared in nearly the entire viral population.

Patient 4

Pt4 was also a partial responder (1.1 log decrease in HIV-1 RNA) in whom virological suppression was not achieved during 48 weeks of raltegravir containing therapy. Sanger sequencing revealed only very small (< 20%) populations of Q148H/Q148R and N155H and resistant variants never dominated the population (Fig. 1d). Before raltegravir therapy, mutation Y143H was detected at low frequency (0.2%) by NGS but did not appear in any of the later samples during raltegravir treatment. In samples 8 and 81 days after start of raltegravir therapy also no significant raltegravir resistance mutations were observed (Fig. 2a). However, in the sample 123 days after raltegravir therapy initiation several primary and secondary raltegravir resistance mutations occurred. N155H was found in 14.4%, of the population, Q148H in 7.5% of which approximately half (4.1%) in combination with G140S and the remainder in combination with E157D (Fig. 2b). Q148R was present in 2.4% of the population and G163R in 2.7%.

Patient 5

In addition to a compromised backbone, this patient also had sub-therapeutic levels of raltegravir levels due to a drug–drug interaction with rifampicin (an interaction unknown at time of raltegravir prescription). Despite the suboptimal levels of raltegravir, switching to a raltegravir containing regimen quickly resulted in complete viral suppression (< 26 days, 3.3 log decrease in HIV-1 RNA) but viral load rebounded within 21 weeks (Fig. 1e). Population sequencing (sample 457 days after initiation of raltegravir) demonstrated that virus with integrase substitutions G140S + Q148H completely dominated the viral population.

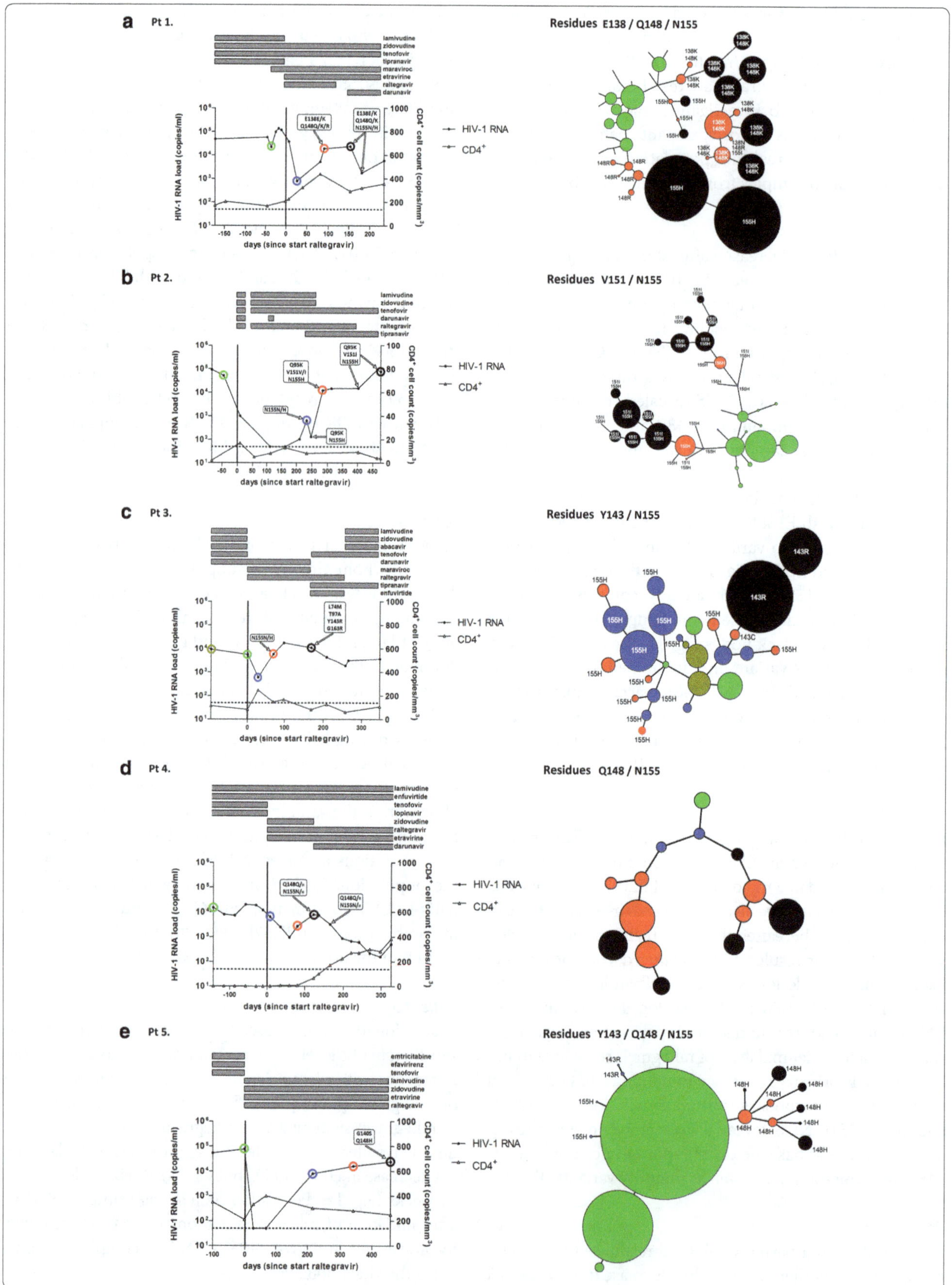
a Pt 1.
Residues E138 / Q148 / N155
b Pt 2.
Residues V151 / N155
c Pt 3.
Residues Y143 / N155
d Pt 4.
Residues Q148 / N155
e Pt 5.
Residues Y143 / Q148 / N155
HIV-1 RNA
CD4+
HIV-1 RNA load (copies/ml)
CD4+ cell count (copies/mm3)
days (since start raltegravir)

(See figure on previous page.)

Fig. 1 Development of raltegravir resistance during raltegravir containing cART. Left hand panels: therapy history, HIV-1 RNA load, CD4^{+} cell count and resistance mutations detected by population sequencing of five patients receiving raltegravir therapy. All viral load measurements are marked by a solid black circle. The CD4^{+} cell counts are represented by open triangles. Samples analyzed by 454 deep sequencing are marked by colored circles. Resistance mutations detected by Sanger population sequencing are indicated in boxes. Only raltegravir resistance associated mutations are given. Right hand panels: evolution of resistance pathways, deep sequence analysis of the integrase core domain. Data was obtained by 454 pyrosequencing. Relevant resistance mutations are indicated at the respective nodes. Figures were generated using the nucleotide sequences and the redundancy-level for calling a variant was set at 80. No mutation information indicates wild-type amino acids. The size of each node is scaled to reflect the relative abundance and viral load at each time point and patient. Time points are indicated by color and correspond to the colored circles in the left hand panels: green is the baseline sample, blue the first sample after raltegravir therapy initiation; red is the second sample after raltegravir therapy initiation; black is the final sample after raltegravir therapy initiation. In patient 3 gold is another baseline sample predating the green sample

No major resistance mutations were detected at baseline by NGS. Interestingly, in the first sample during therapy failure (blue nodes Fig. 1e) mutations from all three major resistance pathways including double mutants with secondary mutations were observed with G140S + Q148H being the most frequently occurring double mutant (Fig. 2). In this sample the variant with mutation N155H was the dominant variant but was outcompeted in the subsequent sample (red nodes Fig. 1e) by double mutant G140S + Q148H.

Evidence for elevated diversification following extinction from drug pressure on large viral populations

To get an impression of the evolution of the viral population during raltegravir pressure we analyzed how the number of derived sequences and unique variants related to the viral load. To allow for easy comparisons, each of these measures were normalized on a 0–1 scale and plotted in the same graph (Fig. 3, left panels). In all patients, the total number of derived sequences and the number of unique variants detected correlated with the viral load

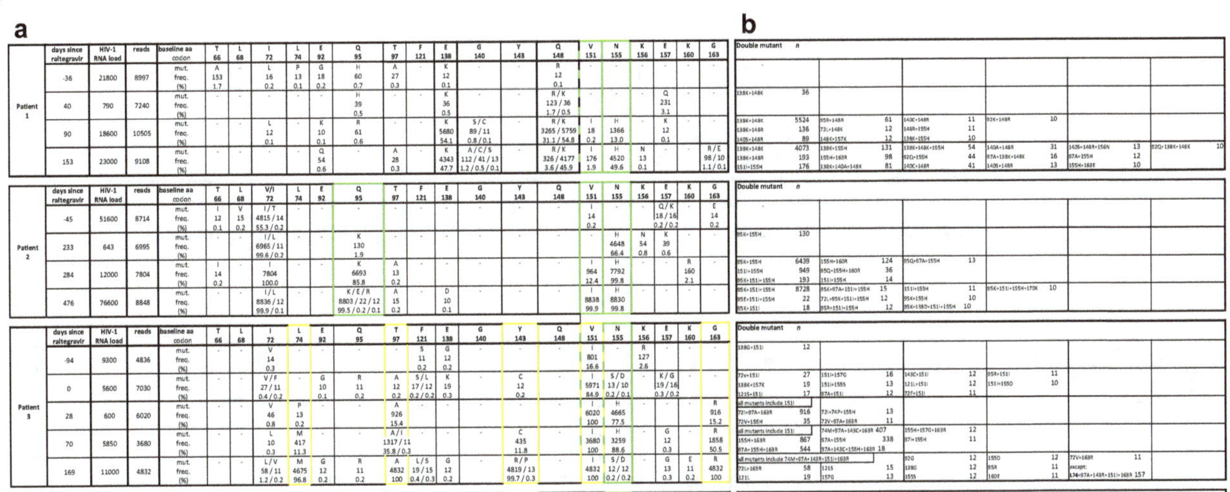

Fig. 2 Analysis of the longitudinal 454 deep sequencing data. **a** Analysis of the frequency of all non-synonymous mutations detected by deep sequencing at the 18 codons associated with raltegravir resistance. Only unique variants with a minimum of 10 reads were included in the analysis. Total number of reads and the proportion of reads containing the denoted mutations relative to the total number of reads are given. Mutations of interest are highlighted by colored boxes. Similar colored boxes are mutations that appeared to be on the same genome. Red boxes indicate mutations from the Q148 pathway, yellow boxes indicate the Y143 pathway and green boxes the N155 pathway. **b** Sequences containing multiple mutations are shown. Double mutants are sorted by frequency

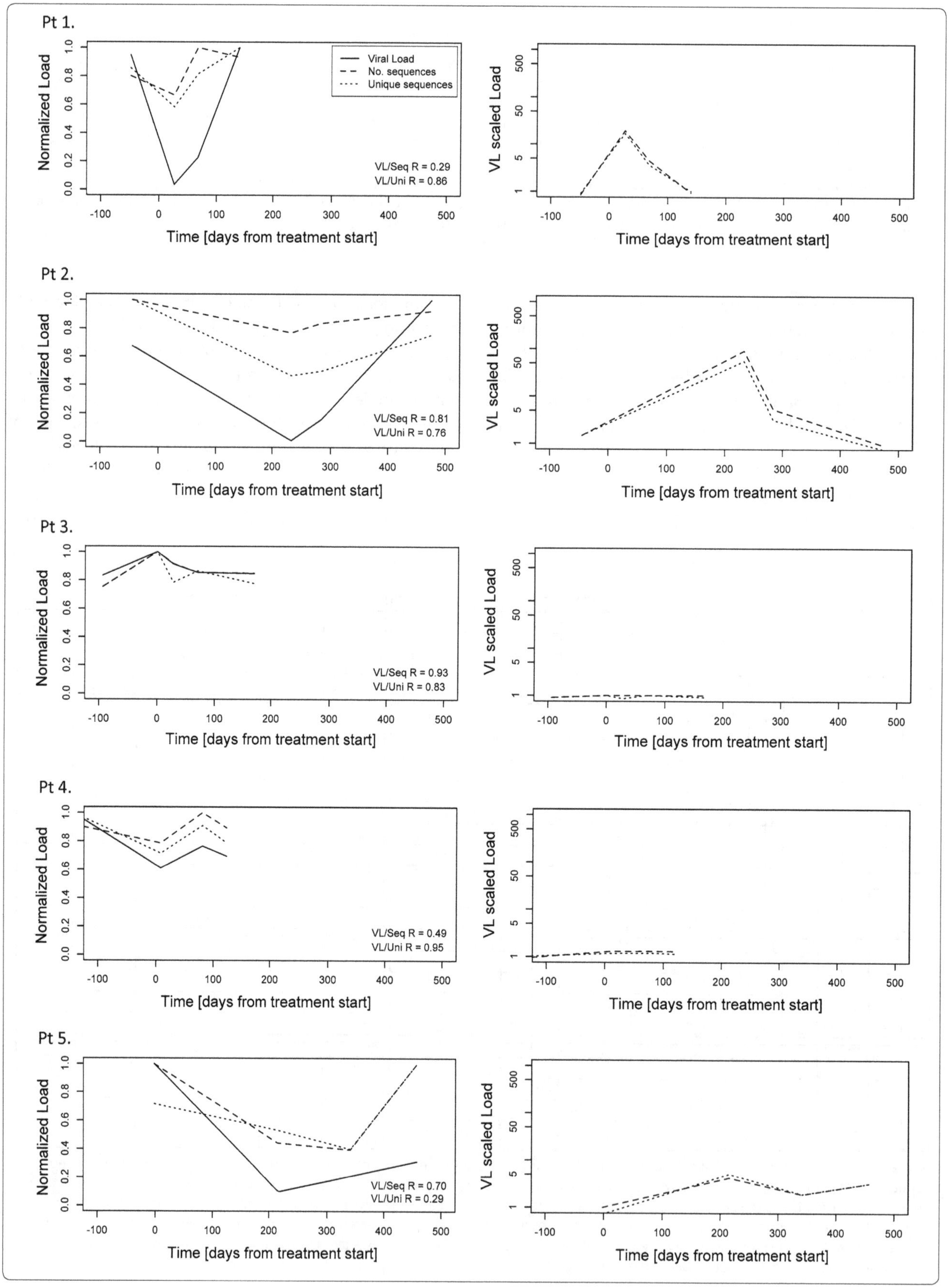
Pt 1.
Viral Load
No. sequences
Unique sequences
VL/Seq R = 0.29
VL/Uni R = 0.86
Pt 2.
VL/Seq R = 0.81
VL/Uni R = 0.76
Pt 3.
VL/Seq R = 0.93
VL/Uni R = 0.83
Pt 4.
VL/Seq R = 0.49
VL/Uni R = 0.95
Pt 5.
VL/Seq R = 0.70
VL/Uni R = 0.29
Normalized Load
VL scaled Load
Time [days from treatment start]

(See figure on previous page.)

Fig. 3 Viral load versus total number of sequences and number of unique variants. Left panels: the normalized (range 0–1) viral load, number of detected sequences and unique number of variants. The correlation coefficient, R, is indicated for the comparisons to the viral load. Right panels: the relative number of derived sequences and the number of unique variants relative to the viral load. Figures were generated using the nucleotide sequences and the redundancy-level for calling a variant was set at 80

(R = 0.29–0.95). However, in a few cases either the number of sequences detected (Pt1 and Pt4) or the number of unique variants (Pt5) showed weaker correlations and thus we also investigated the population diversity relative to the viral load. Therefore, the normalized number of sequences and unique variants were divided by the normalized viral load respectively (Fig. 3, right panels). Interestingly, in patients 1, 2 and 5, the proportion of unique sequence variants appeared to increase when the viral load dropped. This suggests a sudden diversifying pressure on the viral population. In contrast, in patients 3 and 4 there appeared to be no correlation between the relative number of unique variants and the viral load which coincided with the smallest reductions in viral load during raltegravir containing cART (Fig. 1c, d). These data indicate that the stronger the bottleneck is (i.e. largest reduction in viral load), the larger the effect is on subsequent diversification. Thus, extinction due to antiretroviral treatment appears to induce diversification.

The genetic background is not paramount for the emerging INSTI resistance pathway

To investigate the role of the viral genetic background in determining the raltegravir resistance pathway we cloned the integrase coding region from pre-raltegravir therapy samples from two patients (Pt1 and Pt2) in an HXB2 reference background. These recombinant viruses were derived from amplicons of pre-raltegravir samples with a viral load of 21,800 and 51,600 c/ml respectively, creating libraries containing thousands of patient-derived sequences for both samples.

In these patients, raltegravir resistance developed through different pathways during raltegravir therapy. In Pt1, resistance developed initially through E138K + Q148K and later a second variant emerged with N155H (Fig. 1a). In Pt2, raltegravir resistance developed initially through N155H and was complemented by two secondary mutations, Q95K and V151I, which made up nearly 100% of the viral population during prolonged therapy failure (Fig. 1b). With the recombinant viruses of these raltegravir baseline samples and HXB2 reference virus (molecular clone from pHXB2AF, therefore a single sequence input per replicate) multiple in vitro selections were performed in parallel.

All cultures were maintained for 10 serial passages to a final concentration of 1024 nM raltegravir. HXB2 virus predominantly selected mutation Q148K (in three out of the five independent cultures), N155H was selected once and in one culture only secondary resistance mutations emerged (Fig. 4). Two of the Q148K mutations were

		Integrase																
	position HXB2 aa	68 L	72 V	74 L	92 E	121 F	138 E	140 G	143 Y	148 Q	151 V	155 N	163 G	230 S	232 N	233 P	265 A	280 C
n = 5																		
# 1	HXB2	-	-	-	-	-	-	A	-	K	-	-	-	-	-	-	-	-
# 2	HXB2	-	-	M	-	Y	-	-	-	-	I	-	R	-	-	-	-	-
# 3	HXB2	-	V/I	-	-	-	K	-	-	K	-	-	-	-	-	-	-	-
# 4	HXB2		-	-	-	-	-	A	-	K	-	-	-	-	-	-	-	-
# 5	HXB2	-	-	-	Q	-	-	-	-	-	I	H	-	-	-	-	-	-
n = 5	baseline aa		I												D		V	
# 1 pt1	(148 pathway)	-	I	-	-	-	K	C	-	K	-	-	-	-	D	-	V	-
# 2 pt1	(148 pathway)	-	I	L/M	-	-	-	-	C	-	-	H	R	S/K/R/N	D	-	V	-
# 3 pt1	(148 pathway)	-	I	-	-	-	-	-	R	-	-	H	-	-	D	S	V	-
# 4 pt1	(148 pathway)	-	I	-	-	-	-	S	-	R	-	-	-	-	D	-	A	-
# 5 pt1	(148 pathway)	-	I	-	-	-	-	S	-	H	-	-	-	-	D	-	V	-
n = 5	baseline aa		V/I												D			
# 1 pt2	(155 pathway)	-	I	-	-	-	-	S	-	H	-	-	R	-	D	-	-	-
# 2 pt2	(155 pathway)	-	-	-	-	-	-	S	-	H	-	-	-	-	H	-	-	Y
# 4 pt2	(155 pathway)	V	I	-	-	-	-	-	-	-	-	-	-	-	D	-	-	-

Fig. 4 Raltegravir in vitro selections with patient-derived integrase recombinant viruses. Raltegravir concentration was doubled in each serial passage to a final concentration of 1024 nM of raltegravir in passage 10. All differences from the HXB2 reference sequence are given and mutations emerging during the raltegravir in vitro selections in the viral integrase coding region are indicated in red. The indicated mutations were detected by population sequencing of viral RNA in the culture supernatants from passage 10. Cultures 3 and 5 from Pt2 were discontinued due to failed virus propagation

accompanied by G140A, the third by E138K. G140S was probably not selected because it required two nucleotide changes in this HXB2 background; G140A and E138K required just one. The N155H-virus additionally acquired mutations V151I and later E92Q. Amino acid substitutions at residue 143 were not observed. In vitro selections with Pt1 recombinant virus yielded amino acid substitutions at all three major resistance positions. Again, Q148 mutations dominated (3/5 cultures). Interestingly, in the two other cultures, N155H emerged in combination with amino acid substitutions at position 143 (one with Y143C and one with Y143R, Fig. 4). The mutations selected in vitro differed remarkably from what was observed in vivo. The difference between in vivo and in vitro resistance was even more profound for Pt2. In vivo, only mutations relating to the N155H pathway were observed and no other significant mutations were detected by deep sequencing. In contrast, in vitro only mutations from the Q148 pathway were detected (Fig. 4). Cultures #1 and #2 both developed raltegravir resistance through G140S + Q148H. Culture #1 initially selected G140S + Q148R but later switched to Q148H (not shown). Two of the five cultures were not able to replicate at higher raltegravir concentrations. When an earlier passage of both cultures was sequenced, no mutations in the viral integrase were found so these cultures were discontinued. A third culture only selected an L68V substitution, but this virus did not demonstrate phenotypic resistance when tested (data not shown).

Discussion

We investigated the impact of the genetic background and viral population size on the development and evolution of raltegravir resistance in vitro and in vivo. Deep sequencing revealed presence of major raltegravir resistance mutations at baseline in 3/5 patients (patients (1, 3 and 4, Fig. 2a) at very low frequencies ($\leq$ 0.2%). Only the major resistance mutation detected before therapy in Pt3, Y143C (0.2% in 2nd sample on day 0), ultimately became the dominant variant in the resistant population although it surprisingly disappeared from the subsequent sample to reappear in the 4th sample (11.8% on day 70). In the final sample (169 days) Y143C was not detected anymore but virus with mutation Y143R completely dominated the viral population. In the majority of subtype B viruses, the Y143R substitution requires two nucleotide changes (Y143 = TAC → TGC = 143C → CGC = 143R). So it appears that the Y143C variant detected at baseline facilitated selection of the more resistant Y143R variant [27] in this patient. In Pt 1, secondary resistance mutation E138K was detected at baseline (0.1%) and appeared to have acquired Q148K as in the subsequent sample a double mutant with E138K + Q148K was detected (0.5%) which persevered in the population (44.7% in the final sample). This suggests that minor variants present at baseline can play a role in the development and evolution of raltegravir resistance but are not essential for the emergence of resistance [40–42], which is also seen for other drug classes [43, 44].

Remarkably, in Pt1 a variant with major mutation N155H and no apparent secondary resistance mutations emerged after and then co-existed alongside the E138K + Q148K double mutant, each comprising roughly 50% of the viral population. This suggests a fitness advantage for the N155H single mutant over the E138K + Q148K double mutant. This is unexpected considering the in vitro observations regarding resistance and replication of these mutants [17, 45, 46]. A possible explanation could be the so-called hitchhiking effect; this particular variant had a fitness advantage that was located outside the investigated region. For instance, resistance against any of the other antiretrovirals in the therapy regimen (e.g. efavirenz) could have been present in this variant but not in the other variants. While the chances that this occurs in one of the minority variants and not in any of the dominant species are small, the possibility cannot be excluded. Another possible explanation is that the variant with the N155H mutation is more fit than the E138K + Q148K mutant in the presence of raltegravir in this particular setting due to other factors (e.g. specific viral genetic background, immunological host factors, etc.).

We also observed a correlation between the number of unique sequence variants emerging and the magnitude of the virological response. Good virological response in Pts 1, 2 and 5 (> 2 log reductions in HIV-1 RNA) coincided with the number of unique sequence variants observed; the proportion of unique sequence variants increased when the viral load dropped. Pts 3 and 4 demonstrated moderate responses with viral load drops of around 1 log and showed no change in the relative frequencies of unique variants over the course of sampling. This suggests that in Pts 1, 2 and 5, treatment with raltegravir induced an elevated diversification to escape drug pressure while in Pts 3 and 4 the pressure on the population seemed to occur to a much lesser extent. This observation reminds of the explosive diversifications on a macro-evolutionary scale observed in other fields of biology after ice ages and the cataclysmic extinction of dinosaurs [47, 48]. However, regardless of treatment efficacy, i.e. both in patients with less dramatic and more severe virus load reductions, raltegravir resistance mutations developed and multiple resistance pathways were observed in 4/5 patients. The large extinction opens up previously occupied niches for new virus variants, (1) such that all new mutations are accepted and not compete for resources until the population regains a size limited by the carrying capacity of the system, or (2) most of the diversity pre-existed as a permanent, extremely low frequency pool of highly diverse viruses persisting in the shadow of more fit and high frequency variants. Once the high frequency variants are eradicated (e.g. by newly introduced drugs) a glimpse of that diversity surfaces. The larger the impact on the high frequency variants (i.e. reduction in viral load) the larger the proportion of the low frequency pool appears (i.e. number of unique variants increases when the viral load drops). Subsequently, one or two of the low frequency variants gain fitness in the new environment and become dominant, lowering the mean diversity again. The same can be argued from the point of viral escape. If the viral population doesn't have to go that low to find fit variants (in the new environment), the impact on viral load is also less severe.

The raltegravir in vitro selections in which we mimicked population bottlenecking by using a low MOI allowing all resistant variants that are generated to emerge, clearly indicate that the resistance pathway that is selected to escape raltegravir pressure is not predisposed in the genetic background of the integrase CDS. Evaluation of the combined in vivo and in vitro data indicates that stochastic selection plays a major role during the initial development of raltegravir resistance.

In conclusion, the development and evolution of raltegravir resistance can be separated in multiple phases/components: (1) minority variants present at baseline can contribute to the emergence of raltegravir resistance but this is not preordained and seems to occur arbitrarily; (2) during the viral load drop due to drug pressure a burst of new sequence variants emerges creating diversifying selection; (3) these new variants usually include multiple raltegravir resistant variants (from multiple resistance pathways) that can pass the imposed bottleneck; (4) competition of these resistant variants determines the ultimate shape of the viral population. The resistant population can be the product of a single variant that outcompetes all others and only one variant represents the population or multiple variants with a similar fitness emerge that coexist in the viral population.

Further investigation is needed to better assess the exact impact of baseline minority resistance variants and the population size on the development and evolution of raltegravir resistance and determine the clinical implications of these factors.

Conclusions

Emergence of resistance against integrase inhibitor raltegravir in HIV-1 patients is generally associated with selection of one of three distinct resistance pathways. The mechanisms that drive selection of a specific pathway are still poorly understood. Using deep sequencing we observed an inverse correlation between the virological response and the relative amount of unique sequence variants emerging, suggesting diversifying selection during drug pressure. In 4/5 patients multiple signature mutations representing different resistance pathways were observed. In addition, in vitro selections revealed that identical viral clones were also able to select mutations from different resistance pathways indicating that raltegravir resistance is not predisposed in the genetic background of the viral integrase. Importantly, raltegravir resistance develops progressively and discontinuation during early phases of therapy failure is justified to preserve future options with second-generation INSTIs.

Methods

Genotypic analysis

Population sequencing

HIV-1 RNA was isolated using the Nuclisens Islolation kit (BioMérieux, Boxtel, The Netherlands). Briefly, 100 µl of sample was mixed with 900 µl lysisbuffer and 40 µl silica and incubated for 10 min at room temperature to allow binding of the nucleic acid to the silica particles. Unbound material was removed by several washing steps after which the RNA was eluted at 56 °C with 100 µl of 40 ng/µl poly-A RNA. The isolated viral RNA was used to reverse transcribe and amplify the viral integrase coding region in a single-step reaction using the Titan One Tube RT-PCR kit (Roche). The RT-PCR was conducted with primers 5′INoutF1 (5′-GGA ATC ATT CAA GCA CAA CCA GA-3′; 4059–4081) and 3′INoutR1 (5′-TGT ATG CAG ACC CCA ATA TGT TG-3′; 5262–5241). The amount of amplified product was further enhanced in a second PCR using the Expand High fidelity kit (Roche) with primers 5′INinF1 (5′-TAT CTG GCA TGG GTA CCA GCA C-3′; 4143–4164) and 3′INinR1 (5′-TAG TGG GAT GTG TAC TTC TGA AC-3′; 5217–5195). All PCR-amplified products were purified using the QIAquick PCR purification kit (Qiagen, Leusden, The Netherlands). Sequence analysis was performed with the BigDye Terminator v3.1 Cycle Sequencing Kit (Applied Biosystems, Foster City, CA, USA). Integrase sequences were obtained using six primers: Intseq1 (5′-ATT GGA GGA AAT GAA CAA GT-3′; 4173–4192), Intseq2 (5′-AGC AGA AGT TAT TCC AGC AG-3′; 4484–4503), INT-3 (5′-TTC GGG TTT ATT ACA G-3′; 4897–4912), INT-4 (5′-CTT GTA TTA CTA CTG C-3′; 4986–4971), Intseq-5 (5′-CTG GCT ACA TGA ACT GCT AC-3′; 4470–4452) and 3′INinR2 (5′-GCT TTC ATA GTG ATG TCT ATA AAA CC-3′; 5178–5153). Sequence editing and contig assembly were performed using SeqScape v2.6 (Applied Biosystems) with HXB2 as a reference sequence.

Next-generation sequencing and data analysis

To examine the mutation frequencies within the viral integrase by pyrosequencing, RNA was reverse transcribed and amplified in a single step in a touch-down PCR using bar-coded primers to enable 454 pyrosequencing with pooled amplicons. The integrase core domain from amino acid position 53 to amino acid 180 was analyzed. All mutations in the viral integrase at 18 different codons associated with raltegravir resistance [16, 20, 23, 45, 49–52] were evaluated and included amino acids: T66, L68, V72, L74, E92, Q95, T97, F121, E138, G140, Y143, Q148, V151, N155, K156, E157, K160 and G163.

All amplicons were purified with AMPure magnetic beads (Agencourt, Beckman Coulter, Krefeld, Germany), quality checked and quantified using an Agilent 2100 Bioanalyser (Agilent Life Sciences, Waldbronn, Germany) and picogreen using the fluorometer Fluostar Optima (BMG Labtech, Offenburg, Germany), respectively. After equimolar pooling of the amplicons, emulsion PCRs were performed. Pyrosequencing was done using primers A and B (Titanium emPCR kit Lib-L v2; Roche-454 Life Sciences, Branford, CT, USA). After bead recovery and enrichment, approximately 250,000 beads per pool were loaded on one region of a GS FLX PicoTiter plate subdivided with a four-lane gasket. Pyrosequencing was performed on a Genome Sequencer FLX (Roche-454 Life Sciences). Sequence readings from the 454 pyrosequencing run were extracted directly from the Standard-Flowgram-Files (sff). Reads were pair-wise aligned against the integrase sequence of reference strain HXB2. Multiple mutations present in a single read were assumed to originate from the same genome.

The dynamics of the sequence populations were investigated on de-aligned sets to avoid artifacts due to inconsistent alignment gap placements using statistical functions in R [53]. For the analyses of the dominant variants in the populations, the redundancy-level for calling a variant was set at 80, i.e. only variants detected at least 80 times were considered. This level resulted in a strict filter that removed all known 454-sequencing artifacts [54–56]. The resulting dominant variants were aligned using MAFFT [57] and distance matrices were estimated using the R library ape [58] under a F84 substitution model (the choice of substitution model had no effect on the subsequent analyses). The R library sna [59] was used to construct minimum spanning trees (MSTs) based on the distance matrices of patient's HIV population. Nodes were scaled according to both viral load and the relative abundance of each detected variant at time of sampling. We tracked known resistance mutations on the edges and edge lengths were drawn arbitrary in order to make the resulting graphs easy to look at.

Construction of deletion clone HXB2ΔINT

An HXB2 molecular clone (pHXB2AF) was used to construct a molecular deletion clone lacking the integrase coding region. pHXB2AF is derived from pHXB2WT [60], which expresses the full length HIV-1 sequence HXB2 (9719 bp, Genbank accession number K03455.1), with all bacterial sequences non-essential for bacterial expression and replication removed.

The NdeI site present in Gp120 (at HXB2 nt 6404) was inactivated to create a unique NdeI site at the 3′ end of the integrase CDS. Therefore pHXB2AF was digested with NcoI (Roche Diagnostics, Almere, The Netherlands) and NheI (New England Biolabs, Ipswich, MA, USA) to remove a 1586 bp fragment containing the NdeI site. PCR, using Vent$_R$ ® DNA polymerase (New England BioLabs) was performed on pHXB2AF with primers, NcoI-out (5′ CAC TAG AGC TTT TAG AGG AGC TTA AGA-3′; 5614–5640), NheI-out (5′-TTT TAT TAT TTC CAA ATT GTT CTC TTA-3′; 7296–7270) and NdeI-KO (5′-TCA <u>G</u>AT GCT AAA GCG TAT GAT ACA G-3′; 6390–6414). Primer NdeI-KO contained one silent nucleotide change (underlined) to inactivate the NdeI site in the PCR fragment. The amount of amplified product was further increased and enriched by performing a second amplification using Vent$_R$ ® DNA polymerase with primers NcoI-in (5′-GAG CTT TTA GAG GAG CTT AAG AAT GAA-3′; 5619–5645), NheI-in (5′-ATT GTT CTC TTA ATT TGC TAG CTA TCT-3′; 7281–7255) and NdeI-KO. This PCR fragment was then digested with NcoI and NheI and ligated with the digested pHXB2AF, resulting in pHXB2AFNdeIKO which was confirmed by sequence analysis of the complete fragment.

Subsequently, the integrase coding region was removed from pHXB2AFNdeIKO. Therefore pHXB2AFNdeIKO was digested with MluNI (Roche) and NdeI (New England BioLabs). The fragment between MluNI and the 5′ end of integrase was restored by performing a PCR on pHXB2AF using primers RT19 (5′-GGA CAT AAA GCT ATA GGT ACA G-3′; 2454–2472) and NgoMIV-INTlinker (5′-TAA TAT CAT ATG GAC AGC GTC <u>GCC GGC</u> ACT GAC TAA TTT ATC TAC TTG TTC-3′). Primer NgoMIV-INTlinker contained two silent nucleotide changes with respect to pHXB2AFNdeIKO, thereby introducing a unique NgoMIV site (underlined 4209–4214) in the PCR fragment. In addition to these nucleotide changes the primer NgoMIV-INTlinker contains a linker sequence with a unique NdeI site and an AspI site. AspI was used to prevent re-ligation of the vector and the linker. The amount of amplified product was further increased and enriched by performing a second amplification using Vent$_R$ ® DNA polymerase with primers RT19new2 (5′-GGA CCT ACA CCT GTC AAC ATA ATT GG-3′; 2484–2509) and NgoMIV-INTlinker. The

PCR fragment was then digested with MluNI and NdeI and ligated with the digested pHXB2AFNdeIKO, resulting in a molecular deletion clone lacking the integrase coding region (pHXB2AFΔINT), which was confirmed by sequencing the complete fragment.

Generation of recombinant virus

To generate recombinant viruses, the second PCR as described in the population sequencing section, was performed with a different primer pair: forward primer NgoMIV-Int2 (5′-TTA GTC AGT GCC GGC ATC AGG AAA G-3′; 4200–4224) which contains a NgoMIV restriction site (underlined) and reverse primer 3INinR2. The obtained integrase fragment and vector pHXB2AFΔINT were digested with restriction enzymes NgoMIV (New England BioLabs) and NdeI. The PCR product and vector pHXB2AFΔINT were ligated overnight at 4 °C using the Rapid Ligation System (Promega, Benelux, Leiden, The Netherlands). After ligation, plasmids were digested with AspI and subsequently transformed into competent cells. Bacteria were cultured overnight at 37 °C. Plasmid isolation was performed using the QIAgen Plasmid Mini kit (Qiagen).

Viruses were generated by transfecting 293T cells with 10 ug plasmid DNA using Lipofectamine 2000 reagent (Invitrogen) according the manufacturer's protocol. Two days after infection cell free virus was harvested and the viral infectivity ($TCID_{50}$) was determined using end-point dilutions in MT2 cells.

Viral and cell culture

Cells

SupT1 and MT-2 cells were maintained in RPMI 1640 medium with L-glutamine (Lonza, Verviers, Belgium) supplemented with 10% FBS (FBS; Sigma-Aldrich, Zwijndrecht, The Netherlands) and 10 µg/ml gentamicin (Invitrogen, Breda, The Netherlands). 293T cells were maintained in DMEM with L-glutamine (Lonza) supplemented with 10% FBS and 10 µg/ml gentamicin.

In vitro selection experiments

The raltegravir in vitro selection experiments with two recombinant viruses that contained patient-derived integrase CDS and HIV-1 reference strain HXB2 were each performed 5 times. The in vitro selections were initiated by infecting 2×10^6 SupT1 cells at a multiplicity of infection (MOI) of 0.001. The initial raltegravir concentration was 2 nM Raltegravir. Cultures were monitored daily for cytopatic effect (CPE) and twice a week half of the medium was replaced by fresh medium supplemented with raltegravir. When full blown CPE was observed, cell free virus was harvested. The raltegravir concentration was raised in each passage to a final concentration of 1024 nM in passage 10. After passage 10, HIV-1 RNA was isolated from all cultures for genotypic analysis. As a control, in vitro evolution experiments (10 passages) were performed with HXB2 reference strain (5 individual cultures) to monitor the evolution of the integrase CDS during culture in absence of inhibitor.

Phenotypic drug susceptibility analysis

Drug susceptibility was determined by a multiple cycle cell-killing assay [61]. MT-2 cells (4×10^4 in 200 µl RPMI 10% FBS per well) were plated in 96-well microplates. Sample virus and reference virus were inoculated for five days on a single 96-well plate in the presence of threefold dilutions of raltegravir. Both sample virus and reference virus were inoculated at multiple MOIs to adjust for any differences in viral RC. Fold change (FC) values were calculated by dividing the mean 50% inhibitory concentration (EC_{50}) for a sample virus by that of the HXB2 reference strain.

Authors' contributions

AF, LV, AW and MN designed the study. MD, BB, IH and AW collected patient information and provided samples or sequences. AF, MD, AT and PS generated the data. AF, TL, AW and MN analyzed and interpreted the data and wrote the manuscript. All authors read and approved the final manuscript.

Author details

[1] Department of Medical Microbiology, Virology, University Medical Center Utrecht, Heidelberglaan 100, HP G04.614, 3584 CX Utrecht, The Netherlands. [2] Theoretical Biology and Biophysics, Los Alamos National Laboratory, Los Alamos, NM, USA. [3] Department of General Internal Medicine and Infectious Diseases, Ghent University Hospital, Ghent, Belgium. [4] Institute of Immunology and Genetics, Kaiserslautern, Germany. [5] Max Planck Institute for Informatics, Saarbrücken, Germany. [6] Pediatric Clinic, University Medical Center Mannheim, Mannheim, Germany. [7] Department of Internal Medicine and Infectious Diseases, University Medical Center Utrecht, Utrecht, The Netherlands. [8] Department of Internal Medicine, Rijnstate Hospital, Arnhem, The Netherlands.

Acknowledgements

Raltegravir was obtained through the NIH AIDS Reagent Program, Division of AIDS, NIAID, NIH: Raltegravir (Cat # 11680) from Merck & Company, Inc.

Competing interests

The authors declare no competing financial or other interests.

Consent for publication

Not applicable.

Funding

This work was supported by (AF) the Dutch AIDS Fund (Project Number 2006028), (MN) the Netherlands Organization for Scientific Research (NOW VIDI Grant 91796349) and TL was supported by the National Institutes of

Health (NIH Grant R01AI087520). The funders had no role in study design, collecting, analyzing or interpreting the data, or in the preparation of the manuscript.

References

1. Palella FJ Jr, Delaney KM, Moorman AC, Loveless MO, Fuhrer J, Satten GA, Aschman DJ, Holmberg SD. Declining morbidity and mortality among patients with advanced human immunodeficiency virus infection. HIV Outpatient study investigators. N Engl J Med. 1998;338:853–60.
2. Boyer JC, Bebenek K, Kunkel TA. Unequal human immunodeficiency virus type 1 reverse transcriptase error rates with RNA and DNA templates. Proc Natl Acad Sci USA. 1992;89:6919–23.
3. Ji JP, Loeb LA. Fidelity of HIV-1 reverse transcriptase copying RNA in vitro. Biochemistry. 1992;31:954–8.
4. Menendez-Arias L. Molecular basis of fidelity of DNA synthesis and nucleotide specificity of retroviral reverse transcriptases. Prog Nucleic Acid Res Mol Biol. 2002;71:91–147.
5. Bonhoeffer S, May RM, Shaw GM, Nowak MA. Virus dynamics and drug therapy. Proc Natl Acad Sci USA. 1997;94:6971–6.
6. Nowak MA, Bonhoeffer S, Shaw GM, May RM. Anti-viral drug treatment: dynamics of resistance in free virus and infected cell populations. J Theor Biol. 1997;184:203–17. https://doi.org/10.1006/jtbi.1996.0307.
7. Nijhuis M, Boucher CA, Schipper P, Leitner T, Schuurman R, Albert J. Stochastic processes strongly influence HIV-1 evolution during suboptimal protease-inhibitor therapy. Proc Natl Acad Sci USA. 1998;95:14441–6.
8. Duffy S, Shackelton LA, Holmes EC. Rates of evolutionary change in viruses: patterns and determinants. Nat Rev Genet. 2008;9:267–76. https://doi.org/10.1038/nrg2323.
9. Alexander HK, Bonhoeffer S. Pre-existence and emergence of drug resistance in a generalized model of intra-host viral dynamics. Epidemics. 2012;4:187–202. https://doi.org/10.1016/j.epidem.2012.10.001.
10. Espeseth AS, Felock P, Wolfe A, Witmer M, Grobler J, Anthony N, Egbertson M, Melamed JY, Young S, Hamill T, et al. HIV-1 integrase inhibitors that compete with the target DNA substrate define a unique strand transfer conformation for integrase. Proc Natl Acad Sci USA. 2000;97:11244–9. https://doi.org/10.1073/pnas.200139397.
11. Hare S, Gupta SS, Valkov E, Engelman A, Cherepanov P. Retroviral intasome assembly and inhibition of DNA strand transfer. Nature. 2010;464:232–6. https://doi.org/10.1038/nature08784.
12. Krishnan L, Li X, Naraharisetty HL, Hare S, Cherepanov P, Engelman A. Structure-based modeling of the functional HIV-1 intasome and its inhibition. Proc Natl Acad Sci USA. 2010;107:15910–5. https://doi.org/10.1073/pnas.1002346107.
13. ISENTRESS™ (raltegravir) for treatment of HIV (NDA 22145 S036). FDA May 31 2017. https://www.accessdata.fda.gov/drugsatfda_docs/label/2017/022145s036,203045s013,205786s004lbl.pdf.
14. Messiaen P, Wensing AM, Fun A, Nijhuis M, Brusselaers N, Vandekerckhove L. Clinical use of HIV integrase inhibitors: a systematic review and meta-analysis. PLoS ONE. 2013;8:e52562. https://doi.org/10.1371/journal.pone.0052562.
15. Rockstroh JK, DeJesus E, Lennox JL, Yazdanpanah Y, Saag MS, Wan H, Rodgers AJ, Walker ML, Miller M, DiNubile MJ, et al. Durable efficacy and safety of raltegravir versus efavirenz when combined with tenofovir/emtricitabine in treatment-naive HIV-1-infected patients: final 5-year results from STARTMRK. J Acquir Immune Defic Syndr. 2013;63:77–85. https://doi.org/10.1097/QAI.0b013e31828ace69.
16. Cooper DA, Steigbigel RT, Gatell JM, Rockstroh JK, Katlama C, Yeni P, Lazzarin A, Clotet B, Kumar PN, Eron JE, et al. Subgroup and resistance analyses of raltegravir for resistant HIV-1 infection. N Engl J Med. 2008;359:355–65.
17. Miller MD, Danovich R, Ke Y, Witmer M, Zhao J, Harvey C, Nguyen BY, Hazuda DJ: Longitudinal analysis of resistance to the HIV-1 integrase inhibitor raltegravir: results from P005 a phase 2 study in treatment experienced patients. In: XVII international HIV drug resistance workshop, Sitges, Spain. 2008.
18. Johnson VA, Brun-Vezinet F, Clotet B, Gunthard HF, Kuritzkes DR, Pillay D, Schapiro JM, Richman DD. Update of the drug resistance mutations in HIV-1: December 2010. Top HIV Med. 2010;18:156–63.
19. Fransen S, Gupta S, Danovich R, Hazuda D, Miller M, Witmer M, Petropoulos CJ, Huang W. Loss of raltegravir susceptibility by human immunodeficiency virus type 1 is conferred via multiple nonoverlapping genetic pathways. J Virol. 2009;83:11440–6.
20. Malet I, Delelis O, Valantin MA, Montes B, Soulie C, Wirden M, Tchertanov L, Peytavin G, Reynes J, Mouscadet JF, et al. Mutations associated with failure of raltegravir treatment affect integrase sensitivity to the inhibitor in vitro. Antimicrob Agents Chemother. 2008;52:1351–8.
21. Nakahara K, Wakasa-Morimoto C, Kobayashi M, Miki S, Noshi T, Seki T, Kanamori-Koyama M, Kawauchi S, Suyama A, Fujishita T, et al. Secondary mutations in viruses resistant to HIV-1 integrase inhibitors that restore viral infectivity and replication kinetics. Antivir Res. 2009;81:141–6.
22. Quercia R, Dam E, Perez-Bercoff D, Clavel F. Selective-advantage profile of human immunodeficiency virus type 1 integrase mutants explains in vivo evolution of raltegravir resistance genotypes. J Virol. 2009;83:10245–9.
23. Fun A, Van Baelen K, van Lelyveld SF, Schipper PJ, Stuyver LJ, Wensing AM, Nijhuis M. Mutation Q95K enhances N155H-mediated integrase inhibitor resistance and improves viral replication capacity. J Antimicrob Chemother. 2010;65:2300–4.
24. Blanco JL, Varghese V, Rhee SY, Gatell JM, Shafer RW. HIV-1 integrase inhibitor resistance and its clinical implications. J Infect Dis. 2011;203:1204–14.
25. Anstett K, Brenner B, Mesplede T, Wainberg MA. HIV drug resistance against strand transfer integrase inhibitors. Retrovirology. 2017;14:36. https://doi.org/10.1186/s12977-017-0360-7.
26. Malet I, Delelis O, Soulie C, Wirden M, Tchertanov L, Mottaz P, Peytavin G, Katlama C, Mouscadet JF, Calvez V, Marcelin AG. Quasispecies variant dynamics during emergence of resistance to raltegravir in HIV-1-infected patients. J Antimicrob Chemother. 2009;63:795–804.
27. Reigadas S, Anies G, Masquelier B, Calmels C, Stuyver LJ, Parissi V, Fleury H, Andreola ML. The HIV-1 integrase mutations Y143C/R are an alternative pathway for resistance to Raltegravir and impact the enzyme functions. PLoS ONE. 2010;5:e10311. https://doi.org/10.1371/journal.pone.0010311.
28. da Silva D, Van Wesenbeeck L, Breilh D, Reigadas S, Anies G, Van Baelen K, Morlat P, Neau D, Dupon M, Wittkop L, et al. HIV-1 resistance patterns to integrase inhibitors in antiretroviral-experienced patients with virological failure on raltegravir-containing regimens. J Antimicrob Chemother. 2010;65:1262–9. https://doi.org/10.1093/jac/dkq099.
29. Mukherjee R, Jensen ST, Male F, Bittinger K, Hodinka RL, Miller MD, Bushman FD. Switching between raltegravir resistance pathways analyzed by deep sequencing. AIDS. 2011;25:1951–9. https://doi.org/10.1097/QAD.0b013e32834b34de.
30. Goethals O, Vos A, Van Ginderen M, Geluykens P, Smits V, Schols D, Hertogs K, Clayton R. Primary mutations selected in vitro with raltegravir confer large fold changes in susceptibility to first-generation integrase inhibitors, but minor fold changes to inhibitors with second-generation resistance profiles. Virology. 2010;402:338–46.
31. Metifiot M, Vandegraaff N, Maddali K, Naumova A, Zhang X, Rhodes D, Marchand C, Pommier Y. Elvitegravir overcomes resistance to raltegravir induced by integrase mutation Y143. Aids. 2011;25:1175–8.
32. Castagna A, Maggiolo F, Penco G, Wright D, Mills A, Grossberg R, Molina JM, Chas J, Durant J, Moreno S, et al. Dolutegravir in antiretroviral-experienced patients with raltegravir- and/or elvitegravir-resistant HIV-1: 24-week results of the phase III VIKING-3 study. J Infect Dis. 2014;210:354–62. https://doi.org/10.1093/infdis/jiu051.
33. Akil B, Blick G, Hagins DP, Ramgopal MN, Richmond GJ, Samuel RM, Givens N, Vavro C, Song IH, Wynne B, Ait-Khaled M. Dolutegravir versus placebo in subjects harbouring HIV-1 with integrase inhibitor resistance associated substitutions: 48-week results from VIKING-4, a randomized study. Antivir Ther. 2015;20:343–8. https://doi.org/10.3851/IMP2878.
34. Hofstra LM, Nijhuis M, Mudrikova T, Fun A, Schipper P, Schneider M, Wensing A. Use of dolutegravir in two INI-experienced patients with multiclass resistance resulted in excellent virological and immunological responses. J Int AIDS Soc. 2014;17:19755. https://doi.org/10.7448/IAS.17.4.19755.
35. Seki T, Kobayashi M, Wakasa-Morimoto C, Yoshinaga T, Sato A, Fujiwara T, Underwood MR, Garvey EP, Johns BA: S/GSK1349572 is a potent next generation HIV integrase inhibitor and demonstrates a superior resist

ance profile substantiated with 60 integrase mutant molecular clones. In: 17th conference on retroviruses and opportunistic infections, San Francisco, CA, USA. 2010.
36. Brenner BG, Wainberg MA. Clinical benefit of dolutegravir in HIV-1 management related to the high genetic barrier to drug resistance. Virus Res. 2017;239:1–9. https://doi.org/10.1016/j.virusres.2016.07.006.
37. Laskey SB, Siliciano RF. Quantitative evaluation of the antiretroviral efficacy of dolutegravir. JCI Insight. 2016;1:e90033. https://doi.org/10.1172/jci.insight.90033.
38. Malet I, Gimferrer Arriaga L, Artese A, Costa G, Parrotta L, Alcaro S, Delelis O, Tmeizeh A, Katlama C, Valantin MA, et al. New raltegravir resistance pathways induce broad cross-resistance to all currently used integrase inhibitors. J Antimicrob Chemother. 2014;69:2118–22. https://doi.org/10.1093/jac/dku095.
39. Brenner BG, Thomas R, Blanco JL, Ibanescu RI, Oliveira M, Mesplede T, Golubkov O, Roger M, Garcia F, Martinez E, Wainberg MA. Development of a G118R mutation in HIV-1 integrase following a switch to dolutegravir monotherapy leading to cross-resistance to integrase inhibitors. J Antimicrob Chemother. 2016;71:1948–53. https://doi.org/10.1093/jac/dkw071.
40. Liu J, Miller MD, Danovich RM, Vandergrift N, Cai F, Hicks CB, Hazuda DJ, Gao F. Analysis of low-frequency mutations associated with drug resistance to raltegravir before antiretroviral treatment. Antimicrob Agents Chemother. 2010;55:1114–9. https://doi.org/10.1128/AAC.01492-10.
41. Charpentier C, Laureillard D, Piketty C, Tisserand P, Batisse D, Karmochkine M, Si-Mohamed A, Weiss L. High frequency of integrase Q148R minority variants in HIV-infected patients naive of integrase inhibitors. AIDS. 2010;24:867–73. https://doi.org/10.1097/QAD.0b013e3283367796.
42. Codoner FM, Pou C, Thielen A, Garcia F, Delgado R, Dalmau D, Santos JR, Buzon MJ, Martinez-Picado J, Alvarez-Tejado M, et al. Dynamic escape of pre-existing raltegravir-resistant HIV-1 from raltegravir selection pressure. Antivir Res. 2010;88:281–6. https://doi.org/10.1016/j.antiviral.2010.09.016.
43. Johnson JA, Geretti AM. Low-frequency HIV-1 drug resistance mutations can be clinically significant but must be interpreted with caution. J Antimicrob Chemother. 2010;65:1322–6. https://doi.org/10.1093/jac/dkq139.
44. Li JZ, Paredes R, Ribaudo HJ, Svarovskaia ES, Metzner KJ, Kozal MJ, Hullsiek KH, Balduin M, Jakobsen MR, Geretti AM, et al. Low-frequency HIV-1 drug resistance mutations and risk of NNRTI-based antiretroviral treatment failure: a systematic review and pooled analysis. JAMA. 2011;305:1327–35. https://doi.org/10.1001/jama.2011.375.
45. Kobayashi M, Nakahara K, Seki T, Miki S, Kawauchi S, Suyama A, Wakasa-Morimoto C, Kodama M, Endoh T, Oosugi E, et al. Selection of diverse and clinically relevant integrase inhibitor-resistant human immunodeficiency virus type 1 mutants. Antivir Res. 2008;80:213–22.
46. Hu Z, Kuritzkes DR. Effect of raltegravir resistance mutations in HIV-1 integrase on viral fitness. J Acquir Immune Defic Syndr. 2010;55:148–55. https://doi.org/10.1097/QAI.0b013e3181e9a87a.
47. Weir JT, Haddrath O, Robertson HA, Colbourne RM, Baker AJ. Explosive ice age diversification of kiwi. Proc Natl Acad Sci USA. 2016;113:E5580–7. https://doi.org/10.1073/pnas.1603795113.
48. Halliday TJD, Goswami A. Eutherian morphological disparity across the end-Cretaceous mass extinction. Biol J Lin Soc. 2016;118:152–68.
49. Charpentier C, Karmochkine M, Laureillard D, Tisserand P, Belec L, Weiss L, Si-Mohamed A, Piketty C. Drug resistance profiles for the HIV integrase gene in patients failing raltegravir salvage therapy. HIV Med. 2008;9:765–70.
50. Goodman D, Hluhanich R, Waters J, Margot NA, Fransen S, Gupta S, Huang W, Parkin N, Borroto-Esoda K, Svarovskaia ES, et al: Integrase inhibitor resistance involves complex interactions among primary and secondary resistance mutations: a novel mutation L68V/I associates with E92Q and increases resistance. In: XVII international HIV drug resistance workshop, Sitges, Spain. 2008.
51. Markowitz M, Nguyen BY, Gotuzzo E, Mendo F, Ratanasuwan W, Kovacs C, Prada G, Morales-Ramirez JO, Crumpacker CS, Isaacs RD, et al. Rapid and durable antiretroviral effect of the HIV-1 integrase inhibitor raltegravir as part of combination therapy in treatment-naive patients with HIV-1 infection: results of a 48-week controlled study. J Acquir Immune Defic Syndr. 2007;46:125–33.
52. Sichtig N, Sierra S, Kaiser R, Daumer M, Reuter S, Schulter E, Altmann A, Fatkenheuer G, Dittmer U, Pfister H, Esser S. Evolution of raltegravir resistance during therapy. J Antimicrob Chemother. 2009;64:25–32.
53. Team RDC: R: a language and environment for statistical computing. R Foundation for Statistical Computing, Vienna, Austria. http://www.R-project.org. 2008.
54. Brodin J, Mild M, Hedskog C, Sherwood E, Leitner T, Andersson B, Albert J. PCR-induced transitions are the major source of error in cleaned ultra-deep pyrosequencing data. PLoS ONE. 2013;8:e70388. https://doi.org/10.1371/journal.pone.0070388.
55. Hedskog C, Mild M, Jernberg J, Sherwood E, Bratt G, Leitner T, Lundeberg J, Andersson B, Albert J. Dynamics of HIV-1 quasispecies during antiviral treatment dissected using ultra-deep pyrosequencing. PLoS ONE. 2010;5:e11345. https://doi.org/10.1371/journal.pone.0011345.
56. Tsibris AM, Korber B, Arnaout R, Russ C, Lo CC, Leitner T, Gaschen B, Theiler J, Paredes R, Su Z, et al. Quantitative deep sequencing reveals dynamic HIV-1 escape and large population shifts during CCR5 antagonist therapy in vivo. PLoS ONE. 2009;4:e5683. https://doi.org/10.1371/journal.pone.0005683.
57. Katoh K, Asimenos G, Toh H. Multiple alignment of DNA sequences with MAFFT. Methods Mol Biol. 2009;537:39–64. https://doi.org/10.1007/978-1-59745-251-9_3.
58. Paradis E, Claude J, Strimmer K. APE: analyses of phylogenetics and evolution in R language. Bioinformatics. 2004;20:289–90.
59. Butts SF, Owen C, Mainigi M, Senapati S, Seifer DB, Dokras A. Assisted hatching and intracytoplasmic sperm injection are not associated with improved outcomes in assisted reproduction cycles for diminished ovarian reserve: an analysis of cycles in the United States from 2004 to 2011. Fertil Steril. 2014;102(1041–1047):e1041. https://doi.org/10.1016/j.fertnstert.2014.06.043.
60. van Maarseveen NM, Huigen MC, de Jong D, Smits AM, Boucher CA, Nijhuis M. A novel real-time PCR assay to determine relative replication capacity for HIV-1 protease variants and/or reverse transcriptase variants. J Virol Methods. 2006;133:185–94. https://doi.org/10.1016/j.jviromet.2005.11.008.
61. Boucher CA, Keulen W, van Bommel T, Nijhuis M, de Jong D, de Jong MD, Schipper P, Back NK. Human immunodeficiency virus type 1 drug susceptibility determination by using recombinant viruses generated from patient sera tested in a cell-killing assay. Antimicrob Agents Chemother. 1996;40:2404–9.

15

Potent suppression of HIV-1 cell attachment by Kudzu root extract

S. Mediouni[1†], J. A. Jablonski[1†], S. Tsuda[1], A. Richard[1], C. Kessing[1], M. V. Andrade[4], A. Biswas[1], Y. Even[3], T. Tellinghuisen[1,5], H. Choe[1], M. Cameron[2], M. Stevenson[4] and S. T. Valente[1*]

Abstract

There is a constant need to improve antiretrovirals against HIV since therapy is limited by cost, side effects and the emergence of drug resistance. Kudzu is a climbing vine from which the root extract (*Pueraria lobata*), rich in isoflavones and saponins, has long been used in traditional Chinese medicine for a variety of purposes, from weight loss to alcoholism prevention. Here we show that Kudzu root extract significantly inhibits HIV-1 entry into cell lines, primary human CD4^{+}T lymphocytes and macrophages, without cell-associated toxicity. Specifically, Kudzu inhibits the initial attachment of the viral particle to the cell surface, a mechanism that depends on the envelope glycoprotein gp120 but is independent from the HIV-1 cell receptor CD4 and co-receptors CXCR4/CCR5. This activity seems selective to lentiviruses since Kudzu inhibits HIV-2 and simian immunodeficiency virus, but does not interfere with Hepatitis C, Influenza, Zika Brazil and adenovirus infection. Importantly, depending on the dose, Kudzu can act synergistically or additively with the current antiretroviral cocktails against HIV-1 and can block viruses resistant to the fusion inhibitor Enfuvirtide. Together our results highlight Kudzu's root extract value as a supplement to current antiretroviral therapy against HIV.

Background

Current human immunodeficiency virus (HIV) antiretroviral therapy (ART) has dramatically benefited HIV-1 infected individuals, by reducing circulating virus to very low levels; however, it fails to completely eliminate infection and the emergence of drug resistance is an ongoing problem [1, 2]. It is highly desirable to develop additional anti-HIV-1 agents with superior efficacy and safety profiles.

HIV entry into target cells is a multistep process. The envelope glycoprotein is composed of the surface gp120 and the transmembrane gp41. The gp120 first attaches to the target cell through interactions with the negatively charged cell-surface heparan sulfate proteoglycans [3], α4β7 integrins [4, 5], the dendritic cell–specific intercellular adhesion molecular 3-grabbing non-integrin (DC-SIGN) [6] or the mannose binding C-type lectin receptors (MCLR) [7]. These attachment receptors bring gp120/gp41 into close proximity with the virus receptor CD4 and the co-receptors CXCR4 or CCR5 on the cell surface, increasing infection efficiency. Following this initial contact, interaction of gp120 with CD4 induces a conformational change in gp120, exposing the co-receptor binding site to allow interaction with the co-receptors. This also leaves gp41 in an active form, resulting in fusion of virus and cell membranes and release of the nucleocapsid into the cytoplasm [8–10]. Soon after infection, the viral population is mainly composed of R5 strains, defined by their use of the CCR5 co-receptor. As the disease progresses, the R5 strains evolve to dual X4R5 or just X4 strains, defined by their use of the CXCR4 co-receptor.

Virus attachment and fusion (collectively termed "virus entry") have long been the target for drug development against HIV [11, 12]. Only three Food and Drug administration (FDA) approved drugs currently block the entry pathway. Enfuvirtide (Fuzeone) targets the HIV-1 envelope glycoprotein gp41 to prevent viral and host membrane fusion. It is mainly used as salvage therapy in multidrug resistance cases, given that it is dosed

*Correspondence: svalente@scripps.edu
†S. Mediouni and J. A. Jablonski have contributed equally to this work
[1] Department of Immunology and Microbiology, The Scripps Research Institute, 130 Scripps Way, 3C1, Jupiter, FL 33458, USA
Full list of author information is available at the end of the article

twice-daily by subcutaneous injection [13–15]. Maraviroc, is a selective and reversible CCR5 antagonist [16, 17], and genotypic tropism testing for co-receptor usage is usually recommended before administration. Finally, Ibalizumab is an antibody that targets the CD4 receptor. It is administrated intravenously every 2 weeks, approved for multidrug resistant cases [18]. The gp120 is critical for viral entry, but unfortunately drug discovery against this protein has remained unsuccessful so far, possibly due to the incredible ability of this protein of mutating and evading drug-selective pressure.

Natural products have always been a valuable resource for the pharmaceutical industry and have been of great benefit in virtually all-clinical therapeutic areas. Kudzu (also named *Pueraria lobata*, Kudzu vine, foot-a-night vine, vine-that-ate-the-South, and Ko-hemp) belonging to the genus Pueraria, is a fast-growing evergreen vine originating from China. Kudzu was introduced in the United States from Japan at the Philadelphia Centennial Exposition in 1876. Initially it was used as forage crop and ornamental plant, but later in the 1930s, many Southern farmers were encouraged to plant Kudzu for erosion control. Kudzu quickly became invasive, and today an estimated 2 million acres of forestland in the southern United States are covered with Kudzu.

Kudzu's root extract (or radix Puerariae or Gegen) has been traditionally used in Asian countries (since at least 200 BC) as an herbal remedy to treat colds, headaches and diarrhea. The entire plant is edible, however the health supporting benefits of Kudzu comes from its flowers and roots, which contain many natural products including isoflavones and saponins. In Asia, Kudzu root extract is often found in food products, beverages, as well as soaps and lotions as an antimicrobial agent [19, 20]. Recently, in a clinical trial, Kudzu has been found to reduce alcoholism [21]. The chemical composition of Kudzu differs depending on geographic origin and extraction method (Additional file 1: Table 1S) [20, 22, 23]. Chemical analysis of Kudzu root extract revealed several chemical entities including Daidzein (antimicrobial and anti-inflammatory agent), Daidzin (prevents development of cancer), Genistein (anti-bacterial and anti-leukemic agent), and is also a distinct source of the isoflavone puerarin (Additional file 1: Table 1S). Kudzu root extract has also been linked to improvements in glucose metabolism, neuron regeneration, stroke prevention, inflammation, oxidative stress reduction, migraines and cluster headaches reduction [24, 25].

Given the known general antimicrobial activity of Kudzu, we investigated its activity against HIV-1 replication. Here we demonstrate that Kudzu root extract significantly inhibits HIV-1 entry into target cells. Kudzu inhibits both R5 and X4 tropic strains of HIV-1, HIV-2 and simian immunodeficiency virus (SIV). This activity seems selective since Kudzu has no activity against viruses from other families. Kudzu root extract specifically blocks the attachment of HIV gp120 to the target cell. Importantly, Kudzu acts synergistically or additively depending on the dose, with current cocktails of antiretrovirals (ARVs), and can block viruses resistant to the commercial entry inhibitor, Enfuvirtide. Overall, these results define Kudzu root extract as a potential HIV-1 inhibitor that, combined to ART, could improve therapy outcomes by acting at a novel target site of the viral replication cycle.

Results and discussion

Kudzu extract inhibits HIV-1 replication

HIV-1 susceptibility to Kudzu root extract was assayed using a reporter cell line that stably expresses the β-galactosidase (β-Gal) gene (LacZ) [HeLa-CD4-LTR-LacZ cells]; LacZ expression is driven by the 5′LTR promoter of HIV-1 and responds to Tat expressed by an incoming virus. HeLa-CD4-LTR-LacZ cells were infected with HIV-1 isolate NL4-3 (clade B, X4 tropic), in the presence of increasing dilutions of Kudzu, and β-Gal activity was measured 72 h later (Fig. 1a). The inhibition of Tat-dependent transcription of LacZ was dose-dependent, presenting an IC_{50} of $1{:}5263 \pm 6 \times 10^{-5}$ dilution of the Kudzu extract. More than 90% inhibition was observed at dilution 1:1600, a concentration that does not impact HeLa-CD4-LTR-LacZ cells viability (Additional file 1: Fig. 1SA). The IC_{50} of currently used ARVs was included for comparison (Additional file 1: Fig. 2SA).

Kudzu root extract is a tincture comprised of 33% of dried kudzu root and 66% of solvent (a mix of ethanol and glycerol). To verify that Kudzu activity was not aspecifically linked to these solvents, both solutions, at different percentages, were tested in infectivity assays (Additional file 1: Fig. 2SB and C). No viral inhibition was observed at the highest and effective dilution of Kudzu (1:200), which corresponds to 0.3% glycerol or ethanol.

The activity of Kudzu was also assessed by measuring p24 capsid in the supernatant by p24 ELISA, revealing an IC_{50} of 1:1556 (Fig. 1b). The differences in IC_{50} between the β-Gal and p24 ELISA assays most likely reflect the ability of p24 ELISA to detect all p24 production, whether the protein is incorporated into virions or not. Similar results were obtained with Kudzu purchased from another company (data not shown), suggesting that Kudzu's activity is consistent between brands.

Kudzu extract blocks the first steps of HIV-1 entry into target cells

To investigate the mechanism by which Kudzu suppresses HIV-1, we first monitored the integration of proviral DNA into the genome of HeLa-CD4-LTR-LacZ cells 24 h post-infection by Alu-PCR, followed by quantitative real time PCR (qPCR). Kudzu (1:200) significantly inhibited the integration of proviral HIV DNA, similar to Efavirenz (a reverse transcriptase inhibitor, 200 nM), Raltegravir (an integrase inhibitor, 200 nM) and AMD3100 (a CXCR4 antagonist, 10 nM), used as positive controls. As expected, Saquinavir, a protease inhibitor (200 nM) did not inhibit HIV integration during this 24 h assay (Fig. 1c). Dimethyl sulfoxide (DMSO) was used as negative control since the ARVs are solubilized in 0.001 or 0.002% DMSO. Furthermore, ethanol and glycerol, at the highest concentrations of Kudzu, did not interfere with HIV replication (Additional file 1: Fig. 2SB and C).

To insure Kudzu's activity was not cell line dependent, we also assessed the activity of Kudzu on primary human $CD4^{+}T$ cells isolated from blood of 3 individuals. Cells were infected for 24 h with NL4-3 in the presence of the most potent but non-toxic dilutions of Kudzu (1:400 and 1:200; Additional file 1: Fig. 1SB), or a cocktail of ARVs (Raltegravir 200 nM, Efavirenz 100 nM and AZT 180 nM) or Enfuvirtide (1 μg/ml). Total viral DNA was measured by qPCR (Fig. 1d). Kudzu inhibited the infection of primary human $CD4^{+}T$ cells equally well as Enfuvirtide or a cocktail of ARVs. Together these results suggest that Kudzu activity is not cell type dependent and targets an early event of the HIV-1 life cycle.

We next measured the early and late HIV-1 reverse transcription (RT) products by qPCR 10 h post-infection of HeLa-CD4-LTR-LacZ cells (Fig. 1e). Efavirenz and AMD3100 (200 nM and 10 nM respectively) used as controls, inhibited early RT products by approximately 40% and 60% respectively, while the late RT products were decreased by approximately 84%, consistent with the literature [26]. Treatment of the cells with Kudzu resulted in a similar reduction of early and late RT products to controls (60% and 75% respectively). Saquinavir (200 nM), as expected, did not impact the production of late RT products. Altogether these results suggest that Kudzu inhibits an early event occurring before or at the reverse transcription step.

To further understand which step was blocked by Kudzu, we performed time-of-addition assays as previously described [27], using entry and RT inhibitors as controls. We infected HeLa-CD4-LTR-LacZ cells with NL4-3, then added Kudzu at different time points post-infection (from 1 to 6 h), and measured β-Gal activity 72 h later (Fig. 1f). Both dilutions of Kudzu (1:800 and 1:400) displayed similar inhibitory kinetics to the entry inhibitor AMD3100 (4 nM). When Kudzu or AMD3100 were added 3 h post infection, their inhibitory activity started to decline, showing almost no activity if added 6 h later. As expected, Efavirenz (10 nM) displayed stronger inhibition when added at later time points. These results suggest that Kudzu inhibits the entry step of HIV-1 into the target cell.

Kudzu extract inhibits HIV-1 infection independently of tropism

The interaction between gp120 and CD4, followed by interaction with CXCR4 or CCR5 is determinant for HIV-1 entry into the target cell. To determine if the coreceptors usage is determinant for Kudzu-mediated inhibition of HIV-1, we tested the susceptibility of R5 tropic viruses. We infected GHOST-CCR5 cells with JRCSF or YU2 viruses, in the presence of Kudzu concentrations that did not impact cell viability (1:400 and 1:200; Additional file 1: Fig. 1SC), and assessed p24 in the supernatant 72 h post infection (Fig. 2a). Raltegravir (100 nM) and Emtricitabine (a reverse transcriptase inhibitor, 100 nM) were used as positive controls. We observed a dose-dependent inhibition of both R5 viruses, suggesting that Kudzu inhibits HIV-1 entry independently of coreceptor usage.

(See figure on next page.)

Fig. 1 Kudzu inhibits HIV-1 replication of X4 tropic viruses. **a, b** Activity of Kudzu in acute infection of HeLa-CD4-LTR-LacZ cells with NL4-3. HeLa-CD4-LTR-LacZ cells were infected with HIV-1 NL4-3 strain in the presence of different dilutions of Kudzu. **a** β-Gal activity was measured 72 h later. Untreat.: untreated. The mean ± SD of 5 independent experiments is represented. **b** Viral supernatants recovered 72 h post infection were assayed for their p24 antigen content using a sandwich ELISA kit. Untreat.: untreated. Data is a mean ± SD of 2 independent experiments. **c** Impact of Kudzu on HIV-1 integration. HeLa-CD4-LTR-LacZ were infected with NL4-3 in presence of compounds for 24 h. Next day, DNA was extracted and provirus integration was quantified by Alu-PCR followed by qPCR. Saquinavir (Saq., a protease inhibitor, 200 nM), Efavirenz (Efav., a reverse transcriptase inhibitor, 200 nM), Raltegravir (Ralt., an integrase inhibitor, 200 nM), and AMD3100 (an entry inhibitor, 10 nM) were used as controls. Kudzu was used at 1:200. The mean ± SD of 5 independent experiments is represented. **d** Impact of Kudzu on HIV-1 integration in primary $CD4^{+}T$ cells 24 h post-infection and treatment. A cocktail of antiretrovirals (ARVS: 180 nM AZT, a reverse transcriptase inhibitor, 100 nM Efavirenz and 200 nM Raltegravir) and Enfuvirtide (1 μg/ml) were used as controls. Shown is the mean ± SD of 3 independent experiments. **e** Activity of Kudzu on reverse-transcription products of HIV-1 10 h post-infection. HeLa-CD4-LTR-LacZ were infected with NL4-3 in presence of compounds for 10 h. Next, DNA was extracted and early and late RT products were measured by qPCR. Kudzu was used at 1:200. Error Bars from qPCR (n = 3) ± SD from 2 independent experiments for early products. The mean ± SD of 4 independent experiments is represented for late products. **f** Time-of-addition experiment of kudzu in HeLa-CD4-LTR-LacZ infected with NL4-3 strain. Kudzu or control compounds (4 nM AMD3100 and 10 nM Efavirenz) were added at 1, 2, 3, 4, 5 and 6 h postinfection. β-Gal activity was measured 72 h later. Data shown is representative of 2 independent experiments. The two-tailed paired *t* test was used for statistical comparisons. *: p value < 0.05, ***: p value < 0.0005

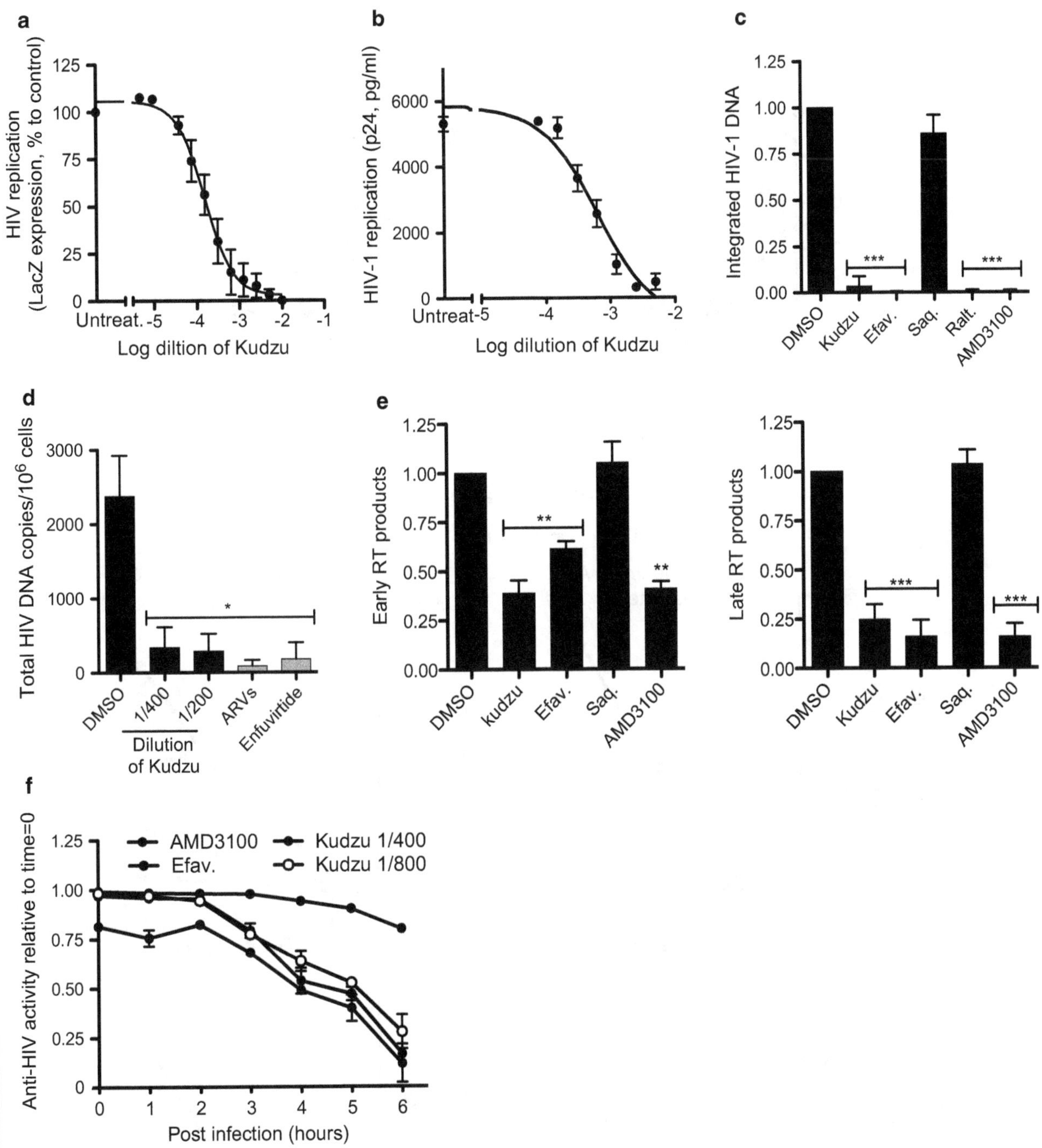

(See figure on next page.)

Fig. 2 Kudzu blocks HIV-1 replication of R5 tropic viruses. **a** Activity of Kudzu on acute replication of GHOST-CCR5 cells with JRCSF and YU2 viruses. Cells were infected with both viruses in presence of 2 potent dilutions of Kudzu. Viral supernatants recovered 72 h post-infection from cells were assayed for their p24 antigen content. Raltegravir (Ralt, an integrase inhibitor, 100 nM) and Emtricitabine (Emt., a reverse transcriptase inhibitor, 100 nM) were used as controls. The mean ± SD of 2 independent experiments is shown. **b** Impact of Kudzu on acute replication of primary Human macrophages infected with macrophage-tropic stains ADA and 5002 M. Cells were infected with both viruses in presence of 1:100 dilution of Kudzu or 5 μg/ml of Enfurvirtide (a fusion inhibitor). Viral supernatants recovered at different times post-infection were assayed for their p24 antigen content

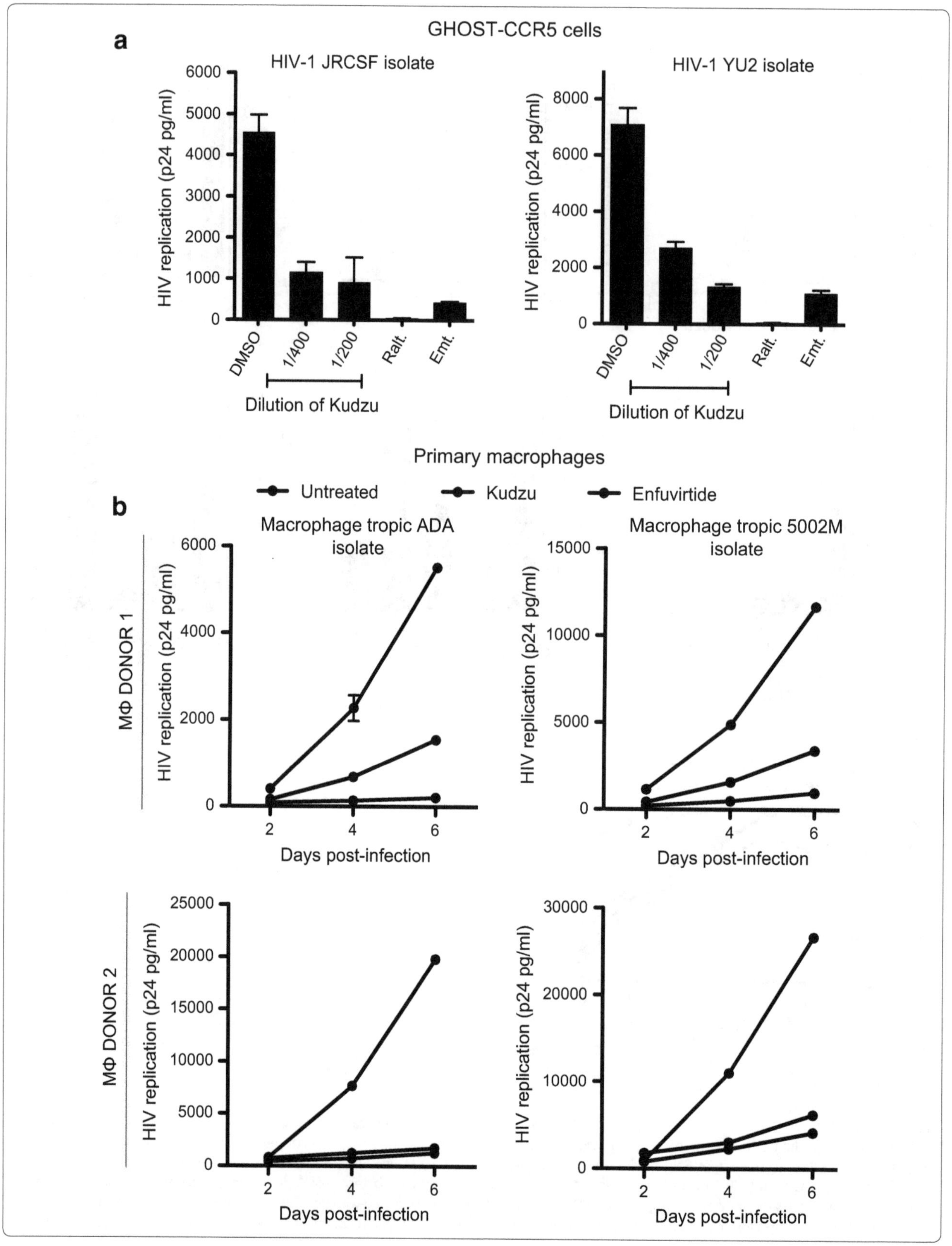
a
GHOST-CCR5 cells
HIV-1 JRCSF isolate
HIV-1 YU2 isolate
HIV replication (p24 pg/ml)
DMSO
1/400
1/200
Ralt.
Emt.
Dilution of Kudzu
Primary macrophages
Untreated
Kudzu
Enfuvirtide
b
Macrophage tropic ADA isolate
Macrophage tropic 5002M isolate
MΦ DONOR 1
MΦ DONOR 2
Days post-infection

Next, we assessed the susceptibility of Kudzu against infection of primary macrophages from two donors, using two R5 macrophages tropic viruses (ADA and 5002 M). We measured p24 production in the supernatant at different times post-infection. Kudzu (1:100) drastically inhibited HIV-1 replication of both viruses in macrophages from the two donors, with very similar kinetics as the fusion inhibitor Enfuvirtide (5 μg/ml, Fig. 2b). In sum, Kudzu seems to be able to block HIV-1 independently of tropism and cell type, important features as HIV-1 changes receptor usage during disease progression.

Kudzu-mediated inhibition of viral entry is specific to lentiviruses

We set out to evaluate the breath of Kudzu activity against other lentiviruses such as HIV-2 and SIV or viruses from other families such as Hepatitis C virus (HCV), Influenza A (H1N1), Zika virus (Zika Brazil) and adenovirus serotype 5 (ADV5) (Fig. 3).

HIV-2, initially found in West Africa, has spread to Europe, other parts of Africa, India and the United States. HIV-2 is naturally resistant to nonnucleoside reverse transcriptase inhibitors, to some protease inhibitors [28, 29] and to Enfuvirtide [30]. We assessed Kudzu's (1:300 and 1:150) activity against the highly cytopathic isolate HIV-2 CBL-20 strain (X4 tropic virus) [31], using TZM-bl cells expressing the luciferase gene under the control of the HIV LTR. Luminescence counts were quantified 48 h later and normalized to total protein concentration (Fig. 3a). Kudzu significantly reduced HIV-2 replication similarly to AMD3100 (10 nM) or Raltegravir (100 nM) used as controls.

Next, we investigated the effect of Kudzu on the replication of SIV (Fig. 3b). Primary macaque $CD4^{+}T$ cells from three independent rhesus macaques (RhM A, B and C) were infected with $SIV_{mac}239$. Six days post-infection,

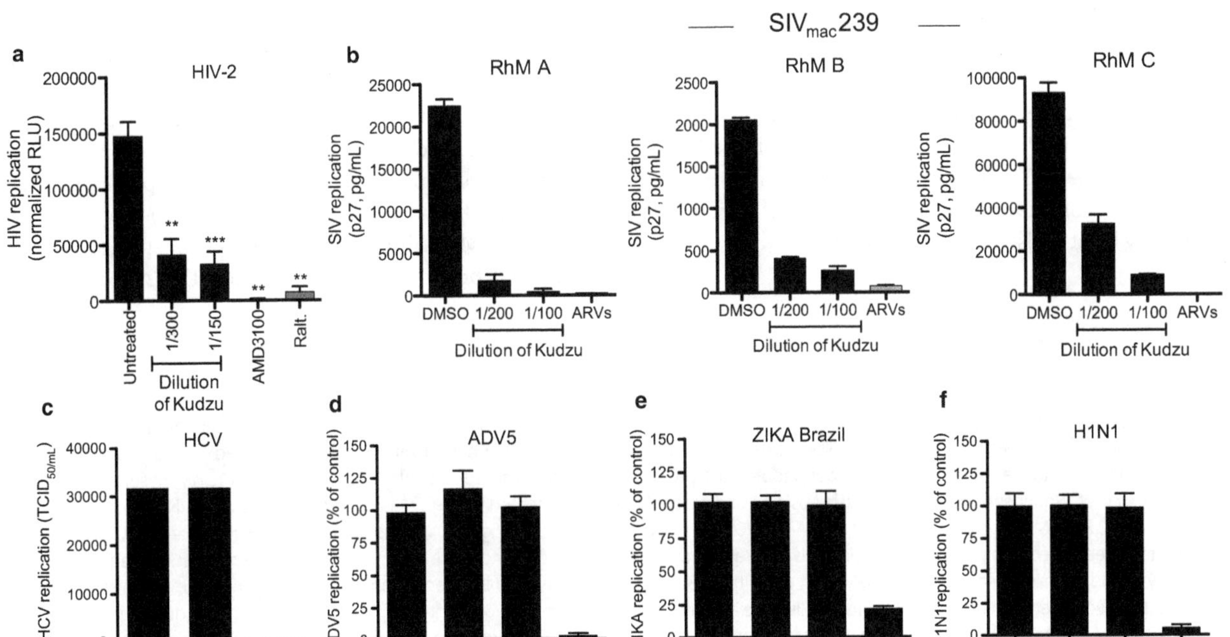

Fig. 3 Activity of Kudzu on diverse viruses. **a** Activity of Kudzu in acute infection of TZM-bl with HIV-2 CBL-20 strain. TZM-bl cells were infected with HIV-2 in presence of different concentrations of Kudzu for 24 h. Luminescence normalized to total protein was measured 48 h later. AMD3100 and Raltegravir used at 10 nM and 100 nM respectively. Data is the mean ± SD of 3 independent experiments. **b** Activity of Kudzu on SIV-infected primary rhesus macaque cells 6 days post-infection. Virus in the supernatant measured by capsid p27 ELISA. A cocktail of antiretrovirals (ARVs, Emtracitabine, Raltegravir and Tenofovir, 200 nM) was used as control. **c** No effect of Kudzu on HCV infection in Huh 7.5 cells. 2′-Cmethyladenosine (2′-C-methyl., 10 μM) was used as a control. Results represent the mean ± SD of 2 independent experiments. **d–f** Hela-CD4-LTR-LacZ were infected with the indicated viruses in the presence of the indicated compounds. The cells were stained 24 h later using virus-specific antibodies to assess the levels of infection and analyzed by flow cytometry. **d** Kudzu has no activity on ADV5 virus infection of HeLa-CD4-LTR-LacZ cells. Heat inactivated virus (H.I) and Raltegravir (Ralt., 100 nM) were used as controls. Results represent the mean ± SD of 3 independent experiments. **e** No activity of Kudzu on ZIKA Brazil virus infection of HeLa-CD4-LTR-LacZ cells. Cabozantinib (Caboz., 1 μM) and Raltegravir (Ralt., 100 nM) were used as controls. Results represent the mean ± SD of 3 independent experiments. **f** No impact of Kudzu on H1N1 virus infection of HeLa-CD4-LTR-LacZ cells. Aleuria Aurantia Lectin (AAL, 100 nM) and Raltegravir (Ralt., 100 nM) were used as controls. Results represent the mean ± SD of 3 independent experiments. Kudzu was used at the dilution 1:200 in **c–f**. The two-tailed paired *t* test was used for statistical analysis. **: *p* value < 0.005, ***: *p* value < 0.0005

p27 capsid in the supernatant was measured by ELISA. A cocktail of ARVs (Emtracitabine, Tenofovir and Raltegravir, 200 nM) was used as control. Kudzu extract inhibited $SIV_{mac}239$ replication in a dose-dependent manner at concentrations (1:200 and 1:100) that did not alter cell viability (Additional file 1: Fig. 1SE). Altogether, these results suggest that Kudzu has a broad activity against lentiviruses, likely because these share similar target cell entry mechanisms.

HCV replication was investigated using Huh-7.5 cells in the presence of either Kudzu (1:200) or 2′-C-methyladenosine (10 μM), an RNA-dependent RNA polymerase inhibitor known to suppress HCV replication. Unlike the latter, Kudzu had no effect on HCV infection (Fig. 3c). The replication of H1N1, Zika or ADV5 was also assessed in HeLa-CD4-LTR-LacZ cells in the presence of Kudzu or known inhibitors of each virus. The Aleuria Aurantia lectin (100 nM) binds the envelope hemagglutinin glycoprotein of H1N1 and blocks its entry [32]. Cabozantinib (1 μM), a tyrosine-kinase inhibitor, suppresses Zika infection [33]. ADV5 infection was controlled by heat inactivation. While all our positive controls proved able to inhibit these viruses, Kudzu (1:200) showed no activity (Fig. 3d–f).

Collectively, these results suggest that Kudzu is specific to lentiviruses, HIV-1, HIV-2 and SIV. This large breath of activity suggests Kudzu targets an aspect of viral entry that is very conserved across lentiviruses, which may potentially limit resistance development.

The anti-HIV-1 activity of Kudzu depends on gp120

The interaction between gp120 and CD4, followed by interaction with CXCR4 or CCR5 is determinant for HIV-1 entry into the target cell. After determining that Kudzu does not affect CD4, CXCR4 and CCR5 cell surface expression (Additional file 1: Fig. 3S), we investigated whether Kudzu affects the virus or the cells. As such, we either pre-incubated Kudzu with HeLa-CD4-LTR-LacZ cells for 6 h, then washed the cells and added NL4-3 virus; or, incubated kudzu and NL4-3 at the same time with HeLa-CD4-LTR-LacZ cells. β-Gal activity was measured 72 h later (Fig. 4a). Kudzu (1:400) was significantly active only when added to the cells simultaneously with the virus. Similar result was obtained with the negative control Raltegravir (100 nM). These results suggest that Kudzu mainly affects the virus, and is not the result of an activity on the target cell.

Given the key role of gp120 and gp41 proteins in the entry process, we assessed whether gp160 (precursor of gp120 and gp41) was necessary for Kudzu's activity by comparing susceptibility of HIV virus-like particles pseudotyped with either the vesicular stomatitis virus G (VSV-G) envelope protein or the X4 gp160 envelope from HXB2 strain. We infected HeLa-CD4-LTR-LacZ cells with VSV-G-NL4-3 or HXB2 gp160-NL4-3 in the presence of Kudzu or a series of control compounds, and measured p24 production in the supernatant 72 h later (Fig. 4b and c). Efavirenz (200 nM) and Raltegravir (200 nM), which inhibit HIV-1 post-entry events, decreased both VSV-G-NL4-3 and HXB2 gp160-NL4-3

(See figure on next page.)

Fig. 4 Kudzu blocks HIV-1 attachment. **a** Kudzu specifically affect the virus and not the cells. Kudzu (1:400) was either preincubated for 6 h with the HeLa-CD4-LTR-LacZ cells then washed and NL4-3 virus was added to the cells ("preincubation with compound/washed") or simultaneously added with the virus to the cells ("virus + compound together"). β-Gal activity was measured 72 h later. Ralt.: Raltegravir, 100 nM. Results represent the mean ± SD of 3 independent experiments. **b** Absence of activity of Kudzu on infected HeLa-CD4-LTR-LacZ cells with VSV-G-NL4-3 pseudovirus. Efavirenz (Efav., 200 nM), Raltegravir (Ralt., 200 nM), AMD3100 (10 nM) and Enfuvirtide (1 μg/ml) were used as controls. Kudzu: 1:400 dilution. Viral supernatants recovered 72 h postinfection from cells were assayed for their p24 antigen content. The mean ± SD of 4 experiments is represented. **c** Activity of Kudzu on infected HeLa-CD4-LTR-LacZ cells with HXB2 gp160 pseudotyped NL4-3 env-. Same controls and Kudzu dilution as in **b**. Viral supernatants recovered 72 h post-infection from cells were assayed for their p24 antigen content. The mean ± SD of 4 independent experiments is presented. **d** No impact of Kudzu on interaction of monomeric YU2 gp120 and CD4-Ig receptor in ELISA. s: soluble. Emtricitabine (Emt., 500 nM) and soluble gp120 or CD4 were used as controls. The mean ± SD of 4 independent experiments is represented. **e** Kudzu does not affect fusion of HeLa-CD4-LTR-LacZ and HL2/3 cells. Emtricitabine (Emt., 100 nM), Enfuvirtide (1 μg/ml), cells only and untreated cells were used as controls. β-Gal activity was measured 48 h later. Data is the mean ± SD of 3 independent experiments. **f** No activity of Kudzu on shedding of transfected JRCSF gp160 in HEK293T cells, revealed by western blot. Data is representative of 3 independent experiments. CD4-Ig (45 μg) was used as a control. **g** No activity of Kudzu on shedding of NL4-3 gp120 from NL4-3 virus. HIV-1 directly loaded on gel (HIV-1+) was used as positive control of the gp120 protein. The other samples were filtered through a column and a fraction of the flow-through was analyzed by western blot. CD4 (30 μg) was used as control of the shedding. Data is representative of 2 independent experiments and quantification of the 2 independent experiments is shown below the blot. **h–j** Kudzu significantly reduces the attachment of HIV-1 and HIV-2 to TZM-bl cells. **h** Schematic of the attachment assay. The attachment assay is performed by adding virus to cells in presence of the different compounds for 3 h at 4 °C. After this period of incubation at 4 °C, the cells are washed with cold PBS and then allowed to incubate at 37 °C for 72 h before luciferase is quantified. **i** Kudzu inhibits HIV-1 attachment. Enfuvirtide (a fusion inhibitor, 1 μg/ml) and Heparin (competes with Heparan sulfate proteoglycans binding to gp120, 1 mg/ml) were used as controls. The data is the mean ± SD of 3 independent experiments. **j** Kudzu inhibits HIV-2 CBL-20 attachment. AMD3100 (10 nM) and Heparin were used as controls. The data is the mean ± SD of 4 independent experiments. The two-tailed paired *t* test was used for statistical comparisons. **: p value < 0.005, ***: p value < 0.0005

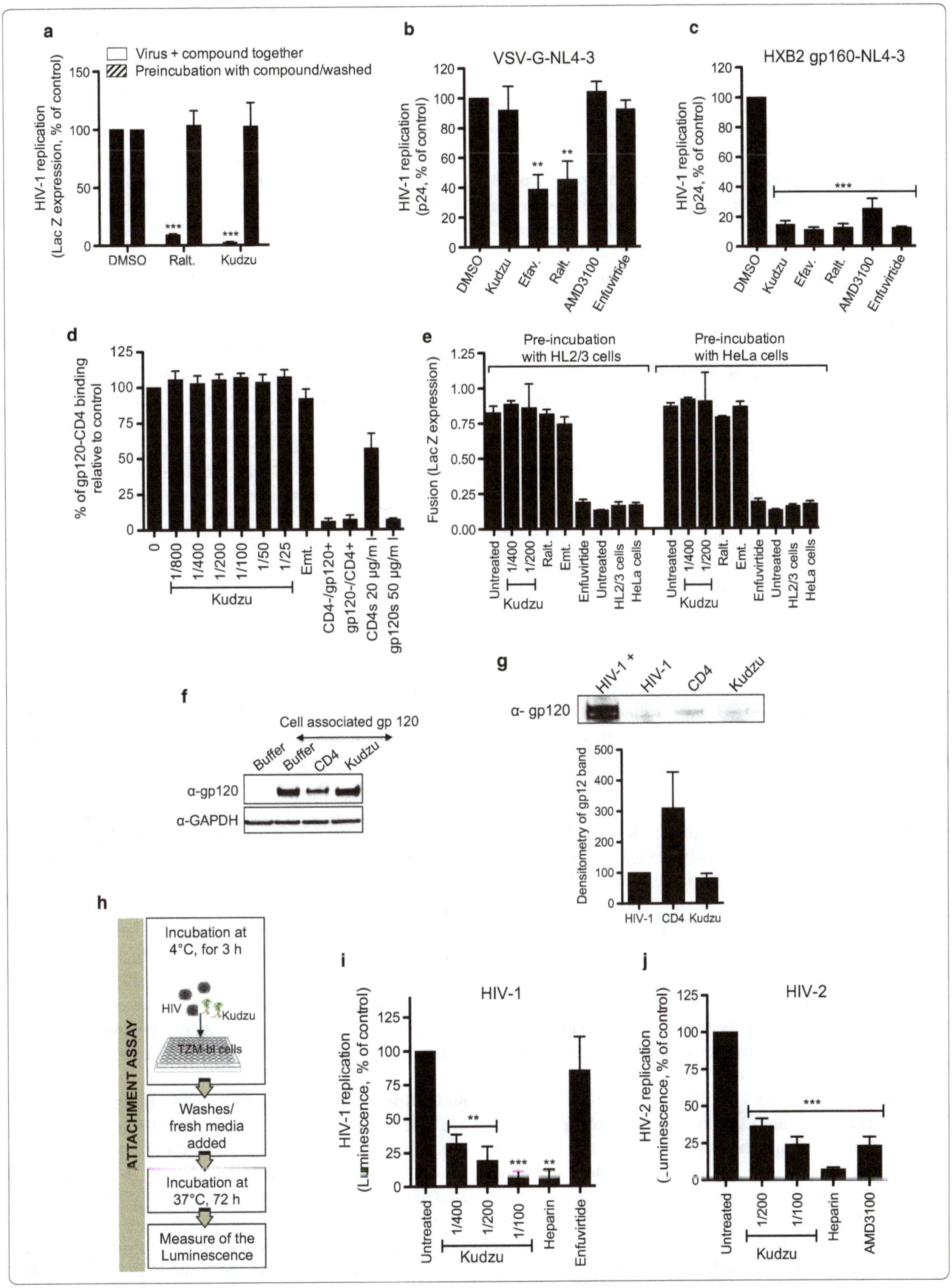
a
Virus + compound together
Preincubation with compound/washed
HIV-1 replication (Lac Z expression, % of control)
DMSO
Ralt.
Kudzu
b
VSV-G-NL4-3
HIV-1 replication (p24, % of control)
DMSO
Kudzu
Efav.
Ralt.
AMD3100
Enfuvirtide
c
HXB2 gp160-NL4-3
HIV-1 replication (p24, % of control)
d
% of gp120-CD4 binding relative to control
Kudzu
Emt.
CD4-/gp120+
gp120-/CD4+
CD4s 20 µg/m l
gp120s 50 µg/m l
e
Pre-incubation with HL2/3 cells
Pre-incubation with HeLa cells
Fusion (Lac Z expression)
Untreated
Kudzu
Ralt.
Emt.
Enfuvirtide
HL2/3 cells
HeLa cells
f
Cell associated gp 120
Buffer
Buffer
CD4
Kudzu
α-gp120
α-GAPDH
g
HIV-1 +
HIV-1
CD4
Kudzu
α- gp120
Densitometry of gp12 band
h
ATTACHMENT ASSAY
Incubation at 4°C, for 3 h
HIV
Kudzu
TZM-bl cells
Washes/ fresh media added
Incubation at 37°C, 72 h
Measure of the Luminescence
i
HIV-1
HIV-1 replication (Luminescence, % of control)
Untreated
Kudzu
Heparin
Enfuvirtide
j
HIV-2
HIV-2 replication (Luminescence, % of control)
Untreated
Kudzu
Heparin
AMD3100

infection. However, the entry inhibitor AMD3100 (10 nM), the fusion inhibitor Enfuvirtide (1 μg/ml) and Kudzu (1:400) significantly inhibited HXB2 gp160-, but not VSV-G-driven infection of the cells. Collectively, these data suggest that Kudzu targets HIV entry, and is dependent on virus-target cell entry mechanisms mediated by gp120 and/or gp41.

To investigate whether Kudzu was targeting gp120 directly, we first assessed its ability to interfere with the interaction between gp120 and CD4. For this, we coated monomeric recombinant gp120 on an ELISA plate and measured its interaction with CD4-Ig (as in [34]), in the presence of increasing concentrations of Kudzu (Fig. 4d). The intensity of the absorbance is directly proportional to the binding of gp120 to CD4-Ig. Kudzu did not interfere with the gp120-CD4-Ig interaction, while soluble gp120 and soluble CD4 both acted as competitors. Emtricitabine (500 nM) was used as a negative control. The background of the assay was evaluated by measuring the signal generated in buffer-coated wells (instead of gp120), as well as buffer-treated wells (instead of CD4-Ig). Together these results suggest that Kudzu does not interfere with gp120 interaction with CD4.

We further investigated whether Kudzu could interfere with the fusion between gp160-expressing cells and target cells that express both CD4 and CXCR4 (Fig. 4e). HL2/3 cells express viral Gag, Env, Tat, Rev, and Nef proteins. Their co-culture with HeLa-CD4-LTR-LacZ cells results in cell fusion, and subsequent activation of 5′-LTR-driven LacZ expression by HL2/3-produced Tat protein. In this assay, we individually pre-incubated HL2/3 cells or HeLa-CD4-LTR-LacZ cells with compounds for 2 h before co-culture for another 48 h, followed by measurement of β-Gal activity. As shown in Fig. 4e, while Enfuvirtide (1 μg/ml) inhibited cell fusion, Kudzu (1:400 and 1: 200) had no inhibitory activity. Raltegravir (100 nM), Emtracitabine (100 nM), untreated and cells alone were used as negative controls. Together these results suggest that Kudzu does not interfere with the fusion between virus and cell membranes, especially with gp120 interaction with CXCR4 and gp41mediated fusion.

We then evaluated the effect of Kudzu on the shedding of gp120 (Fig. 4f), as previously described [35]. The gp120 was ectopically expressed in HEK293T cells after transfection of JR-FL gp160Δtail constructs. After 48 h, cells were incubated for 4 h with Kudzu (1:400), or soluble CD4-Ig (45 μg) used as a positive control. Cells were washed, lysed, and the remaining gp120 attached to the cell surface was detected by immunoblotting using anti-gp120 serum. As shown in Fig. 4f, the amount of gp120 remaining after Kudzu treatment was similar to the control, while soluble CD4-Ig decreased the presence of gp120 on the cell surface. To confirm this result, we exposed concentrated NL4-3 virus to Kudzu (1:200) or CD4-Ig (30 μg) for 6 h, and then passed the solution through a 300-kDa filter, to recover shedded gp120 in the flow-through. A fraction of the flow-through was then analyzed by immunoblotting with an anti-gp120 serum. The gp120 bands and corresponding densitometry are shown in Fig. 4g. Kudzu did not increase shedding of gp120 from the virus surface as opposed to CD4. Altogether these results suggest that the inhibition of HIV-1 entry by Kudzu is not mediated by gp120 shedding.

Kudzu inhibits HIV attachment to the target cell

As we mentioned above, gp120 first attaches to the target cell by interacting with cell-surface heparan sulfate proteoglycans, α4β7 integrins, DC-SIGN or MCLR. These attachment receptors bring gp120/gp41 into close proximity to CD4 and CXCR4 or CCR5, increasing infection efficiency. As such, we monitored the involvement of Kudzu in HIV-1 attachment to the target cell, as previously described [36, 37]. Pre-chilled TZM-bl cells were incubated with NL4-3 and increasing concentrations of Kudzu for 3 h at 4 °C (Fig. 4h). This allows only attachment of the virus to the cell, but not fusion and entry of the virus. After washing with cold PBS to remove unattached virions and/or drugs bound to virions, fresh medium (without compounds) was added. The cells were then further incubated at 37 °C for 72 h, before luciferase activity was measured. The peptide fusion inhibitor Enfuvirtide (1 μg/ml), which binds to gp41, was used as a negative control, while soluble heparin (1 mg/ml), which reduces attachment of the virus to the target cell by competing with the cell surface heparan sulfate proteoglycans [38], was used as a positive control. We observed a dose-dependent inhibition of virus attachment in the presence of Kudzu compared to the untreated control (Fig. 4i). We also verified that Kudzu did not affect TZM-bl cell viability at the indicated concentrations (Additional file 1: Fig. 1SD left). Similar results were obtained with HIV-2 (Fig. 4j). Kudzu inhibited HIV-2 attachment to the same extent as the entry inhibitor AM3100, which tightly binds to CXCR4 on the cell surface. Taken together our results suggest that Kudzu root extract blocks attachment of gp120 to the cell surface, a process independent of CD4, thus inhibiting the first step of the HIV entry life cycle.

Kudzu does not inhibit HIV-1 late replication events

We also investigated the ability of Kudzu to inhibit late replication events, i.e., after integration of the viral DNA into the host cell genome. First, we tested the activity of Kudzu on viral transcription by analyzing the viral production from chronically-infected HeLa-CD4 cells for a period of 72 h (Fig. 5a). We used the Tat inhibitor didehydro-Cortistatin A (dCA) as a control [39]. While dCA

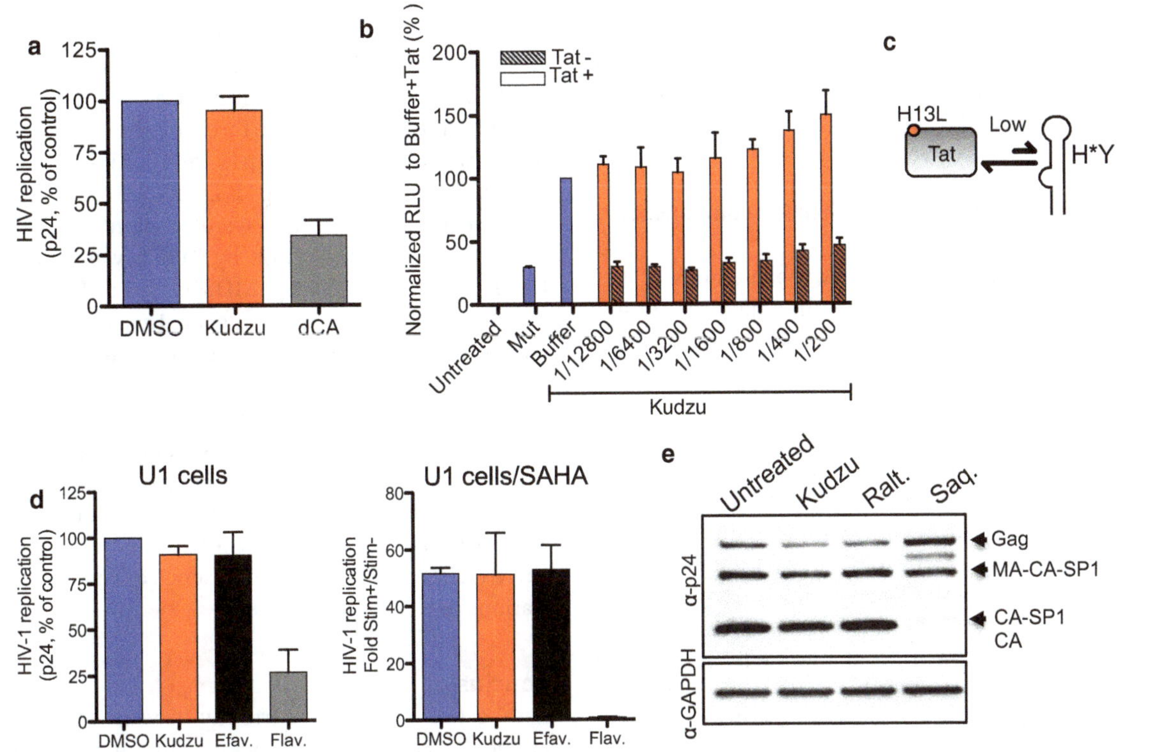

Fig. 5 Kudzu does not interfere with HIV-1 late replication events. **a** Kudzu has no activity on HeLa-CD4 cells chronically infected with NL4-3. Kudzu used at 1:400 dilution. Didehydro-Cortistatin A (dCA, a Tat inhibitor, 100 nM) was used as a control. Viral supernatants recovered 72 h post-infection from cells were assayed for their p24 antigen content. Data is the mean ± SD of 2 independent experiments. **b** Kudzu does not impact the transactivation activity of transfected Tat protein in TZM-bl cells. Tat mutated in the basic domain, Tat Mut, and buffer were used as negative controls. Luciferase activity per protein concentration was determined 48 h later. The data represents the mean ± SD of 2 independent experiments. **c** Schematic describing the mutation in Tat protein in U1 cells. **d** Kudzu has no activity on chronically-infected U1 cells before and after stimulation with SAHA (an histone deacetylase inhibitor, 1 μM). Kudzu used at 1:400 dilution. Flavopiridol (Flav., a CDK9 inhibitor, 100 nM) and Efavirenz (Efav., 100 nM) were used as controls. Viral supernatants recovered 72 h post-infection from cells were assayed for their p24 antigen content. Results are the mean ± SD of 3 independent experiments. **e** Kudzu has no activity on maturation and assembly of HIV-1 capsid. HIV-1 capsid originates from a 55 kDa Gag precursor proteolyzed into three folded proteins [matrix (MA), capsid (CA) and nucleocapsid (NC)] and 3 small peptides [spacer peptides 1 and 2 (SP1 and SP2) and p6]. HEK293T cells were transfected with NL4-3 in presence of compounds, and 72 h later, cell lysates were analyzed by western blot with anti-p24 antibody. Kudzu used at the dilution 1:400. Raltegravir (Ralt., 100 nM) and Saquinavir (300 nM) used as controls. The result are representative of 3 independent experiments

inhibited viral production from these cells, Kudzu (1:400) showed no activity. We further confirmed this result by measuring the activity of Kudzu during Tat transactivation of the HIV-1 promoter in TZM-bl cells (Fig. 5b). A dominant negative variant of Tat mutated in the basic domain ("Tat Mut"), and unable to transactivate HIV-1 promoter, was used as a negative control. We confirmed that Kudzu had no impact on Tat-mediated transcription of HIV-1 while not altering the viability of TZM-bl cells (Additional file 1: Fig. 2SD right). We also investigated the activity of Kudzu on basal transcription using U1 cells, a model of HIV-1 latency where a H13L mutation in Tat abolishes the activity of the protein (Fig. 5c and d). The controls performed as expected, with a CDK9 inhibitor, Flavopiridol, inhibiting basal transcription, but not Efavirenz (Fig. 5d left). In this case again, Kudzu (1:400) did not alter basal transcription. We observed the same results after stimulating the cells with Suberoylanilide Hydroxamic acid (SAHA) for 24 h, a histone deacetylase inhibitor (Fig. 5d right). Collectively, these data suggest that Kudzu does not interfere with the transcription of HIV-1 genome.

To investigate whether Kudzu could interfere with late events of the viral life cycle, such as Gag expression and maturation, we transfected HEK293T cells with NL4-3 molecular clone in the presence of Kudzu or controls, and measured Gag expression in cell lysates by immunoblot, using an anti-p24 antibody (Fig. 5e). The expression of Gag protein and its maturation were similar in Kudzu (1:400), Raltegravir (100 nM), and untreated samples, while the protease inhibitor, Saquinavir (300 nM), blocked the maturation of Gag into Matrix (MA) and Capsid (CA) proteins. Taken together, our results strongly suggest that Kudzu does not impact the late events of HIV-1 life cycle.

Individual antiviral activity of the chemical components of Kudzu

The geographic origin, as well as the methods of culture and root extraction, dictate the chemical composition of Kudzu extract. Several reports ascribe more than 40 chemical entities to Kudzu root extract (Additional file 1: Table 1S). In an effort to identify the compound responsible for Kudzu's activity, we tested its main components and as many as of the commercially available less abundant ones (Fig. 6a). To our surprise, none of the most abundant components presented antiviral activity, including Puerarin (60% of the total isoflavones in Kudzu), Daidzein and Daidzin, which have been previously reported to have anti-microbial activity [25]. In addition, the combination of the highly represented compounds still did not display antiviral activity (data not shown). Moreover, we assessed whether the activity of Kudzu was potentially due to protein traces in the extract. For this, we denatured Kudzu by boiling and tested its activity in infectivity assays using HeLa-CD4-LTR-LacZ cells. Figure 6b shows that the denaturation did not affect the activity of Kudzu. Furthermore, the denatured Kudzu was ran on a denatured electrophoresis gel, transferred to a nitrocellulose membrane, and no proteins were observed after Coomassie blue staining (data not shown). Altogether, these results suggest that Kudzu antiviral activity most likely derives from a component present at a very low concentration or it results from an additive/synergistic effect of several compounds.

Activity of Kudzu in combination with current ART and activity against viruses resistant to Enfuvirtide

We assessed the activity of Kudzu in combination with two different cocktails: ATRIPLA® (Efavirenz, Tenofovir and Emtricitabine) as a single tablet regime, and ALE, Combivir (AZT and Lamivudine) combined with Sustiva (Efavirenz). We first determined the IC_{50} of each ARV cocktail and Kudzu alone. Next, Kudzu and the cocktail at their IC_{50} were mixed together at different ratios (Kudzu: cocktail, 1:1, 4:1, 8:1) to obtain an IC_{50} of the mix. The data was plotted as concentration of ARVs cocktails per Kudzu dilution (Fig. 7a). The mixes of Kudzu with ATRIPLA® at the different ratios presented IC_{50} concentrations lower than the trend line (dotted line between Individual IC_{50}), which is associated with synergistic activity [40]. With the ALE cocktail, the activity of Kudzu was additive when used at ratio of 1:1, and synergistic when used at doses higher than 1:1.

A number of characterized HIV-1 isolates are resistant to the entry inhibitor Enfuvirtide. We evaluated the susceptibility to Kudzu of three of these viral isolates, which carry mutations in gp41 (gp41 D36G V38E N42S, gp41 D36G V38A and gp41 D36G V38A N42D) [41]. As expected, these viruses were resistant to Enfuvirtide (150 ng/ml). The inhibition by Kudzu was as potent as Raltegravir (100 nM) and AMD3100 (10 nM).

Collectively these results suggest that Kudzu does not compete with current ART, but acts additively/synergistically with ART, rationalizing its use as a complement to current HIV-1 therapy. Kudzu may also be helpful in cases of Enfuvirtide resistance, highlighting its potential to complement HIV-1 salvage therapy.

Importantly, we did not observe development of resistance after passaging the NL4-3 strain every 4 days, onto naïve HeLa-CD4 cells in the presence of sub IC_{50} concentrations over a 54 days period.

Conclusion

Only a few agents are currently used in the clinic that target the HIV-1 entry process (Enfuvirtide, Maraviroc and Ibalizumab). There is a continuing need for development of novel antiretroviral drugs and regimens in order to address tolerability, long-term safety concerns, the immune dysfunction mediated by HIV and the emergency of drug resistance. The cost of current treatment and accessibility is also a concern to a significant portion of the HIV affected population. Traditional medicine is very popular in many countries, and it has been used to replace ARVs or off-set side effects from antiretroviral medication. Such medicinal plants have not been properly identified and documented probably due to the lack of collaborations between scientists and traditional healers. Kudzu is a safe and very well-known natural product that has been used for hundreds of years. Here we demonstrated its potent activity against the attachment of HIV-1 gp120 to receptors on the host cell membrane, blocking HIV entry into cell lines, primary $CD4^{+}$T cells and macrophages. To date, no small molecule compound has been described that targets this step of the life cycle. Kudzu's efficacy against HIV-2 is further appreciated given that HIV-2 is naturally resistant to Enfuvirtide, nonnucleoside reverse transcriptase inhibitors, and to some protease inhibitors. Moreover, Kudzu has also been shown to reduce inflammation and oxidative stress, an affliction that is often identified in HIV infected individuals. So far, we have not been able to identify the chemical entity of Kudzu extract that is responsible for the antiretroviral activity. In future studies, we will use a combination of HPLC fractionation and 1H-NMR to identify

a

Compounds	Structure	IC_{50} (μM)	CC_{50} (μM)
Daidzin	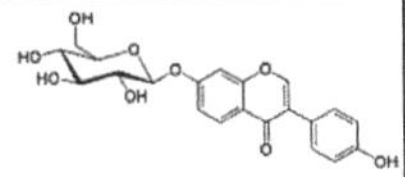	> 500 ± 0.0	> 500 ± 0.0
Daidzein	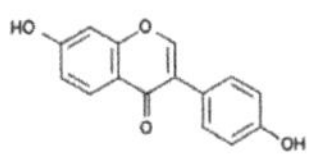	81.6 ± 4.3	151.6 ± 25.5
Genistein	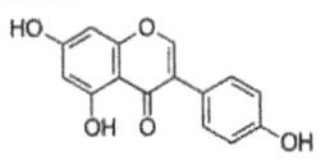	100.9 ± 15.2	135.4 ± 22.8
Puerarin	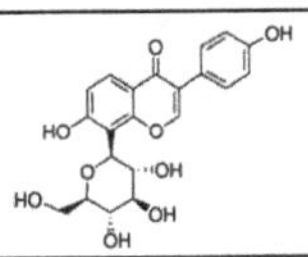	> 500 ± 0.0	> 500 ± 0.0
Lupeol	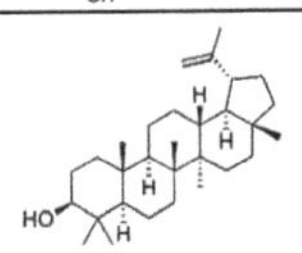	> 56.3 ± 3.9	56.3 ± 3.9
Ononin	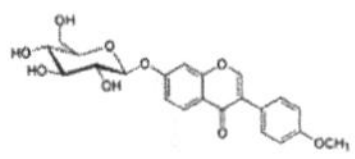	> 500 ± 0.0	> 500 ± 0.0
β- Sitosterol	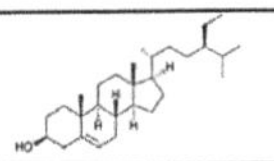	> 500 ± 0.0	> 500 ± 0.0
Lupenone	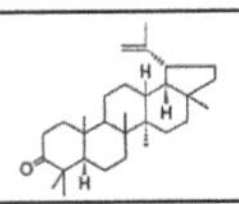	> 500 ± 0.0	> 500 ± 0.0
Allantoin	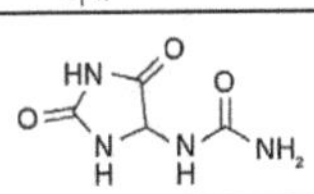	> 500 ± 0.0	> 500 ± 0.0
Formononetin	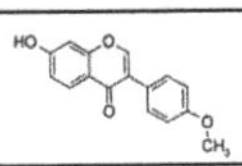	> 500 ± 0.0	> 500 ± 0.0
Coumestrol	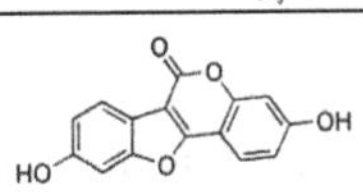	> 182.4 ± 5.6	182.4 ± 5.6
Diisobutyl phthalate	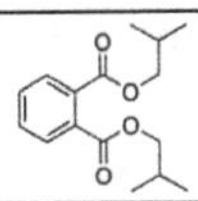	> 500 ± 0.0	> 500 ± 0.0
Bis 2-(ethylhexyl) phthalate	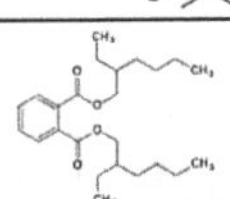	> 500 ± 0.0	> 500 ± 0.0
Sissotrin	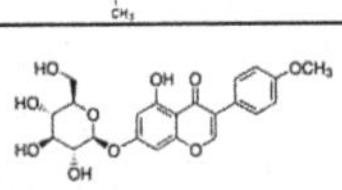	> 500 ± 0.0	> 500 ± 0.0

b

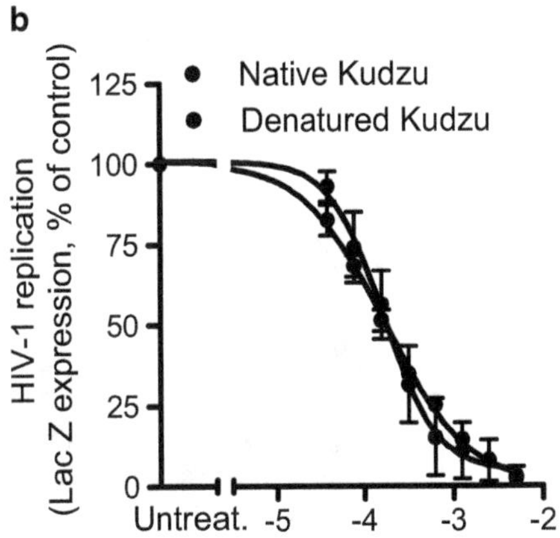

Fig. 6 Activity of the different components of Kudzu on acute replication of HIV-1 of X4 tropic virus in HeLa-CD4-LTR-LacZ cells. **a** HeLa-CD4-LTR-LacZ cells were infected with HIV-1 NL4-3 strain in presence of different concentrations of the indicated compounds. β-Gal activity and viability assays were performed 72 h later. Results represent the mean ± SD of 3 independent experiments. **b** The activity of Kudzu does not involve proteins. HeLa-CD4-LTR-LacZ cells were infected with HIV-1 NL4-3 strain in presence of denatured Kudzu by heat or native Kudzu. β Gal activity was measured 72 h later. Data is the mean ± SD of 4 independent experiments. Untreat.: untreated

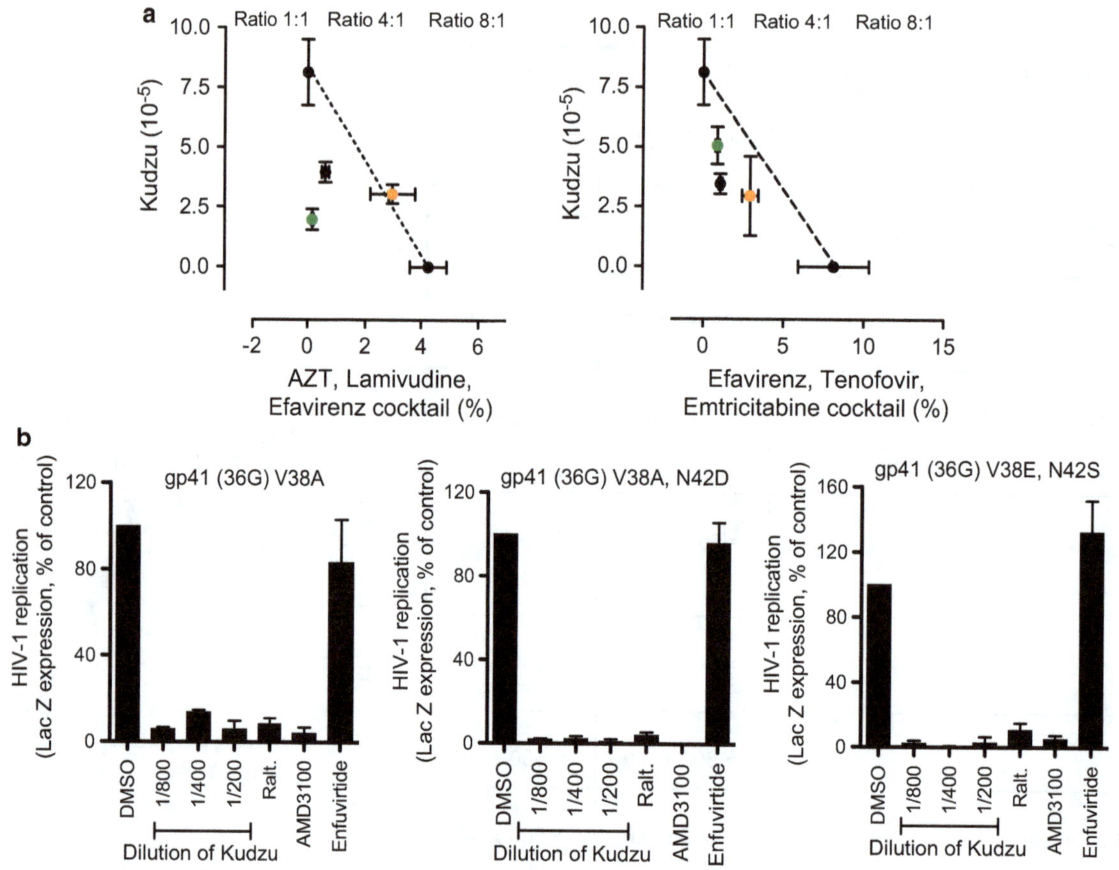

Fig. 7 Activity of Kudzu in combination with antiretrovirals. **a** Combination of Kudzu with different cocktails of antiretrovirals prescribed to HIV-1 patients. HeLa-CD4-LTR-LacZ cells were infected with HIV-1 NL4-3 strain in presence of different amounts of Kudzu or the indicated antiretrovirals. β-Gal activity was measured 72 h later. The IC_{50} of the cocktails or Kudzu alone (black) or the mix of both at different ratios (orange, green, and blue) are represented. Results represent the mean ± SD of 3 independent experiments. **b** Activity of Kudzu in HeLa-CD4-LTR-LacZ cells infected with viruses resistant to Enfuvirtide. HeLa-CD4-LTR-LacZ cells were infected in presence of different dilutions of Kudzu. β-Gal activity was measured 72 h later. AMD3100 and Enfuvirtide used at 10 nM and 0.15 µg/ml respectively. Data is the mean ± SD of 2 independent experiments

active compounds. In sum, given Kudzu's low cost, safety, oral bioavailability, tissue distribution, activity with ART and potent activity against HIV gp120 attachment to host cell, it should be considered as a promising supplement to current HIV therapeutic strategies.

Methods

Kudzu

Kudzu (Hawaii pharm LLC), was used in all experiments. Its activity was confirmed with Kudzu from Secrets of the Tribe (Nevada pharm LLC). Both Kudzus are a tincture solution with 33% of Kudzu and 66% of ethanol-glycerol-water.

Viral replication of HIV-1 in cell lines detected by p24 ELISA

HeLa-CD4-LTR-LacZ cells (5×10^3/well in a 96-well-plate) were first incubated with Kudzu or the indicated controls (Raltegravir: 100 nM, Efavirenz: 100 nM, AMD3100: 10 nM, Enfuvirtide: 1 µg/ml,) then infected overnight (ON) with wild-type HIV-1 isolate NL4-3 (3.2 ng/well). Cells were washed and further cultured in fresh media (DMEM supplemented with 5% Fetal bovine serum (FBS), L-glutamine (292 µg/ml) and antibiotics (100 units/ml penicillin and streptomycin)) containing compounds for another 72 h. HIV-1 replication was assessed by measuring the viral capsid in the supernatant with a p24 ELISA kit (Advances Bioscience Laboratories). Similar protocol was used for the infection of GHOST-CCR5 cells (gift from Dr. Michael Farzan) with JRCSF (21.3 ng/well, NIH AIDS Reagent Program, cat# 394) or YU2 (2.32 ng, NIH AIDS Reagent Program, cat# 1350) viruses. The amount of virus was chosen to obtain similar p24 capsid outputs between the different cell lines.

Viral replication of HIV-1 in cell lines, revealed by chlorophenol red-β-d-galactopyranoside assay (CPRG assay)

HeLa-CD4-LTR-LacZ cells (5×10^3/well in a 96-well-plate) were first incubated with compounds (Kudzu, Kudzu's components, indicated ARVs controls, Glycerol, Ethanol, or denatured Kudzu boiled for 10 min at 90 °C). NL4-3 virus (6.6 ng/well) or viruses resistant to Enfuvirtide (gifts from NIH AIDS Reagent Program, cat# 9490 (5.4 ng/well), cat# 9488 (2 ng/well), cat# 9496 (0.6 ng/ml)) were then added for 72 h. Cells were disrupted with lysis buffer (60 mM Na2HPO4, 40 mM NaH2PO4, 10 mM KCl, 10 mM MgSO4, 2.5 mM EDTA, 50 mM ß-mercaptoethanol, 0.125% Nonidet P-40) for 1 h at 4 °C and a quantitative CPRG-based (Boehringer Mannheim) assay was performed per manufacturer's instructions. The cell extracts were incubated in reaction buffer (0.9 M phosphate buffer [pH 7.4], 9 mM MgCl2, 11 mM ß-mercaptoethanol, 7 mM CPRG) until a red color developed and measured with an LP400 (Becton–Dickinson) plate reader at 572 nm. The amount of virus used for infection aimed to produce a similar OD at approximately 20 min of reaction.

Viral replication of H1N1, ZIKV Brazil and ADV5 viruses

Influenza A (Virginia/ATCC1/2009 H1N1), Zika virus (PB81, Brazil strain), and adenovirus serotype 5 (ADV5) were produced in MDCK, Vero and HEK293T cells, respectively. Each virus was mixed with the indicated compounds, and added to Hela-CD4-LTR-LacZ cells. Aleuria Aurantia lectin (100 nM), an inhibitor of H1N1 entry, cabozantinib (1 μM), a tyrosine kinase inhibitor known to suppress Zika infection, and heat inactivation of ADV5 for 30 min at 70 °C, were used as positive controls. After 1 h at 37 °C, the inoculum was removed, and the cells further grown in DMEM containing 10% FBS and the indicated compounds. Twenty-four hours after the inoculation, the cells were collected, fixed with PBS containing 2% of paraformaldehyde, permeabilized using PBS containing 2% of saponin, and stained with either the monoclonal mouse anti-H1N1/H2N2 antibody clone C179 (H1N1), or the monoclonal mouse pan anti-flavivirus antibody clone 4G2 (Zika), or the mouse monoclonal anti-adenovirus antibody clone 2/6 (ADV5), followed with an Alexa647-conjugated, anti-mIgG antibody (Jackson ImmunoResearch). The cells were then washed, fixed, and analyzed using flow cytometry (BD Biosciences C6 Accuri and IntelliCyt HyperCyt sampler powered by FlowCyt software). Infection levels were normalized to those of cells infected in the absence of any compound.

Viral replication of pseudotyped viruses

Hela-CD4-LTR-LacZ cells (1×10^5/in a 6-well-plate) were infected ON with VSV-G-NL4-3 (70.2 ng or 5.6 ng/well) or with HXB2 gp160-NL4-3 (25.4 ng/well) pseudotyped viruses (constructs gifted by Dr. Michael Farzan), in the presence of Kudzu (1:400) or controls (Raltegravir: 200 nM, Efavirenz: 200 nM, AMD3100: 10 nM, Enfuvirtide: 1 μg/ml). Cells were washed and fresh media containing compounds was added for 72 h. The capsid p24 in the supernatant was assessed by ELISA.

Viral replication of HIV2 CBL-20 strain

TZM-bl cells (4×10^4/well in a 6-well-plate) were incubated with different concentrations of Kudzu or the indicated controls, then infected ON with HIV-2 CBL-20 virus (NIH AIDS Reagent Program, cat# 600, 0.2 ng/well). Media was then removed and fresh media (DMEM supplemented with 5% FBS, L-glutamine (292 μg/ml) and antibiotics (100 units/ml penicillin and streptomycin)) containing compounds were added for an additional 48 h. Luciferase activity was then determined and reported per protein concentration of each sample as previously described [39].

Viral replication of HIV-1 and SIV in primary human and rhesus macaque CD4$^+$T cells

CD4$^+$T cells were isolated and expanded as previously described [42]. Briefly, PBMCs were extracted from 3 healthy human donors (purchased from One Blood-Florida) and 3 healthy rhesus macaques (obtained from the Wisconsin National Primate Research Center). Total CD4$^+$T cells were isolated using positive selection kit (StemCell Technologies). Cells were expanded with 1 μg/ml of PHA (Sigma Aldrich), 100 U/ml of IL-2 (Roche) and irradiated feeder PBMCs (OneBlood) and cultured in RPMI and human serum for 3 days. Primary human CD4$^+$ T cells were treated with DMSO, kudzu or a cocktail of ARVs (AZT: 180 nM, Efavirenz: 100 nM, Raltegravir: 200 nM) then infected with NL4-3 (100 ng). After 24 h, cells were pelleted, DNA extracted, followed by PCR (see below). Primary macaque CD4$^+$T cells were treated with DMSO, Kudzu or a cocktail of ARVs (Emtracitabine, Raltegravir and Tenofovir, 200 nM) then infected with SIV$_{mac}$239 (100 ng). SIV replication was assessed 6 days later, by measuring the viral capsid in the supernatant with p27 ELISA (Advances Bioscience Laboratories).

Viral replication of ADA and 5002M viruses in primary macrophages

Human monocyte-derived macrophages were isolated from elutriated blood PBMCs by Ficoll-Hypaque density gradient centrifugation (GE Healthcare). Monocytes were negatively selected using magnetic particles from EasySep™ Human Monocyte CD14+without CD16 depletion Enrichment kit (Stem Cell Technologies). Monocytes were differentiated in DMEM (Gibco) supplemented with 10% heat inactivated human serum (Sera Care Life Sciences), 2 mM L-glutamine (Gibco), 10 µg/ml gentamicin (Sigma–Aldrich) and 10 ng/ml MCSF (R&D System) and cultured for 7 days at 37°C with 5% CO_2. Macrophages were then seeded onto 48-well plate (1×10^5 cells/well), infected with 1×10^6 CPMs of macrophage-tropic strains ADA or 5002M in the presence of kudzu (1:100) or Enfuvirtide (5 µg/ml). Viral inoculum was removed, and media replaced with fresh compounds 24 h post-infection. Compounds were added every day to the media, and half media was replaced every 2 days. An aliquot of the supernatant was harvested every day for 6 days, for capsid p24 quantification (ClonTech).

Viral replication of HCV, assessed by immunohistochemistry

Huh 7.5 cells were infected with HCV JC1 virus and Kudzu (at the dilution 1:200) was subsequently added. A blank (without compounds, "untreated") or 2′-C-methyladenosine (10 µM) [43] was used as control. After 48 h, cell supernatants/lysates were harvested, cell lysates prepared with 2 cycles of freeze thawing, and stored (−80 °C) for infection of naïve cells for titration. Naïve Huh-7.5 cells (cultured in 96-well plates) were inoculated with the supernatant/lysates for limiting dilution virus titration assay. After 48 h, cells were washed, fixed in methanol, and then probed for NS5A expression using the 9E10 monoclonal antibody. $TCID_{50}$ was calculated as previously described [44].

Latently infected cells

NL4-3 latently infected HeLa-CD4 cells (2.5×10^5/well) and U1 cells (5×10^5/well) were treated with compounds for 72 h, and p24 in the supernatant was assessed by ELISA. HeLa-CD4 cells were treated with 1:400 dilution of Kudzu or 100 nM of dCA. U1 cells were treated with 1:400 dilution of Kudzu, 100 nM of Efavirenz or 100 nM of Flavopiridol. U1 cells reactivation was performed with 1 µM of SAHA for 24 h.

Time-of-addition experiment

HeLa-CD4-LTR-LacZ cells (5×10^3/well in a 96-well-plate) were infected with NL4-3 (6.6 ng/well). Kudzu and ARVs controls were added to the infected cells, at the indicated concentrations, at time 0, 1, 2, 3, 4, 5, and 6 h post-infection. β-Gal activity was measured 72 h later.

Attachment assay in TZM-bl cells

TZM-bl cells (2×10^4) were seeded in a 96-well plate. Next day, cells were incubated for 1 h at 4 °C. In parallel, compounds (Kudzu: 1:400–1:100, Heparin: 1 mg/ml, Enfuvirtide: 1 µg/ml and AMD3100: 10 nM) were mixed with NL4-3 (16.6 ng/well) or HIV-2 CBL-20 (0.8 ng/well) for 1 h at 4 °C. Compounds with virus were then added to the cells and incubated at 4 °C for 3 h. Cells were washed twice with ice cold PBS and incubated with fresh media for 72 h at 37 °C, 5% CO_2. The infection was assessed by measuring the luminescence with Bright Glo (Promega) according to the manufacturer's protocol.

Transactivation assay in TZM-bl cells

Cells were seeded at 2×10^6 in a 10 cm^2 tissue culture dish and transfected with 5 µg of the constructs expressing Tat (PGK-Flag-Tat) or Tat Mut (PGK-Flag-Tat Mut) driven by the murine phosphoglycerate kinase-1 (PGK) promoter, with TransIT-LT1 transfection reagent (Mirus Bio LLC) according to the manufacturer's protocol. The cells were split 24 h post transfection and treated with Kudzu at different concentrations. Luciferase activity per protein concentration of each sample was determined 48 h later as previously described [39].

Quantification of early and late reverse transcription product and provirus integration

In HeLa-CD4-LTR-LacZ cells

Cells were plated at 1×10^5 cells per well in a six-well plate. Twenty-four hours later, cells were infected with NL4-3 virus (66 ng) in the presence of DMSO, Kudzu (1:200), Efavirenz (200 nM), Saquinavir (200 nM), Raltegravir (200 nM) and AMD3100 (10 nM). At 10 h and 24 h post-infection, genomic DNA was prepared and early and late viral DNA products and the integrated proviruses were quantified as previously described [45].

In primary CD4$^+$T lymphocyte cells

Cell pellets were digested at 133 µl/million cells with lysis buffer (10 mM Tris–HCL, 50 nM KCl, 400 µg/ml proteinase K (Invitrogen) to extract DNA. Total HIV or SIV DNA content was quantified by PCR amplification with Taq polymerase, 1X Taq Buffer (Invitrogen), $MgCl_2$, dNTPs, HIV/SIV and CD3 primers for 12 cycles. The second amplification involved a nested PCR using 1:10 dilution of amplification product, HIV/SIV and CD3 primers, SsoAdvanced Universal Probe Supermix (Biorad) using qPCR. A standard curve was established using ACH2 cell lysates (ATCC) that contain a single copy of

HIV per cell or $SIV_{mac}239$ plasmid. Primers used are listed in Additional file 1: Table S2.

gp120-CD4 interaction by ELISA

Recombinant YU2 gp120 protein (Immune Tech, cat# IT-001-0027p, 1.5 μg/ml) was coated onto a 96-well ELISA plate ON at 4 °C. Next day, wells were washed twice with 150 μl of PBS supplemented with Tween 0.05%, and saturated with 5% milk for 1 h at 37 °C. CD4-Ig (gift from Dr. Michael Farzan, 5 μg/ml) in presence or not of increasing concentrations of Kudzu was then added and the plate was incubated at 37 °C for 1 h. Secondary anti-human antibodies were then added at the dilution 1:5000 for 1 h at 37 °C. Then, 50 μl of TMB (Immunochemistry) was added at room temperature and the reaction was stopped with TMB stop solution. The absorbance was read at 450 nm. Several washes were performed after each step except after saturation. Emtracitabine (500 μM), soluble recombinant YU2 gp120 protein (Immune Tech, cat# IT-001-0027p) and soluble mouse CD4 (gift from Dr. Michael Farzan) were used as controls in the assay.

Expression of CD4 and CXCR4 in HeLa-CD4-LTR-LacZ cells analyzed by flow cytometry

Cells were seeded at 1×10^6 in a 6-well plate. The next day, compounds (Kudzu 1:400, PMA 100 nM or buffer) were added. Cells were then collected at different time points, stained with PE/Cy7 anti-human CD184 (for CXCR4, Biolegend) or PE antihuman CD4 (Biolegend) and fixed with 2% formaldehyde in PBS. Cells were analyzed on a LSRII flow cytometer (BD Bioscience).

Expression of CCR5 in Ghost-CCR5 cells analyzed by flow cytometry

Cells were seeded at 1×10^6 per 6-well plate. The following day, compounds (Kudzu 1:400 and 1:200) were added. Cells were then collected at 6 h post treatment, stained with PE anti-human CD195 (Biolegend) or PE control (Biolegend). Flow cytometry was afterwards performed (BD Biosciences C6 Accuri and IntelliCyt HyperCyt sampler powered by FlowCyt software).

Fusion of HeLa-CD4-LTR-LacZ cells with HL2/3 cells

Similar protocol to [27] was performed. Briefly, HL2/3 cells (gift from AIDSreagent program), expressing the HXB2 envelope, Tat, Gag, Rev, and Nef proteins, were co-cultured with HeLa-CD4-LTR-LacZ cells at 1:1 density ratio (2.5×10^4 cells/well in a 96-well-plate) for 48 h in presence of different compounds (Kudzu: 1:400 and 1:200, Raltegravir: 100 nM, Emtricitabine: 100 nM, Enfuvirtide: 1 μg/ml). Upon fusion of the two cell lines, Tat protein from HL2/3 cells activates β-gal expression, measured by a CPRG assay.

Shedding of gp120 from transfected gp160

Performed as previously reported [35]. Briefly, HEK293T cells were transfected with the constructs JR-FL gp160Δtail (1.5 μg) and pCMV Rev (0.5 μg) (gifts from Dr. Michael Farzan) for 48 h with TransIT-2020 transfection reagent (Mirus Bio LLC) according to the manufacturer's protocol. JR-FL gp160Δtail was used to achieve higher envelope trimer expression [46]. CD4-Ig (45 μg, a gift from Dr. Michael Farzan) or Kudzu (1:400 dilution) were incubated with cells for 4 h. Cells were then washed several times with PBS 10 mM EDTA and lysed with RIPA buffer supplemented with a cocktail of protease inhibitor (Roche). Protein concentration was determined by Bradford assay. The cell lysates were analyzed on a reducing gel 4–20% gradient gel (Bio-rad) and subsequently blotted onto a nitrocellulose membrane (Biorad). gp120 content was determined by immunoblotting using a gp120 serum at 1:500 dilution (a gift from Dr. Michael Farzan) and goat anti-human antibody (1:5000, Santa Cruz).

Shedding of gp120 from NL4-3 virus

Kudzu and CD4-Ig (gift from Dr. Michael Farzan, 30 μg) were incubated with NL4-3 (44.8 ng) for 6 h. Samples were then filtered through 300 kDa column (Vivaspin, Sartorius). A fraction of the flow trough was analyzed by immunoblot as described above.

Maturation of HIV-1 capsid

HEK293T cells were transfected with pNL4-3 encoding plasmid (12.5 μg) with TransIT-2020 transfection reagent (Mirus Bio LLC) according to the manufacturer's protocol. Six hours later, cells were washed, trypsinized and split between the conditions tested. Cells were treated with either buffer, Kudzu (1:400), Raltegravir (100 nM), Saquinavir (300 nM) and incubated at 37 °C for 72 h. Next, cells were lysed in RIPA buffer supplemented with a cocktail of protease inhibitors (Roche) and analyzed by immunoblotting with anti-p24 antibody (1:2500, NIH AIDS Reagent Program cat# 3537). Anti-GAPDH antibody (1:500, Santa cruz) was used as loading control.

Assessment of Kudzu's cytotoxicity

MTT (3-[4,5-dimethylthiazol-2-yl]-2,5-diphenyltetrazolium bromide) assay (ATCC) or cell titer Glo luminescent cell viability (Promega) was performed in the presence of increasing concentrations of Kudzu or Kudzu components according to the manufacturer's protocol.

Authors' contributions

AM, JJA, STV and YE involved in conceptualization. SM, JJA, ST, AR, CK, AB, MC performed experiments and analyzed data. TT, CH, MC, STV responsible for methodology, supervision, validation and in funding acquisition. All authors read and approved the final manuscript.

Author details

[1] Department of Immunology and Microbiology, The Scripps Research Institute, 130 Scripps Way, 3C1, Jupiter, FL 33458, USA. [2] Department of Molecular Therapeutics, The Scripps Research Institute, Jupiter, FL, USA. [3] The Botanist's Beach Farm, Jupiter, FL, USA. [4] University of Miami Miller School of Medicine, Miami, FL, USA. [5] Present Address: Roche, Basel, Switzerland.

Acknowledgements

This work was supported by The Scripps Research institute. We thank the NIH AIDS Reagent Program for reagents and Dr. Michael Farzan, Dr. Guillaume Mousseau, Dr. Travis Grim for advice and helpful discussions.

Competing interests

The authors declare that they have no competing interests.

Funding

This work was supported with funding from The Scripps Research Institute.

References

1. Clavel F, Hance AJ. HIV drug resistance. N Engl J Med. 2004;350:1023–35.
2. Wainberg MA, Zaharatos GJ, Brenner BG. Development of antiretroviral drug resistance. N Engl J Med. 2011;365:637–46.
3. Saphire AC, Bobardt MD, Zhang Z, David G, Gallay PA. Syndecans serve as attachment receptors for human immunodeficiency virus type 1 on macrophages. J Virol. 2001;75:9187–200.
4. Arthos J, Cicala C, Martinelli E, Macleod K, Van Ryk D, Wei D, Xiao Z, Veenstra TD, Conrad TP, Lempicki RA, McLaughlin S, Pascuccio M, Gopaul R, McNally J, Cruz CC, Censoplano N, Chung E, Reitano KN, Kottilil S, Goode DJ, Fauci AS. HIV-1 envelope protein binds to and signals through integrin alpha4beta7, the gut mucosal homing receptor for peripheral T cells. Nat Immunol. 2008;9:301–9.
5. Cicala C, Martinelli E, McNally JP, Goode DJ, Gopaul R, Hiatt J, Jelicic K, Kottilil S, Macleod K, O'Shea A, Patel N, Van Ryk D, Wei D, Pascuccio M, Yi L, McKinnon L, Izulla P, Kimani J, Kaul R, Fauci AS, Arthos J. The integrin alpha4beta7 forms a complex with cell-surface CD4 and defines a T-cell subset that is highly susceptible to infection by HIV-1. Proc Natl Acad Sci USA. 2009;106:20877–82.
6. Geijtenbeek TB, Kwon DS, Torensma R, van Vliet SJ, van Duijnhoven GC, Middel J, Cornelissen IL, Nottet HS, KewalRamani VN, Littman DR, Figdor CG, van Kooyk Y. DC-SIGN, a dendritic cell-specific HIV-1-binding protein that enhances trans-infection of T cells. Cell. 2000;100:587–97 **(see comments)**.
7. Gummuluru S, Rogel M, Stamatatos L, Emerman M. Binding of human immunodeficiency virus type 1 to immature dendritic cells can occur independently of DC-SIGN and mannose binding C-type lectin receptors via a cholesterol-dependent pathway. J Virol. 2003;77:12865–74.
8. Clapham PR, Reeves JD, Simmons G, Dejucq N, Hibbitts S, McKnight A. HIV coreceptors, cell tropism and inhibition by chemokine receptor ligands. Mol Membr Biol. 1999;16:49–55.
9. Kwong PD, Wyatt R, Sattentau QJ, Sodroski J, Hendrickson WA. Oligomeric modeling and electrostatic analysis of the gp120 envelope glycoprotein of human immunodeficiency virus. J Virol. 2000;74:1961–72.
10. Moore JP, Binley J. HIV. Envelope's letters boxed into shape. Nature. 1998;393:630–1 **(news; comment)**.
11. Caffrey M. HIV envelope: challenges and opportunities for development of entry inhibitors. Trends Microbiol. 2011;19:191–7.
12. Didigu CA, Doms RW. Novel approaches to inhibit HIV entry. Viruses. 2012;4:309–24.
13. Hardy H, Skolnik PR. Enfuvirtide, a new fusion inhibitor for therapy of human immunodeficiency virus infection. Pharmacotherapy. 2004;24:198–211.
14. Duffalo ML, James CW. Enfuvirtide: a novel agent for the treatment of HIV-1 infection. Ann Pharmacother. 2003;37:1448–56.
15. Tang MW, Shafer RW. HIV-1 antiretroviral resistance: scientific principles and clinical applications. Drugs. 2012;72:e1–25.
16. Dorr P, Westby M, Dobbs S, Griffin P, Irvine B, Macartney M, Mori J, Rickett G, Smith-Burchnell C, Napier C, Webster R, Armour D, Price D, Stammen B, Wood A, Perros M. Maraviroc (UK-427,857), a potent, orally bioavailable, and selective small-molecule inhibitor of chemokine receptor CCR5 with broad-spectrum anti-human immunodeficiency virus type 1 activity. Antimicrob Agents Chemother. 2005;49:4721–32.
17. Hunt JS, Romanelli F. Maraviroc, a CCR5 coreceptor antagonist that blocks entry of human immunodeficiency virus type 1. Pharmacotherapy. 2009;29:295–304.
18. Fessel WJ, Anderson B, Follansbee SE, Winters MA, Lewis ST, Weinheimer SP, Petropoulos CJ, Shafer RW. The efficacy of an anti-CD4 monoclonal antibody for HIV-1 treatment. Antiviral Res. 2011;92:484–7.
19. Zhong Y, Li Y, Zhao Y. Physicochemical, microstructural, and antibacterial properties of beta-chitosan and kudzu starch composite films. J Food Sci. 2012;77:E280–6.
20. Li G, Zhang Q, Wang Y. Chemical constituents from roots of *Pueraria lobata*. Zhongguo Zhong Yao Za Zhi. 2010;35:3156–60.
21. Penetar DM, Toto LH, Lee DY, Lukas SE. A single dose of kudzu extract reduces alcohol consumption in a binge drinking paradigm. Drug Alcohol Depend. 2015;153:194–200.
22. Liu B, Kongstad KT, Qinglei S, Nyberg NT, Jager AK, Staerk D. Dual high-resolution alpha-glucosidase and radical scavenging profiling combined with HPLC-HRMS-SPE-NMR for identification of minor and major constituents directly from the crude extract of *Pueraria lobata*. J Nat Prod. 2015;78:294–300.
23. Du G, Zhao HY, Zhang QW, Li GH, Yang FQ, Wang Y, Li YC, Wang YT. A rapid method for simultaneous determination of 14 phenolic compounds in *Radix puerariae* using microwave-assisted extraction and ultra high performance liquid chromatography coupled with diode array detection and time-of-flight mass spectrometry. J Chromatogr A. 2010;1217:705–14.
24. Zhou YX, Zhang H, Peng C. Puerarin: a review of pharmacological effects. Phytother Res. 2014;28:961–75.
25. Zhang Z, Lam TN, Zuo Z. *Radix puerariae*: an overview of its chemistry, pharmacology, pharmacokinetics, and clinical use. J Clin Pharmacol. 2013;53:787–811.
26. Abram ME, Ferris AL, Das K, Quinones O, Shao W, Tuske S, Alvord WG, Arnold E, Hughes SH. Mutations in HIV-1 reverse transcriptase affect the errors made in a single cycle of viral replication. J Virol. 2014;88:7589–601.
27. Lara HH, Ayala-Nunez NV, Ixtepan-Turrent L, Rodriguez-Padilla C. Mode of antiviral action of silver nanoparticles against HIV-1. J Nanobiotechnology. 2010;8:1.
28. Ren J, Bird LE, Chamberlain PP, Stewart-Jones GB, Stuart DI, Stammers DK. Structure of HIV-2 reverse transcriptase at 2.35-A resolution and the mechanism of resistance to non-nucleoside inhibitors. Proc Natl Acad Sci USA. 2002;99:14410–5.
29. Witvrouw M, Pannecouque C, Switzer WM, Folks TM, De Clercq E, Heneine W. Susceptibility of HIV-2, SIV and SHIV to various anti-HIV-1 compounds: implications for treatment and postexposure prophylaxis. Antivir Ther. 2004;9:57–65.
30. Poveda E, Briz V, Soriano V. Enfuvirtide, the first fusion inhibitor to treat HIV infection. AIDS Rev. 2005;7:139–47.
31. Schulz TF, Whitby D, Hoad JG, Corrah T, Whittle H, Weiss RA. Biological and molecular variability of human immunodeficiency virus type 2 isolates from the Gambia. J Virol. 1990;64:5177–82.
32. Sato Y, Morimoto K, Kubo T, Sakaguchi T, Nishizono A, Hirayama M, Hori K. Entry inhibition of influenza viruses with high mannose binding lectin ESA-2 from the red alga *Eucheuma serra* through the recognition of viral hemagglutinin. Mar Drugs. 2015;13:3454–65.
33. Rausch K, Hackett BA, Weinbren NL, Reeder SM, Sadovsky Y, Hunter CA, Schultz DC, Coyne CB, Cherry S. Screening bioactives reveals nanchangmycin as a broad spectrum antiviral active against zika virus. Cell Rep. 2017;18:804–15.
34. Fellinger CH, Gardner MR, Bailey CC, Farzan M. Simian immunodeficiency virus SIVmac239, but not SIVmac316, binds and utilizes Human CD4 more efficiently than rhesus CD4. J Virol. 2017;91(18):e00847.

35. Ruprecht CR, Krarup A, Reynell L, Mann AM, Brandenberg OF, Berlinger L, Abela IA, Regoes RR, Gunthard HF, Rusert P, Trkola A. MPER-specific antibodies induce gp120 shedding and irreversibly neutralize HIV-1. J Exp Med. 2011;208:439–54.
36. Rapista A, Ding J, Benito B, Lo YT, Neiditch MB, Lu W, Chang TL. Human defensins 5 and 6 enhance HIV-1 infectivity through promoting HIV attachment. Retrovirology. 2011;8:45.
37. Saphire AC, Bobardt MD, Gallay PA. Host cyclophilin A mediates HIV-1 attachment to target cells via heparans. EMBO J. 1999;18:6771–85.
38. Vives RR, Imberty A, Sattentau QJ, Lortat-Jacob H. Heparan sulfate targets the HIV-1 envelope glycoprotein gp120 coreceptor binding site. J Biol Chem. 2005;280:21353–7.
39. Mousseau G, Clementz MA, Bakeman WN, Nagarsheth N, Cameron M, Shi J, Baran P, Fromentin R, Chomont N, Valente ST. An analog of the natural steroidal alkaloid cortistatin A potently suppresses Tat-dependent HIV transcription. Cell Host Microbe. 2012;12(1):97–108.
40. Tallarida RJ. Drug combinations: tests and analysis with isoboles. Curr Protoc Pharmacol. 2016;72:9.19.1–19.19.
41. Pan C, Cai L, Lu H, Qi Z, Jiang S. Combinations of the first and next generations of human immunodeficiency virus (HIV) fusion inhibitors exhibit a highly potent synergistic effect against enfuvirtide- sensitive and -resistant HIV type 1 strains. J Virol. 2009;83:7862–72.
42. Mousseau G, Kessing CF, Fromentin R, Trautmann L, Chomont N, Valente ST. The tat inhibitor didehydro-Cortistatin A prevents HIV-1 reactivation from latency. MBio. 2015;6:e00465.
43. Carroll SS, Tomassini JE, Bosserman M, Getty K, Stahlhut MW, Eldrup AB, Bhat B, Hall D, Simcoe AL, LaFemina R, Rutkowski CA, Wolanski B, Yang Z, Migliaccio G, De Francesco R, Kuo LC, MacCoss M, Olsen DB. Inhibition of hepatitis C virus RNA replication by 2′-modified nucleoside analogs. J Biol Chem. 2003;278:11979–84.
44. Lindenbach BD, Evans MJ, Syder AJ, Wolk B, Tellinghuisen TL, Liu CC, Maruyama T, Hynes RO, Burton DR, McKeating JA, Rice CM. Complete replication of hepatitis C virus in cell culture. Science. 2005;309:623–6.
45. Thenin-Houssier S, de Vera IM, Pedro-Rosa L, Brady A, Richard A, Konnick B, Opp S, Buffone C, Fuhrmann J, Kota S, Billack B, Pietka-Ottlik M, Tellinghuisen T, Choe H, Spicer T, Scampavia L, Diaz-Griffero F, Kojetin DJ, Valente ST. Ebselen, a small-molecule capsid inhibitor of HIV-1 replication. Antimicrob Agents Chemother. 2016;60:2195–208.
46. Binley JM, Cayanan CS, Wiley C, Schulke N, Olson WC, Burton DR. Redox-triggered infection by disulfide-shackled human immunodeficiency virus type 1 pseudovirions. J Virol. 2003;77:5678–84.

Permissions

All chapters in this book were first published in RETROVIROLOGY, by BioMed Central; hereby published with permission under the Creative Commons Attribution License or equivalent. Every chapter published in this book has been scrutinized by our experts. Their significance has been extensively debated. The topics covered herein carry significant findings which will fuel the growth of the discipline. They may even be implemented as practical applications or may be referred to as a beginning point for another development.

The contributors of this book come from diverse backgrounds, making this book a truly international effort. This book will bring forth new frontiers with its revolutionizing research information and detailed analysis of the nascent developments around the world.

We would like to thank all the contributing authors for lending their expertise to make the book truly unique. They have played a crucial role in the development of this book. Without their invaluable contributions this book wouldn't have been possible. They have made vital efforts to compile up to date information on the varied aspects of this subject to make this book a valuable addition to the collection of many professionals and students.

This book was conceptualized with the vision of imparting up-to-date information and advanced data in this field. To ensure the same, a matchless editorial board was set up. Every individual on the board went through rigorous rounds of assessment to prove their worth. After which they invested a large part of their time researching and compiling the most relevant data for our readers.

The editorial board has been involved in producing this book since its inception. They have spent rigorous hours researching and exploring the diverse topics which have resulted in the successful publishing of this book. They have passed on their knowledge of decades through this book. To expedite this challenging task, the publisher supported the team at every step. A small team of assistant editors was also appointed to further simplify the editing procedure and attain best results for the readers.

Apart from the editorial board, the designing team has also invested a significant amount of their time in understanding the subject and creating the most relevant covers. They scrutinized every image to scout for the most suitable representation of the subject and create an appropriate cover for the book.

The publishing team has been an ardent support to the editorial, designing and production team. Their endless efforts to recruit the best for this project, has resulted in the accomplishment of this book. They are a veteran in the field of academics and their pool of knowledge is as vast as their experience in printing. Their expertise and guidance has proved useful at every step. Their uncompromising quality standards have made this book an exceptional effort. Their encouragement from time to time has been an inspiration for everyone.

The publisher and the editorial board hope that this book will prove to be a valuable piece of knowledge for researchers, students, practitioners and scholars across the globe.

List of Contributors

Franziska K. Geis, Melanie Galla, Dirk Hoffmann, Johannes Kuehle, Daniela Zychlinski, Tobias Maetzig, Juliane W. Schott and Adrian Schwarzer
Institute of Experimental Hematology, Hannover Medical School, Carl-Neuberg-Str. 1, Hannover, Germany
REBIRTH Cluster of Excellence, Hannover Medical School, Hannover, Germany

Christine Goffinet
Institute of Experimental Virology, TWINCORE, Centre for Experimental and Clinical Infections Research, Hannover, Germany

Stephen P. Goff
Department of Biochemistry and Molecular Biophysics, Columbia University Medical Center, New York, NY, USA
Department of Microbiology and Immunology, Columbia University Medical Center, New York, NY, USA
Howard Hughes Medical Institute, Columbia University Medical Center, New York, NY, USA

Axel Schambach
Institute of Experimental Hematology, Hannover Medical School, Carl-Neuberg-Str. 1, Hannover, Germany
REBIRTH Cluster of Excellence, Hannover Medical School, Hannover, Germany
Division of Hematology/Oncology, Children's Hospital Boston, Harvard Medical School, Boston, MA, USA

Tadasuke Naito, Hiroe Sejima, Hiroshi Ushirogawa and Mineki Saito
Department of Microbiology, Kawasaki Medical School, 577 Matsushima, Kurashiki, Okayama 701-0192, Japan

Jun-ichirou Yasunaga
Laboratory of Virus Control, Institute for Frontier Life and Medical Sciences, Kyoto University, Kyoto, Japan

Kazumasa Shirai and Kousuke Hanada
Department of Bioscience and Bioinformatics, Kyushu Institute of Technology, Fukuoka, Japan

Yuetsu Tanaka
Department of Immunology, Graduate School of Medicine, University of the Ryukyus, Okinawa, Japan

Tatsufumi Nakamura
Department of Social Work, Faculty of Human and Social Studies, Nagasaki International University, 2825-7Huis Ten Bosch Machi, Sasebo, Nagasaki 859-3298, Japan

Masahiro Fujii
Division of Virology, Niigata University Graduate School of Medical and Dental Sciences, Niigata, Japan

Masao Matsuoka
Laboratory of Virus Control, Institute for Frontier Life and Medical Sciences, Kyoto University, Kyoto, Japan
Department of Hematology, Rheumatology and Infectious Disease, Faculty of Life Sciences, Kumamoto University, Kumamoto, Japan

Yuichi Mitobe
Laboratory of Virus Control, Institute for Frontier Life and Medical Sciences, Kyoto University, Kyoto, Japan
Division of Gene Regulation and Signal Transduction, Research Center for Genomic Medicine, Saitama Medical University, Saitama, Japan

Jolien Blokken, Stéphanie De Houwer, Lieve Dirix, Frauke Christ and Zeger Debyser
Laboratory for Molecular Virology and Gene Therapy, Department of Pharmaceutical and Pharmacological Sciences, KU Leuven, Kapucijnenvoer 33, VCTB +5, Bus 7001, 3000 Leuven, Flanders, Belgium

Hugo Klaassen and Arnaud Marchand
Center for Innovation and Stimulation of Drug Discovery (CISTIM), Leuven, Belgium

Jonas Demeulemeester
Laboratory for Molecular Virology and Gene Therapy, Department of Pharmaceutical and Pharmacological Sciences, KU Leuven, Kapucijnenvoer 33, VCTB +5, Bus 7001, 3000 Leuven, Flanders, Belgium
The Francis Crick Institute, London, UK

Patrick Chaltin
Center for Innovation and Stimulation of Drug Discovery (CISTIM), Leuven, Belgium
Center for Drug Design and Development (CD3), KU Leuven R&D, Leuven, Belgium

Art F. Y. Poon
Department of Pathology and Laboratory Medicine, Western University, London, ON, Canada

Jessica L. Prodger
Department of Microbiology and Immunology, Western University, London, ON, Canada
Department of Medicine, Johns Hopkins School of Medicine, Baltimore, MD, USA

Briana A. Lynch
Laboratory of Immunoregulation, National Institute of Allergy and Infectious Diseases, National Institutes of Health, Baltimore, MD, USA

Jun Lai and Adam A. Capoferri
Department of Medicine, Johns Hopkins School of Medicine, Baltimore, MD, USA

Thomas C. Quinn and Andrew D. Redd
Laboratory of Immunoregulation, National Institute of Allergy and Infectious Diseases, National Institutes of Health, Baltimore, MD, USA
Department of Medicine, Johns Hopkins School of Medicine, Baltimore, MD, USA

Steven J. Reynolds
Department of Medicine, Johns Hopkins School of Medicine, Baltimore, MD, USA
Rakai Health Sciences Program, Kalisizo, Uganda

Jingo Kasule
Rakai Health Sciences Program, Kalisizo, Uganda

Susanna L. Lamers and Christopher W. Rodriguez
Bio- InfoExperts, LLC, Thibodaux, LA, USA

Daniel Bruno, Stephen F. Porcella and Craig Martens
Genomics Unit, Research Technologies Branch, Rocky Mountain Laboratories, National Institute of Allergy and Infectious Diseases, National Institutes of Health, Hamilton, MT, USA

Lingzhen Ma, Jiayong Wen, Xuanxuan Zhang, Fangyuan Lai, Shuwen Liu and Lin Li Shuwen Yin
Guangdong Provincial Key Laboratory of New Drug Screening, Guangzhou Key Laboratory of Drug Research for Emerging Virus Prevention and Treatment, School of Pharmaceutical Sciences, Southern Medical University, 1838 Guangzhou Avenue North, Guangzhou 510515, Guangdong, China

Ruxia Ren
Guangdong Provincial Key Laboratory of New Drug Screening, Guangzhou Key Laboratory of Drug Research for Emerging Virus Prevention and Treatment, School of Pharmaceutical Sciences, Southern Medical University, 1838 Guangzhou Avenue North, Guangzhou 510515, Guangdong, China
Department of Pharmacy, The Seventh Affiliated Hospital of Sun Yat-sen University, Shenzhen 518107, China

Baolong Lai
Department of Pharmacy, The Seventh Affiliated Hospital of Sun Yat-sen University, Shenzhen 518107, China

Xiaocao Ma, Qifei Hu, Feng Huang, Junsong Zhang, Ting Pan, Chao Liu and Hui Zhang
Institute of Human Virology, Zhongshan School of Medicine, Sun Yat-sen University, Guangzhou 510080, China
Key Laboratory of Tropical Disease Control of Ministry of Education, Zhongshan School of Medicine, Sun Yatsen University, Guangzhou 510080, China

Cancan Chen
Institute of Human Virology, Zhongshan School of Medicine, Sun Yat-sen University, Guangzhou 510080, China
Key Laboratory of Tropical Disease Control of Ministry of Education, Zhongshan School of Medicine, Sun Yatsen University, Guangzhou 510080, China
Department of Pathology, The First Affiliated Hospital, Sun Yat-sen University, Guangzhou 510080, China

Xinghua Li and Jinyu Xia
Department of Infectious Diseases, The Fifth Affiliated Hospital, Sun Yat-sen University, Zhuhai 519000, China

Brett D. Anderson, Seyed Arad Moghadasi, Amber St. Martin and William L. Brown
Department of Biochemistry, Molecular Biology and Biophysics, Masonic Cancer Center, Center for Genome Engineering, Institute for Molecular Virology, University of Minnesota, Minneapolis, MN, USA

Terumasa Ikeda and Reuben S. Harris
Department of Biochemistry, Molecular Biology and Biophysics, Masonic Cancer Center, Center for Genome Engineering, Institute for Molecular Virology, University of Minnesota, Minneapolis, MN, USA
Howard Hughes Medical Institute, University of Minnesota, 2231 6th St. S.E., Minneapolis, MN 55455, USA

Hayato Murakoshi, Chengcheng Zou, Nozomi Kuse, Tomohiro Akahoshi, Takayuki Chikata and Masafumi Takiguchi
Center for AIDS Research, Kumamoto University, 2-2-1 Honjo, Chuo-ku, Kumamoto 860-0811, Japan

Hiroyuki Gatanaga and Shinichi Oka
Center for AIDS Research, Kumamoto University, 2-2-1 Honjo, Chuo-ku, Kumamoto 860-0811, Japan
AIDS Clinical Center, National Center for Global Health and Medicine, Tokyo, Japan

Tomáš Hanke
International Research Center of Medical Sciences, Kumamoto University, Kumamoto, Japan
The Jenner Institute, University of Oxford, Old Road Campus Research Building, Roosevelt Drive, Oxford, UK

Matthew F. Chersich, Michelle A. Moorhouse, Willem D. Venter and Samanta T. Lalla-Edward
Wits Reproductive Health and HIV Institute, Faculty of Health Sciences, University of Witwatersrand, Johannesburg, South Africa

Peter Zuithoff and Diederick E. Grobbee
Julius Global Health, Julius Center for Health Sciences and Primary Care, University Medical Center Utrecht, Utrecht University, Utrecht, The Netherlands

Roos E. Barth
Department of Internal Medicine and Infectious Diseases, University Medical Center Utrecht, Utrecht University, Utrecht, The Netherlands

Alinda G. Vos
Wits Reproductive Health and HIV Institute, Faculty of Health Sciences, University of Witwatersrand, Johannesburg, South Africa
Julius Global Health, Julius Center for Health Sciences and Primary Care, University Medical Center Utrecht, Utrecht University, Utrecht, The Netherlands
Department of Internal Medicine and Infectious Diseases, University Medical Center Utrecht, Utrecht University, Utrecht, The Netherlands

Kerstin Klipstein-Grobusch
Julius Global Health, Julius Center for Health Sciences and Primary Care, University Medical Center Utrecht, Utrecht University, Utrecht, The Netherlands
Division of Epidemiology and Biostatistics, School of Public Health, Faculty of Health Sciences, University of the Witwatersrand, Johannesburg, South Africa

Andrew Kambugu
Infectious Diseases Institute, Makerere University, Kampala, Uganda

N. Kumarasamy
YRG CAR E Medical Centre, CART Clinical Research Site, Voluntary Health Services, Chennai, India

Hongyun Wang, Jiping Song, Rui Zhang and Qingling Du
Laboratory of Structural Virology and Immunology, Institute of Biophysics, Chinese Academy of Sciences, Beijing, China

Zene Matsuda
Laboratory of Structural Virology and Immunology, Institute of Biophysics, Chinese Academy of Sciences, Beijing, China
Research Center for Asian Infectious Diseases, Institute of Medical Science, The University of Tokyo, 4-6-1 Shirokanedai, Minato-ku, Tokyo 108-8639, Japan

Dehua Liu
Laboratory of Structural Virology and Immunology, Institute of Biophysics, Chinese Academy of Sciences, Beijing, China
The Chinese University of Hong Kong, Sha Tin, Hong Kong, China

Yasushi Kawaguchi
Research Center for Asian Infectious Diseases, Institute of Medical Science, The University of Tokyo, 4-6-1 Shirokanedai, Minato-ku, Tokyo 108-8639, Japan
Division of Molecular Virology, Department of Microbiology and Immunology, Institute of Medical Science, The University of Tokyo, Tokyo, Japan

Mizuki Yamamoto
Research Center for Asian Infectious Diseases, Institute of Medical Science, The University of Tokyo, 4-6-1 Shirokanedai, Minato-ku, Tokyo 108-8639, Japan
Division of Cellular and Molecular Biology, Institute of Medical Science, The University of Tokyo, Tokyo, Japan

Jun-ichiro Inoue
Division of Cellular and Molecular Biology, Institute of Medical Science, The University of Tokyo, Tokyo, Japan

Damien Bonnard, Julie Brias, Francis Chevreuil, François Moreau, Sophie Chasset, Erwann Le Rouzic and Benoit Ledoussal
Biodim Mutabilis, 93230 Romainville, France

Céline Amadori
Biodim Mutabilis, 93230 Romainville, France
INSERM, U1016, Institut Cochin, Paris, France
CNRS, UMR8104, Paris, France
University Paris Descartes, Sorbonne Paris Cité, Paris, France

Stéphane Emiliani
INSERM, U1016, Institut Cochin, Paris, France
CNRS, UMR8104, Paris, France
University Paris Descartes, Sorbonne Paris Cité, Paris, France

Yme Ubeles van der Velden, Nikki van Bel and Ben Berkhout
Laboratory of Experimental Virology, Department of Medical Microbiology, Center for Infection and Immunity Amsterdam (CINIMA), Academic Medical Center, University of Amsterdam, Amsterdam, The Netherlands

Igor Orlov, Daniele Spehner, Marc Ruff and Bruno P. Klaholz
Centre for Integrative Biology, IGBMC, CNRS, INSERM, University of Strasbourg, Strasbourg, France

Laia Miralles, Felipe García, José Gatell and Montserrat Plana
AIDS Research Group, IDIBAPS, Hospital Clinic, Barcelona, Spain

Luzia Mayr and Christiane Moog
INSERM U1109, Strasbourg, France

Alessia Zamborlini
CNRS, UMR7212, INSERM U944, Université Paris Diderot, Conservatoire National des Arts et Métiers, Paris, France

Richard Benarous
Biodim Mutabilis, 93230 Romainville, France
19 rue de Croulebarbe, 75013 Paris, France

Nathalie Désiré, Vincent Calvez and Anne-Geneviève Marcelin
INSERM, Institut Pierre Louis d'Epidémiologie et de Santé Publique (iPLESP), Sorbonne Université, 75013 Paris, France
AP-HP, Hôpital Pitié-Salpêtrière, Virologie, 75013 Paris, France

Lorenzo Cerutti and Stefan Emler
SmartGene Services, EPFL Innovation Park, 1015 Lausanne, Switzerland

Diane Descamps, Quentin Le Hingrat, Marine Perrier and Benoit Visseaux
IAME, UMR 1137, Université Paris Diderot, INSERM, Paris, France
AP-HP, Hôpital Bichat, Virologie, 75018 Paris, France

Stéphane Hué
Department of Epidemiology & Population Health, London School of Hygiene and Tropical Medicine, London WC1E 7HT, UK

Paul D. Bieniasz
Laboratory of Retrovirology and Howard Hughes Medical Institute, The Rockefeller University, New York, NY, USA

Robert J. Gifford
MRC-University of Glasgow Centre for Virus Research, Glasgow, UK

Daniel Blanco-Melo
Department of Microbiology, Icahn School of Medicine at Mount Sinai, New York, NY, USA

Pauline J. Schipper, Axel Fun and Monique Nijhuis
Department of Medical Microbiology, Virology, University Medical Center Utrecht, Heidelberglaan 100, HP G04.614, 3584 CX Utrecht, The Netherlands

Annemarie M. J. Wensing
Department of Medical Microbiology, Virology, University Medical Center Utrecht, Heidelberglaan 100, HP G04.614, 3584 CX Utrecht, The Netherlands

Thomas Leitner
Theoretical Biology and Biophysics, Los Alamos National Laboratory, Los Alamos, NM, USA

Linos Vandekerckhove
Department of General Internal Medicine and Infectious Diseases, Ghent University Hospital, Ghent, Belgium

Martin Däumer
Institute of Immunology and Genetics, Kaiserslautern, Germany

Alexander Thielen
Max Planck Institute for Informatics, Saarbrücken, Germany

Bernd Buchholz
Pediatric Clinic, University Medical Center Mannheim, Mannheim, Germany

Andy I. M. Hoepelman
Department of Internal Medicine and Infectious Diseases, University Medical Center Utrecht, Utrecht, The Netherlands

Elizabeth H. Gisolf
Department of Internal Medicine, Rijnstate Hospital, Arnhem, The Netherlands

S. Mediouni, J. A. Jablonski, S. Tsuda, A. Richard, C. Kessing, A. Biswas, H. Choe and S. T. Valente
Department of Immunology and Microbiology, The Scripps Research Institute, 130 Scripps Way, 3C1, Jupiter, FL 33458, USA

T. Tellinghuisen
Department of Immunology and Microbiology, The Scripps Research Institute, 130 Scripps Way, 3C1, Jupiter, FL 33458, USA
Roche, Basel, Switzerland

M. Cameron
Department of Molecular Therapeutics, The Scripps Research Institute, Jupiter, FL, USA

Y. Even
The Botanist's Beach Farm, Jupiter, FL, USA

M. V. Andrade and M. Stevenson
University of Miami Miller School of Medicine, Miami, FL, USA

Index

www.ingramcontent.com/pod-product-compliance
Lightning Source LLC
LaVergne TN
LVHW080702210926
848485LV00003B/403